International Workshop on Neutron Optics and Detectors

(NOP&D 2013)

Journal of Physics: Conference Series Volume 528

Munich, Germany
2-5 July 2013

Editor:

Alexander Ioffe

ISBN: 978-1-63439-309-6
ISSN: 1742-6588

Printed from e-media with permission by:

Curran Associates, Inc.
57 Morehouse Lane
Red Hook, NY 12571

Some format issues inherent in the e-media version may also appear in this print version.

Printed by Curran Associates, Inc. (2014)

For permission requests, please contact the Institute of Physics
at the address below.

Institute of Physics
Dirac House, Temple Back
Bristol BS1 6BE UK

Phone: 44 1 17 929 7481
Fax: 44 1 17 920 0979

techtracking@iop.org

Additional copies of this publication are available from:

Curran Associates, Inc.
57 Morehouse Lane
Red Hook, NY 12571 USA
Phone: 845-758-0400
Fax: 845-758-2634
Email: curran@proceedings.com
Web: www.proceedings.com

TABLE OF CONTENTS

MONOCHROMATORS

NEUTRON GUIDES AND FOCUSING

POLARIZATION

DETECTORS

Journal of Physics: Conference Series **528** (2014) 011001 doi:10.1088/1742-6596/528/1/011001

Preface

Every two–three years scientists involved in developments of neutron optics gather together for the International Workshop on Neutron Optics (NOP). Neutron optics has always been considered very important for the development of new neutron instrumentation. The limited brilliance of existing or future neutron sources requires the more effective usage of emitted neutrons. Indeed, improvements of the neutron optical system or an optimization of the neutron-optical tracts of instruments can result in a significant enhancement of their performance. This is especially important at present when the neutron scattering community is strongly engaged in developments of new instrumentation around the spallation neutron sources – SNS, ESS, J-PARC and Second Target Station at ISIS.

In 2013 the workshop was organized by the Jülich Centre for Neutron Science of the Forschungszentrum Jülich GmbH and was held at the Conference Centre in Ismaning next to Munich on July 2-7, 2013 on the eve of the ICNS-2013 in Edinburg. It carried on the series of Neutron Optics workshops held in Villigen (1999, 2007), Tokyo (2004) and Alpe d'Huez (2010). This time it is also aimed to compliment the International Conference on Neutron Scattering in Edinburgh (ICNS-2013) by providing a platform for detailed discussions on the latest developments in the field of neutron optics.

The scope of the workshop was extended to the neutron detectors (in a way similar to the NOP-2004 held in Tokyo) and was labelled as the International Workshop on Neutron Optics and Detectors, NOP&D-2013. However, in contrast to the Tokyo workshop, the focus of discussions was not the detector technologies (which are the subject of many dedicated meetings), rather than the use of detectors for the purpose of the design of modern instrumentation aiming to inform detector developers about real detectors requirements for new advanced instrumental concepts.

The three-full-days workshop gathered a record number of participants, more than 120, that well exceeded usual numbers (around 70) and even exceeded the limit capacity of the conference hall booked for the workshop! It even forced the organizers to stop the registration before the deadline and to establish a kind of waiting list. Such attendance of scientists representing Australia, Japan, India, Saudi Arabia, Russia, USA, Canada and 10 European countries, actually raised the level of this event from Workshop to Conference.

Discussions at the workshop were devoted but not limited to new concepts in neutron instrumentation, focusing optics, neutron detection in the light of requirements imposed by neutron instruments, neutron polarization and polarization analysis and simulation packages.

The current proceedings are representative of about a half of oral and poster presentations made at the workshop and provide a reader with possibility to outlook the current status and perspectives in the field of neutron optics and related detector developments.

Alexander Ioffe Editor and Chairman of NOP&D-2013

The additional financial support by ESS Scandinavia, SwissNeutronics, Mirotron, S-DH Heidelberg and ASTRIUM is gratefully acknowledged.

We gratefully acknowledge the support of:

Journal of Physics: Conference Series **528** (2014) 011003 doi:10.1088/1742-6596/528/1/011003

Peer review statement

All papers published in this volume of *Journal of Physics: Conference Series* have been peer reviewed through processes administered by the proceedings Editors. Reviews were conducted by expert referees to the professional and scientific standards expected of a proceedings journal published by IOP Publishing.

The first prototype diamond monochromator at the Institut Laue-Langevin

P Courtois[*1], M T Fernandez-Diaz[1], G Nenert[1], K H Andersen[2], A K Freund[1,3], S Gsell[4], M Fischer[4], M Schreck[4], P Link[5] and M Meven[6]

[1] Institut Max von Laue - Paul Langevin, Grenoble, France
[2] European Spallation Source ESS, Lund, Sweden
[3] Via Cordis - Consulting and Movement, Bordeaux, France
[4] Institut für Physik, Universität Augsburg, Augsburg, Germany
[5] Forschungs-Neutronenquelle Heinz Maier-Leibnitz (FRM II), Technische Universität München, Garching, Germany
[6] RWTH Aachen, Institut für Kristallographie, Aachen and Forschungszentrum Jülich GmbH, Jülich Centre for Neutron Science (JCNS), Outstation at MLZ, Garching, Germany.

E-mail: courtois@ill.fr

Abstract. We report on the performance of the first diamond neutron monochromator built at the ILL. It has been designed for the hot neutron diffractometer D9 with the aim of improving significantly the instrument performance in particular for short wavelengths in the 0.3-0.9 Å wavelength range. Diamond crystal plates with dimensions of 1.5 x 1.5 x 0.18 cm^3 and an average mosaic spread of 0.15° have been synthesized at the University of Augsburg. They exhibited excellent neutron diffraction properties when examined on a neutron double-crystal test setup. Sufficiently thick diamond elements with a controlled mosaic spread of 0.25° have been obtained by stacking several of these crystals. First tests runs carried out at the ILL confirmed the predicted high reflectivity of the diamond stacks. The diamond prototype monochromator uses the (220) reflection in transmission geometry replacing the Cu (220) monochromator on D9 that has the same *d*-spacing. The final performance studies on D9 showed that the diamond device did not perform better than the original copper crystal. This unexpected result could be explained by significant optical aberrations caused by non-uniformities of both the angular and spatial mosaic distribution in the individual diamond crystals, as revealed by a detailed characterisation study using high-energy X-ray diffraction.

1. Introduction

Thanks to its very favourable crystalline and nuclear properties, diamond offers theoretically the highest performance of all existing materials for neutron monochromator applications [1]. It has a very compact structure with a small unit cell (a = 3.5668 Å) composed of 8 carbon atoms having a big coherent scattering length ($b = 0.665\mathrm{x}10^{-12}$ cm) and very low incoherent and absorption cross-sections

[*] To whom any correspondence should be addressed

for thermal neutrons. At a wavelength of 1.8 Å σ_{inc} = 0.001x10^{-24} cm^2 and σ_{abs} = 0.0350x10^{-24} cm^2, respectively. It has been shown that a sufficiently thick diamond crystal with an appropriate mosaic spread would outperform by far existing materials such as copper and germanium mosaic crystals, especially for the monochromatisation of hot neutrons, in the range of 0.3-1 Å [1,2]. However, when the present project started there we neither crystals of appropriate size available (at reasonable costs) nor existed a concept how to generate a defined mosaic spread in big diamond single crystals. Thus, the controlled synthesis of large area diamond crystals with sufficient thickness and appropriate mosaic spread represented quite a big challenge in material science.

This problem was tackled by a research group at the University of Augsburg that has succeeded in producing mosaic diamond crystals using plasma chemical vapour deposition (MWPCVD). Heteroepitaxial diamond layers are grown in the <001> direction on a 150 nm thick iridium (001) film coated onto a thin layer of yttria-stabilised zirconia deposited on a silicon (001) wafer [3,4]. The surface of crystal plates is nearly parallel to (100) crystallographic planes, typically a few degrees off while both edges are parallel to (110)-type directions. In the meantime it is possible to obtain good quality crystal plates that are up to 0.18 cm thick with lateral dimensions of 1.5 x 1.5 cm^2. Neutron double-crystal diffraction studies of a great number of samples has shown a high diffraction efficiency yielding values of the peak reflectivity close to the theoretical predictions [4,5]. The potential of heteroepitaxial diamond crystals suitable for neutron monochromators has thus been demonstrated unequivocally. Following this promising result, we decided to build a full-scale prototype monochromator for the hot neutron single crystal diffractometer D9 at the ILL [6]. It should produce the highest performance ever achieved at short wavelengths and would thus represent a significant improvement for reactor based hot neutron instruments.

In the following sections we report on the design, describe the different steps involved in the construction and report on the performance of the first diamond monochromator.

2. Monochromator layout

The diffractometer D9 is presently equipped with a copper (220) monochromator in transmission geometry composed of 3 flat, 0.8 cm thick mosaic crystal plates characterised by a mosaic distribution of 0.25°. With an active area of 6 x 6 cm^2, it has been optimised to provide neutrons in the wavelength range from 0.3 to 0.9 Å. The monochromator is flat in order to achieve polychromatic pseudo-focusing at the sample position according to the geometry of the incoming white beam [6].

The prototype diamond monochromator was designed specifically for D9 to replace the existing copper monochromator with the aim of improving significantly the performance of this instrument [6]. To compare directly both monochromators in terms of neutron flux and resolution, we have chosen diamond crystals with characteristics similar to that of the copper crystals currently used on D9, *i.e.* identical mosaic spread, *d*-spacing and active area.

3. Neutron diffraction properties of the diamond crystals

To provide the material for the construction of the monochromator, more than 50 diamond crystals were grown at the University of Augsburg. The neutron diffraction properties of all these diamond samples were studied on the double-crystal test instrument T13C at the ILL. A perfect Ge (331) monochromator crystal selected a wavelength of 1 Å out of a thermal guide tube. The effective neutron mosaic spread was determined as the full width at half height of (220) rocking curves recorded in transmission geometry. The beam cross-section was 0.2 x 0.2 cm^2. Since the *d*-spacing of Ge (331) is very close to that of diamond (220), d_{Ge331} = 1.296 Å ~ d_{C*220} = 1.261 Å, the double-crystal configuration was parallel and the arrangement nearly dispersion-free. Thus the width of the rocking curve represented directly the mosaic distribution of the sample under study. Moreover, there was no $\lambda/2$-contamination since the curvature of the thermal neutron guide cuts off neutrons of wavelength below 0.8 Å. As a result, the neutron reflectivity could directly be derived from the crystal rocking curve as the ratio $R(\theta) = I_r(\theta)/I_0$, where I_0 is the intensity of the direct beam and I_r is the intensity of the reflected beam. The peak reflectivity is the value of $R(\theta)$ in the peak centre.

A typical neutron rocking curve obtained from a 0.18 cm thick diamond crystal is shown in figure 1. The peak reflectivity is close to 34.5% for a FWHM of 0.18°. It corresponds to 80% of the theoretical peak reflectivity calculated from the Darwin model for ideally imperfect mosaic crystals [7]. The origin of the lower reflectivity could be assigned to primary extinction processes and some crystal inhomogeneities [8].

The individual diamond single crystal platelets exhibited high reflectivity, already comparable to that of the best monochromator elements made of copper mosaic crystals despite the fact that their thickness was still significantly below the theoretical optimum. To use the full potential of diamond the thickness of the crystals had to be increased. At the same time, to meet the instrument requirements, it should be possible to tailor the mosaic spread.

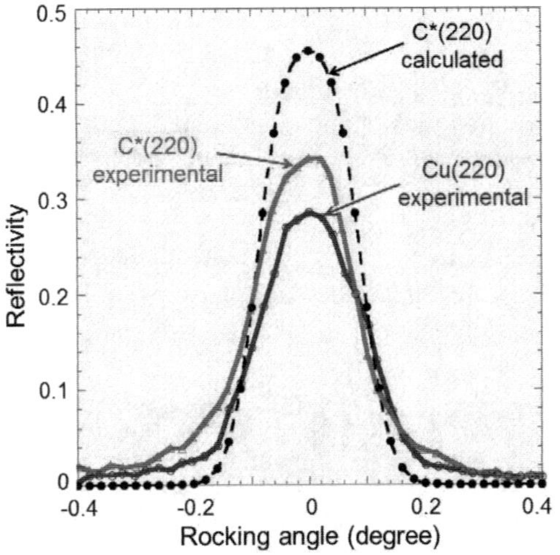

Figure 1. Experimental and theoretical neutron rocking curves at 1 Å of a mosaic diamond (C*) (220) crystal in Laue geometry with a mosaic spread of 0.3° and a thickness of 0.18 cm. For comparison, the experimental rocking curve of the mosaic Cu (220) crystal presently used on D9 as monochromator is shown too.

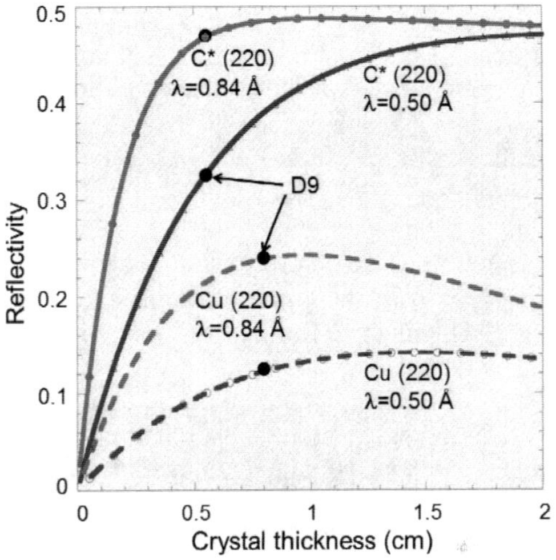

Figure 2. Theoretical neutron peak reflectivity for $\lambda = 0.84$ Å and 0.5 Å as a function of crystal thickness for diamond (220) and Cu (220) in Laue geometry with the same intrinsic mosaic spread of 0.25°. The black dots indicate the theoretically predicted peak reflectivities of Cu and C* as crystal monochromators on D9.

4. Composite diamond crystal systems

First of all we consider the crystal thickness needed to maximise the neutron reflectivity of diamond. Figure 2 shows the theoretical peak reflectivity of the diamond (220) reflection in transmission geometry as a function of crystal thickness for two short wavelengths $\lambda = 0.5$ Å and 0.8 Å and for an intrinsic mosaic spread of 0.25°. It should be mentioned at this point that due to secondary extinction the neutron mosaic spread defined as the FWHM of the diffraction peak is always a little wider than the intrinsic crystal mosaic spread that is given by the average angular spread of the mosaic blocks or the lattice tilts, see [1]. If not stated otherwise, by mosaic spread we mean the neutron rocking curve width. It can be seen that the peak reflectivity reaches a maximum at an optimum thickness, t_{opt}, that depends on the wavelength, see also [9]. Thanks to the very small capture and inelastic scattering cross sections of carbon the maximum peak reflectivity approaches 50 % for both wavelengths. The calculations show that the thickness of the crystals available ($t \sim 0.18$ cm) is too small with respect to the required values as predicted by the theory: $t_{opt} \sim 1$ cm at $\lambda = 0.8$ Å and $t_{opt} \sim 2$ cm at $\lambda = 0.5$ Å.

At this point it should be mentioned that not only capture and inelastic scattering contribute to the attenuation of the neutron beam, see [10,11]. Because carbon is such a strong scatterer, also multiple elastic scattering effects competing with the main diffraction process could significantly affect the reflectivity whereas for copper capture and inelastic scattering dominate the attenuation. Taking this fact into account, the optimum thickness would be smaller than the values given above.

Because the synthesis of 1 cm-thick diamond crystals with a mosaic distribution of 0.25° is currently impossible, we decided to build a composite crystal system by stacking several thin diamond crystal pieces of high quality, each piece being slightly misoriented with respect to its neighbour by an angle α [5]. Note that the term *high quality* refers to crystals of type A as defined in the study described in ref. [8].The effective total mosaic spread of the composite crystal system is the average spread of the n crystal plates plus n-1 times the inclination angle between the plates. Besides the optimisation of the effective thickness for a given wavelength, an important advantage of this technique is that the global mosaic distribution of the crystal stack can be tailored: the misorientation angle must not always be constant within each stack but can be varied if required. This technique called *onion peel method* had been successfully employed earlier for producing highly anisotropic copper crystals with a mosaic spread well adapted for neutron monochromators [12]. It works well if the inclination angle is smaller than or equal to the mosaic spread. Otherwise holes will appear on the angular profile.

Due to the limited number of available crystals, the best compromise was a stack consisting of three plates to obtain an effective thickness of 0.55 cm. Such a stack would theoretically outperform a mosaic copper for the monochromatisation of hot neutrons by more than a factor two as shown in figure 2. The angle α was chosen close to 0.05° in order to achieve a global mosaic spread of the stacks close to 0.25°.

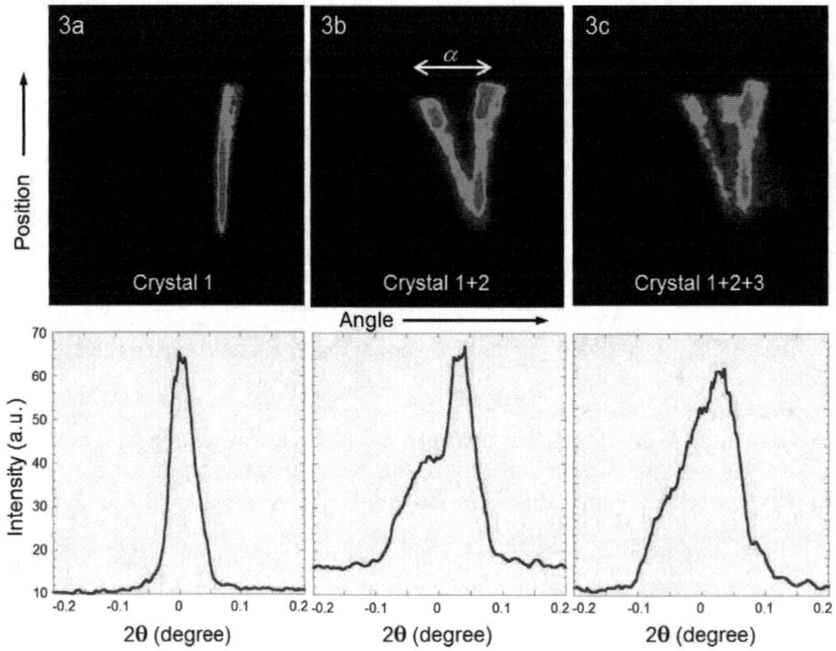

Figure 3. High-energy X-ray diffraction images and peak profiles (C*(220) reflection) showing the evolution of the mosaic distribution during the stacking procedure. A (left): single crystal plate; b (middle): two-crystal stack; c (right): final stack of three crystals (thickness = 0.55 cm). Note that the angle on the images appears to be bigger than the real one due to the projection.

The production of the composite crystal systems was achieved by using high-energy X-ray diffraction for the precise orientation of the slabs [13]. A high-resolution, two-dimensional detector enabled the visualisation of the diffraction image and the determination of the peak position ($2\theta_0$) with an accuracy of 30 arc seconds. Due to the high energy of hard X-rays (100-450 keV), absorption is small compared to that of standard X-rays (30 keV). As a consequence, the whole bulk of the crystals was probed similar to diffraction experiments with neutrons which offered the opportunity to observe simultaneously diffraction peaks from several crystals stacked together. Thus high-energy, white-beam X-ray diffraction was well suited for assembling stacks of diamond single crystals as described in the next paragraphs.

First, a diamond platelet (crystal 1) was mounted onto the sample table that was equipped with x-y-z translation stages and ω–ϕ–κ angular rotation stages. When placed in the white beam, the crystal diffracted X-rays at the Bragg angle θ_0 according to its orientation (figure 3a). The second crystal attached to an aluminium support (crystal 2) was positioned in front of the first one. A specific and very accurate mechanical support had been developed for this purpose at the ILL consisting of a combination of a goniometer and a translation stage. Then crystal 2 was carefully aligned at the desired angle $\theta_0 + \alpha$, moved very close to the first crystal and then bonded on the latter using a carbon based glue (figure 3b). The translation stage had been specially prepared to keep the precise orientation during the transfer operation. Finally, to complete the stack, the third crystal was aligned and glued onto the second one following the same procedure. This technique allowed a precise orientation of crystallographic planes within an accuracy of 0.01° after gluing. In addition, as shown in figure 3c, such an *in-situ* alignment enabled us to control and optimize the peak profile if necessary.

A total of 16 stacks with a neutron mosaic spread of 0.25° ± 0.1° were successfully prepared using this technique. The reflectivity of the stacks was measured using neutrons on the diffractometer T13C at the ILL giving a typical peak reflectivity of 41% at 1 Å. Earlier measurements carried out on D9 at smaller wavelengths studying a test stack on the sample position had confirmed its outstanding performance as shown in figure 4. Even with a bigger mosaic spread the peak reflectivity of the diamond stack exceeded by far that of the copper crystal. Note that the rocking curves were wider than the mosaic spread of the crystals under study because their reflection curves were convoluted with the rocking curve of the D9 monochromator that served as first crystal in this double-crystal arrangement.

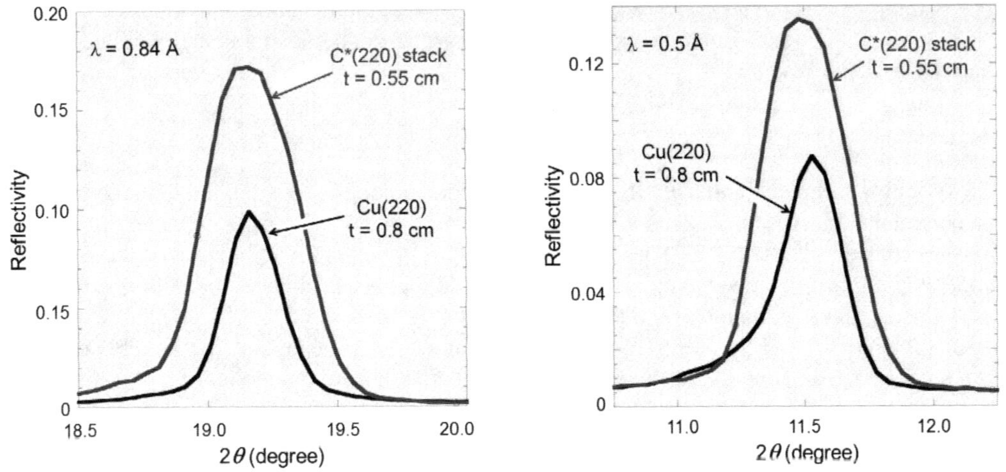

Figure 4. Experimental neutron rocking curves for $\lambda = 0.84$ Å (left) and 0.5 Å (right) obtained with a 0.55 cm thick diamond composite system (FWHM = 0.4°) and a 0.8 cm thick Cu (220) crystal (FWHM = 0.25°).

Journal of Physics: Conference Series **528** (2014) 012001 doi:10.1088/1742-6596/528/1/012001

5. Performance of the diamond monochromator

A photograph of the new diamond (220) monochromator consisting of 4 columns of 4 composite crystal stacks that are approximately 0.55 cm thick with a global mosaic distribution close to 0.25° is shown in figure 5. The crystals were glued onto a backing plate cut from a perfect silicon crystal. Again high-energy X-ray diffraction was used to ensure an alignment of the composite systems with an accuracy of better than 0.05°. The monochromator was installed on D9 in February 2013 and tested in the same configuration as the copper device.

Figure 5. Photograph of the prototype diamond monochromator for the instrument D9 at the ILL.

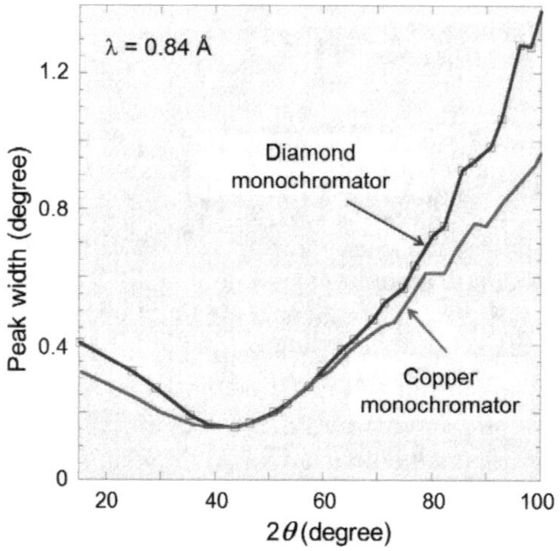

Figure 6. Peak width as a function of scattering angle 2θ measured with a perfect Si crystal on the sample position for $\lambda = 0.84$ Å with both the diamond and the copper monochromator.

5.1. Neutron flux

The neutron flux at 0.84 Å and at 0.5 Å was measured with a neutron monitor at the exit of the D9 monochromator casemate. Surprisingly, it was observed that the diamond monochromator did not provide any gain in neutron flux as compared to copper. The neutron flux was also checked at the sample position by recording the intensity peak profiles of Bragg reflections from a perfect Si crystal as a function of the sample scattering angle 2θ. The integrated peak reflectivity calculated from these measurements confirmed that there was no increase of the neutron flux provided by the diamond monochromator.

5.2. Resolution

Figure 6 shows the resolution curves of the instrument obtained with the diamond and copper monochromators. The two measurements were taken at the same Bragg angle θ_B corresponding to a neutron wavelength of 0.84 Å. As expected, the minimum of the peak width occurred at a scattering angle 2θ close to $2\theta_B$ and the resolution rapidly decreased for $2\theta > 2\theta_B$ [11]. Moreover, the resolution measured with diamond was inferior to that of copper. From figure 6, with the aid of the well-known Cagliotti formula [15] and by setting $\alpha_1 = \alpha_3 = 0$ (no collimation) and $\alpha_2 = 0.75°$, we obtained $\beta_{Cu} = 0.22°$ and $\beta_{C*} = 0.3°$ where α_n are the opening angles of the beam collimation and β is the neutron mosaic spread, respectively. These values were consistent with mosaic spread measured on T13C from rocking curves of single crystal elements.

6. Crystal analysis and discussion

As already mentioned in the previous section, quite unexpectedly the performance of the prototype diamond monochromator in terms of both neutron flux and resolution did not exceed that of the existing copper crystal. Considering the measured high reflectivity of the diamond stacks that composed the monochromator, such a low level of performance suggests that an important part of the diffracted neutron beam did not reach the sample position but was dispersed by a non-uniformity of the crystal mosaic structure. On the double-crystal setup at T13C the detector was wide open and always received all the diffracted intensity.

To validate this assumption, a structural analysis of diamond crystals was performed using the high-energy X-ray diffractometer at the ILL [13] that is a powerful tool to investigate crystals defects on the macroscopic scale, *i.e.* in the mm and sub-mm range. A CCD detector permits to visualize structural defects of the crystal lattice, such as mosaicity, sub grains and curvature. For a perfect silicon crystal, the diffraction spot is a straight and narrow line with a width equal to the generator focus size as shown in figure 7a. The height of the spot corresponds to the crystal size. For a non-perfect crystal, the width of the diffraction spot is expanded as a function of both the mosaicity and lattice plane curvature. As an example, the diffraction image in figure 7b obtained from the (220) reflection of a flat copper crystal currently used on D9 shows a nicely uniform mosaic distribution.

Figure 7c represents a typical diffraction image of the (220) reflection from a diamond crystal. The beam size was 2 x 2 cm^2 so that the full crystal was illuminated by the X-rays. The shape of the image indicates a complex mosaic structure with substantial variations of the peak position in the direction of X-ray scattering (angle 2θ) and also in the vertical direction. Since it is not easy to separate contributions from mosaicity and curvature to the horizontal broadening of the peak width, the crystal was scanned from one end to the other using a narrow beam 0.1 cm wide and 2 cm high so that the effect of horizontal curvature could be neglected. Diffraction images were taken in steps of 0.1 cm along the y-direction of the crystal, see the series of images in figure 8. This scan highlighted relevant information regarding the effect of structural defects in diamond crystals on the reflection properties.

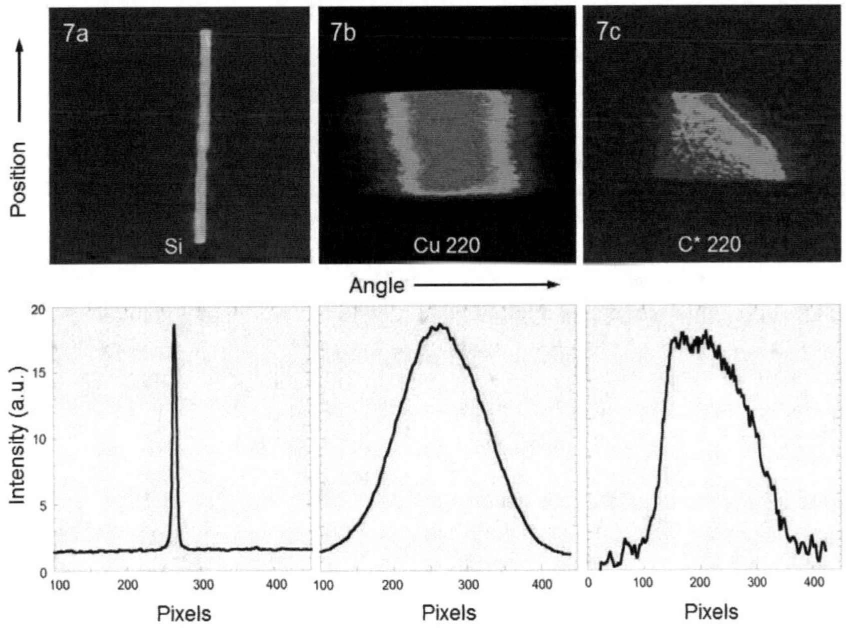

Figure 7. High-energy X-ray diffraction images and peak profiles recorded with a perfect Si crystal (left), with a 0.8 cm thick copper mosaic crystal used on D9 (middle) and with a 0.18 cm thick diamond crystal plate (right). The crystal is fully illuminated by the X-ray beam of cross section 2 x 2 cm^2.

Indeed a peak shift $\Delta\theta_h$ of 0.15° over the length of the plate was observed along the y direction which indicated a bending of the lattice planes perpendicular to the <100> axis. It corresponded to an average radius of curvature of 5.7 m. Such a curvature in diamond crystals had already been reported earlier using neutron and X-ray diffraction [2, 8]. It was confirmed in the present study that most of the diamond crystals exhibited a curvature with a radius in the range from 2 to 5 metres along one (100)-direction. Such a relatively strong curvature can indeed affect the performance of the monochromator because part of the beam is not fully reflected onto the sample. In fact, in the worst case of defocusing, a horizontally bent, 1.5 cm wide crystal with a radius of curvature of 3.5 m produces a fan of 0.5° that after 3 m flight path broadens the beam by 2.6 cm, much bigger than the sample reception width of about 0.5 cm. Thus the pseudo focusing effect that is an important feature at D9 is seriously reduced. This rough estimation already explains the fact that the observed flux was significantly smaller than that expected for a perfectly flat monochromator crystal.

Even more surprisingly, the diffraction image was inclined with respect to the direct beam (z-direction). This indicated a screw type deformation of the crystal lattice along the vertical <220> axis. A maximum twist angle of 0.07°/cm ($\Delta\theta_v \sim 0.1°$) was observed at each end of the crystal plate. In addition, this twist was not constant but varied strongly with the y position and even changed sign at y close to -1 mm as revealed by the inclination of the image with regard to the vertical direction. Such a change of sign and thus sense of the twist suggests that the crystal deformation in the perpendicular direction was quite complex.

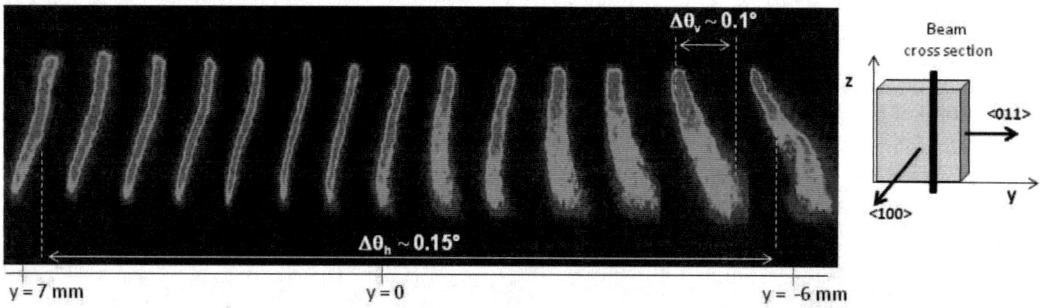

Figure 8. High-energy X-ray diffraction images obtained from a 0.18 cm thick diamond crystal plate with a beam 0.1 cm wide and 2 cm high. The crystal was scanned along the y axis in steps of 0.1 cm.

In addition to these features on the angular scale we observed an important non-uniformity of the spatial mosaic distribution, too. The local (intrinsic) X-ray mosaic spread η_0 was much narrower than the effective mosaic η_{eff} observed from the entire crystal, see figures 7c and 8. The local mosaic distribution η_0 could be determined from the horizontal cross section of the diffraction spot by taking a small region of interest in order to simulate a beam cross section of 1 x 1 mm^2, *i.e.* 3 pixels high. Thus any broadening of the diffraction spot due to bending or twist deformation could be neglected. At the centre of the diamond plate, we found a mosaic η_0 of as small as 0.05°. Such a small value is to be compared to the 0.22° measured for η_{eff} as determined from figure 7c. Moreover, as illustrated in figure 8, η_0 strongly varied not only between the two ends of the plate in the y direction, but also from the top to the bottom in the z direction. For instance, for $y = 7$ mm η_0 was close to 0.1°, twice the value at the centre of the plate.

It is emphasized once more that the neutron beam geometry on D9 had been designed for a flat monochromator in Laue configuration. In the horizontal direction an intermediate neutron source is imaged onto the sample position if a flat mosaic crystal monochromator is used. Also a deformation of the crystal lattice in the vertical plane affects the monochromator performance dispersing the neutron beam and degrading the instrument performance in terms of flux on the sample, although it is smaller than in the scattering plane.

The maximum acceptable angular deviation on D9 can be estimated by considering the beam geometry. The distance between the monochromator and the sample position is close to 3 metres. Usually a collimator of 1 to 6 mm diameter is mounted at a distance of 2.4 m from the monochromator position. Calculations show that an angular deviation of the diffracting planes of 0.1° would already degrade the neutron flux at the sample position. Indeed, it would induce a spatial deviation of the diffracted beam intensity of the order of 10 mm at the collimator position so that a substantial part of the beam is stopped by the shielding. Such an angular limit is smaller than the typical horizontal peak shift of 0.15° observed for most crystals.

Depending on the sense of curvature (concave or convex), both horizontal and vertical crystal curvatures give rise to under- or overfocusing effects increasing the dilution of neutrons at the sample position. The screw type deformation also disperses neutrons leading to a loss of otherwise useful flux. Finally, the non-uniform spatial distribution of the mosaicity causes a degradation of the instrument resolution and may cause a non-uniform intensity distribution over the beam spot received by the sample which might affect the quality of the experiments.

7. Conclusion

The first neutron monochromator made of diamond mosaic crsytals has been built at the ILL with the aim to upgrade its hot neutron diffractometer D9. This prototype device is composed of 16 crystal stacks 0.55 cm thick and 1.5 x 1.5 cm² wide with an effective mosaic spread between 0.25° and 0.30°. Double-crystal experiments conducted at the ILL have shown that such composite systems would outperform existing copper crystal monochromators by more than a factor 2 for the production of short neutron wavelengths below 1 Å. However, the diamond prototype device did not provide any gain over the copper monochromator in terms of both neutron flux at the sample position and resolution.

Detailed characterisation studies using high-energy X-ray diffraction revealed that most of the diamond crystals exhibited quite complex structural deformations of the lattice planes such as curvature and twist. The spatial mosaic distribution was not uniform as well. These non-uniformities of the spatial and angular mosaic block distribution led to even a slight decrease of the diamond monochromator performance compared to that of the existing copper monochromator with undistorted lattice planes and a homogeneous distribution of the mosaic structure.

Acknowledgements

The authors are grateful to ILL staff members involved in the construction and the alignment of the diamond monochromator: Erwin Hetzler (neutron measurements on T13C), Renaud Silvestre (high-energy X-ray characterisation), Jérémie Baudin-Cavallo and Benoît Mestrallet (crystal mounting). We thank John Archer for his help during the experiments performed on the diffractometer D9.

References

[1] Freund A K 2009 *J. Appl.Cryst.* **42** 36-42
[2] Freund A K, Gsell S, Fischer M, Schreck M, Andersen K H, Courtois P, Borchert G and Skoulatos M 2011 *Nucl. Instrum. Methods Phys. Res.* **A634** 28-36
[3] Gsell S, Bauer T, Goldfuss J, Schreck M, Stritzker B 2004 *Appl. Phys. Lett.* **84** 4541-4543
[4] Fischer M, Gsell S, Schreck M, Brescia R and Stritzker B 2008 *Diam. Relat. Mater.* **17** 1035-1038
[5] Courtois P *et al.* 2010 *ILL Annual Report*
[6] *http://www.ill.eu/html/instruments-support/instruments-groups/instruments/d9*
[7] Sears V F 1997 *Acta. Cryst.* **A53** 35-46
[8] Fischer M, Freund A K, Gsell S, Schreck M, Courtois P, Stehl C, Borchert G, Ofner A, Skoulatos M and Andersen K H 2013 *Diam. Relat. Mater.* **37** 41-49
[9] Freund A K 1985 *Nucl. Instr. Methods Phys. Res.* **A238** 570
[10] Freund A K 1983 *Nucl. Instr. Methods* **213** 495
[11] Freund A K 2013 *these proceedings*
[12] Courtois P, Hamelin B, and Andersen K H 2004, *Nucl. Instr. Methods Phys. Res.* **A529** 157
[13] Bastie P, Hamelin B and Courtois P 2000 *J.Phys. IV France* **10** 21
[14] Hewat A W 1975 *Nucl. Instr. Methods* **127** 361
[15] Cagliotti G and Ricci F P 1962 *Nucl. Instrum. Meth.* **15** 155

Effect of crystal shape on neutron rocking curves of perfect single crystals designed for ultra-small-angle scattering experiments

A K Freund[1,*] and C Rehm[2]
[1] Via Cordis Consulting and Movement, Talence, France
[2] Australian Nuclear Science and Technology Organisation, Kirrawee DC, Australia

E-mail: kafreund@free.fr

Abstract. The present study has been conducted in the framework of the channel-cut crystal design for the *Kookaburra* ultra-small-angle neutron scattering (USANS) instrument to be installed at the OPAL reactor of ANSTO. This facility is based on the classical Bonse-Hart method that uses two multiple-reflection crystal systems. The dynamical theory of diffraction by perfect crystals distinguishes two cases: the Darwin case applying to infinitely thick crystals and the Ewald solution for very small absorption taking into account the reflection from the rear face of a plane-parallel crystal reflecting in Bragg geometry. The former is preferable because it yields narrower rocking curves. To prevent the neutrons to "see" the rear face, grooves were machined into the backside of perfect Si test crystals for single reflection and filled with neutron absorbing material. These samples were examined at the S18 instrument of the Institut Laue-Langevin. Unexpectedly the crystals with empty slots showed an increase of the rocking curve width. When filling the slots with an absorber the widths decreased, but without reaching that of the Darwin curve. Understanding the results and achieving a successful crystal design call for the development of a theory that permits to describe neutron diffraction from crystals with a structured back face.

1. Introduction

The present study has been conducted in the framework of the channel-cut crystal design for the new ultra-small-angle neutron scattering (USANS) instrument named *Kookaburra* that is presently under commissioning at the ANSTO OPAL reactor [1]. This facility is based on the classical Bonse-Hart method [2] which consists of using two multiple-reflection crystal systems arranged in non-dispersive geometry to achieve a very steep decrease of the tails of the perfect crystal diffraction curves. This technique permits to detect very small angular deviations of the neutron beam after scattering from a sample placed between the two channel-cut crystals. The state-of-the art of this method and its performance have been reviewed recently by Agamalian *et al.* [3].

The diffraction process of X-rays and neutrons by perfect crystals is described by the dynamical theory. Two cases are distinguished:

i) The Darwin solution for zero absorption that applies to infinitely thick crystals.

ii) The Ewald solution for very small absorption that takes into account the influence of the rear face of a crystal reflecting in Bragg geometry.

[*] To whom any correspondence should be addressed.

Because the former yields narrower rocking curves, for USANS experiments it is preferable to the latter. Several multiple reflection crystal schemes have been designed successfully to achieve the Darwin case [4,5]. However, they cannot be applied to the present instrument because of the big crystal size needed. Therefore we have been looking for a different crystal shape.

The idea we followed was to machine grooves into the backside of the channel-cut crystals and to fill them with neutron absorbing material such as cadmium or boron carbide powder. Indeed, this design prevents the neutrons to "see" the rear face of the crystal. Such crystals have been fabricated for single reflection and examined at the S18 instrument of the Institut Laue-Langevin, Grenoble, by recording rocking curves. Following a brief description of the *Kookaburra* layout, the details of the crystal design and the results will be given.

2. The USANS instrument *Kookaburra* at ANSTO

Figure 1 shows the principle of the Bonse-Hart small angle camera *Kookaburra* installed on a cold neutron guide at ANSTO's Opal reactor. A detailed optimisation including the discussion of several scenarios has been carried out earlier [6].

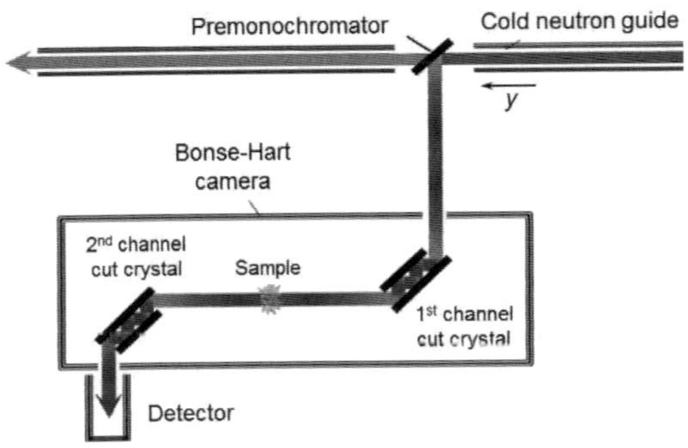

Figure 1. Principle of the USANS facility at ANSTO.

The following design was chosen. A pyrolytic graphite doubly curved premonochromator set at a Bragg angle of 45° reflects simultaneously neutrons of wavelengths 4.74 Å and 2.37 Å from its (002) and (004) lattice planes, respectively.

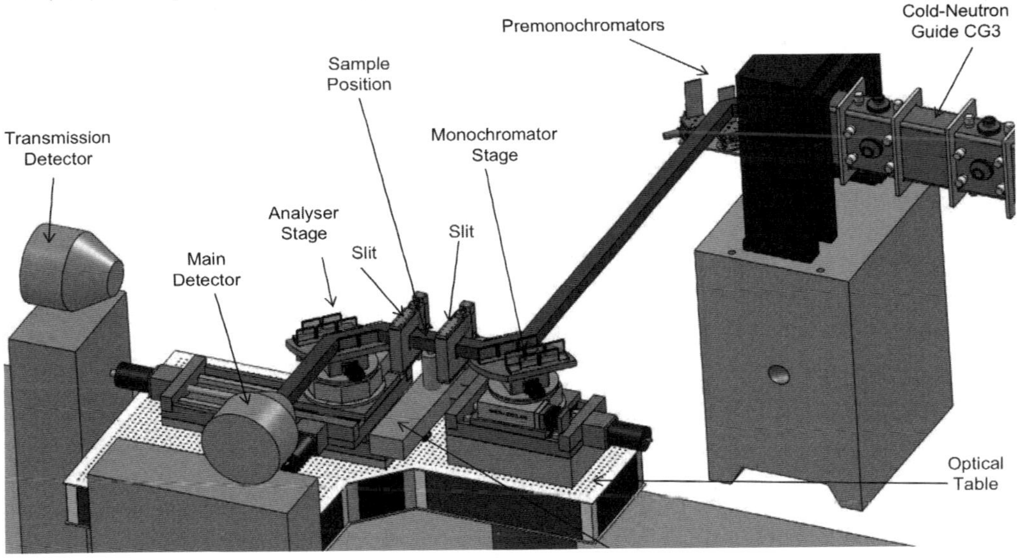

Figure 2. View of *Kookaburra* showing the two sets of monochromator/analyser channel cut crystals.

The deflected beams enter two sets of fivefold reflection channel cut crystals that can be brought into the Bragg condition alternatively (figure 2). The first set of two perfect Si (111) oriented crystals reflects the 4.74 Å neutrons at a Bragg angle of 49.1° while the second set of two Si (311) oriented crystals reflects the 2.37 Å neutrons at a Bragg angle of 46.4°. Figure 3 gives an impression of the size of the channel cut crystals and describes the way how they were cut out of a 100 mm diameter float zone ingot. The <-110> growth axis is typically 6° off the cylinder axis. It is seen that the maximum size is limited by the diameter of the ingot and the angular offset.

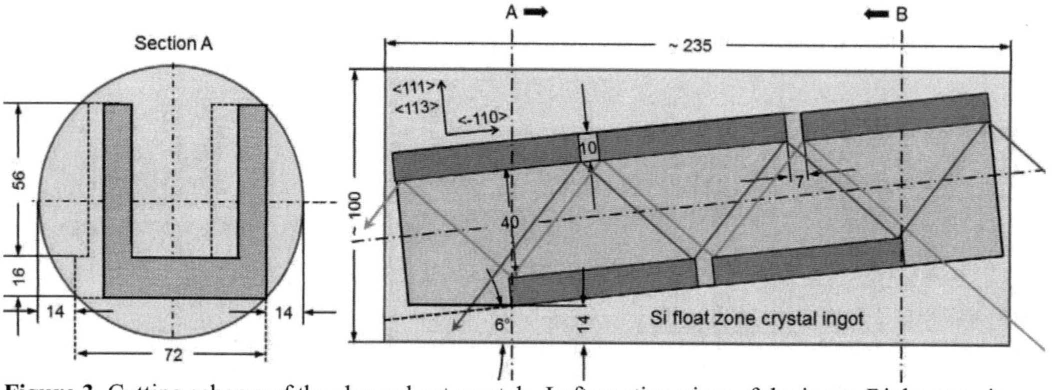

Figure 3. Cutting scheme of the channel cut crystals. Left: section view of the ingot. Right: top view.

3. Theoretical background

In publications referring to USANS and USAXS techniques, see *e.g.* [1], the multiple reflection curves are usually given according to two cases, the Darwin and the Ewald case. The first holds if absorption is so big that reflections from the back face of the crystals can be neglected which is typical for X-rays, or if such reflections can be avoided by special crystal shapes for neutrons that penetrate typically several cm into materials such as Si.

For the Darwin case the diffraction curve is given by

$$R_D(y) = \frac{P_G}{P_e} = \begin{cases} 1 & |y| \leq 1 \\ \left[|y| - \sqrt{(y^2 - 1)} \right]^2 & |y| > 1 \end{cases} \tag{1}$$

whereas the Ewald solution yields

$$R(y) = \frac{P_G}{P_e} = \begin{cases} 1 & |y| \leq 1 \\ 1 - \sqrt{1 - \frac{1}{y^2}} & |y| > 1 \end{cases} \tag{2}$$

where y is a dimensionless parameter that is half the width of the Darwin plateau. On the angular scale the full width, $\Delta\theta_D$, is obtained as

$$\Delta\theta_D = 2[\lambda^2/\sin 2\theta_B][|F(h,k,l)|e^{-W}/\pi V_o] \tag{3}$$

where λ is the wavelength, θ_B the Bragg angle, $F(h,k,l)$ the structure factor containing the coherent scattering length, e^{-W} the Debye-Waller factor where $W = B/(2d_{hkl})^2$, d_{hkl} the lattice plane spacing, B a parameter depending on the Debye temperature and V_o the unit cell volume. Specifically for the Si (311) reflection relevant for this study, a Bragg angle of $\theta_B = 46.4°$ and $\lambda = 2.37$ Å we get $\Delta\theta_D = 5.04$ µrad or about 1 arcsec. In the case of a single reflection we obtain for the full width at half maximum 5.82 and 5.35 µrad for the Ewald and the Darwin case, respectively, i.e. a difference of about 9% between the two solutions. This difference should be easily observable experimentally.

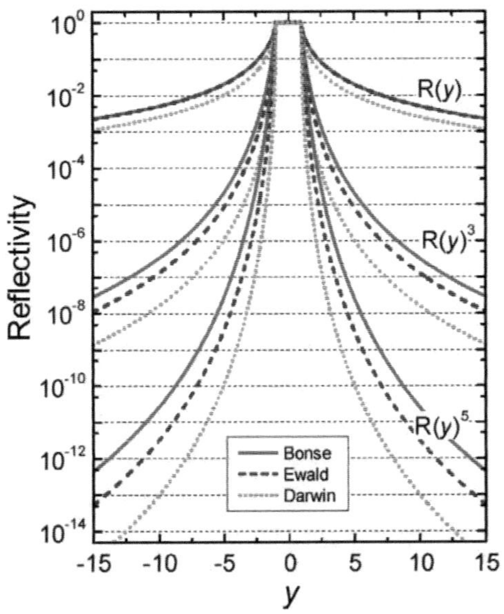

Figure 4. Theoretical perfect crystal reflectivity profiles for single, triple and quintuple bounces. From refs. [7,8].

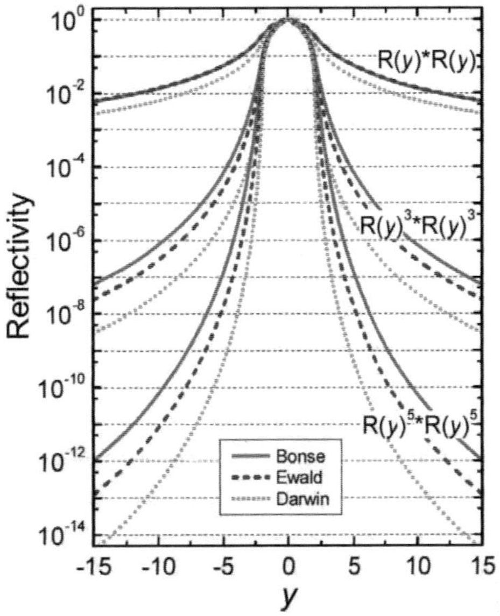

Figure 5. Double crystal rocking curves for single, triple and quintuple bounces. From refs. [7,8].

Figures 4 and 5 show the universal theoretical diffraction profiles and the double-crystal rocking curves, respectively, for perfect crystals and negligibly small absorption. The latter condition holds for neutrons attenuation in perfect Si, because the extinction depth of less than typically 10 μm is very small compared to the absorption depth. In addition to the Darwin and Ewald cases a set of third curves is plotted labelled "Bonse" that corresponds to equations given in [7,8] and is identical to the Ewald solution for single reflections. It is seen that the difference between the Darwin and the Ewald curves increases with the number of reflections and with the distance from the peak centre.

4. Crystal designs

In this section we first recall two channel cut crystal designs that permitted to achieve Darwin type rocking curves for neutrons. Then we describe two crystal designs that should allow us to reach the same goal.

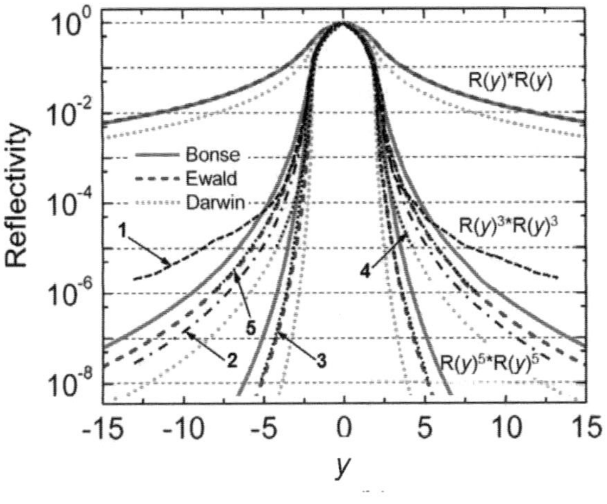

Figure 6. Comparison of some published work. The red (full), blue (broken) and green (dotted) curves are the same as in figure 5. The curve assignments are:
1 - ref. [9], experiment, $n = 3$, $m = 3$, Si 111.
2 - ref. [9], theory, $n = 3$, $m = 4$, Ewald.
3 - ref. [9], theory, $n = 5$, $m = 5$, Ewald.
4 - ref. [4], theory and experiment, $n = 3$, $m = 3$, Si 111, Darwin.
5 - ref. [10], theory, $n = 3$, $m = 3$, Ewald.
The numbers n and m designate the number of reflections in the first and the second crystal, respectively.

An overview of experimental results and a comparison with theory is compiled in figure 6. All the calculated Ewald and Darwin curves for $n = m = 3$ and 5 agree perfectly well with each other, as to be expected. Regarding experimental data, excellent agreement with theory has been obtained for $|y| < 4$, even for the Darwin case [4], but significant deviations occur for $|y| > 4$.

4.1. Existing designs

The two crystal designs achieving the Darwin type rocking curves are shown in figure 7. In the first case (figure 7a) the crystal reflectors were made so thick that the part of the neutron beam hitting the back side of the crystal could not reach the front face after reflection. In this way the term that contributes to reflection from the backside, which gives rise to the broader Ewald profile, was suppressed. Indeed, perfect agreement with the Darwin theory was found.

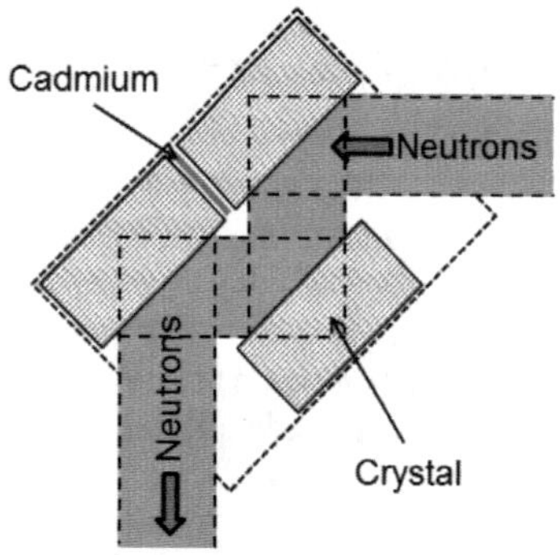

Figure 7a. Crystal design according to ref. [4].

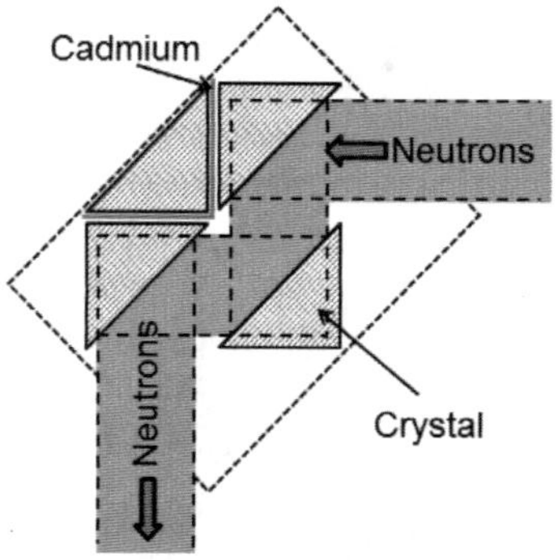

Figure 7b. Crystal design according to ref. [5].

In the second case (figure 7b) the crystals were given a triangular shape so that the neutrons did again not "see" any back side parallel to the front side [5]. Thus the broadening term, see figure 6 in ref. [5], was suppressed as shown experimentally, see figure 10 in ref. [5]. In addition, both geometries permitted to reduce thermal diffuse scattering that otherwise would contribute to the general background.

4.2. Proposed designs

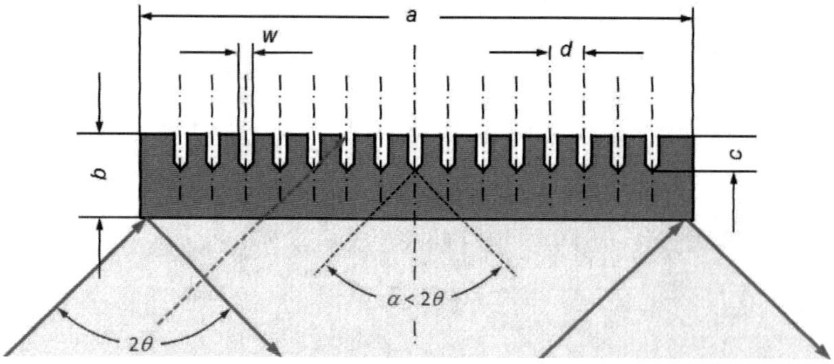

Figure 8. Slotted crystal back to avoid reflections from the rear face. The slots are to be filled with a neutron absorbing material such as cadmium. The neutron beam cannot "see" the back face at all (see broken line). Condition: $c/d > \tan\theta$ where θ is the Bragg angle.

The neutron beam size and the crystal size restrictions for *Kookaburra* shown in figure 3 made it impossible to apply the designs shown in figure 7. Therefore another design was imagined following the principle that the neutrons should not see the back face or, if they see it, they should not be able to interfere with the reflection from the front face. The principles of the two proposals are shown in figures 8 and 9.

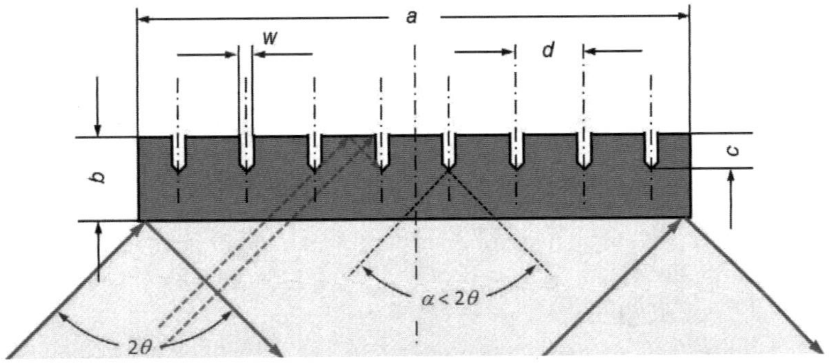

Figure 9. Same design as in figure 9 but with fewer slots. Part of the neutron beam can "see" the back face but cannot escape to the front because of the absorbing material (not shown in the figure) by the cadmium, see the broken line. Condition: $2c/d > \tan\theta$.

For both these designs it is very important that at least 100 μm of Si are removed by chemical etching in order to recover the unperturbed, perfect crystal after cutting and fine grinding. This requirement conditioned the width w of the slots that should be big enough to allow for easy circulation of the etching liquid. A slight rounding of the bottom due to the shape of the cutting tool and the etching occurred that could lead to a small effect on the Darwin profile.

The design shown in figure 9 is a simplified version of the design in figure 8. Here part of the beam hits the back face but cannot escape to the front because it is absorbed. Typical dimensions of the final channel cut reflectors (Si 311, $\theta = 46.4°$) are reflector length a = 82 mm, width b = 12 mm and height h = 59 mm, slot depth c = 5 mm and distance d = 9 mm. The width b is a compromise between mechanical stability including machining risks and the smallest possible crystal volume giving rise to thermal diffuse scattering.

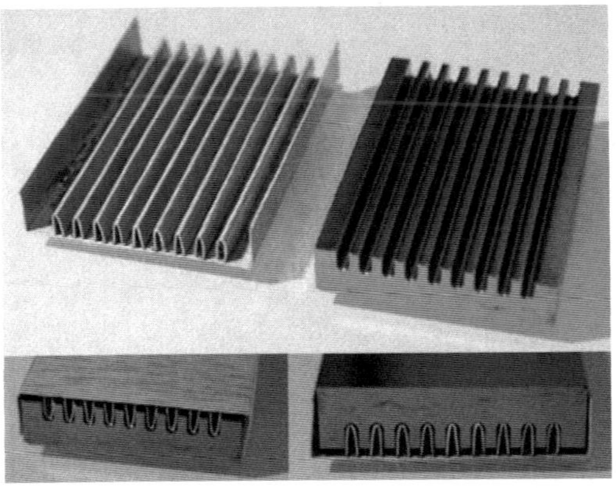

Figure 10. Sample C (9 slots) and the cadmium absorber inserted in the slots and surrounding the back and the sides of the crystal.

The crystals to be prepared for test experiments on the instrument S18 of the ILL [11] should have the same values for b, c, d and w as the final reflector. However, since the beam width at S18 is < 30 mm and its height is < 50 mm, a crystal length of 48 mm and a height of 70 mm were fully sufficient. The three 311 crystals were b = 12 mm thick. Crystal A served as a reference and had no slots. Crystal B had five slots and crystal C 9 slots with a slot distance of 9 mm and 4.5 mm, respectively. The other dimensions were: slot depth c = 5 mm and width w = 2 mm. The crystals were manufactured by the company Andrea Holm, Tann, Germany. The reflecting surfaces were fine ground, deep etched and polished using a strain-free chemical-mechanical process. The other surfaces were deep etched. Photographs of the sample with 9 slots and the cadmium absorber are shown in figure 10. As an alternative to Cd as absorber, in a second test run B_4C powder (24 μm mesh) was filled into the slots and kept in place by adhesive tape.

5. Experiment setup and theoretical prediction

The experimental setup on S18 and a rocking curve obtained with crystal A are shown in figure 11. The Bragg angle θ_B was 45° and the wavelength λ = 2.3158 Å. Because the perfect monochromator crystal was 110 oriented, the 311 reflection was asymmetric from the 111 oriented side-face. To avoid the part of the beam exiting the 110 front face, a slit was positioned intersecting the unwanted neutrons. The reflected beam was 10 mm wide and 25 mm high. The rocking curve was nicely symmetric. It was flat-topped because the reflection curve at the monochromator exit was 1.9 times wider than the symmetric reflection from the sample.

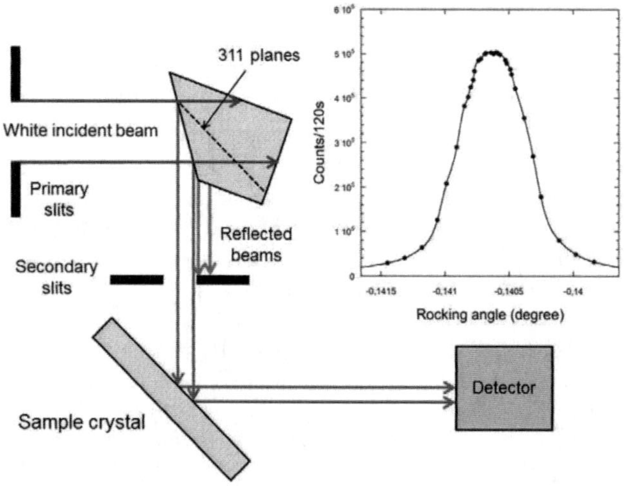

Figure 11. Experimental setup on the instrument S18.

The following numerical data have been calculated from dynamical theory, see equation 3.

<u>*Monochromator*</u>

- Asymmetry angle α = angle between the 311 and the 111 (surface) planes = 29.5°.
- Asymmetry factor: $b^{1/2} = [\sin(\theta_B + \alpha)/\sin(\theta_B - \alpha)]^{1/2}$ = 1.900 for θ_B = 45°.
- Full width of the Darwin plateau, asymmetric geometry: $2y_a = 2y_s \cdot b^{1/2} \Rightarrow \Delta\theta_{Ds} \cdot b^{1/2}$ = 0.523 mdeg.
- Full width at half maximum, ω_a, asymmetric case:
- Ewald case: $\omega_{Ea} = 2\Delta\theta_{Ds} \cdot b^{1/2}/\sqrt{3}$ = 0.604 mdeg.
- Darwin case: $\omega_{Da} = 3\Delta\theta_{Ds} \cdot b^{1/2}/2\sqrt{2}$ = 0.555 mdeg.

<u>*Sample*</u>

- Full width at half maximum, ω_s, symmetric case:
- Ewald case: $\omega_{Es} = 2\Delta\theta_{Ds}/\sqrt{3}$ = 0.318 mdeg.
- Darwin case: $\omega_{Ds} = 3\Delta\theta_{Ds}/2\sqrt{2}$ = 0.292 mdeg.

<u>*Convolution of monochromator and sample curves*</u>

a) Convolution factors, 2 x symmetric and 2 x asymmetric geometry, respectively:

Ewald case: $f_E = \pi/(2\pi - 4)$ = 1.376 => total width Ω_{Es+s} = 0.437 mdeg; Ω_{Ea+a} = 0.830 mdeg.

Darwin case: f_D = 5/4 = 1.250 => total width Ω_{Ds+s} = 0.365 mdeg; Ω_{Da+a} = 0.693 mdeg.

b) Convolution, asymmetric (monochromator) plus symmetric (sample) geometry:
The convolution of two perfectly rectangular distributions of widths w_1 and w_2 has a trapezoidal shape of FWHM equal to the bigger width. Thus, as a first approximation we get for the convolution of the two diffraction curves a total width of $\Delta\theta_{DS}\cdot b^{1/2}$ = 0.523 mdeg, multiplied by a smearing factor due to the fact that the Darwin curves are not perfectly rectangular and then by the convolution factor. For simplicity we assume that the width of the convolution of the two curves is the mean of the asymmetric and symmetric widths.

The final results are then for single reflection curves:
- Ewald case: Ω_{Ea+s} = 0.64 mdeg.
- Darwin case: Ω_{Da+s} = 0.53 mdeg.
- Difference of the FWHM between the two cases: 0.1 mdeg or 17%.

The combination of a first, asymmetric monochromator crystal, Ewald case, with a second, symmetric sample crystal, Darwin case, would give a width somewhere in-between the Ewald and the Darwin case (mixed Ewald-Darwin case), say, in the middle at 0.59 mdeg.

6. Experimental results, discussion and conclusion

Typical examples of rocking curves over a wide angular range and in the central part are given in figure 11. They have been obtained from the reference crystal without slots.

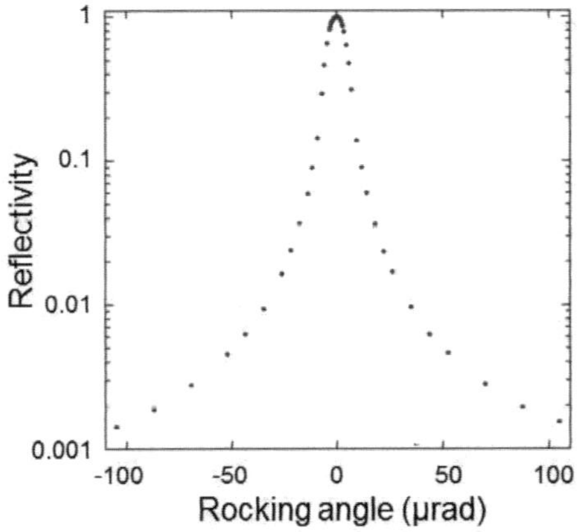

Figure 12a. Rocking curve over a wide angular range for crystal A.

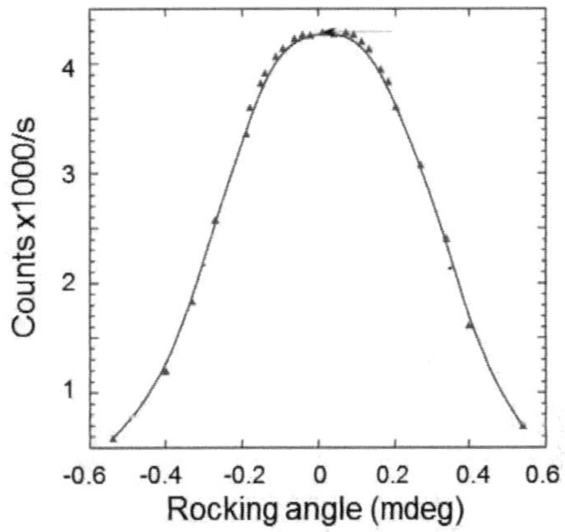

Figure 12b. Central part of the rocking curve shown in figure 12a.

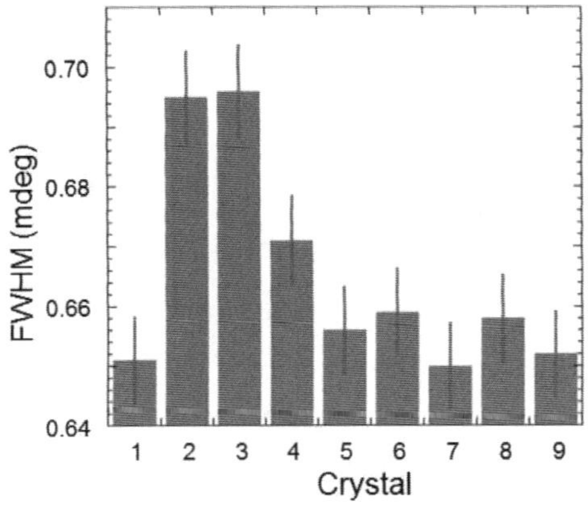

Figure 13. The FWHM of the rocking curves for the following crystal cases:
1. A – no grooves
2. B – 5 grooves, no absorber, first run
3. B – 5 grooves, no absorber, second run
4. C – 9 grooves, no absorber
5. B – 5 slots filled with Cd
6. C – 9 slots filled with Cd
7. B – 5 slots filled with B_4C powder
8. C – 9 slots filled with B_4C powder
9. Backside of a big channel cut crystal

The red vertical lines are the error bars.

The results of the rocking curve full width at half height are summarised in figure 13.

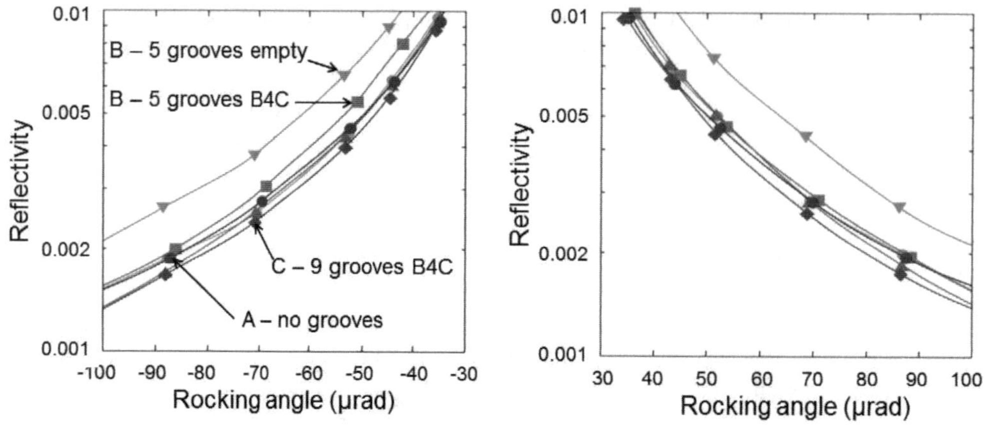

Figure 14. The decrease of the rocking curves in the wings. Left: low-angle side; right: high angle side. The angular scan range of ± 100 µrad corresponds to a range of ± 42 for *y*.

We found very good agreement of the theoretical Ewald FWHM with the experimental result of (0.66 ± 0.01) mdeg observed for crystal A (no slots) showing that the calculation of the convoluted curve and the crystal preparation and mounting have been correct. The smallest value of (0.65 ± 0.01) mdeg for the width of the slotted crystals was obtained for crystal B with B_4C powder in the slots. Within the experimental error this value was identical with the result for crystal A and for the backside of a channel cut crystal. It was above the theoretical result of 0.58 mdeg for the mixed Ewald-Darwin case. For the very far wings crystal C with B_4C in the slots gave the smallest intensities, see figure 14. The 0.01 reflectivity level is crossed at a rocking angle of ± 33 µrad or $y = \pm 14$. This is slightly above the theoretical values for the Ewald case as shown in figure 6 for two single reflections.

The crystals with empty slots exhibited an increase of the rocking curve width outside the error bars above that of the Ewald case, see figure 13. To make sure that there was no experimental error we remeasured the width for crystal B in a second run after dismounting and remounting the crystal and realignment of the vertical tilt angle. The increase was bigger for crystal B (5 slots) than for crystal C (9 slots), which is counterintuitive when supposing that the presence of slots introduces a perturbation of the dynamical wavefield inside the crystals. It could not be ascribed to residual strains. When filling the slots with an absorber, the width of the diffraction profiles decreased, but without reaching that of the Darwin curve. Filling the slots with B4C gave slightly better results than using Cd, which might be due to a better contact of the powder with the groove walls. It is also interesting to note that according to the experiment the reflection curve of the first (monochromator) crystal used on S18 seemed to be of the Ewald type although the neutrons hitting the rear crystal face were eliminated by the slit, see figure 11, and thus the Darwin case should apply.

In conclusion, the results for the widths at half maximum and in the wings did not indicate any shift from the Ewald type diffraction curves towards the Darwin type ones when using the grooved crystals with absorbers. Understanding the experimental results of the present study and determining a successful crystal design for achieving Darwin type rocking curves call for the development of a theory that permits to describe neutron diffraction from crystals with a back face that is different from flat. A theory based on a finite element approach integrated into dynamical theory is presently being developed [12]. First promising results have already been obtained and work is in progress.

Acknowledgments
The authors would like to thank Dr U Kuetgens from the PTB Braunschweig, Germany, for providing the theoretical calculations of the perfect crystal reflectivities. We are grateful to the S18 CRG for kindly providing beamtime for our proposal. We express our sincere thanks to Dr H Lemmel from the Atominstitut Vienna, Austria, for his devoted and very effective assistance during the experiments.

Journal of Physics: Conference Series **528** (2014) 012002 doi:10.1088/1742-6596/528/1/012002

References

[1] Rehm C, Brûlé A, Freund A K and Kennedy S 2013 *J. Appl. Cryst.* In the press
[2] Bonse U and Hart M 1965 *Appl. Phys. Lett.* **7** 238-240
[3] Agamalian M, Carpenter J M and Treimer W 2010 *J. Appl. Cryst.* **43** 900-906
[4] Wagh A G, Rakhecha V C and Treimer W 2001 *Phys. Rev. Lett.* **87**, 125504-1
[5] Villa M, Baron M, Hainbuchner M, Jericha E, Leiner, V, Schwahn, D, Seidl, E, Stahn J and Rauch, H 2003 *J. Appl. Cryst.* **36** 769-773
[6] Freund, A K and Rehm, C 2011 *Nucl. Instrum. Methods Phys. Res. A* **634** S81- S89
[7] Freund, A K, Rehm, C and Kuetgens, U 2011 *ECNS 2011 Conference, Prague,* unpublished
[8] Bonse, U and Graeff, W 1977 *X-Ray Optics,* Queisser H editor, *Topics in Applied Physics* **22** Springer (Berlin, New York)
[9] Agamalian M, Christen D K, Drews A R, Glinka C J, Matsuoka H and Wignall G D 1998 *J. Appl. Cryst.* **31** 235
[10] Treimer W, Strobl M and Hilger A 2002 *Phys. Rev. Lett. A* **305** 87-92
[11] https://www.ill.eu/fr/instruments-support/instruments-groups/instruments/s18/
[12] Ferrero C and Mocella V (2013) *private communication*

Focusing and reflectivity properties of a parallel double bent crystal (+n,-m) setting

P Mikula[1], M Vrána[1], J Pilch[1], J Šaroun[1], B S Seong[2], C Woo[2] and V Em[3]
[1]Nuclear Physics Institute ASCR, v.v.i., 25068 Řež, Czech Republic
[2]Neutron Science Division, Korea Atomic Energy Research Institute, 1045 Daedok daero, Yuseong-gu, Daejeon, 305-353, Republic of Korea
[3]National Research Centre "Kurchatov Institute", Moscow, 123182, Russia

E-mail: mikula@ujf.cas.cz

Abstract. After preliminary results obtained and published recently [1], in our contribution focusing and reflectivity properties of the dispersive double bent-crystal arrangement are presented in much more detail. It has been found that two different bent perfect crystals in (+n,-m) setting can be good candidates for high efficiency neutron microfocusing as well as high-resolution monochromatisation. Due to the (+n,-m) setting of two different bent perfect crystals, a high resolution is expected in both $\Delta(2\theta)$ (2θ is the scattering angle) as well as $\Delta\lambda/\lambda$ (λ is the neutron wavelength). Experimental tests were carried out with the setting employing the bent Si(111) slab and Si(220)-sandwich, which contained either one, or two or four 1.3 mm thin simply stacked slabs. Thanks to a high reflection probability of both bent elements and an easy manipulation with the curvature of the Si(220)-sandwich, an excellently focused intensive monochromatic beam of the width from one to several millimetres was obtained. The properties of the double bent-crystal setting were studied in Rez at the neutron optics diffractometer for the neutron wavelength of 0.162 nm and for various thicknesses and curvatures of the Si(220)-sandwich. It has been also found that besides an excellent focusing and reflectivity properties of the dispersive double bent-crystal setting the obtained monochromatic neutron current is sufficiently high for standard high-resolution diffraction experiments even at the medium power research reactor.

1. Introduction

Nondispersive as well as dispersive double-crystal monochromators are quite common at synchrotron sources. However, in the case of neutrons, at present one can find only several scattering instruments in the world which use nondispersive double-crystal setting for neutron monochromatisation (e.g. HB-1A Triple-Axis Spectrometer at High Flux Isotope Reactor in ORNL, Double Crystal Diffractometer at NIST for USANS, 4F1 and 4F2 spectrometers in LLB Saclay). Such monochromator with two parallel crystals produces monochromatic beam running parallel to incident white beam. For the change of the neutron wavelength such monochromator does not need a movable (rotating) table and the position of the sample is fixed. The problem occurs when using mosaic crystals. Due to the double reflection process monochromatic beam current is much smaller than in the case of one crystal monochromator. For one mosaic monochromator the neutron current is proportional to its peak reflectivity r (r<1). Then for double crystal monochromator it is proportional roughly to r^2 and the situation becomes dramatically inconvenient when $r\approx0.5$ or less

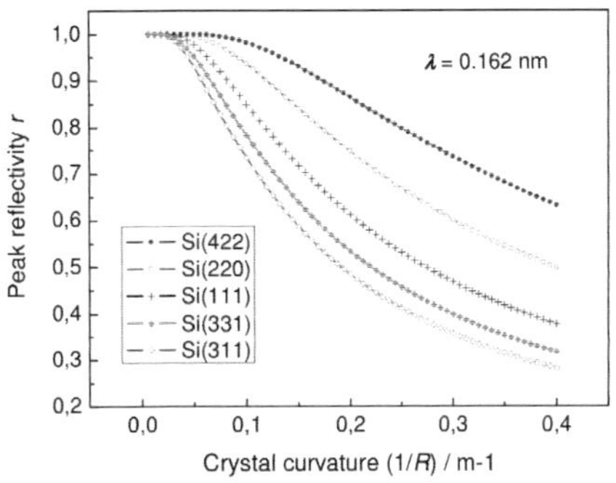

Figure 1. Peak reflectivity vs crystal curvature related to different Si reflections as calculated on the basis of eq. (1) for symmetric reflection geometry (Ψ=0°).

(for details see ref. 1). The situation is even much worse in the case of dispersive double-crystal (1,+1) setting employing mosaic crystals and the obtained monochromatic current is rather small because the individual space-phase elements of the crystals do not match as it is in the case of the nondispersive (n,-n) setting. Therefore, it is commonly accepted opinion that a dispersive double-crystal (DDC) setting is not convenient for neutron monochromatisation due to a low monochromatic neutron current and thus a low luminosity of the scattering instrument which is really true in the case of mosaic crystals. However, in the case of bent perfect crystals (e.g. Si or Ge) the situation is much different and some advantages of the DDC setting can be found. First, one should consider that when the curvature of the crystals is not drastic, the reflection probability of both bent perfect crystals can be close to 1 and there is a small acceptable loss of neutrons due to the double reflection process. The general formula for the reflection probability (often called as peak reflectivity) of neutron passing through a homogeneously deformed perfect crystal has been derived in the form as [2,3]

$$r(R_M) = \{1 - \exp[-Q_{hkl}(\partial\Delta\theta/\partial s_0)^{-1}]\}, \qquad (1)$$

where $(\partial\Delta\theta/\partial s_0)$ is the rate of change of the Bragg angle on the flight path Δs_0 in the crystal in the incident beam direction and Q_{hkl} is the kinematical reflectivity of the crystal unit volume. For cylindrically curved monochromator it has been derived in the general form [3,4] when including also asymmetric diffraction geometry given by the angle ψ of the asymmetry cut of the crystal

$$\partial\Delta\theta/\partial s_0 = (1/R_M)(\cos\psi/\cos\theta_M)[1-(1+\sigma)\sin(\theta_M+\psi)\sin(\theta_M-\psi)], \qquad (2)$$

where R_M is the bending radius, θ_M the Bragg angle and σ the Poisson constant. An example of several calculated peak reflectivities for Si crytal which perfectly fit with experiment [2,3] is shown in figure 1. Furthermore, curved crystal instruments are known by the fact that they do not use Soller collimators and the resolution can be controlled through the bending of the crystals. Possible employment of the focusing does not necessarily imply focusing in real space, but it can also cover focusing in scattering which is very often used in powder diffractometry, namely, for strain/stress measurements in polycrystalline materials [5-9]. Double-crystal arrangements with the bent perfect crystals provide variety of performances which could be exploited in neutron diffractometry. With two bent crystals in successive reflection, the double crystal rocking curve is generally much broader in comparison with the flat perfect crystals. However, when the phase space windows of both bent crystals match, the double crystal rocking curve is narrow because it is related only to the convolution of the reflectivity curves and thus the intensity of the doubly reflected beam is high. This matching of the phase space windows can be even achieved with the bent crystals having different *d*-spacings by a proper adjustment of individual curvatures of the crystals [7,10]. In ref. 7 the authors point out the possibility of a rather sharp focusing (up to 1 mm spot) by means of successive reflections on three Si(111) crystals (mosaic+bent+bent) in the (+n,-n,-n) setting, however, with a rather low luminosity. Another proposal of using the antiparallel (+n,+n) setting of two bent Cu(200) crystals (copper plates of small mosaic spread of about 3′) achieving the spot width of 2.5 mm for residual strain/stress instrument was reported in ref. 11. Very promising results of an extensive investigation of (n,-m) combination of a vertically focusing mosaic Ge(331) crystal in combination with a horizontally bent perfect Si(111) crystal were published in papers [12,13]. Here, it was clearly demonstrated both real space focusing and focusing in scattering on a standard polycrystalline γ-Fe sample (pin of the diameter of 2 mm) at λ=0.183 nm. Furthermore, it was found an excellent property that for an optimum curvature of the bent Si(111), *FWHM*s of the γ-Fe diffraction profiles in the $2\theta_S$-angular scale for both parallel and antiparallel settings were practically the same. This fact can have enormous impact in powder diffractometry and namely for residual strain/stress measurements when permitting to measure two strain components simultaneously. Of course, that in double-crystal combination of mosaic and bent perfect crystal due to a large mosaicity of the mosaic crystal, the matching of the phase space elements is not as good as it can be in the case of two bent crystals. In this paper we deal with the results of the investigation of the properties of the (+n,-m) setting of the bent Si(111)+Si(220)-sandwich for neutron monochromatisation and its possibility to be employed in neutron diffractometry. Some results of the preliminary investigations with two thin (2x1.3 mm thickness) bent Si(220) slabs were published in ref. 1. On the basis of the obtained promising results we continued in extensive investigations of this setting by using a sandwich of four, two and one slab of Si(220), symmetric or asymmetric diffraction geometry and a large range of curvatures. In our case the (+n,-m) setting can be considered as slightly dispersive one.

2. Experimental arrangement

For this experiment we used neutron optics diffractometer installed at the 10 MW research reactor LWR-15. The diffractometer operates at the fixed neutron wave length of λ=0.162 nm which is provided by a cylindrically bent perfect Si(111) crystal of the dimensions of 200x40x4 mm^3 (length x width x thickness). Schematic sketch of the diffractometer performance is drawn in figure 2. Depending on the experimental requirements, the diffractometer can operate in two or three axis mode. First it should be pointed out that the radius of the curvature of the first Si(111) crystal of the DC setting was fixed ($R_I \approx 10$ m) which was not changed during the experiment. The second bent perfect Si(220) monochromator was in fact sandwich of four slabs, two slabs or one slab each of the dimension of 200x40x1.3 mm^3. By using a sandwich instead of a single piece of the same thickness a large range of curvatures without a danger of breaking could be used. Even in the case of diffraction studies with one Si(220) slab, a sandwich of four slabs was used when three additional slabs in the bending device had a different orientation and did not participate in the diffraction process. For neutron detection either a point detector or neutron Imaging Plate (IP) were used. The sample could be put on the third axis of the instrument which is at the distance of 50 cm from the Si(220) crystal. The distance between the Si(111) and Si(220 crystals was fixed and equal to 180 cm. No Soller collimator was installed before and after Si(111) crystal. The focal length of the bent Si(111) crystal was about 130 cm and the convergent rays of the extracted monochromatic beam crossed about 50 cm in front of the Si(220) sandwich. The beam divergence of the incident beam on the Si(111) crystal was 40′ and had no significant impact on the double crystal performance.

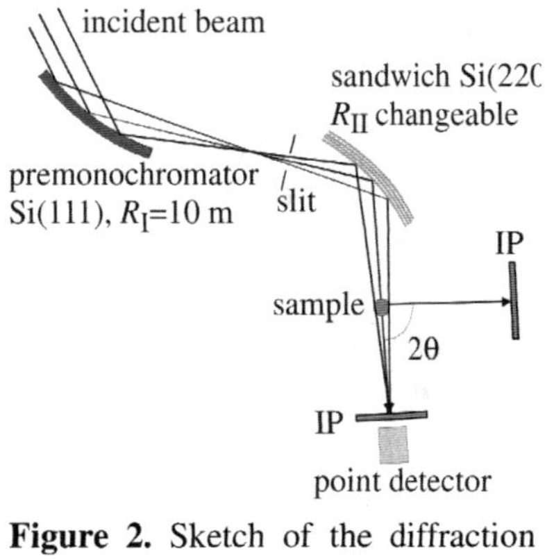

Figure 2. Sketch of the diffraction performance as used for the experimental investigations.

3. Experimental results

3.1. Symmetric diffraction geometry of the Si(220) sandwich

First of all the double crystal rocking curves of the bent Si(220) with respect to the bent Si(111) were investigated. Figure 3 shows a few chosen rocking curves related to different sandwiches, two widths of the incident beam from the bent Si(111) crystal (two widths of the slits) but all for the same radius

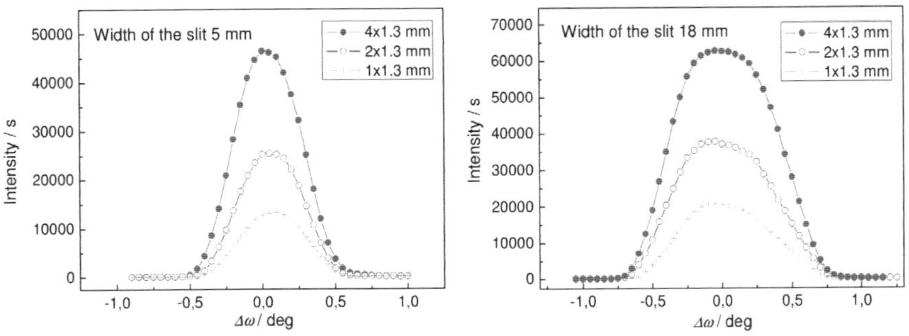

Figure. 3. Double-crystal rocking curves obtained by rocking the Si(220) sandwich with respect to the Si(111) premonochromator for the curvature of the sandwich of 0.28 m^{-1} (R=3.6 m) .

curvature of the bent Si(220) sandwich. It can be seen from figure 3 that the *FWHMs* of the rocking curves related to the 18 mm slit are much larger than that related to the 5 mm slit. However, peak intensities differ substantially in favour of using the 18 mm slit. Furthermore, it can be seen from the rocking curves that the double crystal setting provides very good intensities of monochromatic neutrons (see peak intensities of the rocking curves). As to the reflectivity of the double crystal setting, it can be seen from figure 1 that the peak reflectivity of the bent Si(111) is close to 1 and that of the sandwich is estimated to be about 0.6. It means that the peak reflectivity of the system is still high (r_{111} x $r_{220} \approx 0.6$). For the estimation of the monochromatic neutron current and focusing, the beam profile

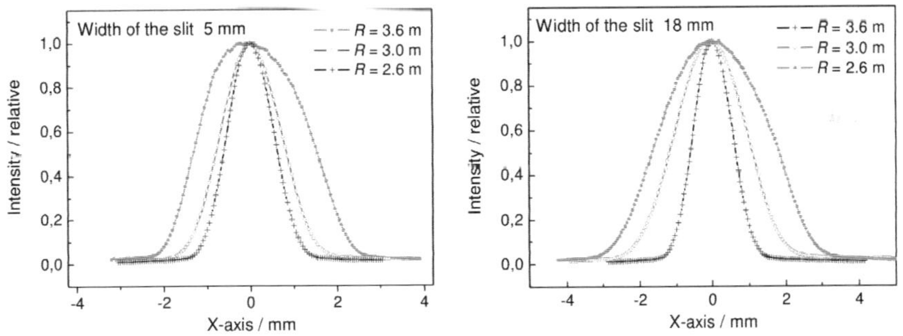

Figure. 4 Beam profiles measured by means of IP situated at the distance of 70 cm from the Si(220) crystal for three radii of curvature and for two widths of the incident premonochromatized beam.

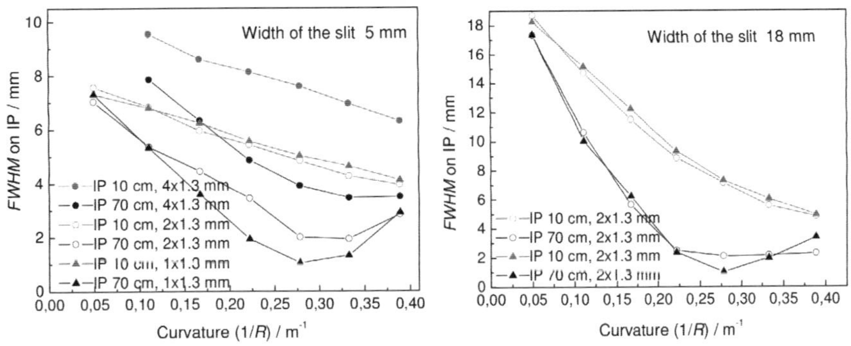

Figure 5. Width of the beam diffracted by the bent sandwich as a function of curvature for two different slits limiting the incident beam on the Si(220) sandwich.

properties were investigated at some distance from the Si(220) sandwich. For the focusing and beam profile studies an Imaging Plate situated either just behind the crystal (distance about 10 cm) or in front of the point detector at the distance of 70 cm. The images of the beam were taken for different radii of curvature of the Si(220) sandwich and again for two widths of the beam impinging on the Si(220) sandwich (5 mm or 18 mm). Figure 4 displays several beam profiles as taken by IP. Thanks to the homogeneous deformation of the crystals, each profile is smooth and has a nearly Gaussian form. Then figure 5 shows the dependence of *FWHM* of the beam profile on the curvature of Si(220) sandwich. It can be seen from figure 5 that a strong focusing effect in the range of curvatures of 0.25 m^{-1} – 0.38 m^{-1} was achieved. The minimum *FWHM* of the diffracted beam of 1 mm and the best focusing effect were found in both cases of the slits with one Si(220) slab in the sandwich for the curvature of 0.28 m^{-1} (R=3.6 m).

Then, the neutron current related to the focused beam obtained by the (+n,-m) double-crystal performance was compared with the neutron current incident on the bent Si(220) crystal for R=3.6 m. For this purpose a 2 mm slit was installed in front of the Si(220) crystal (see figure 2) and the passing beam was imaged by the Imaging Plate at the distance of 10 cm. The peak intensity of the beam profile was not dependent on the width of the slit. Vertically integrated signal registered by the IP was then compared wih the vertically integrated signal registered by the IP situated at the distance of 70 cm behind the Si(220) crystal, just before the point detector (see figure 6). It can be seen from figure 6 that thanks to the strong focusing effect the peak intensities of the obtained profiles are comparable.

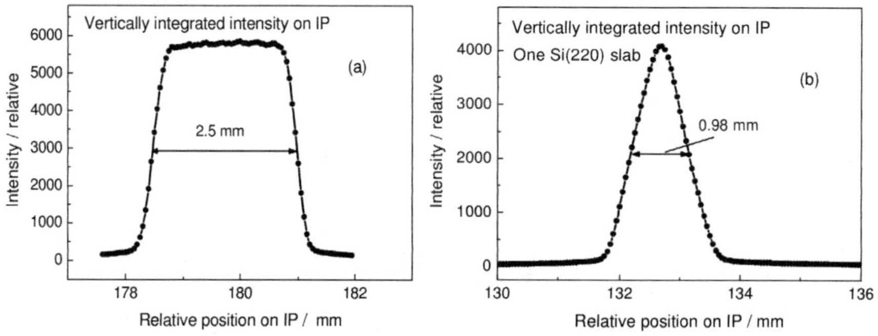

Figure 6. Relative comparison of neutron intensities: (a) - intensity passing through a 2 mm slit measured by IP before the Si(220) crystal, (b) - intensity passing through a 5 mm slit measured by IP at 70 cm behind the Si(220) crystal.

3.2. *Asymmetric diffraction geometry of the Si(220) sandwich with three operating slabs*

The additional slabs in the bending device which were cut with the longest edge parallel to the [001] scattering vector provided us the possibility to study also properties of the Si(220) sandwich in the asymmetric transmission geometry of (see insert in figure 7). Figure 7 shows the dependences of *FWHM* of the beam diffracted by the bent Si(220) sandwich (in this case with three operating slabs) in the asymmetric diffraction geometry as a function of curvature for the slit of 18 mm and two different distances from the sandwich. It can be seen from figure 7 that in this case the minimum *FWHM* (at the distance of 70 cm) achieved in the range of used curvatures was 2 mm. For a larger curvature, *FWHM* can be expected even smaller.

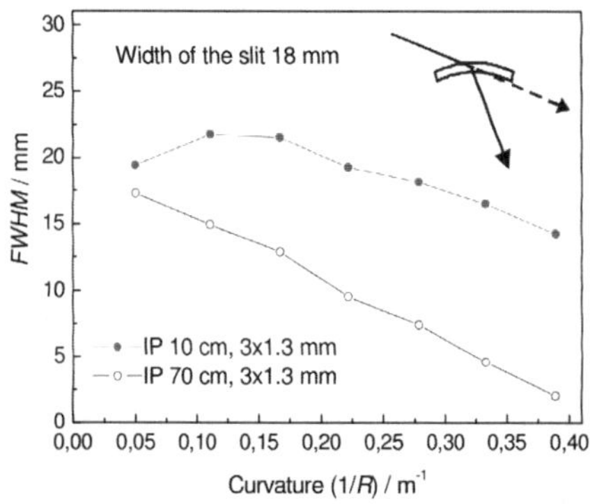

Figure 7. *FWHM* of the beam diffracted by the Si(220) sandwich in the asymmetric diffraction geometry vs its curvature.

3.3 Luminosity of the diffractometer

Of course, the luminosity of the diffraction performance which is determined by the efficiency of the double diffraction process on the bent Si(111)+Si(220) setting also plays an important role. In this way the dependence of the peak intensities of the double-crystal rocking curves of the bent Si(220) slab/slabs on the curvature with respect to the constantly bent Si(111) slab can appear very helpful. The following figure 8 shows an example of the DC rocking curves for one curvature of the sandwich and the dependence of the peak intensities of the DC rocking curves of the individual sandwiches on the curvature of the slabs for symmetric Si(220) reflection as well as of asymmetric transmission geometry. It can be seen from figure 8 that the sandwich with 3 slabs set in asymmetric transmission geometry provides slightly better reflectivity and *FWHM* parameters that the that the sandwich with 4 slabs set in symmetric reflection geometry.

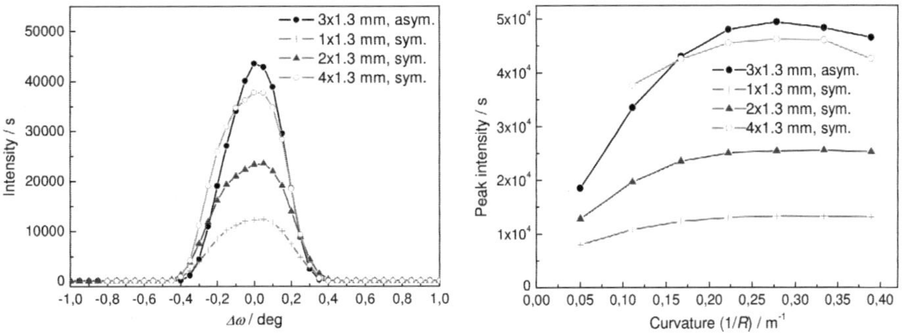

Figure 8. Examples of the rocking curves of the double crystal Si(111)+Si(220) setting for the radius of curvature R=6 m ($1/R$ = 0.167 m^{-1}) – (a) and the dependence of the peak intensities of the rocking curves on the curvature of the sandwich – (b).

3.4 Diffraction by a polycrystalline sample

Finally, in the next step, diffraction profiles on a well annealed standard α-Fe solid polycrystalline pin of 2 mm diameter were taken for different curvatures of the sandwich with one Si(220) crystal and 5 mm slit for the incident beam. The sample was installed at the third axis of the instrument at the distance of 50 cm from the Si(220) crystal and the Imaging Plate was at 45 cm from the sample (see Fig. 2). Scattering angles on the sample for α-Fe(211) and α-Fe(220) reflections were $2\theta_S$=88° and 107°, respectively. Figure 9 shows two of several results taken for different Si(220) crystal curvatures

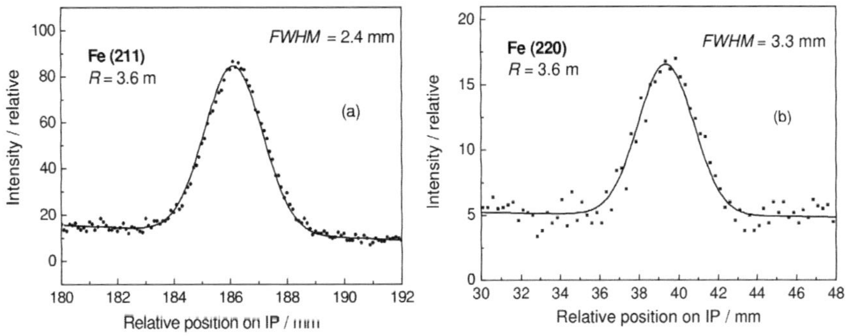

Figure 9. Diffraction profiles related to α-Fe(211) - (a) and α-Fe(220) - (b) reflections obtained by diffraction on a standard polycrystalline pin of the diameter of ϕ=2 mm and imaged by IP.

which provided minimum *FWHMs*. Inspection of figure 9 reveals that the diffracted beams from the α-Fe pin have very low divergence $\Delta(2\theta_S)$ of about 1×10^{-3} rad for 211 reflection and 3×10^{-3} rad for 220 reflection, respectively and document a high resolution property of the double-crystal performance.

4. Conclusion

It can be said from the obtained experimental results shown in figures 2-6 that contrary to a common view, the presented double-crystal monochromator with crystals in the dispersive setting can provide a highly focused neutron beam of a sufficient current at the sample position that can be used in powder diffractometry; of course, at a reasonably high flux neutron source. One should keep in mind that one works with open beams without any Soller collimators and the (+n,-m) setting behaves as a self collimation assembly producing a highly monochromatized beam. Due to the double diffraction on crystal planes of different lattice spacing, a high resolution is thus expected in both $\Delta(2\theta)$ (2θ is the scattering angle) as well as $\Delta\lambda/\lambda$. In the present case, it is necessary to point out that the instrument could not be fully optimized because of using a fixed curvature of the Si(111) crystal. Furthermore, due to the given experimental conditions, the distance between the bent crystals was also fixed and rather large (1.7 m). When all these parameters in connection with a possibility of an employment of vertical focusing are optimized, we believe that the luminosity of the presented experimental setup could be by one order of magnitude higher. Even at our medium power neutron source when using a high resolution position sensitive detector instead of IP (having rather low efficiency), the time expected for the collection of the data related to individual diffraction profile is expected to be less than 20 minutes. For the design and full optimization of such DDC setting, Monte Carlo simulations are desirable [14]. For this purpose one can use a code which has been already published and which permits an evaluation of the resolution function and intensity properties of neutron diffractometer taking into account any spatial configuration of the experimental setup and the monochromator curvature as well as including also a double-crystal monochromator.

Acknowledgements

The related measurements were carried out at the CANAM infrastructure of the NPI ASCR Rez. Bragg diffraction optics investigations are in the Czech Republic supported by GACR project no. 14-36566G, by the ESS project LM2010011: 'Contribution to Partnership in Large Research Infrastructure of Pan-European Importance' as well as by the project of EU-FP7-NMI3 II: "Integrated Infrastructure Initiative for Neutron Scattering and Muon Spectroscopy", 2009-2015.

References

[1] Mikula P, Vrána M, Šaroun J, Em V, Seong B S and Woo W 2012 *Journal of Physics*: Conference Series **340** 012014.
[2] Kulda J 1984 *Acta Cryst.* **A40** 120; 1988 *Acta Cryst.* **A44** 283
[3] Mikula P, Kulda J, Vrána M and Chalupa B 1984 *J. Appl. Cryst.* **17** 189.
[4] Mikula P, Kulda J, Lukáš P, Vrána M and Wagner V 1994 *Nucl. Instrum. Methods in Phys. Research,* A **338** 18
[5] Wagner V, Mikula P and Lukáš P 1994 *Nucl. Instrum. Methods in Phys. Research* A **338** 53
[6] Vrána M, Lukáš P, Mikula P and Kulda J 1994 *Nucl. Instrum. Methods in Phys. Research* A **338** 125-131.
[7] Popovici M and Yelon W B 1995 *J. Neutron Research* **3** 1-25; 1997 *J. Neutron Research* **5** 227.
[8] Wimpory R C, Mikula P, Šaroun J, Poeste T, Li J, Hoffman M and Schneider R 2008 *Neutron News* **19** 16.
[9] Seong B S, Em V, Mikula P, Saroun J and Kang M H 2010 *J. Appl. Cryst.* **43** 654
[10] Popovici M, Stoica A D, Chalupa B and Mikula P 1988 *J. Appl. Cryst.* **21** 258
[11] Popovici M and Yelon W B 1994 *Nucl. Instrum. Methods in Phys. research* A **338** 132.

[12] Mikula P, Vrána M, Choi Y N, Kim S A, Oh H S, Sung K H and Lee C H 2002 *Mater. Sci. Forum* **404-407** 299

[13] Mikula P, Vrána M, Wagner V, Choi Y N, Moon M K, Lee C H and Inoue K 2004 Proc. of 12th Int. Conf. on Experimental Mechanics - ICEM12, 29 August - 2 September, Bari, Italy, McGraw-Hill, ed. C Pappalettere, full paper No. 038 on CD-ROM

[14] Saroun J and Kulda 2009 J Monte Carlo simulation package RESTRAX 5.2.0 and SIMRES 6.0.6, June 2009, http://neutron.ujf.cas.cz/restrax/index.php?chap=3 /

Neutron diffraction studies of a high resolution double crystal (+n,-m) setting containing Si(220) and Si(311) bent perfect crystals in symmetric and fully asymmetric diffraction geometry, respectively

P Mikula[1], M Vrána[1], B S Seong[2], W Woo[2], V Em[3] and D Korytár[4]
[1]Nuclear Physics Institute ASCR, v.v.i., 25068 Rez, Czech Republic
[2]Neutron Science Division, KAERI, 1045 Daedok daero, Daejeon, Yuseong-gu, 305-353, Republic of Korea
[3]National Research Centre "Kurchatov Institute", Moscow, 123182, Russia
[4]Institue of Physics SAS, Dúbravská cesta 9, 845 11 Bratislava, Slovak Republic

E-mail: mikula@ujf.cas.cz

Abstract. In this paper studies of neutron diffraction properties of the double crystal (+n,-m) setting containing Si(220) and Si(311) bent perfect crystals (BPC) in symmetric and fully asymmetric diffraction (FAD) geometry with the output beam expansion (OBE), respectively, are presented. Namely, our attention was focused on the properties of the FAD geometry of the BPC Si(311) crystal slab. It has been found that after a beam expansion this FAD geometry can provide a monochromatic beam of a rather large cross-section and of very small divergence with some possible application use.

1. Introduction

After the first studies of diffraction properties of BPC crystal slabs in FAD geometry which were carried out either in NPI Rez or in PTB Braunschweig in the past [1-10], further studies practically have not appeared for quite a long time. Originally, the motivation for studying the FAD geometry was brought about by the effort to achieve the so called effective mosaicity of the BPC crystals comparable to the mosaicity of the conventional neutron monochromators [1-5] or to enlarge the range of analyzed angles which has been futher used for high resolution double-bent-crystal SANS diffractometers [6-10] operating in the nondispersive (+n,-n) setting. The (+n,-n) setting means that the two step diffraction was carried out on the same set of lattice planes. Basic calculations of the peak and integrated reflectivities as well as the effective mosaicity of the BPC elements, generally in the asymmetric diffraction geometry (including also FAD geometry), were introduced in ref. 1. In this paper also a highly efficient double-crystal monochromator based on two identical BPC slabs in the parallel nondispersive (+n,-n) setting was proposed and then experimentally proved [2]. Also one slab BPC-FAD monochromators in the geometry with the output beam compression were succesfully tested as monochromator for thermal neutrons [3-5]. It should be also mentioned a unique proposal of a possible use of FAD BPC Si slab with the output beam expansion (OBE) geometry as a high resolution TOF analyzer of thermal neutrons which was succesfully tested and desribed in refs. 11,12. Our recent attention has been focused on other FAD properties of the BPC slab, mainly in combination with other BPC slab of a different *d*-spacing in the so called parallel (+n,-m) double-

crystal setting which could provide further application possibilities. In this case the (+n,-m) setting can be considered as something between the dispersive and nondispersive one.
Namely, in the present paper we deal with the studies of some properties of the BPC Si crystal in FAD geometry with the output beam expansion (for a narrow incident beam and spatially expanded diffracted beam) employed as a second crystal but in a (+n,-m) double-crystal setting. The properties of such performance have not been tested yet.

2. Reflecting properties of the bent FAD perfect crystal

Bent perfect crystals (BPC) are usually employed in symmetric or asymmetric diffraction geometry. They are well known namely for their focusing properties in real as well as momentum space with respect to a sample. Generally, the BPC elements are attractive for an employment from the following reasons: predictability and reproducibility of the effective mosaicity, predictability and reproducibility of a rather high peak reflectivity and its uniformity over large areas of the crystal, a rather high peak reflectivity for asymmetric or transmission geometry and predictability of the integrated reflectivity. In the transmission geometry one can also benefit from natural wavelength focusing. In the limiting case - the FAD geometry with OBE - as studied in the present case (see figure 1), some more properties can

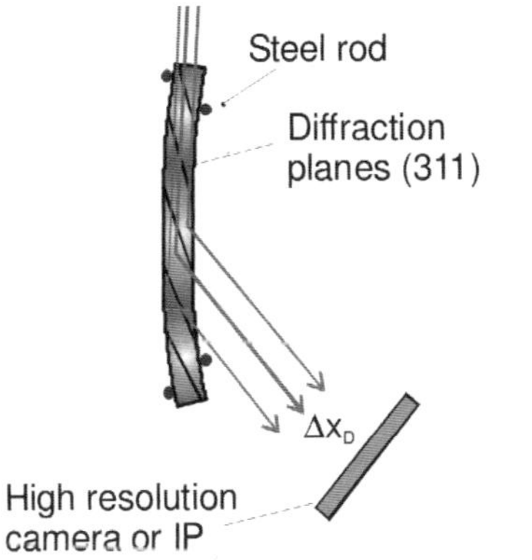

Figure 1. Schematic sketch showing FAD geometry of the Si crystal slab as used in the experiment.

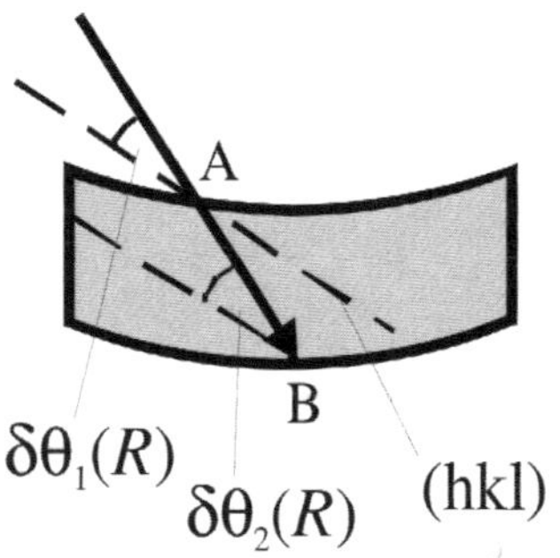

Figure 2. Detail of the incident beam passing through the BPC slab (general diffraction geometry).

be studied which could be possible used for application in neutron diffractometry. As can be seen from figure 1, depending on the curvature of the crystal slab, a narrow incident beam can be spatially enlarged and in such a way one can manipulate with focusing as well as with a spacial distribution (width) of the diffracted beam. Even though one can expect in the diffracted beam some focusing in real space, due to the output beam expansion the focusing is not so pronounced. On the other hand, the expansion of the diffracted beam directly influences separation of the wavelengths of individual neutron rays within the Δx_D distribution (see figure 1) which is, of course, strongly correlated. It is well known that in the case of BPC symmetric and asymmetric diffraction , Bragg law introduces a correlation between the direction and the wave-vector of neutrons, but in the case of FAD geometry there is a new correlation between the position within the Δx_D range and the wave-vector of neutrons. In some experiments this can be successfully exploited in combination with a position sensitive detector [6-10].

Big advantages of BPC elements over mosaic crystals are predictability and reproducibility of their reflectivities as well as the uniformity over large areas. The general formula for the peak reflectivity of neutron passing through a homogeneously deformed perfect crystal (see figure 2) has been derived in a form as [13]

$$r(R) = \{1 - \exp[-Q_{hkl}(\partial\Delta\theta/\partial s)^{-1}]\}, \quad (1)$$

where $(\partial\Delta\theta/\partial s)$ is the rate of change of the Bragg angle θ on the flight path Δs in the crystal in the incident beam direction and $Q_{hkl} = (F_{hkl})^2 \cdot \lambda^3/(\Omega^2 \cdot \sin 2\theta)$ is the kinematical reflectivity of the crystal unit volume. F_{hkl}, λ and Ω are the structure factor, neutron wavelength and the unit-cell volume, respectively [14]. The effective mosaicity $\delta\theta = \delta\theta_1 - \delta\theta_2$ (see figure 2) and the integrated reflectivity $\rho^\theta(R)$ can be defined as $\delta\theta(R)=\Delta s(\partial\Delta\theta/\partial s)$ and $\rho^\theta(R) = \delta\theta(R)\cdot r(R)\cdot A(\mu)$, respectively [1]. $\Delta s=D/\sin(\theta+\psi)$ is the distance between the points A and B (see figure 2), D is the thickness of the crystal slab, ψ is the angle of asymmery and $A(\mu)$ is the attenuation factor. Then, for FAD geometry we can derive simple expressions for the effective mosaicity and the peak reflectivity as

$$\delta\theta(R) = \Delta s\,/R\,, \quad (2)$$

$$r(R) = 1 - \exp(-Q_{hkl}\cdot R)\,. \quad (3)$$

Of course, the effective mosaicity is quite different for incident and diffracted beam. In the present FAD case, the effective mosaicity is $\delta\theta(R) = L/R$, where L is the length of the crystal slab. Figure 3 shows the calculated dependence of the peak reflectivity $r(R)$ for Si(311) FAD diffraction geometry and for the sake of comparison also for Si(220) slab set in symmetric diffraction geometry (both were used in the experiment) on the crystal curvature $(1/R)$. It can be seen from figure 3 that the peak reflectivity corresponding to Si(311) FAD geometry is smaller than the one corresponding to the symmetric Si(220) diffraction geometry, namely due to a difference in $(F_{hkl})^2$ and thus in Q_{hkl} (see eq. (1)) As to the attenuation factor $A(\mu)$, then in our case of low absorbing Si crystal, we can simply use the approximation expression $A(\mu)=\exp(-\mu\cdot L/2)$.

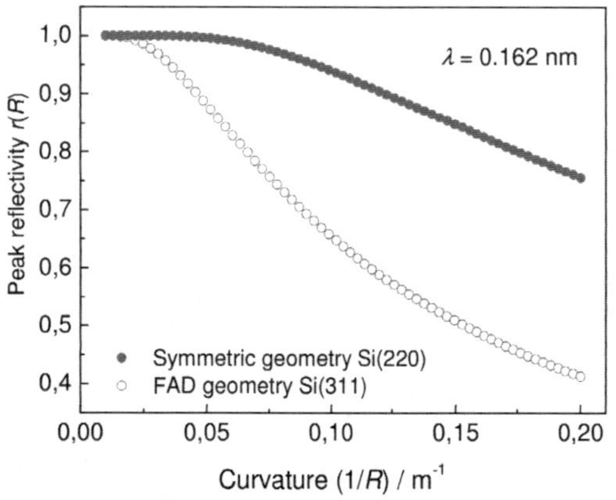

Figure 3. Peak reflectivity vs crystal curvature as calculated according to the formulae (1) and (3) for symmetric Si(220) as well as FAD Si(311) diffraction geometry.

3. Parallel (+n,-m) double-crystal arrangement

The experimental studies of the (+n,-m) double-crystal arrangement with the FAD OBE geometry of the second crystal-analyzer were carried out on the dedicated neutron optics diffractometer installed at the reactor LWR-15 in Řež and operating at the fixed neutron wavelength of 0.162 nm provided by a bent Si(111) (the thickness - 4 mm, the radius of curvature - 10 m) premonochromator. No Soller collimators are used on the beam path from the reactor to the detector. The schematic sketch of the experimental performance is shown in figure 4. Due to a small Bragg angle for Si(111) crystal, the premonochromatic beam is of rather large divergence and the $\Delta\lambda$ spread [14] which is then used for the testing of the next (+n,-m) double BPC setting. The distance between the Si(111) and Si(220) crystals was 1.7 m and the distance between the Si(220) and Si(311) FAD crystal was 0.5 m. As the curvature of both BPC slabs Si(220) and Si(311) were changeable, properties of the beam profiles as registered by a scintillation camera or imaging plate could be studied for different combination of the individual curvatures. The obtained results are shown in the following figure 5. It can be seen from figure 5 that for small curvatures $(1/R_2)$ the area where Bragg condition for neutrons coming from the bent Si(220) crystal is inside the FAD crystal quite large when expanding for R_1=9 m and R_1=12 m up to 20-25 mm (see figures 5a and 5b), though the width of the incident beam entering the FAD crystal was limited by a 3 mm slit.

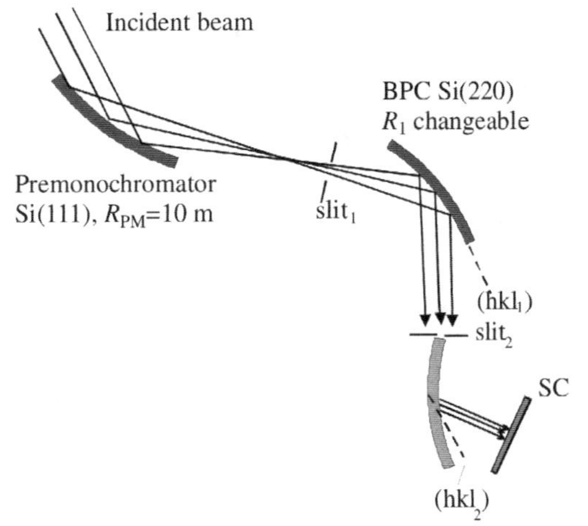

Figure 4. Schematic sketch of the experimental arrangement as used for the studies of the FAD-OBE properties of the BPC Si (311) crystal in the double crystal setting with respect to the Si(220) one. SC - scintillation camera was at 56 cm from the FAD crystal. Dimensions of the crystal slabs: 200x40x4 mm^3 (length x width x thickness).

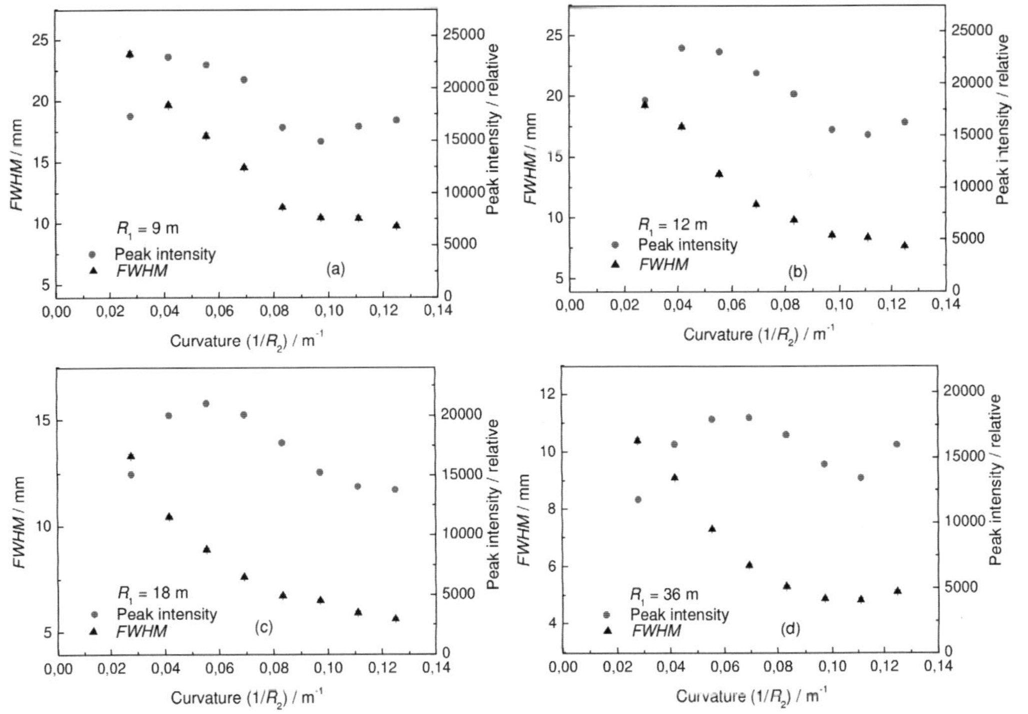

Figure 5. Peak intensity and *FWHM* of the diffraction profile dependence on the FAD crystal curvature for fixed curvatures of the Si(220) one as registered by the scintillation camera: (a) - 0.11 m^{-1}, (b) - 0.083 m^{-1}, (c) - 0.056 m^{-1} and (d) - 0.028 m^{-1}.

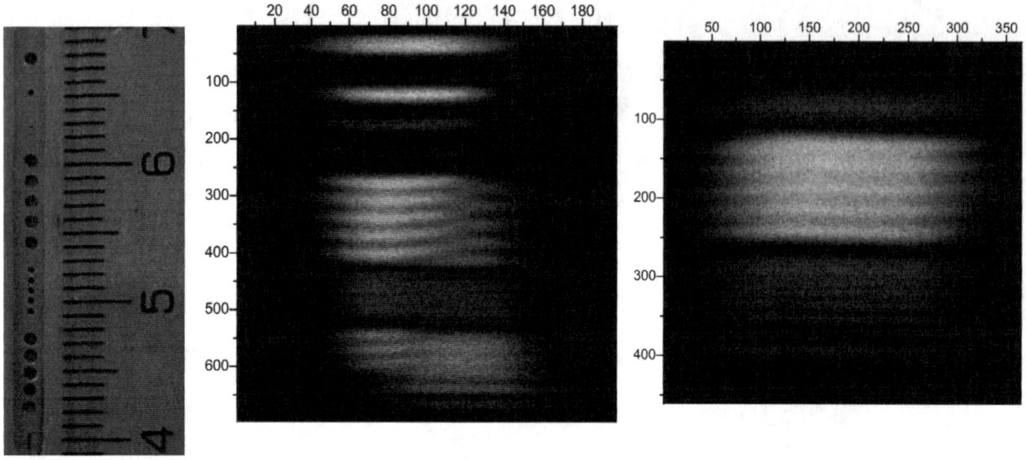

Figure 6. Holes in the Cd absorber vertically situated in front the FAD crystal (a) and the images of them for R_1= 36 m and R_{FAD} = 9 m (b) and R_{FAD} = 36 m (c). IP with the pixel dimension of 50 μm x 50 μm was at the distance of 10 cm from the FAD crystal.

However, in all cases of fixed curvatures of the bent Si(220) crystal slab, *FWHM* of the double diffracted beam decreases with increasing of the curvature of the FAD crystal and at the place of the camera achieves the values of several millimeters. As the corresponding peak intensity of the profiles does not change considerably, it means that the change of *FWHM* is not given directly by focusing in real space but overlapping of the phase space elements of individual crystals. Therefore, one can manipulate with *FWHM* of the double diffracted beam. The expansion of the diffracted beam inside the FAD crystal is clearly demonstrated in the following figure 6 when imaging vertically arranged small holes (of different diameters) in a Cd sheet which was situated just before the front face of the FAD crystal in the place of the slit S_2 (see figure 4). The image of the holes was taken by the imaging plate having a pixel size of 50 μm x 50 μm and put at the distance of 10 cm from the FAD crystal. It can be seen from figure 6 that as expected the smaller urvature ($1/R_2$) of the FAD crystal is applied, the wider image of the holes is obtained. High-resolution properties of the expanded beam obtained by the FAD crystal in scattering plane (in our case it is horizontal plane) was also proved on imaging of some linear objects - two office staples and a hypodermic needle (see photo in figure 7). The obtained images taken by IP are shown in figure 8. Inspection of figure 8 reveals that a typical refraction of neutrons takes place at the edges of the office staples and its observation depends on the curvature of the crystals and a proper distance of IP from the sample. Of course, such refraction effect can be seen only when sufficiently high resolution of the output beam is used. It should be mention that such refraction effects have been already observed on an ordinary neutron diffractometer when employing a multiple reflection monochromator [15]. However, as there are no sharp edges in the case of the hypodermic needle, the corresponding image is blurred.

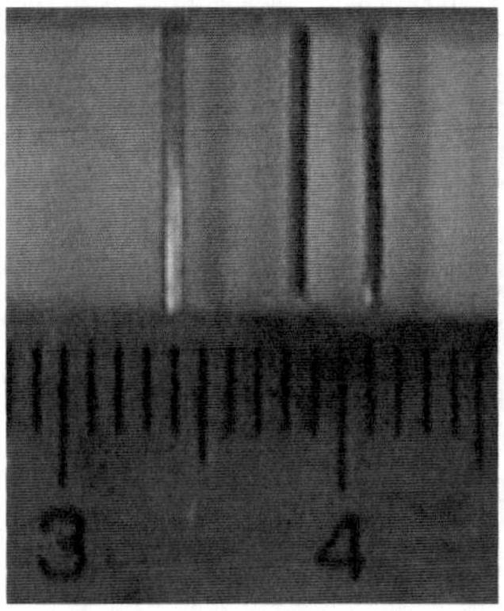

Figure 7. Photo of the hypodermic needle and two office staples.

After a detailed inspection of cylindrical bending of FAD crystal it could be expected that the Bragg angle linearly changes along the longest edge of the curved FAD crystal. This property can be successfully used for investigations of some effects related to scattering to small angles. Figure 9 practically demonstrates this property of the linear change of the Bragg angle when rocking the FAD

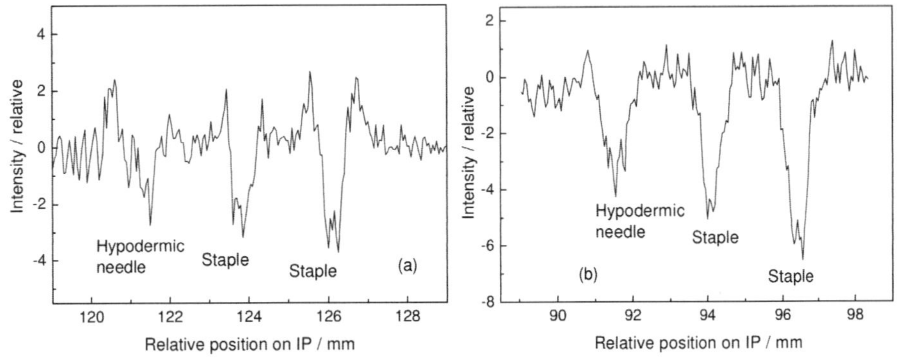

Figure 8. Images of two office staples and a hypodermic needle: (a) - $R_1 = 36$ m, $R_{\mathrm{FAD}} = 36$ m, IP at the distance of 50 cm from the sample, (b), R_1=36 m, R_{FAD} = 72 m, IP 60 cm from the sample.

crystal with respect to the bent Si(220) one. The shift of the diffraction profile on the IP can be expressed by a simple formula $\Delta x = \Delta\theta \cdot (R_{\mathrm{FAD}}\sin(2\theta))$.

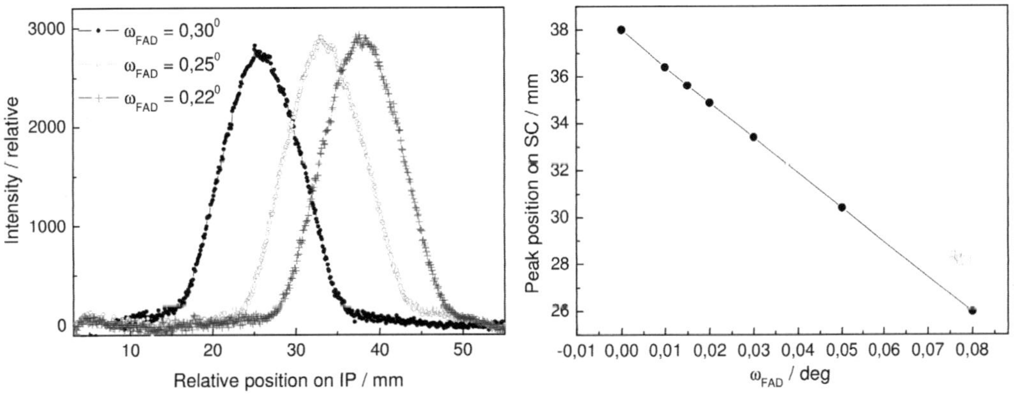

Figure 9. Calibration of the rocking angle of the FAD crystal versus the peak position of the diffraction profile imaged by scintillation counter SC for R_1=36 m, R_{FAD}=36 m: (a) – an example of diffraction profiles for three rocking angles of the FAD crystal, (b) – verification of the linearity of the change of the Bragg angle along the longest edge of the FAD crystal.

4. Summary

Diffraction properties of the FAD geometry of the BPC Si(311) slab as a second crystal in the parallel (+n,-m) double-crystal setting with respect to the BPC Si(220) one were investigated. Monochromatic beam provided by such a flexible double-crystal setting is highly collimated and *FWHM* of the profile could be easily manipulated by the curvature of the FAD crystal. Contrary to earlier conservative meaning, the output monochromatic beam current can be sufficiently high and the presented double bent crystal setting has been found useful for an employment in some special diffraction experiments. Thanks to a high peak reflectivity of the FAD Si(311) and Si(220) crystals (see figure 3), the obtained neutron monochromatic current after double diffraction process corresponds to the resulting phase-space element given by the overlapping of phase space elements of the individual BPC Si(220) and FAD BPC Si(311) crystals in the (+n,-m) setting. As the two step diffraction is carried out on crystals with different lattice spacing, the double diffracted beam has also a narrow $\Delta\lambda/\lambda$ spread of the order 10^{-3} (or less), which is dependent on the curvatures of the individual crystals. It can be said that the

(+n,-m) setting as presented in this paper can be used for some experiments of neutron imaging and for high resolution (in scattering angle as well as $\Delta\lambda/\lambda$) diffraction experiments.

Acknowledgements
The related measurements were carried out at the CANAM infrastructure of the NPI ASCR Rez. Bragg diffraction optics investigations are in the Czech Republic supported by the ESS project LM2010011: 'Contribution to Partnership in Large Research Infrastructure of Pan-European Importance' and by the project EU-FP7-NMI3 II: "Integrated Infrastructure Initiative for Neutron Scattering and Muon Spectroscopy", 2009-2015.

References

[1] Mikula P, Kulda J, Vrána M and Chalupa B 1984 *J. Appl. Cryst.* **17** (1984) 189
[2] Mikula P, Chalupa B, Kulda J, Vrana M, Sedlakova L and Michalec R *J. Appl. Cryst.* 1985 **18** 135
[3] Mikula P, Kulda J, Horalík L, Chalupa B and Lukas P 1986 *J. Appl. Cryst.* **91** 324
[4] Mikula P, Lukas P and Michalec R 1987 *J. Appl. Cryst.* **20** 428
[5] Mikula P, Krüger E, Scherm R and Wagner V 1990 *J. Appl. Cryst.* **23** 105
[6] Mikula P, Lukas P and Eichhorn F 1988 *J. Appl. Cryst.* **21** 33
[7] Saroun J, Lukas P, Mikula P and Strunz P 1993 *Journal de Physique*, Coll. C7, **3** 439
[8] Saroun J, Lukas P, Mikula P and Alefeld B 1994 *J. Appl. Cryst.* **27** 80
[9] Hempel A and Eichhorn F 1994 *BENSC Experimental Reports*, HMI Berlin, p. 350-351.
[10] Lukas P, Mikula P, Šaroun J and Strunz P 1994 *Nucl. Instrum. Methods in Phys. Research* A **338** 111
[11] Mikula P, Vrána M, Furusaka M, Yasushige Y, In Proc. of the Int. Conf. ICANS XVI, May 12–15, 2003, Düsseldorf-Neuss, p. 401
[12] Mikula P, Vrana M, Furusaka M, Wagner V, Choi Y N, Moon M K Em V and Lee C H 2004 *Nucl. Instrum. Methods in Phys. Research* A **529** 138
[13] Kulda J 1984 *Acta Cryst.* A **40** 120
[14] Bacon G E, *Neutron Diffraction*, Clarendon Press, Oxford, 1975, p. 67 and p. 95.
[15] Mikula P, Vrana M, Saroun J, Krejci F, Seong B S, Woo C and Furusaka M 2013 *J. Appl. Cryst.* **46** Part 2 128

"m=1" coatings for neutron guides

C P Cooper-Jensen[1,2], A Vorobiev[2], E Klinkby[3,1], V Kapaklis[2], H Wilkens[4], D Rats[5], B Hjörvarsson[2], O Kirstein[1,2,6], P M Bentley[1,2]

[1] European Spallation Source ESS AB, Lund, Sweden
[2] Department of Physics and Astronomy, Uppsala University, Uppsala, Sweden
[3] DTU Nutech, Technical University of Denmark, Lyngby, Denmark
[4] General AtomicNes, San Diego,Californian, USA
[5] NeoCoat SA, La Chaux de Fonds, Switzerland
[6] School of Engineering, University of Newcastle, Australia

carsten.cooper-jensen@esss.se

Abstract. A substantial fraction of the price for a supermirror neutron guide system is the shielding, which is needed because of the gamma radiation produced as a result of neutron absorption in the supermirror layers. Traditional coatings have been made of nickel-titanium heterostructures, but Ni and Ti also have a fairly high absorption cross section for cold and thermal neutrons. We examine a number of alternatives to Ni as part of a study to reduce the gamma radiation from neutron guides. Materials such as diamond and Be have higher neutron scattering density than Ni, smaller absorption cross section, and when a neutron is absorbed they emit gamma photons with lower energies. We present reflectivity data comparing Ni with Be and preliminary results from diamond coatings showing there use as neutron guide coatings. Calculations show that Be and diamond coatings emit two orders of magnitude fewer gamma photons compared to Ni, mainly because of the lower absorption cross section.

1. Introduction

Neutron guides were introduced in 1963 [1] to increase the neutron flux at the sample position. The guides were evacuated to prevent the neutrons from scattering on the air molecules, and the low-divergent neutrons hitting the neutron guide walls at low grazing angles were reflected and transported more efficiently. Most materials have a refractive index that is slightly below 1 for cold and thermal neutrons. This means that there is a critical angle, θ_c, below which there will be total external reflection of the neutrons, as demonstrated by Fermi and co-workers [2,3] in the mid 1940s. The critical angle is given by

$$sin\theta_c = \lambda\sqrt{Nb/\pi}$$

where N is the atomic number density of the material and b is the average coherent neutron scattering length. Nb is referred to as the neutron scattering length density. Natural nickel (Ni) has one of the highest neutron scattering length densities of the elements in the periodic table. The critical angle for Ni is defined as m=1. To reflect neutrons at still higher angles than that corresponding to m=1, one could use the ^{58}Ni isotope. This has a higher neutron scattering length density compared to Ni, equivalent to m=1.18. Another option is to use supermirrors, proposed by Mezei [4] in 1976, and which today can be made with m-values above m=7. The cost of Ni isotope enrichment is very high, and ^{58}Ni costs at present around 1000 USD per gram. As such, neutron guides with coatings above m=1 use supermirrors as the most economical option.

Neutrons that are not reflected by the coating will be transmitted, and will be absorbed in the coating or in the substrate. Most neutron absorption processes cause the emission of gamma radiation, and the quantity and energy of the photons depend on the absorbing atom.

The substrates for most neutron guides are made of glass that contains boron, and the products "borofloat" and "borkron" are common. Natural B contains 20% of the isotope ^{10}B, which has a very high absorption cross section for cold and thermal neutrons. When a neutron is absorbed in ^{10}B, the majority of the emitted gamma radiation is at a relatively low in energy (0.477 MeV). The other major components in the substrate are Si and O, which both have an absorption cross section that is many orders of magnitude below that of ^{10}B. Consequently, most of the neutrons absorbed in the glass will be due to absorption in ^{10}B. But before the neutrons can be absorbed by the ^{10}B, the neutrons have to pass through the Ni coating. The absorption cross section for Ni is lower than for B, but large enough to cause substantial absorption. When a neutron is absorbed in Ni, a large part of the emitted gamma radiation has energies exceeding 8 MeV.

The most common used shielding materials around neutron guides are iron, different versions of concrete, and sometimes lead. For all these materials, the attenuation coefficient for 0.5 MeV photons is higher then it is for 8 MeV photons. For lead the attenuation coefficient is 1.2 cm^{-1} for 0.5 MeV photons and 0.43 cm^{-1} for 8 MeV photons [5]. Therefore, to get the same decrease in gamma radiation, one needs nearly 3 times more material if the gamma radiation is at 8 MeV than for 0.5 MeV.

The material with the highest neutron scattering length density in the periodic table is carbon in the form of diamond. Thereafter come Ni and Be, which have nearly the same scattering length density. Both C and Be have a very low absorption cross section compared to Ni, and when a neutron is absorbed both emit gamma radiation at lower photon energies than Ni.

The coating for neutron guides is mainly produced by DC magnetron sputtering, since large areas of optical components can be sputtered efficiently with the quality needed. Be can also be coated using DC magnetron sputtering [6], but it requires special safety procedures. Be dust, which will appear in the coating facility during production, is a health hazard and can lead to a chronic lung disease called berylliosis.

Much effort has been spend developing the manufacture of carbon layers with high number of sp^3 bindings compared to number of sp^2 bindings, making coatings that are more like diamond than graphite. This corresponds to an increase in the mass density of the carbon coatings from 1.8 g/cm^3 for amorphous carbon towards the 3.5 g/cm^3 for diamond. This can be achieved by making coatings that are crystalline diamond or coatings that are amorphous but with high mass density, which are called diamond-like carbon (DLC). Current manufacturing processes for DLC coatings result in elevated concentrations of hydrogen in the coating, which reduce the total neutron scattering length density [7] and therefore make them less useful for coatings in neutron guides applications. DLC coatings without hydrogen for ultracold neutron guides have been demonstrated but these coatings have only had 60 % to 70 % sp^3 bindings and a mass density of 2.88 g/cm^3 [8]. This means that the neutron scattering length density is around the same as for Ni. Alternatively one can get close to 100% sp^3 bindings but these diamond coatings are crystalline [9].

In this paper we will compare Be, crystalline diamond, and Ni coatings to be used as m=1 coatings for neutron guides.

For the reflectivity calculation we have used IMD [10] with optical constants for neutrons from Neutron OPtics Scattering (NOP) [11]. To calculate the emitted gamma radiation when a neutron is absorbed we used MCNPX [12]. Three samples are shown to validate the concept that Ni in guides could be replaced by alternative materials, reducing the emitted gamma radiation as a result.

2. Samples and experimental setup

The Be coating was made by DC magnetron sputtering on a 100 mm Si wafer. The coating facility is normally used for coating thick (>150 μm) Be coatings [6]. The Be coating was 276 nm thick and the roughness of the Be surface was measured to be 0.3 nm using 8 keV (Cu kα) X-ray from a rotating anode.

The diamond coating was made by CVD [9] on a 100 mm x 20 mm fused silica substrate that was 10 mm thick. The diamond coating is on average 225 nm thick as measured by weight, but there is some thickness variation over the sample. The 2 mm from the edges of the samples have a thinner coating than the rest of the sample. The grain size was evaluated from SEM measurements to be 50 nm to 70 nm and the roughness of the coating was measured with 8 keV X-ray to be 6 nm. The deposition temperature of the diamond layer was 800°C. The content of sp^3 bindings is higher than 99% and the hydrogen content is around 2%, mainly concentrated at the grain boundaries.

The Ni sample was made using DC magnetron sputtering, on a system with base pressure of $2*10^{-10}$ Torr. The target was 99.995 % Ni and Ar was used as sputtering gas. The substrate was a 20 mm x 20 mm Si sample with a thermal SiO_2 layer.

The neutron reflectivity measurements shown in this paper were preformed using 4.41 Å neutrons at the SuperADAM reflectometer at ILL [13]. The samples were measured using a non-polarizing periodic Bragg mirror that gives a divergence of the neutrons of 0.27 mrad. The detector was a 2D Position Sensitive Detector (PSD). The Be and diamond coatings were measured with a 0.15 mm horizontal slit in front of the sample. The Ni coating was measured with a 0.05 mm horizontal slit because of the smaller sample size. The size of the vertical slit in front of the sample remained constant at 50 mm in all experiments. All data are corrected for variations in the incoming neutron flux by scaling the data to a beam monitor.

3. Results and discussion

Figure 1 compares the reflectivity measurements of the Ni coating and the Be coating. It is apparent that the critical angles for the Ni and the Be coatings are practically the same, as would be expected since they have a neutron scattering length density that is almost equal, as shown in Table 1. It can also be seen that the measured reflectivity is in good agreement with the calculated reflectivity for Ni.

In Figure 2 we show the measured reflectivity curve for the Be and diamond coating together with the calculated reflectivity curve for Ni and ^{58}Ni. In the angular range 0-0.2 degrees the length of the sample and the horizontal slit size define the shape of the reflectivity curve, which is consistent with total external reflection. The diamond data have been scaled to the Be data so the shape in this region is the same since they the samples have the same length and are measured with the same slit settings.

Table 1. The neutron scattering length density is calculated using data from [14] and the absorption cross section is from [14]. The data is for thermal neutrons (2200 m/s).

	Neutron scattering length density ($1*10^{14}$ m^{-2})	Absorption cross section (barn)
Be	9.63	0.0088
C(amorphous)	6.00	0.0035
C(diamond)	11.70	0.0035
Ni	9.41	4.49
^{58}Ni	13.14	4.50

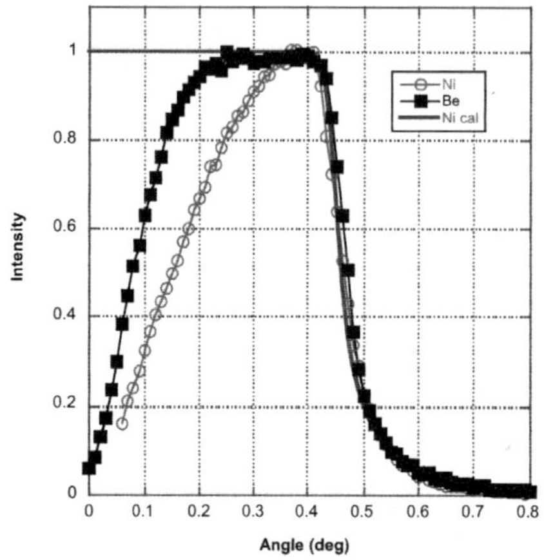

Figure 1. Reflectivity curves of the Ni and the Be samples measured with 4.41 Å neutrons. The thick line is the calculated reflectivity for Ni, the other lines are guides for the eye. The shapes of the experimental curves below 0.3 degrees are caused by the finite sample length and the size of the horizontal slit. Error bars are generally smaller than the marker size.

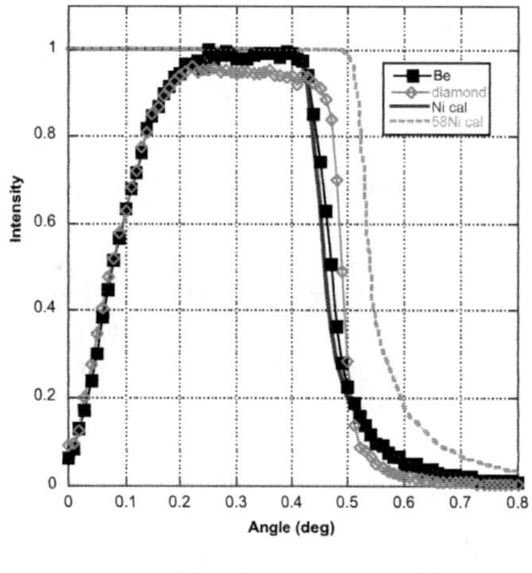

Fig. 2. Reflectivity curves for the Be and the diamond samples measured with 4.41 Å neutrons. The two thick lines are the calculated reflectivity for Ni (solid) and ^{58}Ni (dashed), the thin lines are guides for the eye. The intensity curve for the diamond data is scaled with a constant so the shape of the curve follows that of Be below 0.2 deg. Error bars are generally smaller than the marker size.

From 0.2 degree and to the critical angle of the diamond coating the reflectivity curve is flat as expected for total external reflection, but at a reduced reflectivity of 0.95. This reduction is most likely to be caused by the crystalline diamond coating, which has a high roughness of 6 nm. Furthermore, the sample contains about 2% hydrogen, which are mainly concentrated in the grain boundaries. Hydrogen has a very high incoherent scattering cross section that also contributes to the loss of reflectivity at the sample surface.

We see also that the diamond coating has a higher critical angle than Ni and Be, corresponding to m=1.07, which is nonetheless below the theoretical value, m=1.12, for diamond. There are several reasons for this. The diamond film bindings are likely to not be 100% sp^3 bindings, so the mass density of the coating is lower than for diamond. The hydrogen contamination and roughness would also be expected to contribute to this lower performance.

We have calculated the gamma radiation emitted from a 250 nm thick coating of the different materials using MCNPX [12]. For each coating 10^9 neutrons were calculated. Since the tabulated values for absorption cross sections are for thermal neutrons (λ = 1.80 Å) we used thermal neutrons with an incidence angle of 0.3 degree for these calculations. For thermal neutrons, m=1 corresponds to 0.18 degrees. Therefore, at 0.3 degrees there will be very little reflection and all the neutrons will be transmitted into the coating. In Figure 3, we show the calculated probability for emitted gamma rays per neutron from the various coatings. The emitted gamma rays from Be and C have fewer than 20 different discrete energies. For Ni, the numbers of discrete energies exceeds 500. In order to make the plot easier to read, the data are shown in 0.1 MeV bins. MCNPX does not include the 2 carbon gamma ray lines at energies of 6.1 MeV and 8.2 MeV, however both have a very small probability to occur.

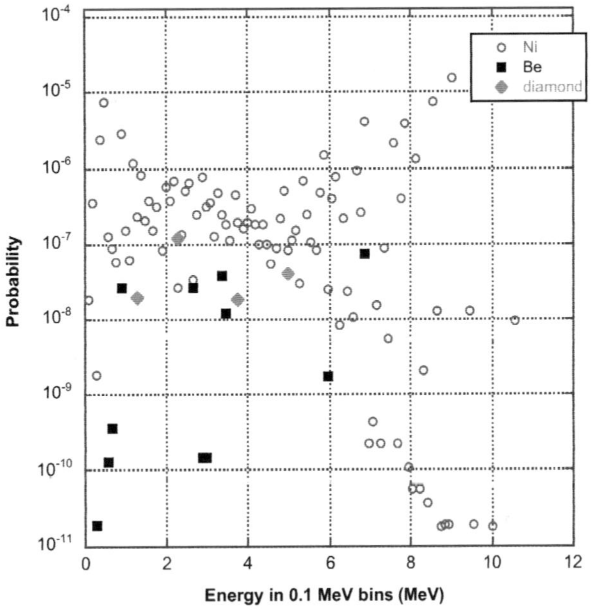

Figure 3. Probability of gamma photon emission from a 250 nm thick films from one grazing incidence neutron as a function of photon energy. The data were calculated using MCNPX as described in the text.

Table 2. The integrated probability for gamma photon emission from one absorbed neutron. The data were calculated using MCNPX as described in the text.

	Gamma ray emitted from one absorbed neutron (Probability)	Total gamma ray emitted from film (Probability)
Be	1.8	$1.8*10^{-7}$
Diamond	1.5	$2.0*10^{-7}$
Ni	1.5	$6.6*10^{-5}$
^{10}B	0.94	-

From Table 2, we see that when a neutron is absorbed the probability for gamma ray emission is the same for Ni and diamond, and for Be the total probability is a little higher. But because the absorption cross section for Ni is order of magnitude larger then for Be and diamond (Table 1), the total number of gamma ray photons from the Ni coating is much larger than for the Be and diamond coating, as shown in Table 2. From Figure 3, we see that the gamma radiation from Ni also has higher energies than that of Be and diamond.

The non-reflected neutrons, which are not absorbed in the coating, will still need to be absorbed, which is why borated substrates are generally used. When a neutron is absorbed in ^{10}B the number of emitted gamma photons is lower than when the neutron is absorbed in Ni, as shown in Table 2. Also the majority of the emitted gamma rays from ^{10}B are at the energy of 0.477 MeV. As a result, by replacing the Ni coating with Be or diamond, one will both reduce the total amount of gamma radiation and shift the radiation to lower energies. Both these two effects will help reduce the amount of shielding required around the guide.

4. Conclusion and future work

Today many new neutron guides have sophisticated shapes compared to traditional straight neutron guides. Especially for long guides there are still parts of the guide where only m=1 coatings are required or where the benefit of using higher m-value coatings do not outweigh the increase in price for the supermirrors. Gamma radiation that is emitted when a neutron is absorbed in the guide has to be attenuated for the safety of the people working in the area and because the gamma radiation can be problematic for some experiments at the end of the neutron guide. Both by reducing the amount of gamma radiation and/or by reducing the energy of the emitted gamma radiation, one will reduce the amount of shielding needed.

With this work we have demonstrated that Be could be a promising coating for neutron guides to replace the traditional Ni coatings. This would be expected to reduce the total emitted gamma ray from the guide, and shift the gamma ray spectrum towards lower energies. Coating the guide optic with Be will be more expensive than coating with Ni, because of the extra precautions that have to be taken during the coating process, but are expected to result in lower overall prices due to the fact that less shielding is needed around the neutron guide. A cost-benefit analysis of this will be made in the near future.

On the other hand, with the high roughness of our diamond sample coating it would not compete as a coating for neutron guides. Instead, we conclude that diamond coatings with a high level of sp^3 bindings will have a m-value that is higher than 1. More work is needed on the diamond-like carbon (DLC) coatings and nano-crystalline diamond coatings, but these could provide performance advantages over both Ni and Be and without the toxicity drawbacks that Be carries.

Acknowledgement

This work was supported under the Cluster of Research Infrastructures for Synergies in Physics (CRISP) program, which is co-funded, by the partners and the European Commission under the 7th Framework Programme Grant Agreement 283745.

References

[1] Maier-Leibnitz H and Springer T 1963 *Reactor Science and Technology* **17** 217
[2] Fermi E and Zinn W H 1946 *Phys.Rev.* **70** 103
[3] Fermi E and Marshall L 1947 *Phys. Rev.* **71** 666
[4] Mezei 1976 *Communications on Physics* **1** 81
[5] Dianoux A J and Lander G 2003 *Neutron Data Booklet, Secound Edition* (Institit Laue-Langevin) p 4.2-1 and 4.2-2
[6] Xu H, Alford C, Chason E, Detor A J, Fuller T, Hamza A V, Hayes J, Moreno K A, Nikroo A, Buuren T, Wang Y, Wu J, Wilkens H and Youngblood K P 2012 *J. Mater. Res.* **27** 822
[7] Grundy M J, Richardson R M, Roser S J, Beamson G, Brennan W J, Howard J, O'Neil M,

Penfold J, Shackleton C and Ward R C 1989 *Thin Solid Films* **172** 269

[8] Heule S, Atchison F, Daum M, Foelske A, Henneck R, Kasprzak M, Kirch K, Knecht A, Kuzniak M, Lippert T, Meier M, Pichlmaier A and Straumann U 2007 *Applied Surface Science* **253** 8245

[9] Mueller C, Haenni W, Binggeli M and Hintermann H E 1993 *Diamond and Related Materials* **2** 1211

[10] Windt D L 1998 *Computers in Physics* **12** 360

[11] Alianelli L, Sánchez del Río M and Felici R 2004 *Physica B* **350** e737

[12] Waters L S, McKinney G W, Durkee J W, Fensin M L, Hendricks J S, James M R, Johns R C and Pelowitz D B 2007 *AIP conference Proceedings* **896** 81

[13] Devishvili A, Zhernenkov K, Dennison A J C, Toperverg B P, Wolff M, Hjörvarsson B and Zabel H 2013 *Review of Scientific Instruments* **84** 025112

[14] www.ncnr.nist.gov/resources/n-lengths/

Optimization of multi-channel neutron focusing guides for extreme sample environments

D D DiJulio[1], E Lelièvre-Berna[2], P Courtois[2], K H Andersen[1] and P M Bentley[1,3]

[1] European Spallation Source ESS AB, P.O. Box 176, SE-221 00 Lund, Sweden
[2] Institut Laue-Langevin, 6 rue Jules Horowitz, BP 156, 38042 Grenoble Cedex 9, France
[3] Department of Physics and Astronomy, Uppsala University, Uppsala, Sweden

E-mail: douglas.dijulio@esss.se

Abstract. In this work, we present and discuss simulation results for the design of multi-channel neutron focusing guides for extreme sample environments. A single focusing guide consists of any number of supermirror-coated curved outer channels surrounding a central channel. Furthermore, a guide is separated into two sections in order to allow for extension into a sample environment. The performance of a guide is evaluated through a Monte-Carlo ray tracing simulation which is further coupled to an optimization algorithm in order to find the best possible guide for a given situation. A number of population-based algorithms have been investigated for this purpose. These include particle-swarm optimization, artificial bee colony, and differential evolution. The performance of each algorithm and preliminary results of the design of a multi-channel neutron focusing guide using these methods are described. We found that a three-channel focusing guide offered the best performance, with a gain factor of 2.4 compared to no focusing guide, for the design scenario investigated in this work.

1. Introduction

The focusing of neutrons is an important technique for many neutron scattering studies. It is particularly true when investigations are to be performed on samples limited to small volumes. In this case, neutron focusing guides play a critical role in determining the feasibility of such an experiment. A number of previous studies have reported on the development of multi-channel supermirror-coated neutron guides for focusing purposes [1, 2]. To benefit the most from the intensity gains of these types of guides, the focusing lens should extend close to the sample position. This is however at odds with the requirements of specific sample environments used in neutron scattering studies. Of particular interest for the current study is the use of cryomagnets to expose a sample to low temperatures and high magnetic fields. One possible solution is to extend the focusing guide into the cryomagnet itself. At the Institut Laue-Langevin (ILL), preliminary tests have been carried out which showed that some supermirrors tolerate both low temperatures and high magnetic field variations. A series of tests at the Paul Scherrer Institute (PSI) have also reported similar results at low temperatures for Ni/Ti supermirrors [3]. These combined results suggest that it is feasible to install a focusing guide into a cryomagnetic sample environment.

We have begun a project to design and build a split, multi-channel neutron focusing guide that begins focusing outside a cryomagnet and completes the focusing optic within the extreme

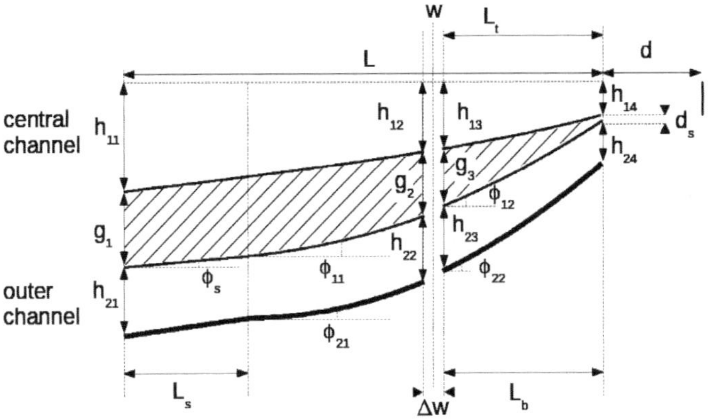

Figure 1. The parameters describing a three-channel guide. The guide is symmetric across its central horizontal axis and thus only the bottom half of the guide is shown. The dashed regions are supermirror-coated mountings. The parameter L corresponds to the length of the central channel, L_t is the length of the top face of the outer channel, and L_b is the length of the bottom face of the outer channel. These lengths can take on different values in the simulations, as described in the text.

sample environment apparatus. This requires optimizing two physically separate multi-channel optics with a gap for the cryomagnet beam window. This work extends the application of the software described in Ref. [4]. The software consists of a Monte-Carlo ray tracing simulation of the multi-channel guide which is coupled to an optimization code in order to find the best guide for a given scenario. We have selected the design of a multi-channel focusing guide for the IN5 instrument [5] at ILL as a test case for the software. The IN5 instrument is a possible candidate for the installation of a Cryogenic Limited [6] cryomagnet, which has already been designed to accept a focusing guide. A similar guide shall also be implemented for a 10T cryogmagnet to boost further the performance of the new ThALES cold-neutron three-axis spectrometer [7]. We therefore present an initial optimization study of this double-optic system, resulting in a candidate design that is capable of extreme vertical focusing elements crossing into a sample environment system. At the same time, we will also benchmark the particle-swarm optimization software used previously [4] against some alternative metaheuristic algorithms that have recently been receiving attention.

2. Multi-channel neutron guides

The geometry of the type of focusing guide that is being investigated is shown in Fig 1. The bottom half of the guide is illustrated as it is symmetric across its central horizontal axis. It can be mentioned that all simulations reported here are carried out in two spatial dimensions. There are no inclined faces in the guide and thus the vertical and horizontal focusing are decoupled. The guide comprises two separate sections, divided by a gap of width Δw centered at position w. This makes it possible to independently design both the section of the guide installed into the sample environment and the section installed outside. The central channel sections follow two separate parabolas, described by the parameters h_{11}, h_{12}, h_{13} and h_{14}, which are the heights at points along the length L of the channel. The outer channel pairs begin with a straight section of length L_s and tilt ϕ_s with respect to the central axis. They then follow circular arcs, starting with the angles ϕ_{11} and ϕ_{21} with respect to the axis of the beam, up to the break in the optics. Likewise, the parameters ϕ_{12} and ϕ_{22} describe the starting angles of the arcs for the second section. The entrance and exit heights h_{21}, h_{22}, h_{23} and h_{24} of the outer channel for the two sections are also adjustable parameters. The channel height at the transition from the straight section to the curved section of a channel is linearly interpolated between the values h_{21} and h_{22}. The vertical offsets of the channel entrances for both sections of the guide along with the offsets of the channel exits of the first section, with respect to the previous channel, are described by the parameters g_1, g_2 and g_3. The vertical offset of the channel exit of the second section is constrained to the value $d_s = 0.45$ mm. This is equal to the thickness of two 0.2 mm pieces of glass separated by a gap of 0.05 mm. The distance d of the end of the guide to the sample position is also an adjustable parameter. Each channel edge, with respect to the previous channel edge closer to the center of the guide, is allowed to extend closer to the sample with the central channel being the furthest from the sample position. In addition, any number of curved outer channel pairs can be defined in the simulation.

3. Population-based optimization algorithms

Population-based algorithms, such as Genetic algorithms (GA) [8] and particle-swarm optimization (PSO) [9], have been shown to be well suited for neutron instrumentation optimization problems [4, 10, 11, 12]. These types of algorithms work with a large number of candidate solutions which evolve over time towards a best solution in a particular parameter space. The evolution is guided by an appropriately selected figure of merit (FOM). In this work, the FOM is the number of neutrons impinging on a sample at the end of the focusing guide. Additional constraints to guide the simulations, such as penalty functions for the divergence at the sample position or beam missing the sample can be included in the simulation if desired.

The above described algorithms are inspired by nature. For example, GA was inspired by the self optimization of biological systems [13] while PSO has its roots in the swarm behavior of social animals, such as the flocking of birds or schooling of fish [9]. Neutronic optimization problems can be difficult to solve using other simple traditional optimization algorithms. This was highlighted in Ref. [14] where a least-squares algorithm, based on partial derivatives, was used to optimize the design of a converging guide exit. In a multi-dimensional search the algorithm was unable to locate the global minimum. In a more recent study, a number of other optimization routines have been proposed for the optimization of neutron scattering instruments [15].

The advantages of population-based algorithms, compared to traditional approaches, are that they have a tendency to avoid local optima, thoroughly sample the parameter space, and are simple to implement partly because they require no partial derivatives. Additionally, these types of methods are less sensitive to random noise, which is important when using the Monte-Carlo method, as discussed in Ref. [15]. In particular, PSO has been shown to have excellent performance for the design of multi-channel focusing guides [4]. To compliment these previous investigations with PSO, we have explored several other popular optimization algorithms for the design of focusing guides. These include artificial bee colony (ABC) [16] and differential evolution (DE) [17], both of which have been compared to PSO in literature; see e.g. Refs. [18, 19, 20]. The PSO method, applied to multi-channel focusing guides, has been described in detail in Ref. [4]. The same algorithm was applied in the current work and is not described here. In the following, the ABC and DE methods will however be briefly outlined.

3.1. Artificial Bee Colony (ABC)

ABC [16] is a population-based algorithm inspired by the intelligent behavior of honeybees in nature. In the algorithm, an artificial bee colony is made up of employed bees, onlooker bees, and scout bees, denoted as n_e, n_o and n_s, respectively. Food sources represent candidate solutions, each characterized by their nectar amounts. The number of food sources is always equal to the number of employed bees. In the first step, a population consisting of a number of food sources is initialized randomly, represented by the quantity SN. Each employed bee will visit an individual food source and determine a new food source based on a comparison with other surrounding sources. A parameter in the new source, represented by v_{ij} where j is a parameter index for the ith source, is produced by the following mutation of the original parameter x_{ij},

$$v_{ij} = x_{ij} + \phi_{ij}(x_{ij} - x_{kj}), \tag{1}$$

where k represents a randomly selected neighbor source and ϕ_{ij} is a random number between -1 and 1. The employed bee will then evaluate the nectar amount of the new source and memorize the source if it produces more nectar than the old one. Otherwise, the bee remembers the old source. After all the employed bees have visited their food sources they return to the hive and share their information with the onlooker bees. In the next step, the onlooker bees select a particular food source to visit based on the probability p_i, given by

$$p_i = \frac{F_i}{\Sigma_{n=1}^{SN} F_n}, \tag{2}$$

where F_i is the nectar amount for the ith source. Sources which produce more nectar are thus more likely to be visited by the onlooker bees. Once an onlooker bee arrives at a particular food source, it searches in the nearby area for a new food source and evaluates the nectar amount of the source, as given by Eq. 1. The onlooker bee memorizes the new source if it produces more nectar otherwise it remembers the old one. If a particular food source cannot be improved after a certain number of iterations, represented by the parameter *limit*, then that food source is forgotten and a scout randomly searches for a new food source. The algorithm is then repeated with the employed bees visiting their food sources and continues for a given number of iterations.

3.2. Differential Evolution (DE)

DE [17] is a type of evolutionary algorithm which operates in continuous parameter spaces. After the population of solutions is initialized and evaluated, each individual solution of the current generation, referred to as a target vector, undergoes mutation. For each ith target vector, a donor vector is generated by the following,

$$\mathbf{v}_i = \mathbf{x}_{r_1} + F(\mathbf{x}_{r_2} - \mathbf{x}_{r_3}), \tag{3}$$

where $\mathbf{v}_i$ represents the donor vector, $\mathbf{x}_{r_1}, \mathbf{x}_{r_2}$ and $\mathbf{x}_{r_3}$ are three different randomly selected solution vectors of the current population and F is a scaling factor. The population diversity is then enhanced through a crossover operation where the donor vector exchanges a number of components with the target vector to create a trial vector. The operation is controlled by a parameter CR, called the crossover constant. For each parameter, crossover is performed if a randomly selected number between 0 and 1 is less than or equal to CR. If the trial vector performs better than or the same as the target vector, it replaces the target vector in the next iteration. In this way, the population of solutions never gets worse but can only get better or stay the same. The algorithm continues for a given number of iterations.

3.3. Comparison of the optimization methods

Two separate tests have been carried out in order to compare the three different optimization methods. These include the design of a single-channel and a three-channel guide based on the IN5 instrument at ILL [5]. The geometry is described in detail below. Single- and three-channel guides were selected for the initial tests. Our preliminary simulations suggest that a

Table 1. The parameter values used in the different optimization methods. The values are taken from the indicated references. See text for the definition of the parameters for ABC and DE. For the PSO parameters, see Ref. [4]

PSO [4]		ABC [19]		DE [18]	
pop. size	50	col. size	100	pop. size	50
d	0.95	n_e	50	CR	0.8
c_L	1	n_o	50	F	0.5
c_G	1	n_s	1		
		limit	$n_e \times D$		

five-channel guide does not provide additional improvement over a three-channel guide for the scenario described below.

The parameters used for the optimization algorithms are given in Table 1. For PSO, the parameters and software previously implemented for the optimization of multi-channel neutron guides were used, as described in Ref. [4]. The population size was set to 50, the damping parameter d to 0.95 and both the control parameters c_L and c_G were set to unity. The parameters for the other two methods were adopted from literature values [18, 19]. For ABC, the colony size was 100 with n_e and n_o both set to 50, the number of scouts n_s was set to 1 and the *limit* parameter was given by $n_e \times D$, where D is the dimension of the problem. In DE, the population size was 50, the crossover constant CR was set to 0.8 and the scaling factor F was equal to 0.5. The optimization software was taken from Ref. [21] for ABC and Ref. [22] for DE. Both are freely available on the internet. The data resulting from each optimization run was analyzed using ROOT [23] based software.

The first test problem was the optimization of a single-channel parabolic neutron guide described by the guide length L and the entrance and exit heights h_{11} and h_{14}. The parameters h_{12} and h_{13} were not adjusted individually during the runs. The entrance to sample position length was fixed to ~856 mm with a 30 mm gap between the inner and outer sections of the focusing guide centered around 591 mm. A rectangular divergence profile, filling the divergence up to $m = 3$, was used as input for the focusing guide. The m-value for the channels of the focusing guide itself was set to $m = 4$. The sample height was selected to be 10 mm. Following discussions with the IN5 instrument team, we have selected to optimize the performance of the multi-channel guide in the wavelength region of $6 - 6.5$ Å, i.e. where the flux starts decreasing and the resolution gets better [24]. In the current work, the wavelength of the neutrons was fixed to $\lambda = 6$ Å.

After setting the parameter boundary conditions, each optimization was carried out 30 separate times and the average best FOM values as a function of the number of guides evaluated were tabulated. The results are indicated in Fig 2. The FOM value represents the number of neutrons, divided by 10^5, impinging on the target position for 10^6 neutrons entering the guide. All three optimization methods converged to the same value. PSO reached a plateau the fastest followed closely by DE. The ABC method however required more guide evaluations to reach the same performance as the other two methods.

The second test problem was the optimization of a three-channel guide for the scenario described above and involved 20 free parameters. The 20 parameters included the heights h_{11}, h_{12}, h_{13}, h_{14} and the length L describing the central channel, the vertical offsets g_1, g_2, g_3 and heights h_{21}, h_{22}, h_{23}, h_{24} of the outer channels, the angles ϕ_{12}, ϕ_{22},ϕ_{23}, ϕ_{24}, the tilt angle ϕ_s and the lengths L_s, L_t and L_b. In a similar fashion as above, after setting the boundary conditions

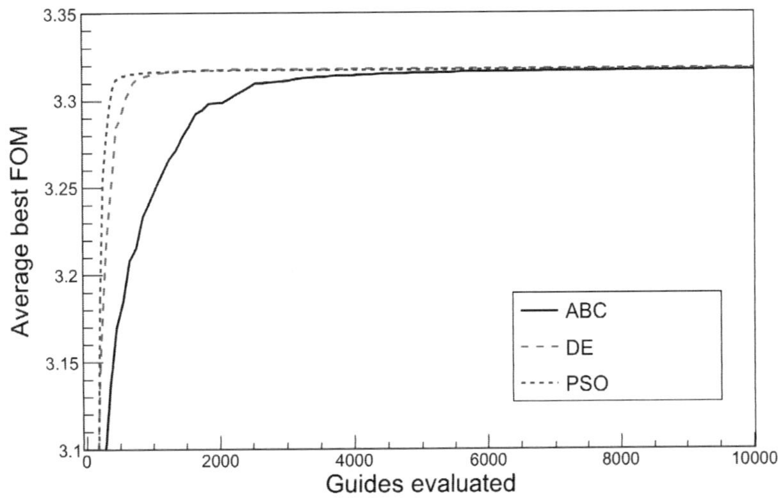

Figure 2. Average best FOM values as a function of the number of single-channel guides evaluated. The three different optimization methods, as discussed in the text, are indicated by the different color lines.

a total of 30 independent optimization runs were carried out for each method. The average best FOM values are shown as a function of the number of guide evaluations in Fig 3. For the ABC method, the FOM value is given as a function of the average number of guides evaluated. The number of guide evaluations per iteration may be different between two independent optimization runs due to the restriction imposed by the *limit* parameter, as discussed in section 3.1. This number however only deviated by a maximum of 14 evaluations for a given iteration in all cases examined here. Interestingly, each of the three optimization methods approached different FOM values. ABC on average found guides with the highest performance followed closely by DE. Furthermore, the results suggest that additional improvements in performance could be reached with more ABC evaluations than indicated in the figure. DE was however able to reach its optimum on average much faster than ABC. PSO on the other hand had a tendency to find single-channel guides, and was unable to escape from the local optimum. While each method on average found different optima, the final FOM values in Fig. 3 differ by no more than 5% between the different methods.

An example of a guide optimized using the DE method is shown in Fig. 4. The left hand panel shows the final geometry settled on by the optimization algorithm and the right hand side shows the divergence of the focused beam averaged over the sample height with and without the focusing guide. The total gain in the number of neutrons at the sample position is a factor of 2.4. The gain in intensity is also accompanied with an increased divergence as indicated in the figure. The FOM used in the simulations purely maximizes the number of neutrons at the sample without considering the shape of the divergence profile. A penalty function taking into account the shape of the divergence profile or the beam missing the sample can also be introduced in the future in the software if desired. It can also be noted that the lateral shift in the position of the mirrors after the break in the optics in Fig. 4 fills out the region of phase space which would otherwise be unfilled due to the gap between the mirrors. Lastly, the three-channel guide shown in Fig. 4 outperforms the single-channel guide with ~10% more neutrons

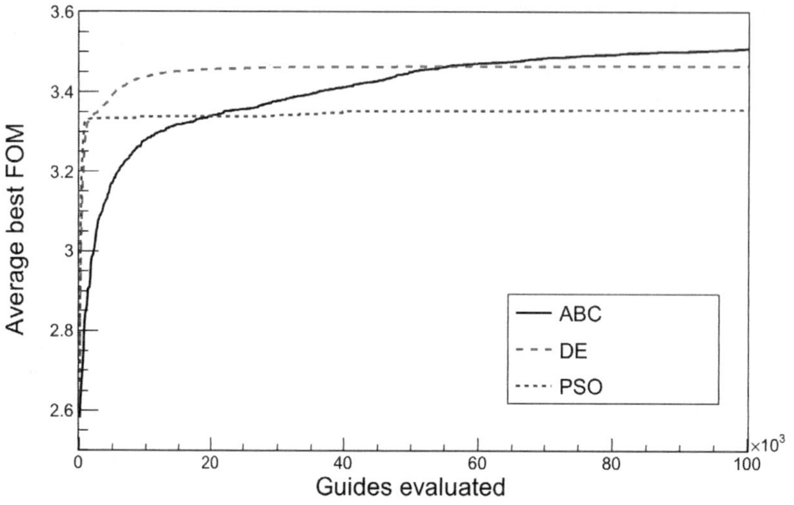

Figure 3. Average best FOM values as a function of the number of three-channel guides evaluated. The three different optimization methods, as discussed in the text, are indicated by the different color lines.

arriving at the sample position. It can however be mentioned that a more detailed investigation of multi-channel guides, including wavelength dependence, number of channels and sample size, just to name a few, is the aim of future work.

4. Conclusion

To summarize, we have developed software for the optimization of multi-channel focusing guides for extreme sample environments. The software includes a Monte-Carlo calculation of the guide coupled to a number of different population-based optimization algorithms. In this work, we investigated the use of PSO, ABC and DE for this purpose. The optimization methods were compared against each other for two different guide designs. For a single-channel guide, it was found that PSO and DE on average performed the best while ABC required many more guide evaluations to reach the same performance. For a three-channel guide, it was however shown that ABC and DE on average found the best guide designs. PSO had a tendency to get stuck in the single-channel local optima. We found that the three-channel focusing guide offered the best performance, with a gain factor of 2.4 compared to no focusing guide, for the test case investigated in this work. These results also highlight the potential benefit of investigating several different optimization routines when designing focusing guides.

5. Acknowledgments

This work was supported by NMI3-FP7 via contract number 283883.

References

[1] Osakabe T and Soyama K 2005 Rev. Sci. Instrum. **76** 073102
[2] Yamada S et al., 2006 Physica B **385-386** 1243
[3] Bartkowiak M, Filges U, and Panzner T 2012 J. Phys.:Conf. Series **340** 012021

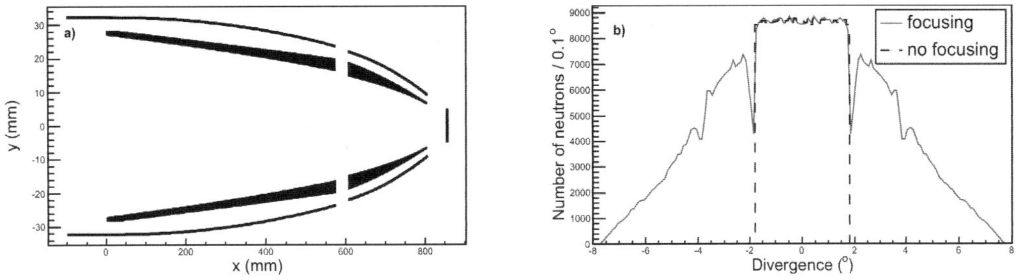

Figure 4. The a) geometry for an example focusing guide and b) the divergence averaged over the sample height for the (solid line) focusing guide and (dashed line) without the focusing guide. The guide design was optimized using the DE method. The shaded regions shown in panel a) are supermirror-coated mountings. The solid vertical line represents the height of the 10 mm sample.

[4] Bentley P M, and Andersen K H 2009 J. Appl. Cryst. **42** 217
[5] Disk chopper time-of-flight spectrometer IN5, http://www.ill.eu/in5
[6] Cryogenic Limited, http://www.cryogenic.co.uk/
[7] Three Axis instrument for Low Energy Spectrometry, http://www.ill.eu/instruments-support/instruments-groups/instruments/thales/description/the-project-in-short/
[8] Holland J T 1975 *Adaption in Natural and Ariticial Systems* (Ann Arbor, MI: University of Michigan Press)
[9] Kennedy J and Eberhart R 1995 Proc. IEEE Int. Conf. Neural Netw. **4** 1942
[10] Hölzel R, Bentley P M, Fouquet P 2006 Rev. Sci. Instrum. **77** 105107
[11] Bentley P M, Pappas C, Habicht K, and Lelièvre-Berna E 2006 Physica B **385-386** 1349
[12] Bentley P M, Fouquet P, Böhm M, Sutton I, Dewhurst C D, and Andersen K H 2011 J. Appl. Cryst. **44** 483
[13] Darwin C 1859 *On the Origin of Species by Means of Natural Selection* London: John Murray
[14] Lieutenant L 2005 J. Phys.:Condens. Matter **17** S167
[15] Farhi E, Debab Y, Willendrup P 2014 J. Neut. Res., **17** 5
[16] Karaboga D 2005 An Idea Based On Honey Bee Swarm For Numerical Optimization, Technical Report-TR06, Erciyes University, Engineering Faculy, Computer Engineering Department
[17] Storn R and Price K 1997 J. Glob. Optim. **11** 341
[18] Krink T, Filipič B, and Fogel G 2004 *Noisy Optimization Problems - A Particular Challenge for Differential Evolution?* Proceedings of 2004 Congress on Evolutionary Computation, IEEE Press, 332
[19] Karaboga D and Basturk B 2008 App. Soft. Comp. **8** 687
[20] Karaboga D and Akay B 2009 Appl. Math. Comp. **214** 108
[21] Karaboga D and Akay B 2009 ABC algorithm coded using C programming language, *Available online* http://mf.erciyes.edu.tr/abc/
[22] Kukkonen S 2007 Basic DE in C (Standard C for Unix/linux/Windows or any other platform), *Available online* http://www.iitk.ac.in/kangal/codes.shtml
[23] ROOT, A Data Analysis Framework, *Available online* http://root.cern.ch/
[24] Mutka H and Ollivier J, private communication

Development and demonstration of a multi-channel spheroidal focusing device for neutron beams

H Hayashida[1], K Soyama[1], D Yamazaki[1], R Maruyama[1] and K Yamamura[2]

[1] J-PARC Center, Japan Atomic Energy Agency, Tokai, Ibaraki 319-1195, Japan
[2] Department of Precision Science & Technology, Osaka University, Suita, Osaka 565-0871, Japan

E-mail: hayashida.hirotoshi@jaea.go.jp

Abstract. A multi-channel neutron focusing mirror is a compact device that can effectively enhance neutron intensity because the multi-channel structure can cover a large divergence of a neutron beam. In this study, we attempted to develop a compact multi-channel spheroidal (MS) neutron focusing device for two-dimensional focusing. A prototype of the MS mirror consists of three spheroidal mirrors of different diameters. The mirrors are fabricated through the copper plating method without supermirror coating and are aligned coaxially using ring-shaped spacers. The MS mirror was demonstrated at beam line 10 NOBORU port at J-PARC, which provides neutron beams with time-of-flight spectra. A gain factor of 6 in neutron intensity was obtained over wavelengths greater than 0.5 nm, and an imaging test with sample scanning could be performed with an exposure time of 10 s. A Gd-patterned standard sample was employed and a 2D image with a spatial resolution of 200 μm was successfully obtained.

1. Introduction

Focusing devices are often used to increase neutron intensity over a minute area, such as a scattering experiment with small samples. Mirrors are the most commonly used focusing devices, and several types have been developed [1-4]. In this study, we attempted to develop a multi-channel neutron focusing mirror that can cover a wide divergence of neutron beams while boasting of a small size. This can be applied to a Wolter mirror, which is already very widely used in microscopy as well as in telescopes. Our approach could also lead to the realization of a compact neutron microscope, which is expected to drastically improve the spatial resolution of the neutron imaging technique. Such a microscope would have sufficient spatial resolution to be a powerful tool in magnetic imaging methods, which have been gaining prominence in recent years [5-7], to visualize magnetic domain structures of the order of micrometers. Neutron spin flippers and neutron spin analyzer are necessary components in these magnetic imaging methods. Consequently, the distance between the sample and detector is at least 500 mm. Even with a detector having a high spatial resolution of the order of micrometer, information on magnetic structure would be lost in such a long flight-path because of the divergence of the neutron beam. On the other hand, a neutron beam with large divergence can be used in the experimental setup comprising such a microscope. Because these magnetic imaging techniques consist of many parts such as a neutron spin flipper, polarizing devices, permalloy magnetic shields, and a guide field coil, sufficient space is not available for installing the microscope. Therefore a compact size is more effective for the microscope. As a first step to overcome these issues, in this study, we attempted to develop a multi-channel spheroidal (MS) mirror for focusing neutron beams.

The MS mirror was demonstrated at beam line 10 (BL10) NOBORU port at J-PARC, and an imaging test with sample scanning was performed using a focused neutron beam. The design of a prototype MS mirror is described in section 2, and a demonstration of the MS mirror is described in section 3 along with the results.

2. Design of the prototype MS mirror

Figure 1 shows a conceptual 2D diagram of the prototype MS mirror. Two focal points f_1 and f_2 are shown, and three ellipses, which represent the full spheroids of the spheroidal mirrors, are indicated with dashed lines. Thick lines along the ellipses indicate the MS mirror, and the prototype MS mirror consists of three spheroidal mirrors of different diameters. The distance between f_1 and f_2 is 1200 mm and the prototype MS mirror is designed to be placed 900 mm downstream of f_1 (300 mm upstream of f_2). The length of the MS mirror is 200 mm, and the diameters on the upstream edge of the three spheroidal mirrors (from inside to outside) are 6 mm, 8 mm and 10 mm. The incident angles of the neutron beam to the three spheroidal mirrors are also shown in Fig. 1. The three spheroidal mirrors are indicated as No. 1, No. 2, and No. 3 (from inside to outside). Incident angles as the upstream and downstream sides of mirror No. 1 are denoted by θ_{1u} and θ_{1d}, respectively. Similarly, θ_{2u} and θ_{2d} indicate those on mirror No. 2, while θ_{3u} and θ_{3d} indicate those on mirror No. 3. The neutron wavelengths reflected by the spheroidal mirrors can be estimated from these incident angles as it is known that copper has 70% the scattering length density of nickel. Table 1 lists the incident angles and neutron wavelengths corresponding to the critical angle for total reflection. The upstream and downstream sides on mirror No. 1 reflect neutron beams with wavelengths greater than 0.46 nm and 0.58 nm, respectively. Therefore, neutron beams with wavelengths greater than 0.58 nm can be reflected off the whole surface of mirror No. 1. In the case of mirrors No. 2 and No. 3, neutron beams with wavelengths greater than 0.77 nm and 0.97 nm, respectively, can be reflected off the whole surface. Although neutron beams with wavelengths shorter than 0.97 nm can be focused by coating the surface of the spheroidal mirrors with a supermirror, such a coating has not been performed in this study.

Figure 2 shows a schematic diagram of the coaxial alignment of the three spheroidal mirrors and an image of the MS mirror. The spheroidal mirrors are fabricated through the copper plating method and are aligned coaxially using ring-shaped spacers formed by winding a kapton film of width 1 mm. The surface figure error is approximately 100 nm, which is determined by the metal mold.

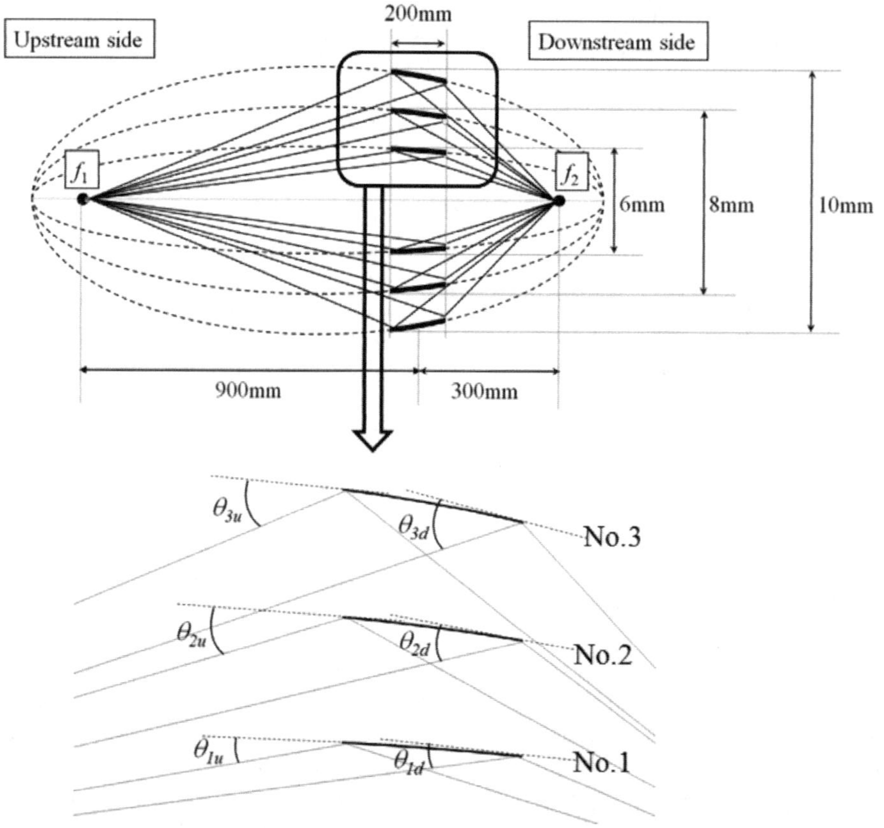

Figure 1. Conceptual 2D diagram of the prototype MS mirror. Schematics of the incident angles are enlarged.

Table 1. Incident angles at the upstream and downstream edges of each spheroidal mirror, along with neutron wavelengths corresponding to the critical angle for total reflection.

Mirror number	θ_u / degree	λ_u / nm	θ_d / degree	λ_d / nm
No.1	0.32	$0.46 \leqq \lambda$	0.41	$0.58 \leqq \lambda$
No.2	0.43	$0.61 \leqq \lambda$	0.54	$0.77 \leqq \lambda$
No.3	0.54	$0.77 \leqq \lambda$	0.68	$0.97 \leqq \lambda$

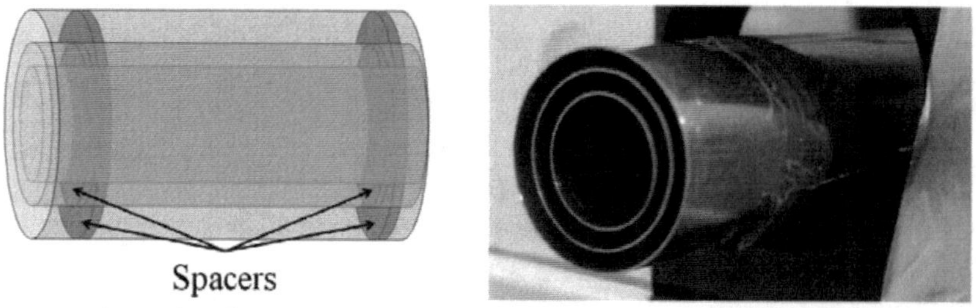

Figure 2. Schematic of the coaxial alignment of the three spheroidal mirrors and an image of the MS mirror. The ring-shaped spacers are used for coaxial alignment.

3. Demonstration and results of the prototype MS mirror

The prototype MS mirror was demonstrated at BL10 NOBORU at J-PARC. Figures 3 (a), (b), and (c) show an image of the experimental setup, and schematic diagrams of this experiment without and with the MS mirror, respectively. A pinhole slit of diameter 100 μm is set at the focal point. The MS mirror is set upstream of the pinhole slit. The distance between the center of the MS mirror and the pinhole slit is 300 mm as shown in Fig. 3(c). The MS mirror can be moved in the horizontal direction, which enables us to switch between the experimental setups with and without the MS mirror. A two-dimensional position sensitive detector (2D detector) is set downstream of the pinhole slit. The distance between the pinhole slit and the surface of the 2D detector is 130 mm as shown in Fig. 3(b). A supermirror with $m = 4$ is set upstream of the MS mirror to shield the 2D detector from a direct high-energy beam. The performance of the MS mirror was evaluated using both experimental setups.

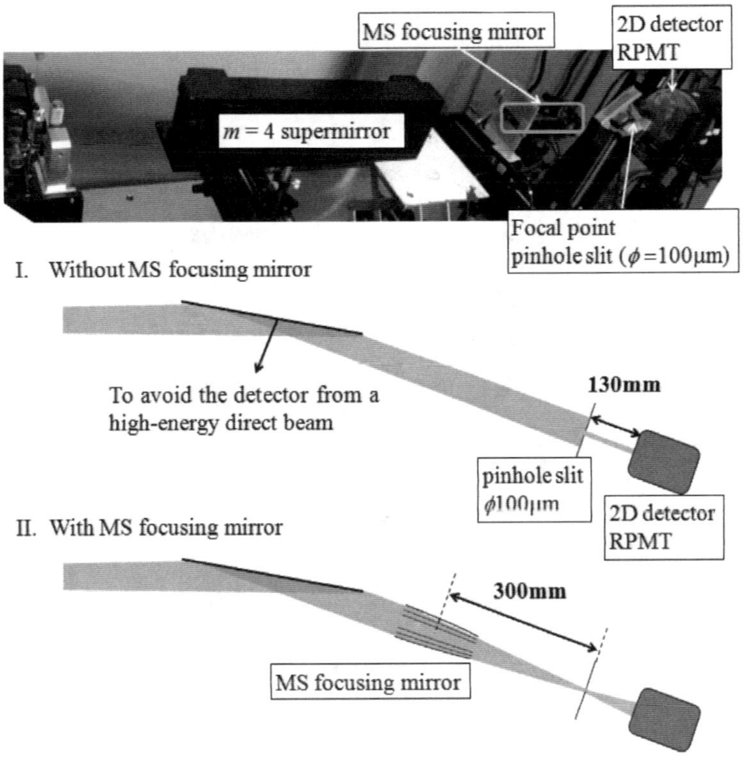

Figure 3. (a) Image of the experimental setup, (b) schematic diagram of the experiment without the MS mirror, and (c) schematic diagram of the experiment with the MS mirror.

Figure 4 shows a 2D image of the neutron beam focused by the MS mirror. A focused neutron beam has a large divergence as compared to that of its unfocused counterpart. Hence, after passing through the pinhole slit, the focused beam spreads into a ring shape at the detector position. The center spot in Fig. 4 shows a direct beam not focused by the MS mirror, and the ring shape around the center spot indicates a neutron beam focused by the MS mirror. Figure 5 shows the time-of-flight (TOF) spectra of the neutron intensity with and without the MS mirror and the ratio of these spectra. The ratio indicates a gain in neutron intensity caused by the MS mirror.

The prototype MS mirror has not been coated with supermirror and mirror No. 3 can reflect neutron beams with wavelengths greater than 0.97 nm, as indicated in Table 1. Mirror No. 3 does not work in this experiment, because the neutron wavelength in the BL10 NOBORU can be used up to 1 nm. Mirrors No. 2 and No. 1 work on neutron beams of wavelengths greater than 0.77 nm and 0.58 nm, respectively, as indicated in Table 1. In the wavelengths of 0.77 nm, mirrors No. 1 and No. 2 work, and the neutron intensity exhibits a gain by a factor of 7.6, which is estimated from the ratio of beam divergences between the two experimental setups. For wavelengths of 0.58 nm, only mirror No. 1 works, and the gain factor can be evaluated as 3.7. However, the gain factors between the two experimental setups for neutron wavelengths from 0.58 nm to 0.77 nm are approximately 6, as shown in Fig. 5, which is inconsistent with geometric estimations. These inconsistent results could be attributed to both setting and coaxial alignment errors. Figure 6 shows a conceptual diagram of misalignment occurring in two typical situations with a spheroidal focusing mirror. The incident angle of the neutron beam on a part of the mirror is less than the designed incident angle, and in the other part of the mirror, it is greater than the designed incident angle. In the former case, neutron beams with wavelengths shorter than the design specification can get focused by the spheroidal mirror surface, whereas in the latter, neutron wavelengths corresponding to the critical angle shift to longer wavelengths. Because of these possibilities, neutron beams with wavelengths different from the design specifications get focused. Hence, the image of the focused neutron beam on the 2D detector does not have a perfect ring shape, as shown in Fig. 4. Therefore, it can be considered that the alignment precision of the setting error and coaxial alignment error lead to results inconsistent with what is expected on the basis of design specifications.

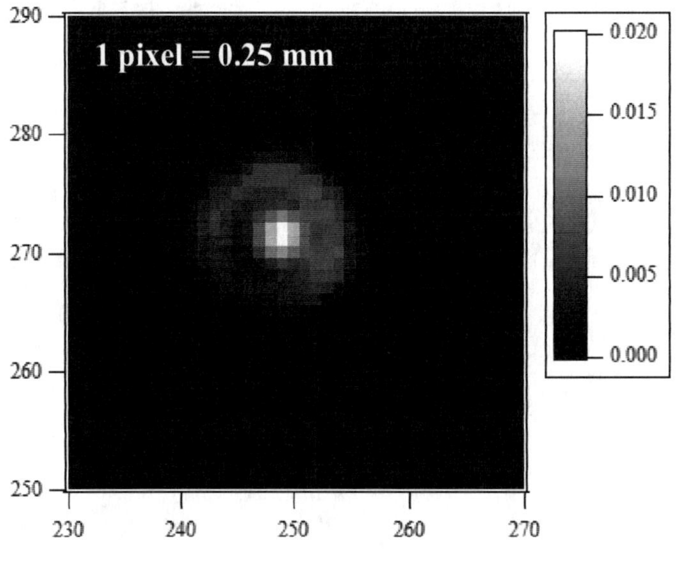

Figure 4. A 2D image of the neutron beam focused by the MS mirror. The spot at the center is the direct beam not focused by the MS mirror. The ring shape around the center spot is the beam focused by the MS mirror.

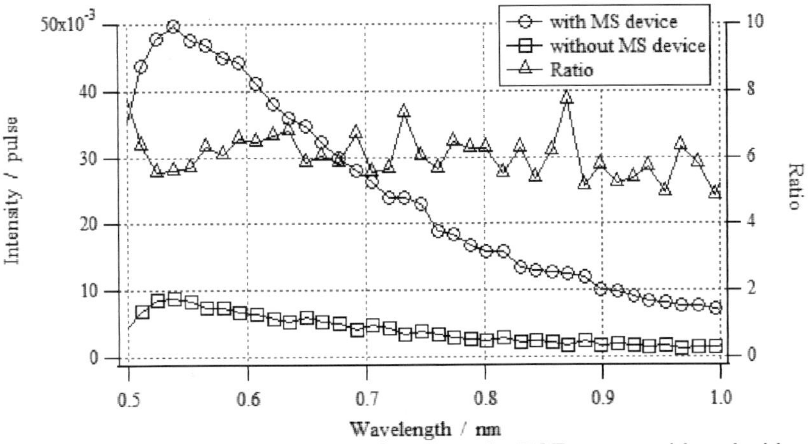

Figure 5. The circles line and the squares line show the TOF spectra with and without the MS mirror, respectively. The triangles represents the ratio of these spectra.

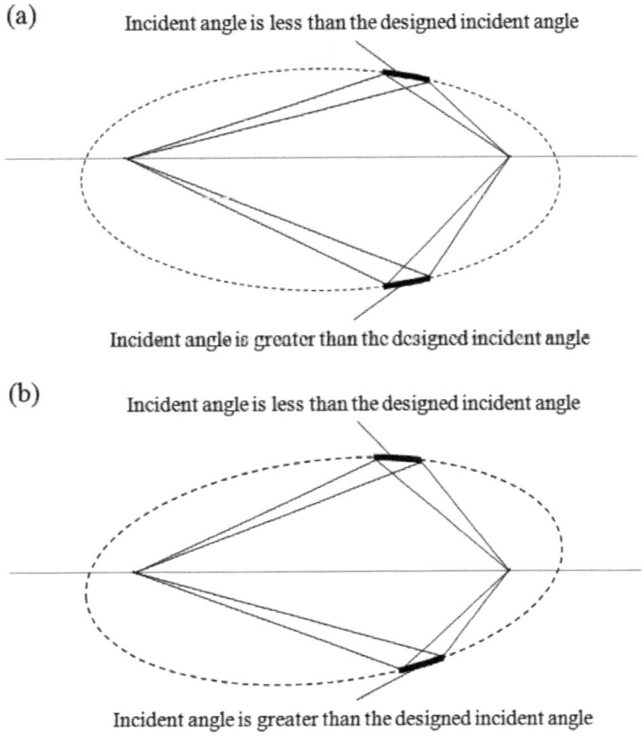

Figure 6. Conceptual diagram of two typical misalignment patterns with a spheroidal focusing mirror.

Although numerous issues remain in the MS mirror, the intensity gain factor of 6, shown in Fig. 5, and the results of the scanning imaging test are promising. Figure 7 shows a schematic of the imaging test, a picture of a test sample, and the resulting 2D image. A Gd-patterned sample, which is a standard sample made by PSI group [8], was used to evaluate spatial resolution. The pitches of the Gd patterns are 500 μm, 400 μm, 300 μm, and 200 μm (from outside), as shown in the central image in Fig. 7. The sample was placed just downstream of the pinhole slit. A scanning pitch of the sample was 100 μm, identical to the pinhole slit size, which corresponds to neutron imaging using a 2D detector with spatial resolution of 100 μm. A 2D image obtained from scanning the sample is shown on the right in fig. 7. The scale bar of the 2D image shows integrated neutron intensity. Gd patterns with a resolution of 200 μm can be visualized. Measurement time of each period was 10 s. The MS mirror enables us to perform the imaging test with a spetial resolution of 200 μm within such a short measurement time.

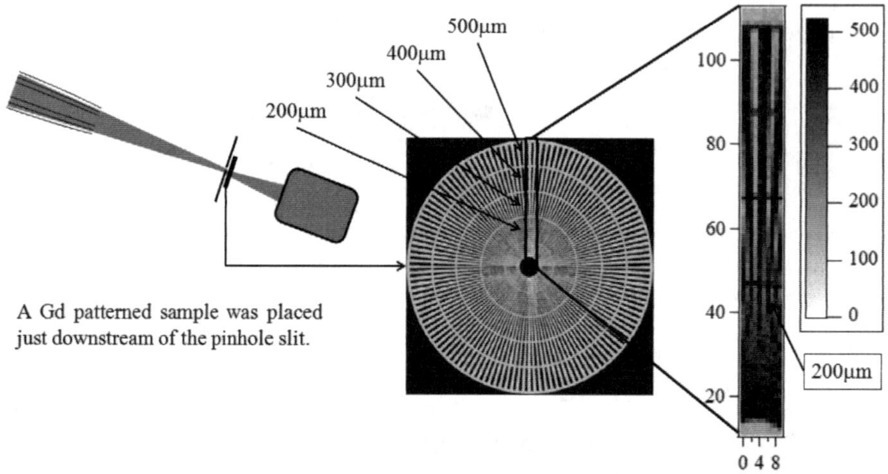

Figure 7. Schematic diagram of the imaging test, a picture of a test sample, and the resulting 2D image. A Gd patterned sample was used as a test sample.

4. Summary

A prototype MS mirror has been developed, and its demonstration for a neutron beam has been performed. The prototype MS mirror consists of three spheroidal mirrors of different diameters fabricated through the copper plating method and aligned coaxially using ring-shaped spacers. Supermirror coating was not performed on the surface of the MS mirror. The length of the MS mirror was 200 mm. The MS mirror was demonstrated at BL10 NOBORU at J-PARC, which provides TOF neutron beams. Since the setting and coaxial alignment precision have a serious effect on the focusing ability in this experiment, these alignment methods have to be improved to obtain a high level of precision. However, the focused beam was successfully observed as a ring shape at the detector, and the MS mirror enabled us to perform an imaging test with a short exposure time of 10 s and a spatial resolution of 200 μm.

Acknowledgments

We are grateful to the staffs of the MLF at J-PARC for their help.

References

[1] Nagano M, Yamaga F, Yamazaki D, Maruyama R, Hayashida H, Soyama K and Yamamura K 2012 *J. Phys. Conference Serries* ***340*** *012016.*

[2] Nagano M, Yamaga F, Yamazaki D, Maruyama R, Hayashida H, Soyama K and Yamamura K 2012 *J. Phys. Conference Serries* ***340*** *012034.*

[3] Liu D, Khaykovich B, Gubarev V. M, Robertson J. L, Crow L, Ramsey D. B and Moncton E. D 2013 *Nature Communications* ***4*** *2556.*

[4] Kardjilov N, Hilger M, Dawson M, Manke I, Banhart M, Strobl M and Boni P, 2010 *Journal of*

Applied Physics ***108*** *034905.*

[5] Kardjilov N, Manke I, Strobl M, Hilger A, Treimer W, Meissner M, Krist T and Banhart J 2008 *Nat. Phys.* ***4*** *399.*

[6] Shinohara T, Sakai K, Ohi M, Kai T, Harada M, Oikawa K, Maekawa F, Suzuki J, Oku T, Takata S, Aizawa K, Arai M and Kiyanagi Y 2011 *Nucl. Instrm. Meth. A* ***651*** *121.*

[7] Hayashida H, Yamazaki D, Ebisawa T, Maruyama R, Soyama K, Tasaki S, Hino M and Matsubayashi M 2011 *Nucl. Instrm. Meth. A* ***634*** *S90.*

[8] Grunzweig C, Frei G, Lehmann E, Kuhne G and David C 2007 *Review of Scientific Instruments* ***78*** *053708.*

The dependence of the gravity effect in elliptic neutron guides on the source size

D Nekrassov[1,2], **C Zendler**[1,2], **K Lieutenant**[1,2]

[1] Helmholtz-Zentrum Berlin für Materialien und Energie GmbH, Hahn-Meitner-Platz 1, 14109 Berlin, Germany

[2] ESS design update programme, Germany

E-mail: daniil.nekrassov@helmholtz-berlin.de

Abstract.
Elliptic neutron guides are expected to be widely used for construction of long neutron beamlines at the future European Spallation Source and other facilities due to their superiour transmission properties compared to conventional straight guides. At the same time, neutrons traveling long distances are subject to the action of gravity that can significantly modify their flight paths. In this work, the influence of gravity on a neutron beam propagating through elliptic guides is studied for the first time in a systematic way with Monte-Carlo simulations. It is shown that gravity leads to significant distortions of the phase space during propagation through long elliptic guides, but this effect can be recovered by a sufficiently large source size. The results of this analysis should be taken into account during design of long neutron instruments at the ESS and other facilities.

1. Introduction

Flight paths of thermalised neutrons at reactor and spallation sources are modified by the action of gravity. Its influence increases with rising wavelength of the neutrons and longer flight paths. For example, a 10 Å neutron traveling a distance of $d = 100$ m is displaced by gravity by

$$\Delta h = 0.5 \times g \times (d/v(\lambda))^2 = 0.5 \times 9.81\,\mathrm{m/s}^2 \times \left(\frac{100\,\mathrm{m} \times 10\,\text{Å}}{3956\,\mathrm{m\text{Å}/s}}\right)^2 \approx 31\,\mathrm{cm}.$$

This effect is significant and needs to be taken into account when designing or operating long neutron beamlines.

The gravity problem has already received some attention in the past. A way to counteract the vertical displacement of long-wavelength neutrons for instruments not comprising neutron guides was found to be a modification of the vertical position of the (virtual) source with respect to the sample and detector position [1], [2]. The influence of gravity on neutrons propagating through a collimation system and a consequential distortion of reflectivity measurements on liquid interfaces was included in an elaborated resolution theory that was confirmed by Monte-Carlo (MC) simulations [3]. On the other hand, the treatment of the gravity effect in neutron guide tubes requires a numerical approach due to the occurrence of reflections. At a time when computing power was limited and thus extensive MC simulations were difficult to carry out, an analytical matrix formalism being less demanding in terms of calculation time was developed to trace phase space during its propagation through straight or curved guides, including the influence of gravity [4]. The calculation of neutron trajectories in and outside guides then became more accessible since the invention of MC software packages like VITESS [5].

At the future European Spallation Source (ESS) facility [6], gravity will play a significant role due to the length of planned instruments that often exceed 100 m. Such long beamlines will need to include ballistic neutron guides to efficiently transport the neutrons to the sample position [7]. In particular, elliptically shaped guides have lately received a significant attention regarding their transmission and focusing properties [8], [9]. Currently, it can be expected that several instruments at the ESS will comprise elliptic guides. However, the influence of gravity on the beam properties after transmission through an elliptic guide has not been studied so far in a systematic way, even though it was found that the focusing ability of elliptic guides might be severely disturbed [8]. For a very long instrument of 300 m using a quite narrow waveband of 0.8 Å around 6.66 Å it appears that incorporating the trajectory curvature into the shape of the elliptic guide allows to remove the direct line-of-sight (LoS) without suffering flux losses and preserve the instrument resolution [10]. The latter study, however, was carried out for a potential ESS backscattering instrument, for which the influence of the beam divergence distribution on the measured resolution is significantly reduced. Hence the present work aims at studying the influence of gravity on the phase space structure for a large neutron waveband after propagation in elliptic guides, in particular concerning the shape of the vertical divergence distribution. This is important for instruments where the divergence distribution has a direct impact on the illumination homogeneity of the sample/detector or on the shape of structures in the scattering spectrum (e.g. for diffraction).

2. Analysis and results

The gravity effect is studied using a simple instrument layout, see Fig 1. A source emitting a constant spectrum as a function of wavelength is followed by an elliptic neutron guide with a square cross section, of which the semi-axes are $a = 75$ m along the instrument axis and $b = 0.15$ m in both directions perpendicular to it. The reflectivity of the guide coating $R(\tilde{m})$ is 0.99 for $\tilde{m} < 1$ (with $\tilde{m} = 10 \times \theta\,[^\circ]/\lambda\,[\AA]$, where θ is the reflection angle), decreases linearly to $R(\tilde{m} = 5.7) = 0.52$ and then drops quickly [11]. The guide is followed by a 1 cm^2 sample. Both the source and the sample are located in the focal points of the ellipse. This study utilises the new elliptic guide module available in the VITESS software from version 3.0, which handles neutron propagation through a perfect ellipse, thus avoiding effects connected with guide segmentation [8], since in such a study the gravity effect should be considered separately from other imperfections. The distance D_0 between the source and the guide entry and between the guide exit and the sample is the same and the total length of the instrument is fixed to 150 m. D_0 is varied between 20 cm and 5 m, thus varying the entry/exit width W_0 between 2.2 cm and 10.8 cm. The source is a square with the edge length X_0, which is varied between 1×1 cm^2 and 12×12 cm^2. **The input divergence is wavelength independent and solely depends on the source size, the distance to the guide entry and the guide entry parameters.** All input parameters are summarized in Tab. 1. **The coordinate system follows the convention used in VITESS, i.e. the x-axis corresponds to the instrument axis, while the y- and z-axis are completing a right-handed coordinate system in horizontal and vertical directions, respectively.**

The goal of the performed simulations is to monitor the beam characteristics at the sample position with regard to gravitational effects. Since it can be expected that gravity modifies the vertical divergence distribution (and thus the distribution in real space at the detector), an asymmetry parameter $\Delta\gamma$ is introduced in order to describe this effect:

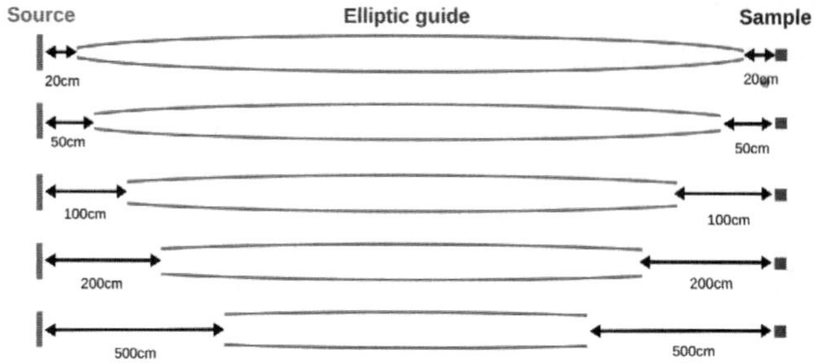

Figure 1. A sketch of the instrument layout used in the present study. The distance between the source and the sample is kept constant at 150 m, whereas the distance between the source and the guide entry (the guide exit and the sample) is varied between 20 cm and 500 cm. The source edge length is varied between 1 cm and 12 cm. See text for further details.

Table 1. Summary of input parameters used in the simulations.

Source	Spectrum: $I(\lambda) = const$
	Continuous wavelengths: 1 Å – 12 Å
	Discrete wavelengths: 2 Å, 6 Å, 10 Å
	Edge length X_0: 1 cm, 2 cm, 4 cm, 6 cm, 8 cm, 10 cm, 12 cm
Guide	Elliptic guide, a = 75 m, b = 15 cm, m=6 coating
	Source-to-guide (guide-to-sample) distance D_0:
	0.2 m, 0.5 m, 1 m, 2 m, 5 m
	Guide entry (exit) W_0: 2.2 cm, 3.5 cm, 4.9 cm, 6.9 cm, 10.8 cm
	Source and sample at focal points
Sample	Size: 1×1 cm^2

$$\Delta\gamma = \left|\frac{I(\gamma)_+ - I(\gamma)_-}{I(\gamma)_+ + I(\gamma)_-}\right| = \left|\frac{\int_0^{\gamma_{\max}}[I(\gamma) - I(-\gamma)]d\gamma}{\int_0^{\gamma_{\max}}[I(\gamma) + I(-\gamma)]d\gamma}\right|, \tag{1}$$

where $I(\gamma)$ is the beam intensity as a function of divergence γ either in vertical or horizontal direction. $\gamma_{\max}$ **can be the full or some selected divergence range.** The simulations are carried out for the waveband ranging from 1 Å to 12 Å and for discrete wavelengths 2 Å, 6 Å and 10 Å to illustrate the different behaviour of short- and long-wavelength neutrons.

The results of the simulations are illustrated in Figs. 2 - 7. A careful consideration of the obtained distributions reveals **the** following major findings:

- Gravity can strongly affect the vertical divergence distribution in long elliptic guides, as opposed to propagation through a straight neutron guide of comparable dimensions, see Fig. 2.
- The magnitude of distortion of the divergence distribution due to gravity depends on two parameters:
  - (i) Wavelength: As it could be expected, the divergence distribution is more distorted with increasing wavelength, see Fig. 3. For short wavelengths (and small sources/samples

[8]), the final divergence distribution is more affected by propagation through an ellipse than by gravity, see Fig. 3 a).

(ii) Source and guide entry size: The **asymmetry** decreases with increasing source size (Fig. 2 (b) **and Fig. 3**) or with decreasing entry width W_0/source-to-guide distance D_0 (Fig. 4).

- Despite the asymmetry in the divergence distributions, elliptic guides still provide a reasonable focusing in space, i.e. the flux is the largest at the sample position. At the same time, the focusing is more strongly reduced for larger X_0 distances in vertical than in horizontal direction (if the source size W_0 is kept constant), see Fig. 5.
- To minimize the distortion of the divergence distribution, the source needs to be of the same size or larger than the guide entrance, see Fig. 7 (a). But – not surprisingly – the actual ratio needed of the source size X_0 to the guide entrance size W_0 is wavelength dependent, see Fig. 7 (b) for a (rough) quantification[1].
- When a symmetric and featureless divergence distribution is reached, a further increase of the source size does not increase the flux at the sample, see Fig. 2 (a), (b) and Fig. 3.

In particular the last two points are important findings. It can be observed that for a given elliptical guide there is a certain (virtual) source size that completely smears out the gravity distortion and features characteristic **for** transmission through elliptic guides. At the same time, the source of this size provides the maximum flux on the sample, a fact deserving serious attention when designing a neutron beamline. This is confirmed by an additional set of simulations using an elliptic guide with the semi-axes $a = 37.5$ m, $b_1 = 0.15$ m and $b_2 = 0.075$ m, a fixed distance between the source and the guide of 1 m and varying the size of the source again between 1×1 cm^2 and 12×12 cm^2. Here it was again observed that the divergence distribution obtains a symmetric shape around zero every time the source is larger than the guide entry, see Fig. 6. The conclusion is that the recovering of a symmetric divergence distribution happens by mixing the neutron trajectories through multiple reflections, which are by far the most dominant transmission regime in elliptic guides [8], such that all inhomogeneities are smeared out. This process is more efficient if the source injects more phase space into the guide.

3. Discussion and conclusions

The simulation results described in the last section clearly show that gravity can play an important role in neutron transport in long elliptic guides, in particular for small sources and long wavelengths. At the same time, it has been shown that these effects can be removed by increasing the size of the (virtual) source such that it exceeds the guide entry dimensions. Such a source is able to smear out the features in the divergence distribution at the sample position coming both from gravity influence and transmission effects. Hence in principle, elliptic guides are able to transport neutrons over long distances and provide a smooth phase space at the sample position, if provided with an adequate input beam. This should be kept in mind for design of instruments that are in need of a smooth phase space at sample/detector position.

Acknowledgments

We thank W. Schweika for initiating this study and L. D. Cussen for fruitful discussions.
This work was funded by the German BMBF under "Mitwirkung der Zentren der Helmholtz

[1] **For each wavelength, this ratio was determined by simply averaging the smallest source-to-entrance size ratios that still led to an asymmetry smaller than a 5% threshold for all source sizes under study.**

[2] **The relatively small $\Delta\gamma$ value for the whole divergence range, despite a visible asymmetry in Fig. 3 for the 2x2 cm^2 source can be increased by restricting $\gamma_{\max}$ to, e.g. 0.5°.**

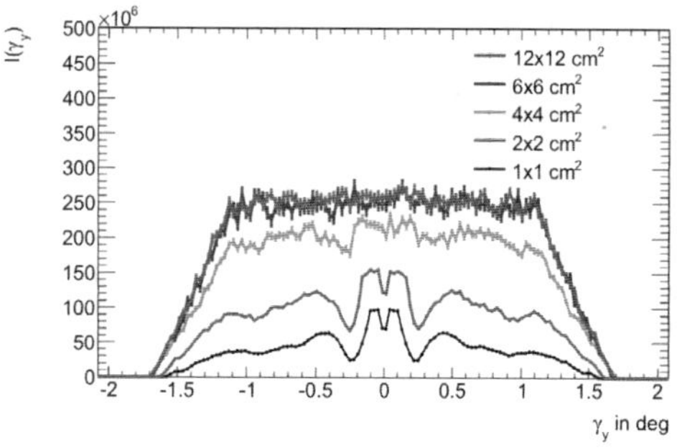

(a) Horizontal divergence distribution at sample position for an elliptic guide

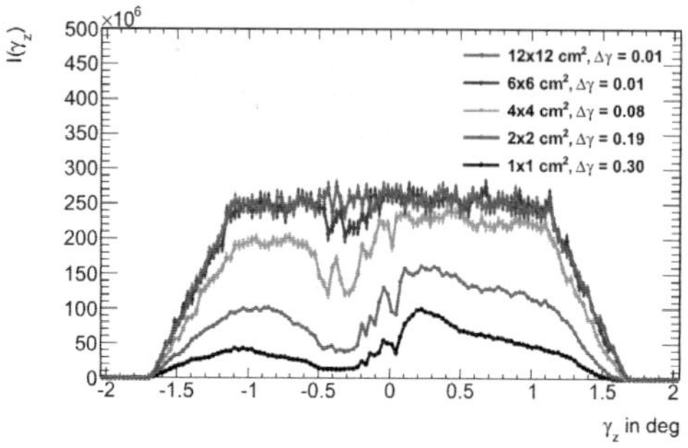

(b) Vertical divergence distribution at sample position for an elliptic guide for varying source size X_0

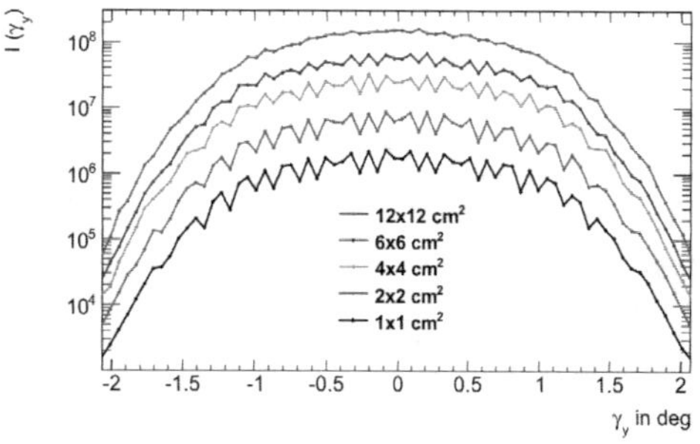

(c) Horizontal divergence distribution at sample position for a straight guide

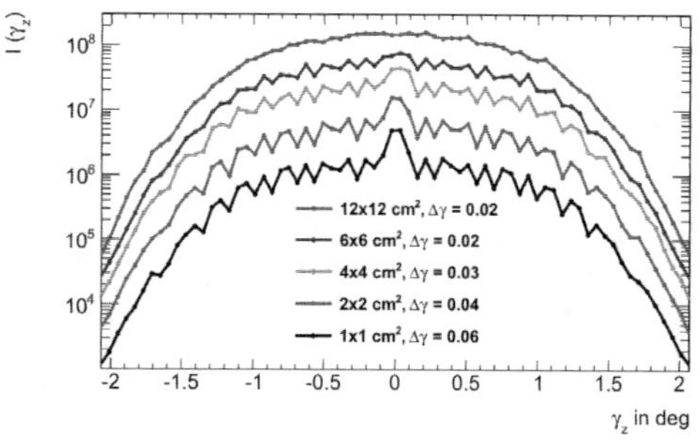

(d) Vertical divergence distribution at sample position for a straight guide

Figure 2. (a) and (b): Divergence distribution at sample position for $W_0 = 4.9$ cm ($D_0 = 1$ m) for an elliptically shaped guide, see Tab. 1, using source sizes from 1×1 cm^2 to 12×12 cm^2 and the full spectrum between 1 Å and 12 Å. (c) and (d): Horizontal and vertical divergence distributions at sample position for a 148 m long guide with a constant 10×10 cm^2 cross section having the same source-to-guide and guide-to-sample distance $D_0 = 1$ m. The y-axis is logarithmic to fit in distributions for small source sizes. The zig-zag structures, which are particularly visible for small source sizes, are a systematic effect and arise due to fractions of the total phase space missing the 1 cm^2 sample, since straight guides lack focusing abilities. **The horizontal divergence distributions also represent the shape of vertical divergence distributions, if gravity would be absent.** Here and in other plots the error bars (mostly too small to be visible) represent the statistical uncertainty due to the number of simulated trajectories.

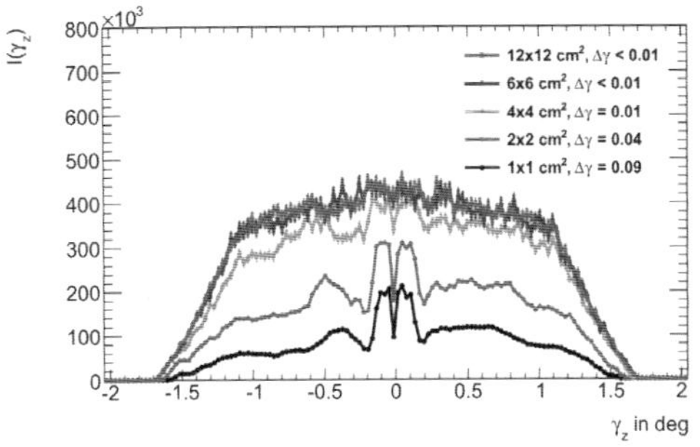

(a) Vertical divergence distribution at sample position for 2 Å neutrons

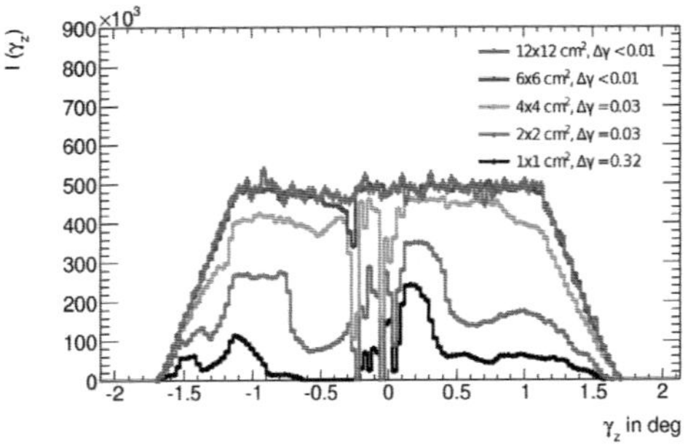

(b) Vertical divergence distribution at sample position for 6 Å neutrons

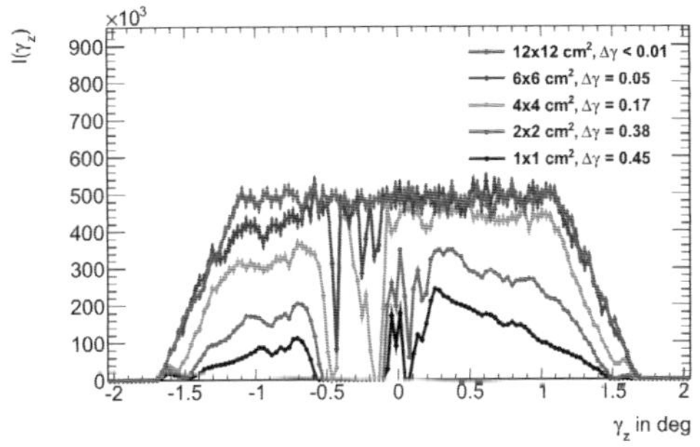

(c) Vertical divergence distribution at sample position for 10 Å neutrons

Figure 3. Vertical divergence distribution at sample position using $W_0 = 4.9$ cm ($D_0 = 1$ m) for 2 Å, 6 Å and 10 Å neutrons. Gravity leads to a modification and an asymmetry $\Delta\gamma$ of the vertical divergence distribution in particular for long wavelengths[2].

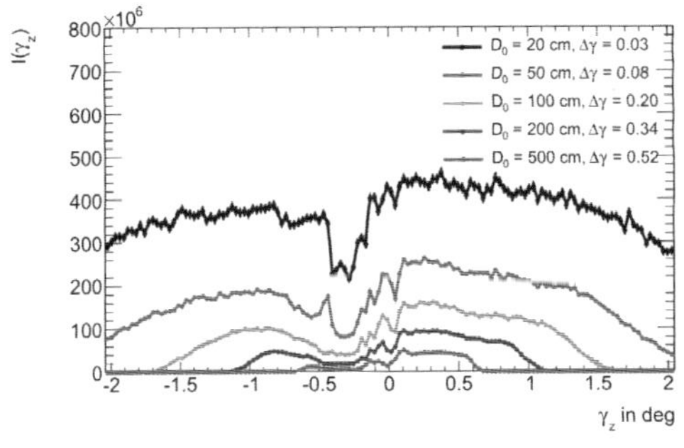

Figure 4. Vertical divergence distribution at sample position for an elliptically shaped guide using a source with $X_0 = 2$ cm and W_0 between 2.2 cm and 10 cm (D_0 between 20 cm and 5 m).

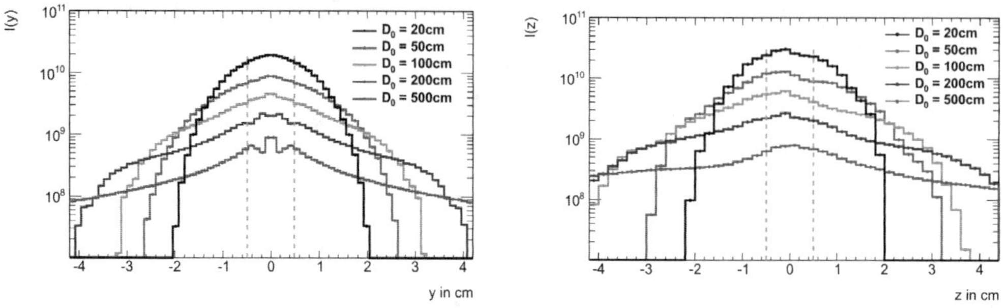

(a) Horizontal position distribution at sample position

(b) Vertical position distribution at sample position

Figure 5. Distribution of neutrons in space for a 2×2 cm^2 source. The 1×1 cm^2 sample indicated by dotted lines is located at the guide symmetry axis that coincides with the location of the highest flux even for a 5 m distance between guide and sample (corresponding to ≈ 11 cm guide exit width).

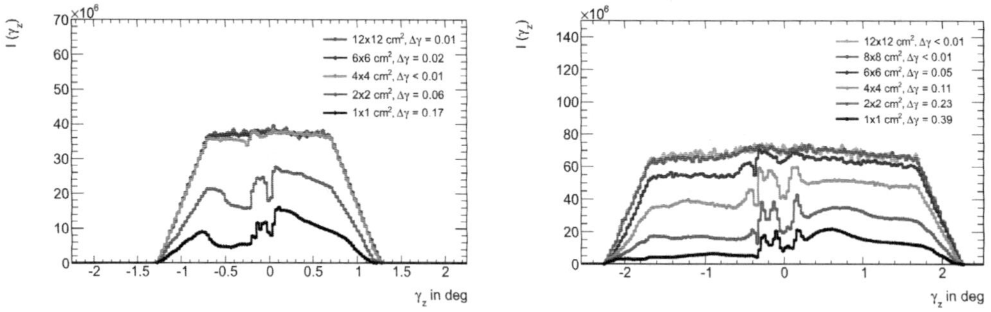

(a) a = 37.5 m, b = 0.075 m

(b) a = 37.5 m, b = 0.15 m

Figure 6. Vertical divergence distribution at the 1×1 cm^2 sample being in the focal point of two different elliptic guides with given parameters. The distance D_0 is 1 m. The guide entry width is 3.44 cm for b = 0.075 m and 6.88 cm for b = 0.15 m, respectively. The divergence distribution saturates for source sizes being larger than the guide entry and the asymmetry parameter $\Delta\gamma$ becomes negligible.

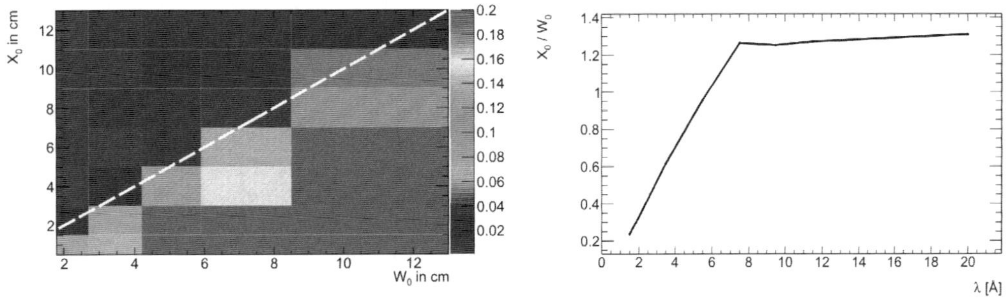

(a) Vertical divergence asymmetry as a function of W_0 and X_0

(b) Size ratio of source to guide entry as a function of λ for $\Delta\gamma_z < 5\%$

Figure 7. (a) The vertical asymmetry parameter $\Delta\gamma$ defined in Eq. 1 as a function of the source edge length X_0 and the entry width W_0 of the elliptic guide for all neutrons. The colour plot is saturated at 0.2, with the maximum asymmetry at $X_0 = 1$ cm and $W_0 = 10.8$ cm. The white line corresponds to the relation $X_0 = W_0$. It is well visible that the gravity effect dominates for $X_0 < W_0$, i.e. for sources being smaller than the guide entrance. (b) **An estimation of the ratio of the source edge length X_0 to the entry width W_0 as a function of the neutron wavelength λ that is needed to achieve a vertical divergence distribution at the sample position, which exhibits an asymmetry of less than 5%.** As expected, larger source sizes are needed for larger wavelengths.

Gemeinschaft und der Technischen Universität München an der Design-Update Phase der ESS, Förderkennzeichen 05E10CB1."

References

[1] J. Copley, *Rev. Sci. Instrum.* **69**, 188 (1996)
[2] R. Cubitt et al., *Nuclear Instruments and Methods in Physics Research A* **665** (2011) 7-10
[3] I. Bodnarchuk et al., *Nuclear Instruments and Methods in Physics Research A* **631** (2011) 121-124
[4] A. Guglielmi, *Nuclear Instruments and Methods in Physics Research A* **325** (1993) 241-254
[5] K. Lieutenant et al., *Proc. SPIE* **5536(1)** (2004) 134-145
[6] European Spallation Source, URL http://www.esss.se
[7] K. Klenø et al., *Nuclear Instruments and Methods in Physics Research A* **696** (2012) 75-84
[8] L. D. Cussen et al., *Nuclear Instruments and Methods in Physics Research A* **705** (2013) 121-131
[9] J. Stahn et al., *Nuclear Instruments and Methods in Physics Research A 634, S12 (2011)*, Proceedings of the International Workshop on Neutron Optics NOP2010
[10] K. Klenø et al., *Nuclear Instruments and Methods in Physics Research A* **634** (2011) 100-103
[11] Swiss Neutronics, URL http://www.swissneutronics.ch/products/coatings.html

Journal of Physics: Conference Series **528** (2014) 012009 doi:10.1088/1742-6596/528/1/012009

Using Parabolic Supermirror Lenses to Focus and De-focus a Neutron Beam

Emmanouela Rantsiou[1], **Tobias Panzner**[1], **Patrick Hautle**[1], **Uwe Filges**[1]

[1]Paul Scherrer Institute, Laboratory for Developments and Methods, 5232 Villigen PSI, Switzerland

E-mail: emmanouela.rantsiou@psi.ch

Abstract. We designed a focus/defocus neutron optics system, in order to investigate the performance, precision, efficiency, and operational and designing challenges of such coupled 2-lens systems, which could potentially find applications where small beam cross sections are beneficial, e.g., virtual neutron source concepts and high efficiency chopper systems. Our particular prototype (as described and discussed in this paper) has already been used in an on-going experiment, involving neutron spin filtering with dynamically polarized protons.

After the designing and construction phases, we continued by performing a long series of simulations and measurements, in order to facilitate the alignment of the lenses, and investigate and understand the behaviour and output of the system. All measurements were performed at the BOA beamline at PSI. The simulations were particularly useful in aligning the lenses: tilts as small as 0.04° could easily be accounted for in our simulations and guide successfully the experimental aligning procedure of the first lens. Although harder to do in the case of two lenses, we were still able to reproduce fairly successfully with our simulations, tilts from both lenses. We have noticed (both in our experiments and simulations) that the sensitivity of such a set-up is $\sim 0.01°$.

1. Introduction

Ever since supermirrors were proposed and built [1, 2, 3], they have found numerous applications in transporting, bending, focusing and polarizing neutron beams. Supermirror technology makes use of multilayer depth-graded coatings (most commonly of Ni and Ti) in order to increase the critical angle of reflection of the mirror by a factor m compared to that of natural ^{58}Ni. Looking at the definition of that factor $m = \theta_c/\theta_c^{Ni}$, with $\theta_c^{Ni} = 0.1°/\AA$, it becomes apparent that the longer the wavelength of the incident neutron, the higher the reflectivity (i.e., the probability of it being reflected), as the divergence tolerance increases with wavelength. Nowadays, supermirrors exhibit reflectivity above 80% for $m \leq 4$, but that value drops to 50% as their coating reaches $m = 7$. A thorough discussion on technicalities of supermirror development can be found in [4].

With technological advances allowing for construction of supermirrors with higher m values, improved reflectivity, and new shapes (elliptical and parabolic), and also ongoing theoretical work on supermirrors, which has been greatly facilitated by simulations, more uses of supermirrors are becoming apparent. The idea of using supermirror lenses to focus and defocus a neutron beam has begun being investigated and tested over the last few years [5, 6]. In this paper we present a proof-of-principle project, carried out at PSI, on the performance and overall behaviour of a 2-lens focus/defocus system.

2. Conceptual Phase and Simulations

The purpose of this project has been to investigate the use of parabolic supermirror neutron lenses in a focus-defocus system, starting from the initial design phase, and following through with the experimental set up and measurements. Besides our interest in the behaviour, sensitivity, efficiency and experimental outcome of such a coupled lens system, we were particularly focused on determining whether simulations can function not only as designing tools, but also reliably guide and correct the alignment of the two lenses during the experiment.

We began by designing the two lenses and determining their characteristics using the Monte Carlo Ray-tracing software McStas [7, 8]. Here we had to take into account that the behaviour and performance of these prototypes should be explored in conjunction with a potential application: a novel neutron spin filter using DNP (Dynamic Nuclear Polarization) [9]. The specifics of that experiment dictated to a certain extent the characteristics and dimensions that the two lenses should have. For example, the defined sample size for use in the DNP experiment was only $\sim 5 \times 5\,\mathrm{mm}^2$, it should be placed as close to the exit of the focusing lens as possible (to maximize intensity), but not closer than $\sim 200 - 250\,\mathrm{mm}$ due to restrictions from equipment placed around the sample. Furthermore, at the sample position the focusing lens (hereafter L1) should provide spatially homogeneous intensity, while the defocusing lens (hereafter L2), placed after the sample, should be able to accept as much as possible of the neutrons coming through the sample and restore the original beam properties. All the above, combined with considerations on costs for the prototype, resulted after several rounds of simulations and optimizations in two lenses whose characteristics are presented on Table 1. L2 is ideally identical to L1 (turned 180° to serve as a defocusing lens), but we ended up constructing only half of it (thus its 0.5 m length), as a full defocusing lens was not really necessary for these preliminary, proof of principle experiments. In our simulations however, we integrated both the half-meter and the one-meter L2 lens, as it is obvious from the results presented on Table 2.

Table 2 shows simulated intensity and divergence values at various crucial positions along the 2-lens set up, for 3 different configurations for L2: i) full one-meter length, placed 0.8 m after L1, ii) full one-meter length, placed 0.5 m after L1, iii) half-meter length, placed 0.5 m after L1. At the first configuration, the focal points of L2 and L1 are at the same position, midway between the two lenses (both lenses have a focal point of 0.4 m). At the second and third configuration (denoted as 'defoc'), L2 is positioned closer to L1, 'looking' 0.3 m before the focal point of L1. Also, in all cases we have used slits before L1 (but also before L2), in order to make sure that a) the full entrance of L1 is illuminated and b) no neutrons pass around L1, contaminating either L2 or making their way onto our final CCD images.

As indicated by the results summarized on Table 2, L1 exhibits a neutron transport of 87% for a white beam ($1 - 10\mathring{A}$) of $\sim 0.5°$ divergence. 22% of those transported neutrons are focused on a $5 \times 5\,\mathrm{mm}^2$ area at the focal point of the focusing lens, with the divergence increased to $\sim 1.3°$. About 37% to 76% of the neutrons transported by L1 are passed through to L2, depending on the position of L2, as it is seen by the two intensity values listed for the entrance of L2. This seemingly low value of intensity transfer between the two lenses can be attributed to the divergence of the direct beam, connected to the relatively large distances, and aberration effects. Aberration alone accounts for up to 30% of intensity loss. As mentioned earlier, our lenses were optimized for providing homogeneous intensity distribution on a $5 \times 5\,\mathrm{mm}^2$ target at 200 mm after L1, and not as a stand alone focus-defocus lens, in which case, the losses in neutrons transferred among the two lenses could be lowered to $10 - 15\%$.

Finally, the length and position of L2 change significantly (as expected) its intensity output and the outcoming neutrons' divergence. For example, increasing the length of L2 restores the initial divergence of the beam, on the expense of intensity. On the other hand, moving L2 closer to L1 (configurations ii and iii) boosts up intensity, but doesn't quite manage to lower the divergence as efficiently as the third configuration.

We mention here that the option to 'defocus' L2 came out of the optimization of the system for the needs of the DNP experiment: we opted for higher intensity input for L2, at the expense of the restored divergence of the beam.

Table 1. Characteristics of the two prototype lenses.

	Focusing Lens (L1)	Defocusing Lens (L2)
Shape	(trully-shaped) parabolic	(trully-shaped) parabolic
Length [mm]	1000	500
Entrance size [mm^2]	25×25	13.363×13.363
Exit size [mm^2]	13.363×13.363	20.04×20.04
m	3.6	3.6
focal point [mm]	400 (after L1 exit)	400 (before L2 entrance)
gain factor[1]	$7.5 - 11$	–

Table 2. Characteristic intensity and divergence values from simulations

	Intensity [$n/s/mA$]	Divergence [$^\circ$] (horizontal/vertical)
L1 entrance	2.34×10^8	0.5/0.4
L1 exit	2.04×10^8	1.3/1.2
200mm after L1 [2]	5.4×10^7	1.3/1.2
Focal point[2]	4.55×10^7	1.3/1.2
L2 entrance	7.08×10^7	1.3/1.2
L2 entrance (defoc)	1.55×10^8	1.3/1.2
(full) L2 exit	5.17×10^7	0.55/0.45
(full) L2 exit (defoc)	8.68×10^7	0.7/0.7
(half) L2 exit (defoc)	1.11×10^8	0.9/0.9

3. Experimental Phase: Measurements and Simulations

After the construction of the lenses was complete, we proceeded to perform experiments at the BOA beamline at PSI. Our experimental procedure was divided in two steps: i) the placement and alignment of the lenses in the beamline one ofter the other, and ii) testing and investigating the performance and outcome of the complete 2-lens system for various configurations, i.e. varying the relative lens distances and orientations, usage of slits, etc. The lenses were placed on separate motorized stages that could move along the direction of the beam path, and also rotate horizontally and vertically with respect to the axis of the lens. Furthermore, both lenses were equipped with microscrews mounted on their holders, enabling for very small inclination

[1] Wavelength dependent

[2] On an area of $5 \times 5mm^2$

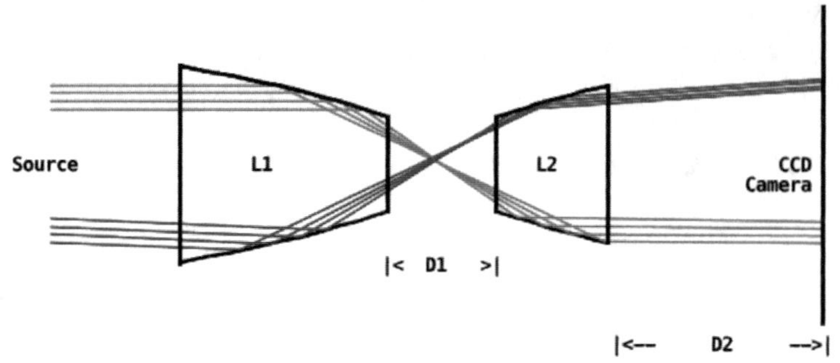

Figure 1. A schematic representation of our experimental setup. The focusing (L1) and defocusing (L2) lenses are shown, along with two groups of rays: parallel rays (upper group), and divergent rays (lower group). The distance D1 between the two lenses as well as the distance D2 between the defocusing lens and the CCD camera, were varying in our simulations and experiments. Relevant scales, ratios and sizes have been exaggerated for illustration purposes.

adjustments. A CCD camera was mounted on a third motorized stage, with the ability to move along the beamline, positioning the camera at various distances from the lenses. All results (experimental and simulated) presented here are for a white neutron beam. During all steps of the experimental procedure, extended simulation runs were performed in order to verify and guide our experiments.

3.1. Measurements and Simulations for the Focusing Lens

We began the experiments by placing and aligning the focusing lens L1 into the beamline. After an initial 'approximate' alignment with the use of a laser beam and a common light bulb, we finalised the alignment of L1 using the full neutron beam, performing scan measurements along the flightpath of the neutrons and correcting the position of L1 accordingly. McStas simulations were particularly helpful in the alignment procedure as the following representative example demonstrates.

At some point during the alignment of L1, the 1D intensity profile of the neutron beam in the Y-direction (up-down direction with respect to the path of the neutrons) as captured by the CCD camera, featured an asymmetry (Fig. 2(a)). Using McStas, we were able to reproduce that feature in our simulations, by introducing an upward tilt of the lens of 0.06°. Fig. 2(b) shows the simulated profiles (X- and Y-direction) for a perfectly aligned lens, and Fig. 2(c) shows the simulated profiles for the tilted lens. We were then able to correct our experimental set-up by tilting L1 0.06° downwards, and finalise its alignment.

Fig. 3 shows the 2-dimensional neutron intensity distributions after the alignment of L1 was completed, at two positions after L1: 200mm (panels (a) and (c)) and at the focal point (400mm) (panels (b) and (d)). Plots (a) and (b) correspond to the experimental images and plots (c) and (d) are result of simulations. We notice an excellent agreement between the experimental and simulated images, also highlighted on the results presented in Fig. 4: after we were convinced that L1 was aligned to a satisfactory level (through extensive measurements and comparison to simulations), we performed some measurements where we introduced a tilt of the lens (either in the horizontal of vertical direction), in order to quantify the sensitivity of our first lens to such misalignments. The upper row of Fig. 4 shows the 2D neutron intensity (experimental (a) and simulated (b)) distribution, 200mm after L1 and with L1 tilted by 0.2° to the right, relative the the neutrons' path. We notice a dramatic change of the intensity distribution profiles, due to a relatively small misalignment when we compare Figs. 4(a) and 4(b) to Figs. 3(a) and 3(c) respectively. Plot (c) of Fig. 4 shows the one-dimensional horizontal profiles of the 2D distributions. The agreement between simulation and experiment is excellent: besides the match in the shape of the two curves, we notice that the length-scales of the two lines agree very well, and so do their relative intensity changes.

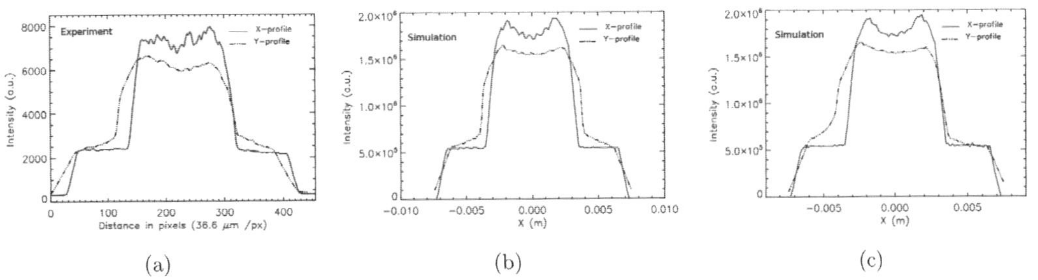

Figure 2. (a) Experimental 1D intensity profiles of L1 in X- and Y-directions. The Y-profile features an asymmetry: a bulging and a shift to the left, due to an accidental tilt of L1. (b) Simulated 1D intensity profiles of L1 in X- and Y-directions, for a perfectly aligned lens. (c) Simulated 1D intensity profiles of L1 in X- and Y-directions after introducing a 0.06° upward tilt of L1. Measurement and simulations correspond to a distance of 200mm after L1.

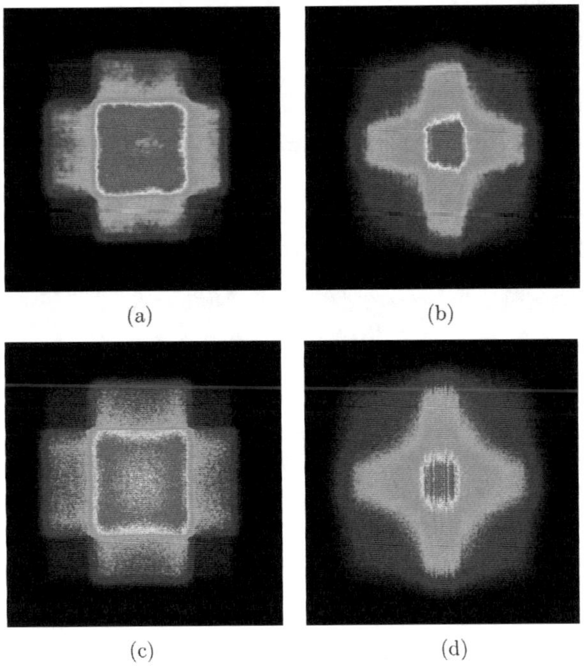

Figure 3. Experimental (panels (a) and (b)) and simulated (panels (c) and (d)) 2D Neutron intensity distributions at 200mm ((a) and (c)) and 400mm - focal point ((b) and (d)) after L1.

3.2. Focusing and De-focusing Lenses Setup

With L1 nicely aligned, we introduced the defocusing lens L2 into the beamline and began its alignment. Simulations were again successfully used to correct and guide the procedure. With both lenses aligned we performed scan measurements along the beam path with the CCD camera covering a range of distances between 300mm-2000mm from the exit of L2. We also performed inclination scans, were L2 was rotated vertically (up-down) and horizontally (left-right) from its aligned position, in order to quantify the effect of tilts of L2 on our set up.

Fig. 5 shows experimental (rows 1 and 3) and simulated (rows 2 and 4) 2D intensity distributions for three different positions after L2 (400mm, 1100mm, and 2000mm). The two lenses are 500 mm apart on the plots of the two upper rows and 800 mm apart on the plots of the two lower rows.

In Fig. 6 we show an example of the defocusing lens: as part of our experiments, L2 was rotated vertically (up-down direction) within a range of angles. Panel (a) of Fig. 6 shows the 1D intensity profile 300 mm after the exit of L2, for a 'downwards tilt' of 0.04°. Panel (b) of the same figure, was out attempt to simulate the effect of such a tilt. The match between experiment and simulation, even for such a small angle, is remarkable.

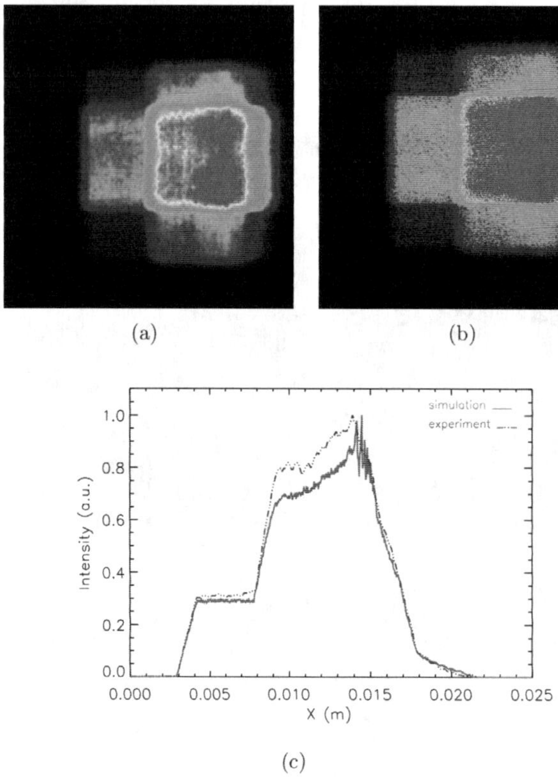

Figure 4. *Up:* 2D neutron intensity distributions, 200mm after L1. The lens was tilted towards the right (relatively to the neutrons' direction) by 0.2°. Panel (a) shows the experimental result and panel (b) the simulated one. *Down:* 1D horizontal profiles of the plots in the upper row. Notice how precisely the simulation matches the experimental result, not only in the overall shape of the 1D profiles, but also the sizes of their features, and the relative increase of the central peak within each plot.

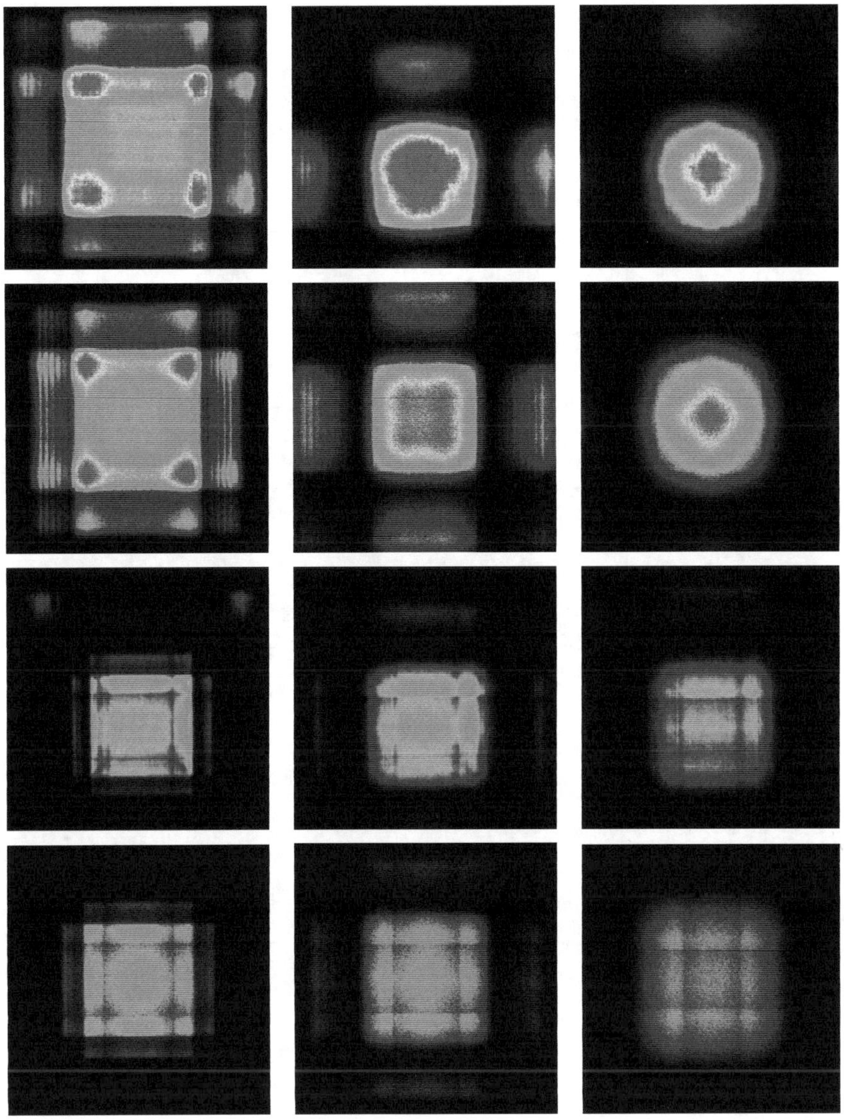

Figure 5. Experimental (rows 1 and 3) and simulated (rows 2 and 4) 2D intensity distributions in various distances from L2: 400mm (left), 1100mm (centre) and 2000mm (right) after L2. L2 was placed 500 mm after L1 for the plots on rows 1 and 2, and the distance was increased to 800 mm for the plots of rows 3 and 4

4. Summary

We have designed, built and tested a focus-defocus, 2-lens system, using parabolic supermirrors. We used McStas not only to simulate the two lenses during the designing phase, but also as a tool to guide us through the experimental alignment and testing. The simulations have proven to be a valuable tool for the alignment procedure, being able to successfully and reliably reproduce effects of misalignments as small as 0.04°. Generally speaking, we noticed that the system of the 2 lenses was sensitive to tilts of the order of 0.01°, making its alignment a tedious, non-trivial

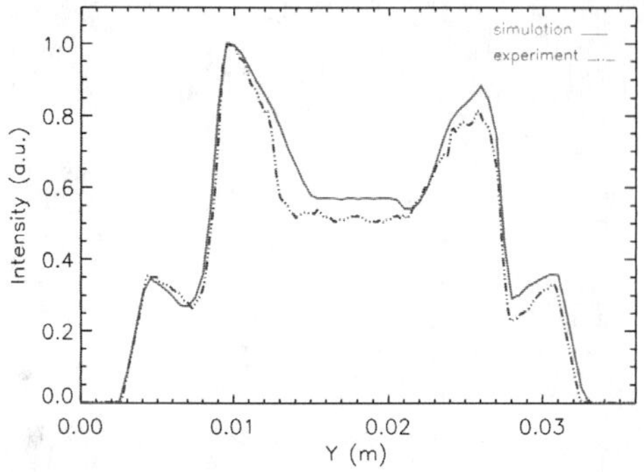

Figure 6. Simulated and experimental 1D intensity profiles in Y-direction, at 300 mm after L2. The two lenses were placed 500 mm apart. In our experimental set up, L2 was tilted downwards by 0.04°. Our simulations were able to reproduce the effect of the tilt very successfully.

task, which however is possible and can greatly benefit from simulations. On this note, as a follow-up to this project, we are currently planning the design of an 'automatic alignment tool', where camera images and simulations are used simultaneously, to provide real-time feedback to the lenses' motors, and thus spead-up and automatize the alignment procedure for coupled lens systems.

Acknowledgments

This project was funded by the European Spallation Source

References

[1] Turchin V F 1967 *At. Energy* 22
[2] Mezei F 1976 *Comm. Phys.* **1** 81
[3] Mezei F and Dagleish P A 1977 *Comm. Phys.* **2** 41
[4] Erko A, Idir M, Krist T and Michette A G (eds) 2008 *Neutron Supermirror Development* (Berlin-Heidelberg: Springer)
[5] Böni P 2007 *NIMPA* **586** 1
[6] Desert S, Panzner T and Permingeat P 2013 *J. Appl. Cryst.* **46** 35
[7] Lefmann K and Nielsen K 1999 *Neutron News* **10** 20
[8] Willendrup P, Farhi E and Lefmann K 2004 *Physica B* **350** 735
[9] Eichhorn T, Niketic N, van den Brandt B, Hautle P and Wenckebach W T *Proc., International Workshop on Neutron Optics and Detectors (NOP&D) 2013 Munich(Ismaning)*

Development of a new neutron mirror made of deuterated Diamond-like carbon

Dai SAKURAI[1] Junsei CHIBA[1], Takashi INO[3], Nobunori KAKUSHO[8], Naokatsu KANEKO[3], Ryo KATAYAMA[5], Masaaki KITAGUCHI[2,9], Kenji MISHIMA[4], (Deceased)Suguru MUTO[3], Kazuhide OZEKI[7], Yoshichika SEKI[8], Hirohiko M SHIMIZU[2], Satoru YAMASHITA[4], Tamaki YOSHIOKA[6], Daiki NISHIMURA[1]

[1]Faculty of Science and Technology, Tokyo University of Science, Noda 278-8510, Japan
[2]Department of Physics, Nagoya University, Nagoya 464-8602, Japan
[3]High Energy Accelerator Research Organization, Tsukuba 305-080, Japan
[4]International Center for Elementary Particle Physics, The University of Tokyo, Tokyo 113-0033, Japan
[5]School of Science, The University of Tokyo, Tokyo 113-0033, Japan
[6]Research Center for Advanced Particle Physics, Kyushu University, Fukuoka 812-8581, Japan
[7]School of Science, Ibaraki University, Hitachi 316-8511, Japan
[8]Institute of Physical and Chemical Research, Wako 351-0198, Japan
[9]Center of Experimental Studies, Nagoya University, Nagoya 464-8602, Japan

E-mail: j6212608@ed.tus.ac.jp

Abstract.
We developed a new neutron mirror made of *Diamond-like carbon* (DLC). DLC is a film of amorphous carbon that has characteristics of both diamond and graphite. We produced DLC mirrors by ionization deposition method which is one of the chemical vapor deposition (CVD). Generally, DLC made by CVD contents a few tens of percentages of hydrogen. It decreases the Fermi potential of the DLC coating because hydrogen has negative Fermi potential. In order to increase the Fermi potential of the coating, we deuterated the DLC by using deuterated benzene for the source gas. The characteristics of the deuterated DLC(DDLC) coating was evaluated by RBS, ERDA, x-ray reflectivity, AFM. As a result, DDLC coating has 243neV due to deuteration, which is the same level as Ni. The RMS of height of the DDLC was 0.6nm so that the DDLC coating can be applied for a focusing mirror or specular transportation of pulsed neutron. Besides, we also develop Hydrogen/Deuterium DLC multiple layer mirror. So far, 4 layers mirror has been succeeded.

1. Introduction

With the development of various neutron mirrors, neutron application is spreading in particle physics and even industrial society. Therefore, better performance of the neutron mirror is getting more important. Generally, it is said that good neutron mirror has high Fermi potential and low off specular reflectivity. The Fermi potential determined the limit of neutron reflection energy and how much amount of neutrons can be collected. The off-specular reflectivity determined how controllable neutron beam is. Ni alloy is well-used as a neutron mirror because of high Fermi potential and low off-specular reflectivity , for example, NiC, NiV. Other typical

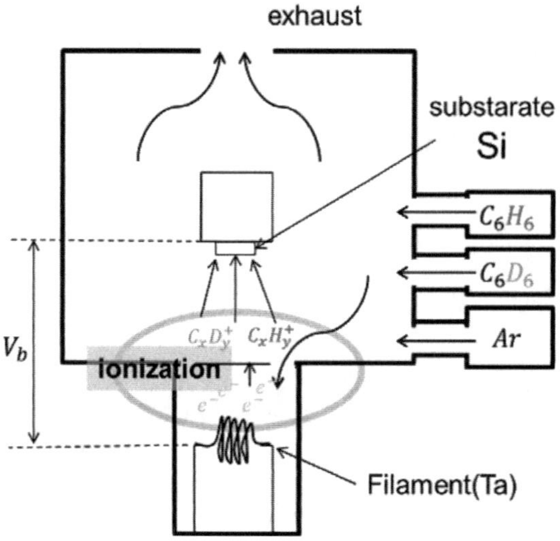

Figure 1. Schematic view of ionized evaporation method. The deposition mechanism is that (a) benzene or deuterated benzen is flowed in the chamber, (b) source gas is ionized by reacting the electron generated from a filament, (c)supettering ionized benzene to substrate by voltage,V_b. Ar gas is used to make substrate's surface smooth before the deposition of DLC layer

materials are Fe for a magnetic mirror, Be and Diamond-Like Carbon(DLC) for an alternative to Ni mirror. However, Be is difficult to handle because of poisonous nature. On the other hand, as the technology of making a film of DLC has recently been developing, DLC is gradually getting attention as a new neutron mirror.
The critical energy of the reflection of neutrons is given by the Fermi V_F potential of the material, which is written as

$$V_F = \frac{2\pi\hbar^2}{m_n}\sum_i N_i b_i \tag{1}$$

where $\hbar$ is plank constant divided by 2π, m_n is the neutron mass, N_i is the number density of the elements in the material and b_i is their scattering lengths. Ni mirror has 243neV in this parameter, which is well-known as m = 1. The diamond is known as the best potential in all materials. Therefore, it is said that DLC is possible to be a mirror with a large Fermi potential. DLC is a film of amorphous carbon that has characteristics of both diamond and graphite. It can be classified by its three components, which are diamond, graphite and hydrogen. At recent work[1], hydrogen free DLC is selected for a neutron guide. This is because hydrogen has a negative factor of it. However, if all hydrogen is replaced by deuterium, its Fermi potential could be large enough to be applicable because deuterium has a positive potential and the small absorption cross-section for neutrons. We report about fabrication and evaluation of our new DLC(DDLC) neutron mirror, comparing with previous researches[2][3].

2. Fabrication

Techniques for producing diamond coatings have changed dramatically over the last few years. It is now possible to coat large areas with a diamond-like material by means of chemical vapor deposition(CVD)[4]. We selected an ionized evaporation method to make a film of DLC, which is classified as a means of CVD. A reason to select this method is to be able to coat on a large and curve area. Another reason is that it is possible to make DDLC and HDLC multiple layer mirror, if hydrogen benzene and deuterium benzene are used. Hydrocarbon gas, benzene, is used for the source material. DDLC has been deposited by using deuterated benzene as source gas to replace hydrogen in the layer. Figure.1 is schematic view of ionized evaporation method. The deposition mechanism is that (a) benzene or deuterated benzen is flowed in the chamber, (b) source gas is ionized by reacting the electron generated from a filament, (c)supettering ionized benzene to substrate by voltage,V_b. Ar gas is used to make substrate's surface smooth before the deposition of DLC layer. Silicon wafers with 3 inch diameter for a substrate. Surface of the wafer was sputtered by argon gas to make its surface smooth before the deposition of DLC layer. The parameters to deposit DLC are voltage to substrate, pressure of source gas, and temperature in the chamber. We produced DLC mirrors with changing these parameters for optimization. The range of parameters were 0 to 3.0kV for substrate voltageV_b, 0.001 to 0.1Pa for the pressure of the source gas, and 200°C to 400°C for the temperature in the chamber.

3. Evaluation of the DDLC mirrors

The characteristics of DDLC as a neutron mirror was evaluated by Rutherford Backscattering Spectrometry(RBS), Elastic Recoil Data Analysis(ERDA), X-Ray Reflectivity(XRR), Atomic Force Microscope(AFM) and Neutron Reflectivity(NR) measurements at J-PARC BL16 SOFIA. From now we report results of these measurements respectively.

3.1. Neutron Reflectivity

Neutron reflectivity(NR) was measured at neutron reflectometer in J-PARC BL16 SOFIA[5]. This is the direct measurement for the Fermi potential. The wavelength of the incident neutron is 0.2 - 0.9nm. By changing two slits, it is able to select the range of neutron beam at a film. The result of typical neutron reflectivity is shown in Figure.2(left). The solid line in this figure is the calculated reflectivity of a layer with the thickness of 75nm and the critical Fermi potential of 243 neV. The filled circle is the reflectivity data of DDLC, the open circle is of HDLC. As shown in this figure, the Fermi potential of DDLC is increased by deuterated, compared with HDLC. Figure.2(right) shows that the correlation between film-forming condition and neutron reflectivity, where high pressure is 0.1Pa, the filled circle, low pressure is 0.01Pa, the open circle. The best condition for the Fermi potential was 0.5kV for the substrate voltage. There was no temperature in the chamber dependence between 200°C to 400°C. Therefore, temperature in chamber is set at about 300°C. On the other hands, Fermi potential of DDLC is high when pressure condition is low.

3.2. Measurement of mass density and chemical compositions

As futher evaluations, mass density and chemical compositions were measured by XRR in RIKEN[6] and RBS/ERDA at Tsukuba university[7].
XRR is a method to measure electron number density and thickness of a film from a spectrum of x-ray reflection. When X-rays are applied to a material's surface, a portion of x-rays is reflected at every interface. Interference of these partially reflected x- ray beams creates a reflectometry pattern. By analyzing it, the parameters are determined. The source of x-ray is CuKα with wavelength of 1.54Å. The two slits set to measure the same position, which was measured by the NR. The evaluation of thickness in XRR is consistent with NR within 3%. RBS/ERDA is a method to measure chemical composition of a film. It is possible to determine the chemical composition and amount of the targets by measuring counts of scattered ions caused by He ions irradiation on the sample. Figure.3 is shematic view of RBS/ERDA. Two detector were used; a forward Silicon Semiconductor Detector(SSD) detects the hydrogen and deuterium. A backward SSD detects the He scattered by carbon and silicon. The energy of incident beam , He+, was

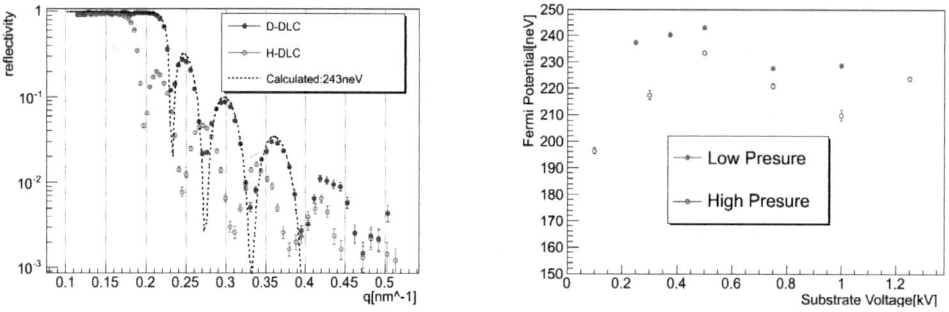

Figure 2. Above figure shows the result of neutron reflectivity measurement (right) neutron reflectivitiy of D-DLC, compared with H-DLC, (left)correlation between Fermi potential and film-forming condition. (right): closed circle is DLC made from Deuterated benzene,we called DDLC, open circle is DLC made from benzene, HDLC, and dashed line is calculated line by two parameters, Fermi potential:243neV and film thickness:75nm. Obviously, the total reflection range of DDLC is much larger than HDLC. (left):open circle is deposited by high pressure condition, 0.1Pa,.On the other hand, closed circle is deposited by low pressure, 0.01Pa.

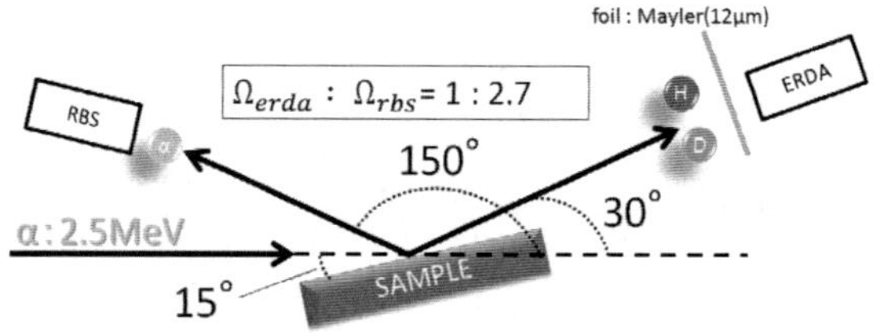

Figure 3. Schematic view of RBS/ERDA. The incident angle of accelerated He is 15°. The setting angles of forward SSD are 30°, whereas backward are 150°. A mylar sheet of 12 μ m is set before forward SSD to eliminated recoiled He.

2.5MeV, and the beam size is 20mm diameter. The incident angle of accelerated He is 15°. The setting angles of forward SSD are 30°, whereas backward are 150°. A mylar sheet of 12 μ m is set before forward SSD to eliminated recoiled He. Though the cross section of He/H and He/D is still uncertain[8], a high molecule polyethylene and deuterated polyethylene are used as a standard sample for a calibration. For analysis RBS/ERDA, SIMNRA program[9] which is the simulation program of RBS/ERDA is used to obtain the components of DDLC. More detail is shown in Ref.[7].

Figure.4 shows the result of relation between (right)Deuterium components and film-forming condition, (left)density of film. The mass density of film was calculated by electron number density measured by XRR and chemical compositions measured by RBS/ERDA. There is a tendency in this figure that deuterium components and mass density of film is high with low pressure, which is reason why Fermi potentital is high when pressure is low. For the sample of 243neV, the parameters are respectively, the mass of density is 2.07(8) g/cm^3, the components are 43(4)% of deuterium and 3.3(6)% of hydrogen in DDLC. Additionally, DDLC still contents 1-5% of hydrogen components.

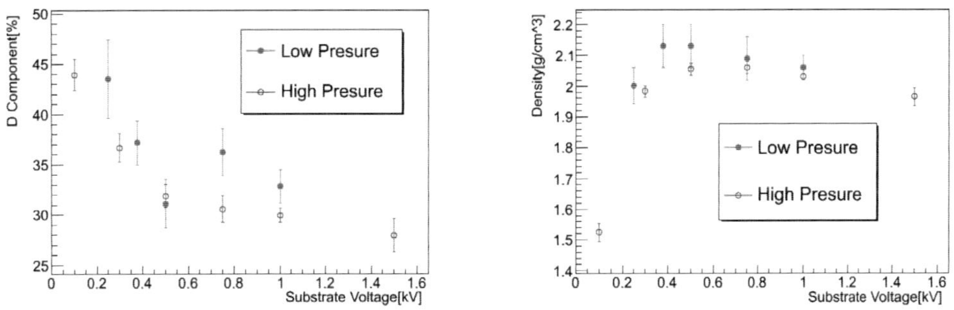

Figure 4. Above figure shows the relation between (right)Deuterium component in film, (left)density of film and film-forming condition.

Table 1. Typical data of various parameters in DDLC and HDLC

material	C[%]	D[%]	H[%]	$\rho[gcm^-3]$	V_F[neV]
DDLC	65(2)	31(2)	3.8(5)	2.07(8)	240(10)
HDLC	64(3)	0.8(1)	35(3)	2.18(4)	150(10)

3.3. Consistency of Fermi potential between neutron reflectivity and RBS/ERDA,XRR
From the equation(1), we get the equation

$$\begin{aligned} V_F &\approx b_C N_C + b_D N_D + b_H N_H \\ N_i &= \rho_M R_i \times \frac{N_A}{A_i} \end{aligned} \quad (2)$$

where N_A is Avogadro constant, A_i is the atomic mass of i material, R_i is the material of chemical composition. NR is the direct measurement of Fermi potential, whereas XRR and RBS/ERDA have the relation with Fermi potential. Therefore, substituting two parameters, mass density and chemical compositions, for Eq.2, Fermi potential is calculated. Figure.5 shows that y axis is Fermi potential of NR measurement, x axis is Fermi potential calculated by ρ_M、R_i. The open circle is hydrogen DLC(HDLC) and closed circle is DDLC. In HDLC data, the value measured by NR tend to be larger than the value calculated by ρ_M、 R_i, while DDLC data has opposite trend of it. There is the consistency with error of 8%. The error of cross section of H/He, D/He caused this error.

3.4. roughness of DDLC
There are two types of refection; one is a specular reflectivity whose incident angle is same as refection angle, the other is off specular reflectivity whose angle not. As a model recently studied[10], roughness causes off-specular reflectivity. As generally, roughness causes off-specular reflectivity. Therefore, firstly, we measured the surface roughness by AFM to evaluate it. The result shows that DDLC surface is very smooth, rms of height is 0.6(2)nm in DDLC, on the other hand, 0.3(1)nm in Si substrate in Fig.6, where b is rms of height. Therefore, low off-specular reflectivity of DDLC can be expected.

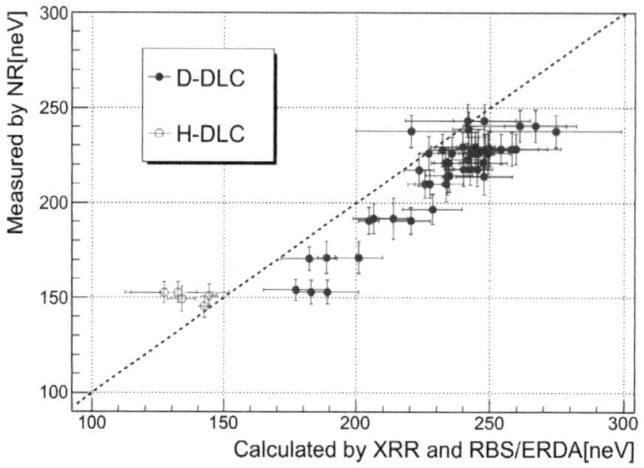

Figure 5. Consistency between neutron measurement and RBS/ERDA,XRR

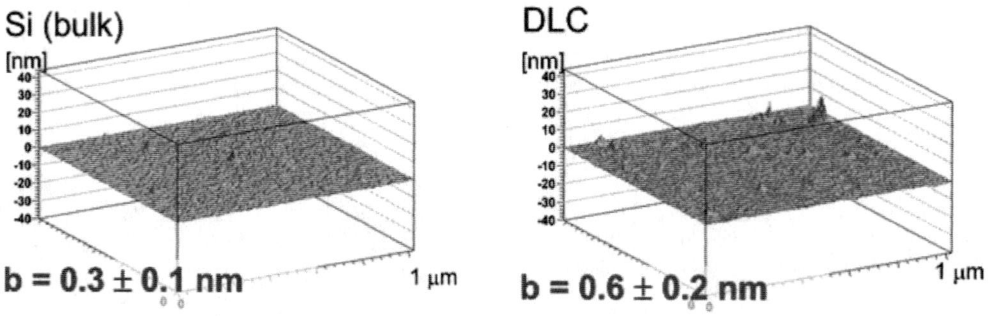

Figure 6. Above figure shows a roughness of (left)Si substrate and (right) DDLC. The thickness of DDLC is 107nm, the parameter of b is the height of Root Mean Square.

3.5. a multiple layers of DDLC/HDLC

As mentioned above, it is possible to produce a DDLC/HDLC multiple layer by changing the source gas between hydrated and deuterated benzene. a multiple layers is a method of exceeding the critical angle of total reflection. It provide reflections from multiple interfaces between different indexes of refraction. Therefore, on trial, we made the forth multiple layer. Figure.7 shows a spectrum of neutron reflectivity of DDLC/HDLC 4 layer(solid line) with a single layer mirror of DDLC(dotted line). There are some sign of multiple reflection at figure.7.

4. conclusion

We fabricated a DDLC coating for a new neutron mirror, and measured its characteristics. Fermi potentials measured by NR were 190-243 neV. The maximum Fermi potential of DDLC gained 10% from the previous report[2][3], which is the same as Fermi potential of NiC, 243neV. The typical surface roughness of DDLC measured by AFM was 0.6 nm as a RMS of height. Therefore, low off-specular reflectivity of DDLC can be expected. The chemical composition was measured by RBS/ERDA. The typical components of hydrogen and deuterium was 3% and 43%, respectively. These results show that performance of DDLC as a neutron mirror is satisfactory. Therefore, DDLC will be useful for neutron guides with complicated shapes

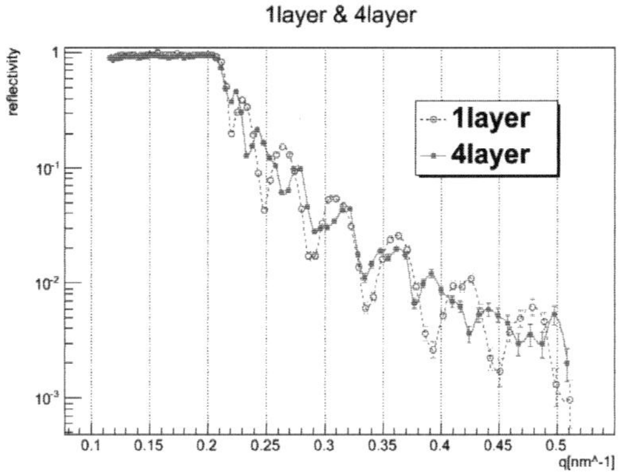

Figure 7. reflectivity of D-DLC/H-DLC multiple layer

because DDLC can be deposited on any surface by CVD. In addition, we successfully made DDLC/HDLC multiple layer mirror.

References

[1] S.Heule, F.Atchison et al.,2007 *Diamond-like carbon coated ultracold neutron guides* (Applied Surface Science) 253 8245-8249

[2] M.G.D. van der Grinten, J.M. Pendlebury et al., 1999 *Characterization and development of diamond-like carbon coatings for storing ultracold neutrons* (Nuclear Instruments and Methods in Physics Research) A423, 421-427

[3] Y. Kawabata, M. Hino et al., 2004 *Characteristics of deuterated diamond-like carbon as neutron mirror* (Nuclear Instruments and Methods in Physics Research) A529, 84-86

[4] J. Robertson, 2002 *Diamond-like amorphous carbon* (Materials Science and Engineering) R37, 129-281

[5] N.L.Yamada N.Torikai et al., 2011 *Design and performance of horizontal-type neutron relectometerSOFIA at J-PARC/MLF* (The European Physical Journal Plus) 126, 108

[6] `http://www.rigaku.com/applications/xray_reflectometry`

[7] D. Sekiba et al., 2012 *Hydrogen isotope tracer experiment in a-C:H film deposition: Reactive RF magnetron sputtering with CH_4 and D_2* (Diamond & Related Materials) 27-28, 60-63

[8] A.F. Gurbich 2010 *Evaluated differential cross-sections for IBA*(Nuclear Instruments and Methods in Physics Research) B268 1703-1710

[9] `http://home.rzg.mpg.de/~mam/`

[10] F. Atchison, M. Daum et al., 2010 *Diffuse reflection of ultracold neutrons from low roughness surfaces* (The European Physical Journal) A44, 23-29

Journal of Physics: Conference Series **528** (2014) 012011 doi:10.1088/1742-6596/528/1/012011

Development of highly-mechanically polished metal-substrate for neutron supermirrors

Shin Takeda[1, 2], Jiang Guo[2], Shinya Morita[2], Hirohisa Ono[1], Tatsuro Oda[3], Junichi Kato[2], Hirotaka Sato[1], Masahiro Hino[4], Yutaka Yamagata[2], Michihiro Furusaka[1]

[1] Graduate School of Engineering, Hokkaido University, Sapporo, Japan
[2] RIKEN Center for Advanced Photonics, RIKEN, Wako, Japan
[3] Department of Nuclear Engineering, Kyoto University, Kyoto, Japan
[4] Research Reactor Institute, Kyoto University, Kumatori, Japan

s-takeda@eng.hokudai.ac.jp

Abstract.

A small-angle neutron scattering instrument using an ellipsoidal focusing mirror has been developed at Hokkaido University and a prototype machine (mfSANS) has been installed at the JRR-3 research reactor at Japan Atomic Energy Agency (JAEA). They are based on an ellipsoidal focusing mirror on a borosilicate glass substrate. It turned out that borosilicate glass was very brittle and difficult to machine or polish it to the required surface-finishing process. In order to improve this situation, we decided to develop a method to create a metal substrate based ellipsoidal focusing mirror. As the first step, test pieces of flat and ellipsoidal surface neutron supermirrors using metal substrates were fabricated. They were first roughly shaped by a conventional numerically controlled (NC) cutting machine and amorphous NiP was plated. They were then machined using an ultrahigh precision cutting (UPC) method and finished by NC polishing techniques. Surface roughness of $R_a \approx 0.78$ nm and surface figure error of sub-micrometer were realized so far.

1. Introduction

A neutron supermirror is an extremely powerful and useful optical device used to handle neutron beams. Polished-glass, float-glass or polished silicon single crystal plates are usually used as neutron supermirror substrates. In only a few trials have neutron supermirrors been fabricated with metal substrate, although metal substrates can be very useful [1]. This is because surface roughness of less than about 0.3 nm is required for the coatings for high-m supermirrors and small d-spacing monochromators, which is not so easy to realize with most metallic materials due to the grain boundaries in their polycrystalline structure. Because NiP is a metal amorphous which has no such boundries, we predict that we can polish the surface as smooth as a glass surface. Namba et al. fabricated a Wolter-type convex mirror for X-ray telescopes with sub-nanometer roughness using metal material [2]. The material used was aluminum alloy with electroless NiP plating on the surface and was fabricated by turning using a diamond tool. For Wolter-type mirrors, turning is applicable since the shape is axisymmetric. However, by using a shaper cutting process, aspherical profiles such as ellipsoidal surfaces can be made. This research was conducted to develop a new fabrication method for aspheric mirror substrates using a metal material, particularly for use in aspheric neutron supermirrors such as used in a mirror focusing small-angle neutron scattering (SANS) instruments or in a focusing reflectometer.

The SANS technique [3] is a powerful tool for the study of various structures in the range of nanometer to micrometer scale. Typical conventional pinhole SANS instruments requires rather long instrument length of about 20-30 m. However, by using focusing optics, the beam diameter on the detector can be reduced to the order of a few millimeters and can therefore reduce the instrument length while maintaining angular resolution. Furthermore, it can be realized without reducing the sample size, therefore not sacrificing intensity. For time of flight (TOF) SANS instruments, shorter instruments result in a wider wavelength band, especially to a longer wavelength range. It is therefore possible to measure a low momentum transfer (Q) region while maintaining intensity.

The mini-focusing small-angle neutron scattering instrument (mfSANS) is a SANS instrument that employed an ellipsoidal neutron focusing mirror with supermirror coating. The basic concept of mfSANS is to make a SANS instrument short, about one-tenth of the conventional SANS instruments, while covering the Q range by application of a focusing optical system. The mfSANS instrument was first developed at Hokkaido University using the Hokkaido University Neutron Source (HUNS). Later, a prototype has been installed at the research reactor (JRR-3) at the Japan Atomic Energy Agency (JAEA). Figure 1 shows a schematic layout of mfSANS.

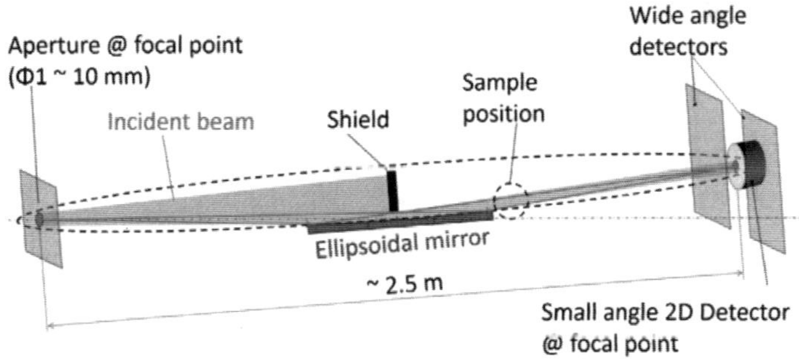

Mirror specifications

Shape	Ellipsoid
Radius	Major: 1250 mm Minor: 20 mm
Size	3 segments each 300 mm × 60 mm × 9 mm (Full length = 900 mm)
Machining method	Grinding + Polishing
Machining time	6 months

Figure 1. Schematic layout of mfSANS: The neutron beam passes through an incident aperture, which is set at one of the focal points. It is then reflected by a focusing mirror which focuses it around the other focal point downstream. It is possible to cover the conventional Q range of a typical SANS instrument together with a larger Q range.

The mfSANS instrument yielded reasonable results, especially in the wide-Q range of up to about 5 nm^{-1} [4]. However, the background level at the small-Q region was rather high due to the diffuse scattering from the mirror. Beam profile deformation caused by misalignment of the mirror segments was also observed. Both problems are related directly to the surface condition of the ellipsoidal mirror. The ellipsoidal mirror installed in the prototype of mfSANS was made of borosilicate glass. Due to the size limitations posed by the coating machine, the mirror was divided into three segments. Each segment was 300 mm long. The surface figure error was 5.0 μm Peak-to-Valley (P–V) for the minor axis direction and 7.5 μm P–V for the major axis direction with 39.7 μm difference in level between the mirror segments. The surface roughness was about 1.0 nm Root-Mean-Square (RMS). This was reasonable for grinding conducted using a #800 grinder. It was planned to use a #1200 grinder, which was cancelled due to machine failure. The machining process was 6 months for grinding and polishing the whole mirror, all three segments in this case, including the time for tool adjustments, such as truing and exchanging polishing tools.

By using a NiP plated metal substrate and applying ultra-high precision cutting (UPC) and polishing methods, there is a hope of making an aspherical mirror while drastically reducing the machining time and maintaining very good surface finish of sub-nm level. By using a metal substrate, it also enables the production of a mechanical fixture directly to the substrate to improve the alignment between the mirror segments. The main objective of this paper is to present the validity of the machining method described above for creating aspherical mirror substrates for neutron mirrors. We fabricated test pieces of two types: a flat and an ellipsoidal test pieces.

2. Fabrication of test pieces

All test pieces were machined from bulk blocks of metal. The UPC method was used, whereby cutting was conducted from only one direction. Surface roughness was characterized by using a white light interferometer, and neutron reflectivity, as described later in this paper.

2.1. Flat test pieces

Stainless steel was used as the substrate for flat test pieces. They were then roughly finished using a conventional numerically controlled machining process with subsequent electroless NiP plating with 100 μm thickness. The dimensions of the test pieces were 50 mm × 50 mm × 13 mm. The plated NiP surface was then roughly finished by turning with a single crystal diamond bite with subsequent machining by UPC. The machine used for the turning and UPC processes was the NIC-300 from NAGASE INTEGREX. As the final process, the surface was finished using a numerically controlled mechanical polishing instrument with colloidal silica as the slurry. The machining time was 13 hours for UPC and 24 hours for mechanical polishing.

2.2. Ellipsoidal mirror test pieces

The base material was reviewed as aluminum alloy for an ellipsoidal mirror test piece. The dimensions of the substrate were 100 mm × 30 mm × 10 mm, as shown in Figure 3(a). The base substrate was roughly shaped using a machining center with a ball-end mill. After creating a rough ellipsoidal surface by milling, the surface was finished by the UPC method. The same machine and machining method was applied as the flat test pieces for the UPC process. Both machining processes required calculation of the tool offset. The machining time was 3 hours for rough shaping and 13 hours for UPC.

3. Surface roughness and surface figure error

The surface roughness was measured using a white light interferometer (new view 7200; Zygo). The measured area was 140 μm × 100 μm, with spatial resolution of 100 nm × 100 nm. The surface profile was measured using a Point Autofocus Probe Surface Texture Measuring Instrument (PF-60; Mitaka Kohki. Co., Ltd.). The laser spot diameter was 1 μm; the pitch between each point was 10 μm. Figure 2 shows the results of surface roughness and the surface profile measured by the interferometer after UPC and mechanical polishing. For the sample finished by UPC, the roughness was 1.3 nm [rms] in the cutting direction and 3.8 nm [rms] for the plane. Horizontal cutting marks were observed. For the sample finished by polishing, the sample roughness after 24 hours of mechanical polishing was 0.78 nm for the plane [rms]. The cutting marks were no longer visible. Figure 3 shows the picture of the ellipsoidal test piece and the surface figure error after the UPC process. The surface figure error was 0.7 μm P–V for the major axis and 3.0 μm P–V for the minor axis.

4. Reflectivity measured by neutrons

In order to determine whether the surface roughness of the flat test piece is adequate, we conducted a neutron reflectometry experiment at the CN3 beam port at the Kyoto University Research Reactor Institute (KUR-CN3) using flat test pieces with different finishing conditions. The samples were flat test pieces with the NiP plated surface finished by (a) UPC and (b) polishing. Sample (a) was set so that the neutron beam is parallel to the cut marks.

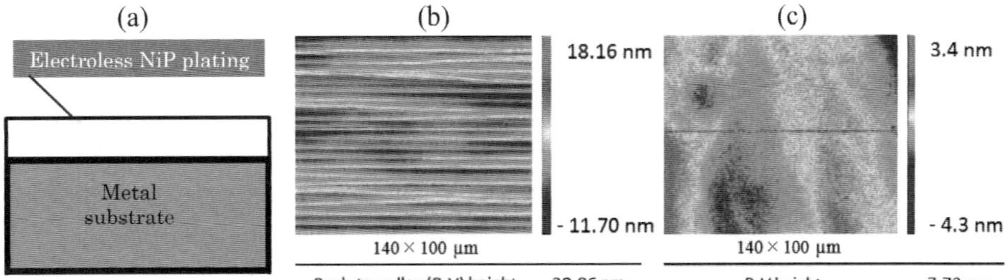

(b)		(c)	
Peak-to-valley (P-V) height	29.86 nm	P-V height	7.72 nm
Roughness [rms]	4.14 nm	Roughness [rms]	0.78 nm

Figure 2. (a) Diagram of the flat test piece. (b) Sample surface profile after UPC. Roughness was 1.3 nm [RMS] in the cutting direction and 4.14 nm [RMS] for the plane. Horizontal cutting marks are visible. (c) Surface profile of the sample after polishing. The surface roughness after 24 hours of mechanical polishing was 0.78 nm [RMS] for the plane. Cutting marks are no longer visible.

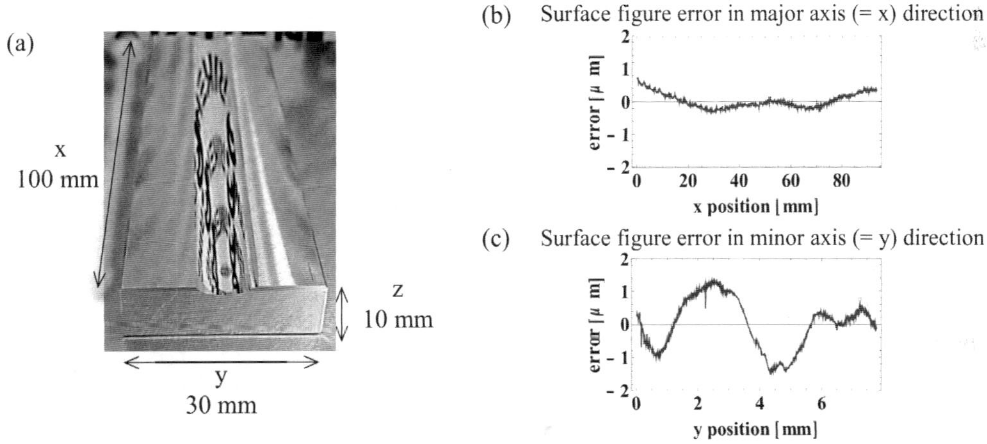

Figure 3. (a) Picture of the ellipsoidal test piece. (b) Surface figure error of the major axis was 0.7 µm P–V. (c) Surface figure error of the minor axis was 3.0 µm P–V.

4.1. Geometry of neutron reflectometry experiment

Figure 4 shows a schematic drawing of the geometry of the reflectometer at KUR-CN3 reactor-based neutron source. Consequently, the energy spectrum of the neutron beam was measured by application of a beam chopper combined with the time of flight (TOF) method and a time sensitive 2D position sensitive detector. The wavelength resolution was 10% for 0.2 nm neutrons. The sample was set on a goniometer used to adjust the incident angle of the beam, which was collimated using a set of vertical slits. The reflectivity and beam profile were measured using the 2D position sensitive detector.

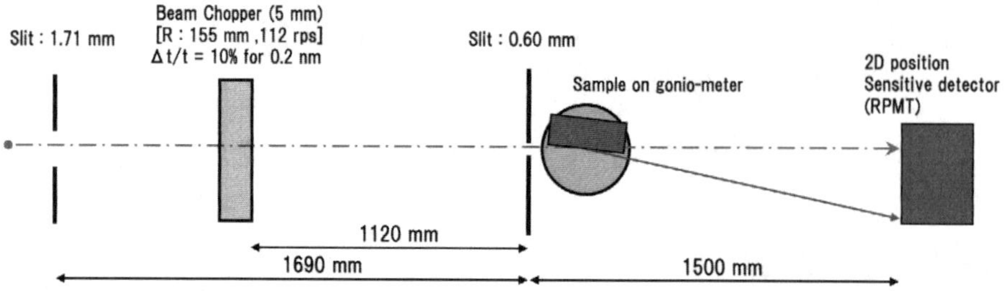

Figure 4. Geometry of the reflectometer mocked up at KUR-CN3 as seen from above: A 2D position sensitive detector was placed at the most downstream of the geometry to measure both the reflectivity and the beam profile.

4.2. Results of neutron reflectometry experiment

4.3. Reflectivity was measured only for flat test pieces. The measurement time was 2 hours. The reflectivity and intensity of diffuse scattering is shown in Figure 5. Diffuse scattering is visible on the figure showing the horizontal projection of the beam profile. Diffuse scattering is visible from 5.0×10^{-2} for (A); the test piece machined only with UPC, but is not visible for (B); the test piece after polishing, within the level of background.

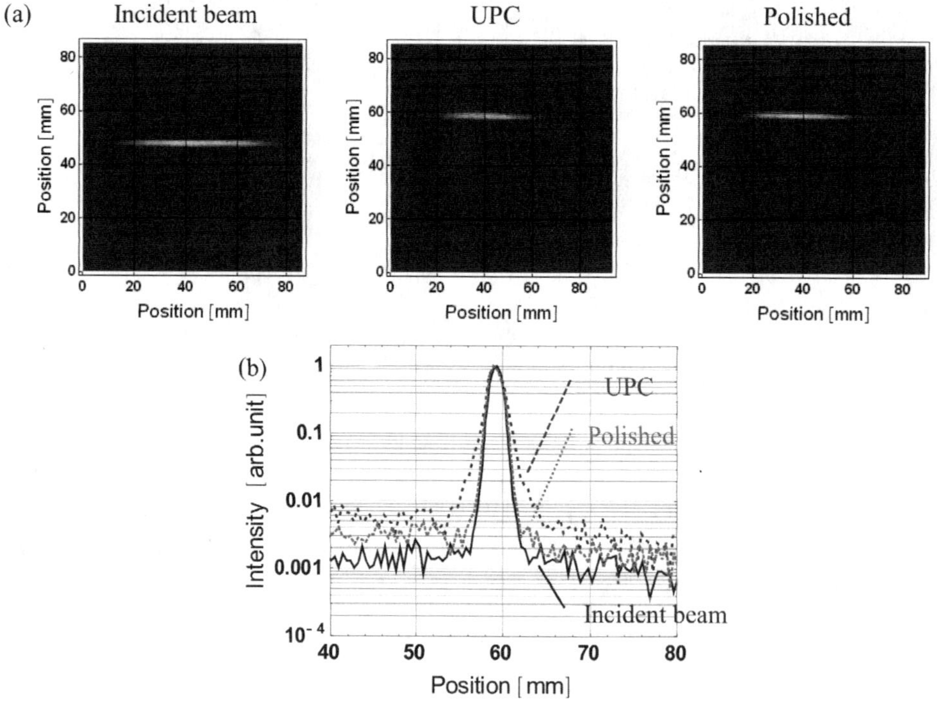

Figure 5. Neutron reflectivity: (a) 2D map of the incident and reflected beam. (b) Vertical projection of the beam profile normalized by peak height.

5. Discussion

The surface figure error in the major axis direction is already better than the previous ellipsoidal mirror, which was in the order of several micrometers. The surface roughness we achieved for the flat test pieces could be improved by optimizing the polishing conditions such as the polishing pressure, type of slurry, and polishing speed. The surface figure error could also be improved using a diamond bite with better shape accuracy. Based on these results, the fabrication method we applied is considered feasible for creating a substrate with better surface figure error and surface roughness compared to the former ellipsoidal mirror used in mfSANS. The intensity of diffuse scattering from the flat test piece was reduced by finishing the surface by polishing, but it was too small to be evaluated quantitatively using our current reflectometer. This resulted from the low signal to background ratio, which was partially attributable to the small incident angle and the beam chopper for monochromization.

6. Conclusion

We fabricated flat test pieces with electroless NiP coating on metal blocks. They were then machined by UPC and polished and the surface roughness we achieved this time was as low as 0.78 nm [rms]. Diffuse scattering was measured by neutron reflectometry. The sample finished by polishing exhibited lower diffuse scattering than the one finished by UPC, but it was not evaluated quantitatively because of its low signal–background ratio of the reflectometer.

We also fabricated ellipsoidal test pieces using the same method. The surface figure error was quite good, 0.7 μm P–V for the major axis and 3.0 μm P–V for the minor axis.

Acknowledgments

S. Takeda acknowledges the support of RIKEN's Junior Research Associate system for providing a good environment for research. All authors are thankful for help in advanced manufacturing support team of RIKEN for technical support in machining. The authors also thank Dr. Homma of Hokkaido University for technical advice and Japan Neutron Optics (JNOP) for experimental support.

References

[1] Schanzer C, Böni P and Schneider M 2010, J. Phys. Conf. Ser. 251, 012082

[2] Namba Y, Shimomura T, Fushiki A, Beaucamp A, Inasaki I, Kunieda H, Ogasaka Y and Yamashita K 2008, *CIRP Annals – Manufacturing Technology*, **57**, 337-340.

[3] Jackson, A. J. 2008, Introduction to Small-Angle Neutron Scattering and Neutron Reflectometry. *NIST Center for Neutron Research.*

[4] Oba Y, Ohnuma M, Ohnuma S, Furusaka M, Koppoju S, Takeda S 2013, *Journal of Magnetism and Magnetic Materials*, pp.45-51

Development of a compact vertical splitting system for the cold neutron beam at JRR-3

I Tamura[1,4] , R Maruyama[2], D Yamazaki[2], K Nakamura[1], K Yamamoto[3], K Aizawa[2] and K Soyama[2]

[1]Department of Research Reactor and Tandem Accelerator, Nuclear Science Research Institute, Tokai Research and Development Center, Japan Atomic Energy Agency, Tokai, Ibaraki 319-1195, Japan
[2]J-PARC Center, Japan Atomic Energy Agency, Tokai, Ibaraki 319-1195, Japan
[3]Nuclear Human Resource Development Center, Japan Atomic Energy Agency, Tokai, Ibaraki 319-1195, Japan

E-mail: tamura.itaru@jaea.go.jp

Abstract. In order to efficiently execute neutron beam experiments at JRR-3, the number of beam ports was increased and neutron beam instruments were rearranged in the C2 cold neutron beam line by using a newly developed compact vertical splitting system. Considering the space available for instruments at the downstream region of the cold neutron beam line, the deflection angles of the compact vertical splitting system were set as 10° and 20°. To branch off the cold neutron beam, a bender with characteristic wavelength of 4 Å was developed using Ni/Ti supermirrors coated on both sides of 0.2-mm-thickness silicon substrates. The channel width of the bender is 0.2 mm. The radius of curvature of the compact bender is 938 mm and the lengths of the benders are 160mm and 320 mm, yielding deflection angles of 10° and 20°, respectively. Neutron flux at the end of the neutron guide installed at the bender exit, the deflection angle of which is 20°, was measured to be 1.58×10^7 $n{\cdot}cm^{-2}{\cdot}s^{-1}$ with a gold foil activation method. We found that the measured flux agrees with the simulated value.

1. Introduction

The Japan Research Reactor 3 (JRR-3) at the Japan Energy Atomic Agency (JAEA) was designed as a multipurpose reactor for the use of neutron beam experiments and material irradiation tests in various fields of basic research and industrial applications. In order to efficiently supply thermal and cold neutron beams, five neutron guides were installed in the beam hall at JRR-3. The two thermal neutron guides are about 60 m long and the three cold neutron guides are 31-51 m long. The characteristic wavelengths of the cold neutron guides C1, C2 and C3 are 4Å, 4Å and 6Å, respectively. Neutron guides are used to simultaneously supply many instruments with intense neutron beams.

There are great demands for a white neutron beam. The Neutron Spin Echo (NSE) spectrometer was installed at branch port C2-2 by deflecting a monochromatic beam using a multilayer mirror. Great benefits are expected if the NSE is relocated to the end of the C2 guide, because white neutrons increase the incident neutron flux and the large space expands the Q region by increasing the detector area [1]. However, instruments such as the Cold Neutron Radiography Facility (CNRF) [2] and

[4] To whom any correspondence should be addressed.

Prompt Gamma Activation Analysis (PGAA) [3] are not permanent installations and therefore share the beam time at the end of the cold neutron beam line (C2-3) with only one or two cycles available per year. Thus, in order to construct new white neutron beam ports, a splitting system has been developed using the neutron benders.

Beam line splitting d

evices have been introduced in some experimental facilities [4-12]. These devices deflect a neutron beam using a bender, which can be used as not only a branching device but also a wavelength filter [13]. It is common to install these splitting devices close to the neutron source. Splitting devices using the conventional benders can bend the neutron beam by several degrees with a radius of curvature of at least tens of meters. They require a large space to install.

Benders with smaller curvature have also been developed [14, 15]. The production of compact benders with a radius of curvature one-tenth of a conventional device allows relocating instruments to the cold neutron beam hall at JRR-3 because the length of the bender can be shortened.

Since the beam path of the compact multichannel bender is short, it can improve the transmissivity of a neutron by using the silicon substrates for bender mirrors. We carried out simulation calculations that take into account the absorption of the silicon substrate in the bender.

In this paper, we describe the experiments performed to characterize the neutron spectrum and flux and examine the performance of the compact vertical splitting system by using McStas simulation code [16, 17].

2. Design and fabrication of the compact neutron splitting system

The compact vertical splitting system consists of neutron guide tubes and two compact benders as shown in Figure 1. The design specifications of the compact benders are described below. The compact bender is a multichannel micro-bender fabricated with a characteristic wavelength of 4 Å. The channels of a multichannel micro-bender function as curved neutron guides.

The design of the micro-bender is based on the neutron optics as follows. The critical angle for total reflection at a supermirror is defined as

$$\theta_c = \lambda \times m \times \gamma_c^{Ni},$$

where λ denotes the wavelength of the neutron, γ_0^{Ni} denotes the critical angle of nickel per unit wavelength (~1.7×10^{-3} rad/Å) and m denotes the relative ratio of the critical angle of total reflection of the supermirror to that of natural nickel [18].

The characteristic wavelength λ^* of the bender is a function of the radius of curvature ρ, and the width d of each channel in the bender, and it is defined as

$$\lambda^* = \frac{1}{m \times \gamma_c^{Ni}} \left(\frac{2d}{\rho} \right)^{1/2}.$$

These supermirrors are stacked with a spacing of 0.2 mm on a pair of curved iron walls as shown in Figure 1. The compact bender yielding deflection angle of 10° consists of a stack of 47 pairs of curved NiC/Ti supermirrors deposited on both sides of silicon-wafer substrates of 0.2 mm thickness. The radius of curvature of the compact bender is 938 mm and the lengths of the benders are 160mm yielding deflection angles of 10°. In order to fabricate the bender yielding deflection angle of 20°, the two benders, which yield deflection angle of 10°, were arranged in tandem along a neutron beam travelling direction. For such arrangement, the length of the benders yielding deflection angles of 20° become 320 mm.

NiC/Ti supermirrors are fabricated with 403 layers deposited by ion beam sputtering [19] onto silicon substrates with a roughness less than 3 Å r.m.s.. We have fabricated 144 pieces of NiC/Ti neutron supermirrors deposited on the both sides of silicon substrate with m = 3.0. The silicon substrate is 40 mm wide, 160 mm long and 0.2 mm thick. Since the substrate is not very thick, the supermirror can be bent to some degree along the curved frame.

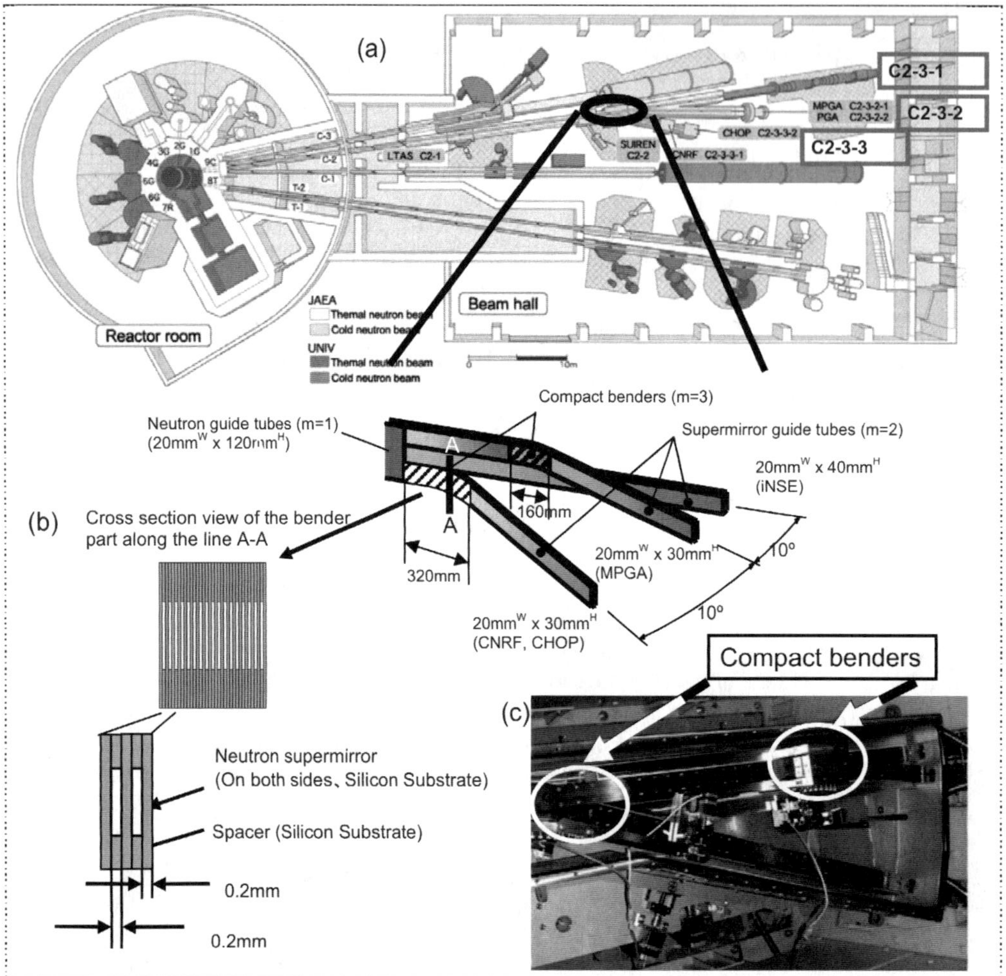

Figure 1. (a) Layout of the instrument and compact vertical splitting system at JRR-3. (b) Cross section view of the compact bender along the line A-A; these supermirrors are stacked with a spacing of 0.2 mm on a pair of curved iron walls. The photograph (c) shows the compact vertical splitting system. Two compact benders and neutron guide tubes are in the vacuum tank.

Figure 2 shows the reflectivity of the supermirror measured using the reflectometer SUIREN [20] at JRR-3. The neutron reflectivity of the front and back surfaces of the supermirrors in the compact bender is 90% at m = 3. These data are used in the simulations for evaluating the compact vertical splitting system.

3. Evaluation of the bender

To characterize the neutron beam, the spatial distribution of the neutron beam intensity after branching by the bender was measured using a neutron image plate as the detector. The plate was placed at the CNRF sample position, which is 150 mm downstream from the C2-3-3 beam port. In this position, the neutron beam size was 20 mm in width and 30 mm in height. The distance between the compact bender exit and neutron image plate is 3960 mm. The measured neutron intensity distribution at CNRF sample position is shown in Figure 3. The distribution of intensity is spatially uniform indicating that the arrangement of bender mirror was successful.

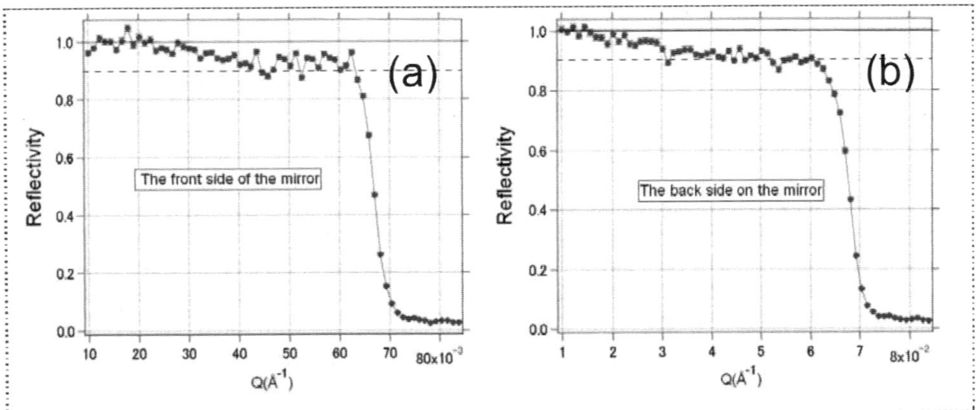

Figure 2. Front surface (a) and back surface (b) reflectivity of a neutron supermirror.

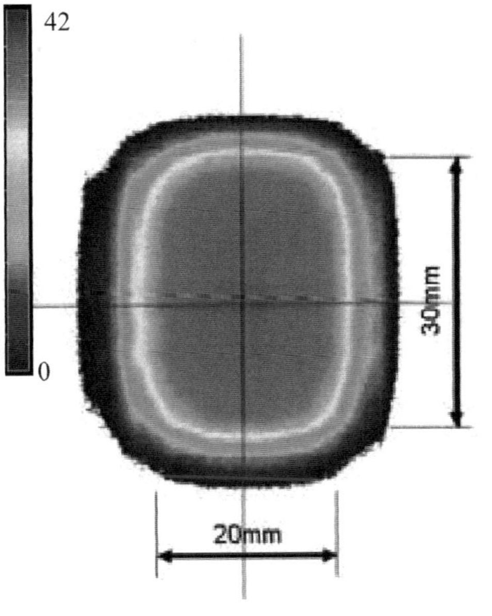

Figure 3. Neutron intensity distribution, seen from the reactor core, at the sample position of CNRF of the C2-3-3 beam port, measured using a neutron image plate.

The neutron spectrum transmitted through the bender was measured using a CHOP (chopper spectrometer) at the cold guide tube. The neutron spectra were measured at the C2-3-3 beam port with the time-of-flight method. A beam collimator in the chopper spectrometer at the neutron guides exit made the neutron beam size of 10 mm in width and 10 mm in height. The neutron chopper has a single slit of 10 mm width and its disk diameter is 700 mm. The chopper rotational speed was 5,000 rpm. The bin width of the multi-channel scale was 1.0×10^{-5} seconds. The flight path between the chopper and the neutron detector was 1.75 m at C2-3-3. The time resolution for 4 Å neutrons is 1.5%. The reactor power was 20MW.

Figure 4 shows the spectrum of transmitted neutrons and the results of the Monte Carlo simulation for the neutron spectrum. The simulation was executed with the same conditions as the measurements. Open circles and open triangles show the measured spectra and the spectra calculated using McStas code, respectively. In this calculation, Maxwellian energy/wavelength spectrum was chosen for the incident neutron source.

Figure 4 shows that the characteristic wavelength of C2-3-3 beam port is 4.8 Å. The neutron spectra are measured and calculated at the detector position of the chopper spectrometer. Measured spectra were scaled for easier comparison. The dips of the spectrum at 3.85 Å, 4.61 Å, 5.11 Å and 5.61 Å are ascribed to the Bragg reflections of the SUIREN monochromator installed at C2-2, polycrystal Al with index 111, single crystal Si with index 111, and the LTAS (Low energy Triple-Axis Spectrometer) monochromator installed at C2-1, respectively. The agreement between the measured dips and the monochromator angles shows the propriety of our measurement. The measured spectrum almost agrees with the simulated spectrum. This indicates that the compact vertical splitting system achieves the expected performance.

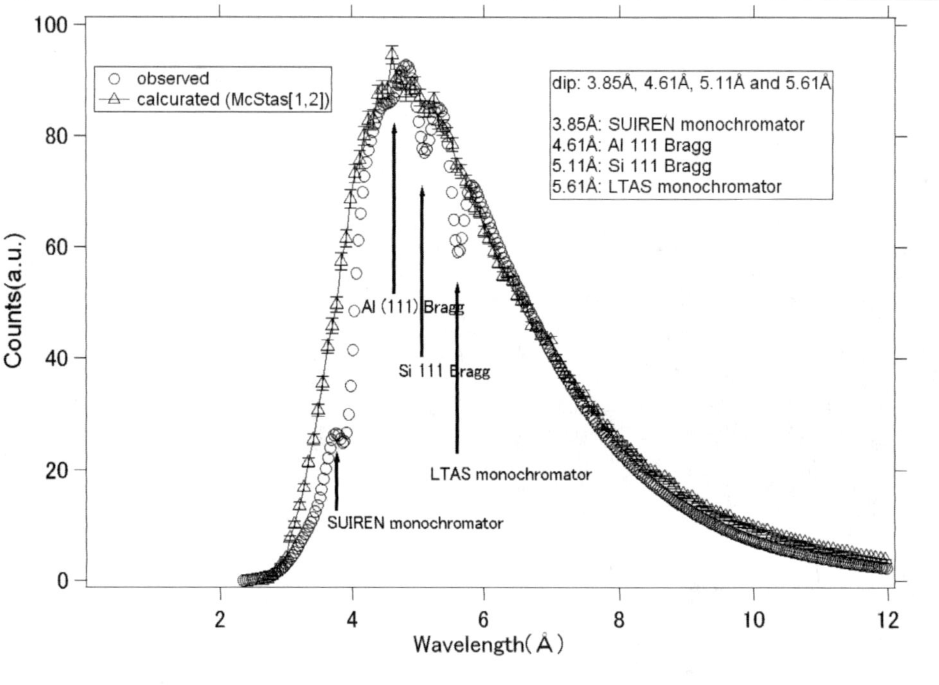

Figure 4. Measured and calculated spectra of C2-3-3 beam port. The dips of the spectrum at 3.85 Å, 4.61 Å, 5.11 Å and 5.61 Å are observed.

Neutron fluxes at the C2-1, C2-2, C2-3-1, C2-3-2, and C2-3-3 beam ports were measured by the gold foil activation method at a reactor power of 20 MW. The gold foil was of area 5 mm^2 and thickness 0.05 mm and the irradiation time was 60 minutes.

Because the neutron absorption cross section of gold depends on the neutron energy, the neutron fluxes were obtained using absorption cross sections averaged over the neutron spectra calculated using McStas at the measuring positions. The measured cold neutron fluxes at each beam port are listed in Table 1. Neutron fluxes integrated over energy were measured. The difference in the measured and calculated ratio fluxes is within 15%.

Table 1. Neutron fluxes at C2-1, C2-2, C2-3-1, C2-3-2, and C2-3-3 beam port (calc. & obs.) The ratio of measured fluxes is in good agreement with that of calculation results.

Beam port	Measured neutron flux (n/cm^2/s)	Ratio (measured)	Ratio(calculated)
C2-1	2.66×10^8	1	1
C2-2	1.47×10^8	0.553	0.630
C2-3-1	9.56×10^7	0.359	0.423
C2-3-2	3.13×10^7	0.118	0.130
C2-3-3	1.58×10^7	0.0594	0.0622

4. Conclusion

To meet the demand for a white neutron beam, we have developed a compact splitting system. This has enabled relocating the NSE spectrometer to the guide end, where white neutrons increased the incident neutron flux, and the large space expanded the Q region by increasing the detector area. Furthermore, the CNRF and PGAA instruments became permanent installations.

We evaluated the performance of the compact splitting system through measurements and simulation. The compact splitting system consists of a straight guide tube and two compact multi channel benders. The deflection angles of the compact benders with the same radius of curvature were set as 10° and 20°.

NiC/Ti supermirrors with m=3 were deposited on both sides of a silicon substrate by an ion beam sputtering apparatus in JAEA to fabricate the compact bender. The neutron reflectivity of the fabricated supermirrors is 90% at m = 3.

Neutron fluxes at the C2-3-1, C2-3-2, and C2-3-3 ports were measured to be 9.56×10^7 n·cm^{-2}·s^{-1}, 3.13×10^7 n·cm^{-2}·s^{-1} and 1.58×10^7 n·cm^{-2}·s^{-1}, respectively. The neutron spectra were measured at the C2-3-3 beam port by the time-of-flight method and a characteristic wavelength of 4.8 Å was found for the C2-3-3 beam port.

We examined the performance of the compact splitting system with a Monte Carlo simulation. The agreement between the calculation and the experimental measurements is good. Therefore, we succeeded in achieving the desired design performance of the compact splitting system.

References

[1] Nagao M, Yamada L N, Kawabata Y, Seto H, Yoshizawa H and Takeda T 2006 *Physica* B **385-386** 1118-21
[2] Matsubayashi M and Tsuruno A 1994 JRR-3 neutron radiography facility *Neutron radiography 4. Including radioscopy and complementary inspection methods using neutrons (Electronic Materials* vol 4*)* ed J P Barton (Yverdon: Gordon and Breach Science Publishers) pp 415–22
[3] Yonezawa C, Magara M, Sawahata H, Hoshi M, Ito Y and Tachikawa E 1995 *J. Radioanal. Nucl. Chem.* **193** 171
[4] Alefeld B, Duppich J, Schärpf O, Schirmer A, Springer T and Werner K, 1988 *SPIE* **983 75-83**
[5] Schirmer A, Heitjans P, Faver W and Samuel F J 1990 *Meas. Sci. Technol.* **1** 654-56
[6] Schärp O 1991 *Physica* B **174** 514-27
[7] Ballot B, Samuel F and Farnoux B 1992 *SPIE* **1738** 159-65
[8] Kawabata Y, Utsuro M and Ebisawa T 1992 *SPIE* **1738** 448-53
[9] Schmidt U, Dubbers D, Raum K, Joeres O and Schärpf O 1996 *J. Neutron Research* **5** 81-8
[10] Mildner D F R, Chen-Mayer H H, Lamaze G P and Sharov V A 1998 *Nucl. Instr. And Meth.* A **413** 341-51
[11] Kawabata Y and Horie T 2004 *Appliel Radiation and Isotopes* **61** 541-6

[12] Suzuki J, Tasaki S, Soyama K and Ebisawa T 2002 *Appl. Phys.*A **74** [Suppl.] S308-10
[13] Rekveldt M T, Verkerk P and Van Well A A 1988 *Nucl. Instr. And Meth.* B **34** 285-9
[14] Soyama K, Suzuki M, Kodaira T, Ebisawa T, Kawabata Y and Tasaki S 1995 *Physica* B **213&214** 951-3
[15] Courtois P, Hamelin B, Humblot H, Alianelli L and Pfeiffer F 2002 *Applied Physics* A **74** [Suppl.] S320-2
[16] Lefmann K and Nielsen K 1999 *Neutron News* **10** 20
[17] Willendrup P, Farhi E and Lefmann K 2004 *Physica* B **350** 735
[18] Sears V F 1989 *Neutron Optics* (Oxford: Oxford University Press)
[19] Maruyama R, Yamazaki D, Ebisama T, Hino M and Soyama K 2006 *Physica* B **386** 1256-58
[20] Yamazaki D, Takeda M, Tamura I, Maruyama R, Moriai A, Hino M and Soyama K 2009 *Physica* B **404** 2557-60

International Workshop on Neutron Optics and Detectors (NOP&D 2013) IOP Publishing
Journal of Physics: Conference Series **528** (2014) 012013 doi:10.1088/1742-6596/528/1/012013

High energy particle background at neutron spallation sources and possible solutions

N Cherkashyna[1,9], K Kanaki[1], T Kittelmann[1], U Filges[2], P Deen[1], K Herwig[3], G Ehlers[4], G Greene[5], J Carpenter[6], R Connatser[1], R Hall-Wilton[1,7], P M Bentley[1,8]

[1]European Spallation Source ESS AB, 221 00 Lund, Sweden
[2]Paul Scherrer Institute, 5232 Villigen PSI, Switzerland
[3]Instrument and Source Division, Oak Ridge National Laboratory, Oak Ridge, TN, 37831, USA
[4]Quantum Condensed Matter Division, Oak Ridge National Laboratory, Oak Ridge, TN, 37831, USA
[5]Physics Division, Oak Ridge National Laboratory, Oak Ridge, TN, 37831, USA
[6]Argonne National Laboratory, Argonne, IL, 60439, USA
[7]Department of Electronics, Mid Sweden University, Sundsvall, Sweden
[8]Department of Physics and Astronomy, Uppsala University, 751 05 Uppsala, Sweden

E-mail: nataliia.cherkashyna@esss.se

Abstract. Modern spallation neutron sources are driven by proton beams ~ GeV energies. Whereas low energy particle background shielding is well understood for reactors sources of neutrons (~20 MeV), for high energies (100s MeV to multiple GeV) there is potential to improve shielding solutions and reduce instrument backgrounds significantly. We present initial measured data on high energy particle backgrounds, which illustrate the results of particle showers caused by high energy particles from spallation neutron sources. We use detailed physics models of different materials to identify new shielding solutions for such neutron sources, including laminated layers of multiple materials. In addition to the steel and concrete, which are used traditionally, we introduce some other options that are new to the neutron scattering community, among which there are copper alloys as used in hadronic calorimeters in high energy physics laboratories. These concepts have very attractive energy absorption characteristics, and simulations predict that the background suppression could be improved by one or two orders of magnitude. These solutions are expected to be great benefit to the European Spallation Source, where the majority of instruments are potentially affected by high energy backgrounds, as well as to existing spallation sources.

[9] To whom any correspondence should be addressed.

1. Introduction

The European Spallation Source combines a powerful spallation source with the instrumental goal of being able to measure weak processes with cold neutrons. Whereas existing spallation sources are either short pulse (microsecond) or continuous, the ESS will be a "long pulse" source, of the order of a few milliseconds, with a low repetition rate. Despite the low repetition rate, an optimisation of the long pulse instrumentation favours measurements across frame boundaries – i.e. when the next proton pulse produces neutrons (neutrons from several pulses are present in the instrument) – on approximately 70% of the instrument suite. The data spike that can be detected as the next pulse arrives is a high energy physics phenomenon that has been called a "prompt pulse", because those neutrons possess such high energy that they appear at virtually zero time-of-flight as measured by the neutron instruments. The "prompt-pulse" has been observed since the first days of neutron spallation sources. The issue about high energy background and the technical challenges, related to it, has been known since those times [1].

In this paper we will give an overview on how the ESS is aiming to reduce the backgrounds on the instruments, present some new shielding solutions, with particular emphasis on reducing the signal from high energy background. This is of particular importance, because although the data spike is generally small, it is sometimes comparable or 1 or 2 orders of magnitude stronger in intensity to the interesting, weak phenomena that the instruments are designed to measure at meV energies.

One example of the high energy background problem is the data measured on HYSPEC instrument, the hybrid spectrometer at the Spallation Neutron Source (SNS, Oak Ridge National Laboratory, USA). The background signal is discussed in HYSPEC technical and status reports [2-3]. The time-dependent signal has a count rate comparable to expected inelastic count rate in some regions of the time window [2]. The background is observed when the instrument shutter is open and closed; it is not significantly affected by the heavy beam shutter, and therefore the source of this signal is believed not to be the instrument's own beamline.

If it was possible to reduce the background signal by two orders of magnitude, the intrinsic background on HYSPEC would be at the level desired for polarized inelastic neutron scattering, for all of incident energies available. It is one of our conclusions that modifications to the shielding strategy may be the best option to reduce this background. The prompt pulse is accompanied by a long tail of fast neutron spectra; which is likely to be caused by energetic neutrons being moderated by instrument and it's surroundings.

Another evidence of the problem is the data measured at CNCS instrument, the cold neutron chopper spectrometer at the Spallation Neutron Source (SNS, Oak Ridge National Laboratory, USA), see figure 1.

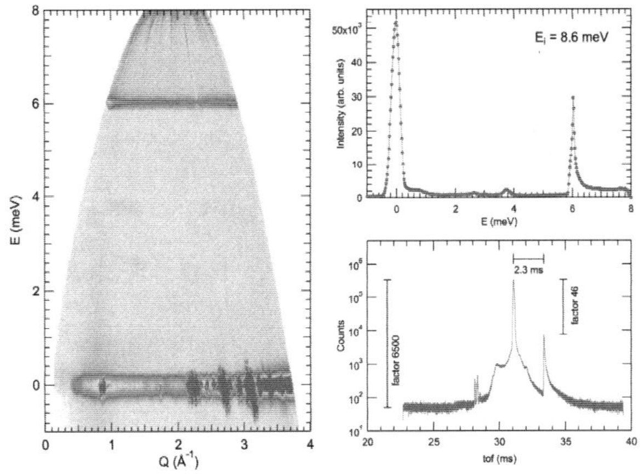

Figure 1. The bottom right shows the raw data as of time-of-flight, the prompt pulse peak is at 33.333 ms and the elastic line is at 31 ms. The top right shows how time-of-flight translates to energy transfer, with the prompt pulse at 6meV energy loss.
The left panel shows the data converted to the (Q,w) plane but otherwise uncorrected. The prompt pulse is at 33.333 because the instrument is in the third frame at this particular energy. The frame length is 16.667, SNS operates at 60Hz.

The high energy background does reduce the functionality of the neutron instruments, by increasing the noise-to-signal ratio. Typically, it causes a degradation in performance by forcing the instruments to discard a region in time-of-flight where the high energy background spike occurs. The spike is systematic – it is time correlated, and it is not a random error that reduces in relative size compared to the measured signal. Attempts to subtract this background have proven to be unsatisfactory.

A prime candidate mechanism in the prompt pulse is the phenomenon of particle showers. These are the cascades of secondary particles produced by high energy particle interactions with dense matter. There are two different mechanisms of particle showers: electromagnetic and hadronic [4]. As shown in figure 2, electromagnetic shower is more localized in a material, while hadronic shower is more spatially extended. Electromagnetic showers are caused by photons and electrons. Electrons create many events in matter by ionization and bremsstrahlung, and photons are able to penetrate quite far through material before they lose energy. The second type, hadronic showers, is caused by hadrons – particles that are made of quarks - and strong nuclear forces are involved into those interactions. The hadronic showers are characterized by ionization and interactions between incident particles and nuclei of the material. Hadronic showers are complex, as many reactions take place and many different particles are produced, and it is essential to use sophisticated modelling packages such as GEANT4 [5] to understand hadronic showers.

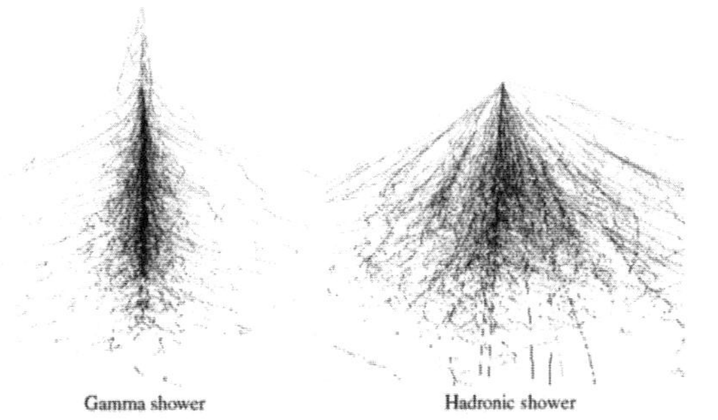

Figure 2. The different character of electromagnetic and hadronic showers [4].

One can distinguish two different types of showers, but typically the hadronic showers include electromagnetic processes (hadronic shower tails are electromagnetic) and both of them take place within the high energy interactions in matter. High energy particles interact with nuclei, secondary particles are produced, and these impact other particles. This process will continue until the particle energies are reduced below the thresholds for generating more secondary particles, at which point the particles are absorbed in matter.

The spatial extent of a particle shower depends on the energy of incoming particles (that dependence is logarithmic) and on the material they hit. This is why examining different materials in order to know how they can produce and/or stop particle showers is important for building a solution which will reduce high energy background.

2. Geant4 simulation idea

Geant4, a reliable simulation toolkit for areas where particle interactions in matter play a main role [6], was used to perform the simulations. That toolkit is a well-known main simulation engine for high energy physics applications. Different materials were investigated in order to understand their shielding properties against high energy particles. Materials studied here include iron, copper, steel, aluminium, regular concrete and tungsten. They were investigated in terms of their high energy shielding properties. The simulations involved blocks of each material of 2x2 meters (length x hight), with thicknesses varying from 10 centimetres to 3 meters. The particle source in Geant4 simulations (particle gun) is placed in front of the centre of the block. 10^6 neutrons with energy 2.5 GeV were emitted along the z-axis, representing a fast neutron emitted with all of the proton energy at an ESS-like target station. The "perfect detector", with a size of 2x2 meters (length x hight), is placed right after the block, as shown schematically in figure 3. The "perfect detector" was used for illustrative purposes in those simulations, but for a full optimisation here the details of sensitivity of the detectors to various particle species and energy would be absolutely key. The physics list *QGSP_BERT_HP*, recommended for shielding applications in the energy range of interest, was chosen for the simulations. This is the set of physical processes which describe the interaction of particles with dense matter [6]. The abbreviation means the following: *QGS* is for quark-gluon string model, *P* means precompound, *BERT* means that Bertini intra-nuclear cascade model is used, and *HP* states that high precision neutron tracking model is also included [7]. The results of simulations are the types of the "surviving" particles after going through the each block and their energies. In other words, for every fast neutron interacting with the shielding block, we determine how many and what kinds of particles are transmitted to the other side, including all generated particles.

Figure 3. A scheme of the simulation setup

3. Simulations and their results

Iron, with a relatively small atomic number, was chosen for the simulations to investigate its shielding properties because iron-based materials are widely used for shielding purposes in neutron scattering facilities. The isotopic composition of iron, the default G4_Fe material, is the following:

Fe54 (Z = 26, N = 54, A = 53.94 g/mole), abundance = 5.84%

Fe56 (Z = 26, N = 56, A = 55.93 g/mole), abundance = 91.75%
Fe57 (Z = 26, N = 57, A = 56.94 g/mole), abundance = 2.12%
Fe58 (Z = 26, N = 58, A = 57.93 g/mole), abundance = 0.28%

As one can see from figure 4, the number of particles developed by particle showers in iron, increases in small (up to 70 cm) blocks of iron. The particle count peak is known as "shower maximum". Note that 50 cm of iron actually "amplifies" the background flux by 1-2 orders of magnitude. This has implications for "T0 Chopper" design at spallation sources that may not have been accounted for in their design at present generation spallation sources. After the peak, it approaches a straight line in a log-plot (or exponential in a linear plot) when the secondary particles' equilibrium is reached. We then observe that the number of particles decreases to the initial number (10^6) of neutrons when the thickness of the block is about 180 cm. After 3 metres of iron, the number of particles is still greater than 10^4; i.e. a shielding factor corresponding to only 2 orders of magnitude. For some materials, neutrons dominate, for some other, as one can see below, the difference in number of neutrons and other particles is significant. However, we're particularly interested in neutron performance of the shielding at the moment.

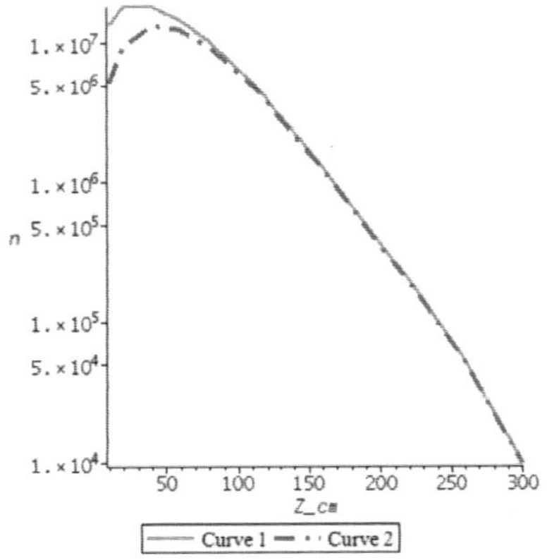

Figure 4. Number of particles detected in the simulation with 10^6 incident neutrons after initial particles are emitted into the blocks of iron as a function of thickness of the block. *n* is the number of particles, *Z_cm* is the thickness of the block. Curve 1 (solid) corresponds to all particles (excluding neutrinos), curve 2 (dashed) corresponds to neutrons.

Copper has been avoided on the past at neutron sources because of the cost, but we are interested in reducing the total cost of the whole shielding system. Copper has been simulated since it has very favourable shielding characteristics. As a result, copper is used in high energy physics laboratories (for example, CMS experiment in CERN [8]). Thus, we should not exclude expensive materials, which offer high performance and can be a part of solution.
The isotopic composition of copper, the default G4_Cu material, is the following:
Cu63 (Z = 29, N = 63, A = 62.93 g/mole), abundance = 69.17%
Cu65 (Z = 26, N = 65, A = 55.93 g/mole), abundance = 30.83%

As one can see from figure 5, the number of particles developed by particle showers in copper has a similar qualitative shape to that of iron, but reaches the initial number (10^6) of neutrons when the thickness of the block is about 110 cm. When it comes to blocks with thickness of blocks about 200 cm and more, the difference between iron and copper is about 2 orders of magnitude. This demonstrates the advantage of copper in stopping high energy particles in comparison with iron. (Please see also figure 7 for that comparison and also for comparison with steel, discussed below).

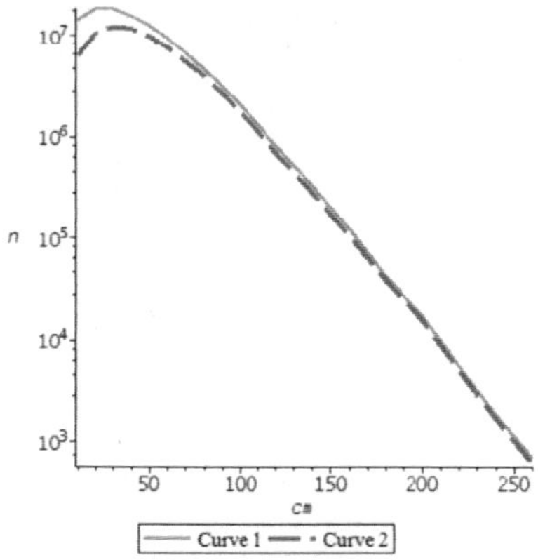

Figure 5. Number of particles detected in the simulation with 10^6 incident neutrons after initial particles are emitted into the blocks of copper as a function of thickness of the block. *n* is the number of particles, *Z_cm* is the thickness of the block. Curve 1 (solid) corresponds to all particles (excluding neutrinos), curve 2 (dashed) corresponds to neutrons.

Steel, as one of the widely used materials for shielding purposes, was also simulated. The type of steel that was used is carbon structural steel, which is used at SNS. It is defined in "ASTM A36 Standard Specification for Carbon Structural Steel" by American Society for Testing and Materials (CFR Section: 46 CFR 160.035-3(b)(2)).

The following composition of materials (the standard requirements for shapes of all thicknesses) was chosen: carbon, 0.26%; phosphorus, 0.04%; sulfur, 0.05%; silicon, 0.4%; copper, 0.2%; iron, 99.05%. The density of the steel is 7.85 g/cm^3. In that particular simulation, the number of neutron emitted into the block is not 10^6 as for all other simulations, but 10^5.

As shown on the figure 6, the structural carbon steel has approximately the same behaviour as iron in terms of blocking the high energy particles, which is not surprising given that the composition is mostly iron. The number of particles developed by particle showers in steel again initially increases in a small (up to 70 cm) blocks up to shower maximum, and decreases back to the initial number (10^5) of neutrons when the thickness of the block is about 170 cm. To see the comparison of number of neutrons as a function of thickness of the block for iron, copper and steel, refer to figure 6, where the values for steel are multiplied by order of magnitude to be within the same scale with values for iron and copper.

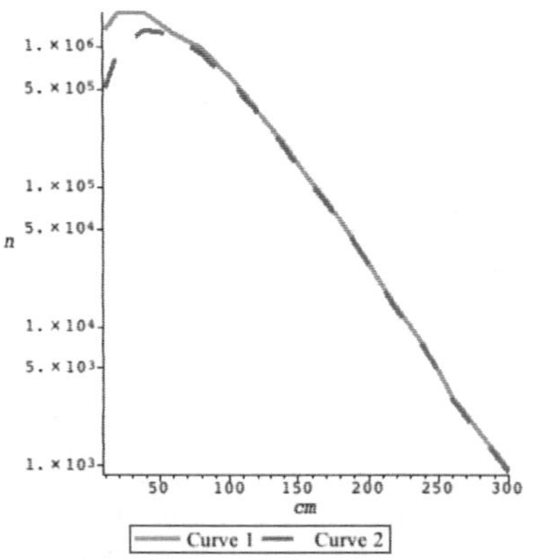

Figure 6. Number of particles detected in the simulation with 10^6 incident neutrons after initial particles are emitted into the blocks of steel as a function of thickness of the block. *n* is the number of particles, *Z_cm* is the thickness of the block. Curve 1 corresponds to all particles (excluding neutrinos), curve 2 corresponds to neutrons.

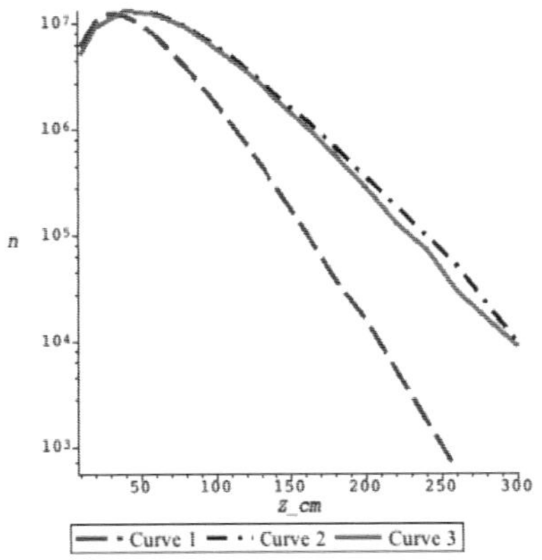

Figure 7. The number of neutrons which are detected in the simulation after initial particles are emitted into the blocks of copper (curve 1, long dash), iron (curve 2, dash-dot) and steel (curve 3, solid) as a function of thickness of the block. Number of incident neutrons is 10^6 for curve 1 and curve 2. Number of incident neutrons is 10^5 for curve 3 and those results are multiplied by 1 order of magnitude. *n* is the number of particles, *Z_cm* is the thickness of the block.

In figure 8, figure 9 and figure 10 one can see the total cross sections as a function of neutron energy in 3 materials, discussed above – iron, copper and steel, correspondingly, which illustrates the performance differences between a copper-based shielding strategy and an iron-based shielding strategy. As one can see, iron and steel are very similar in terms of nuclear resonances and windows of high transparency. Measured cross sections taken from the EXFOR/IAEA Nuclear Data Services website (http://www-nds.iaea.org/nrdc/basics/).

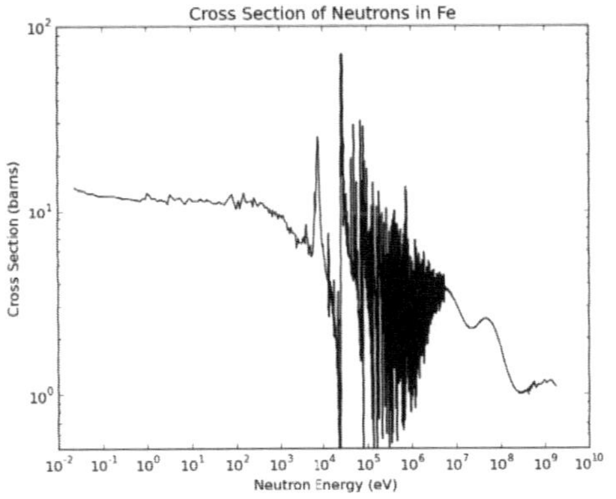

Figure 8. Measured total cross section of neutrons in iron as a function of neutron energy.

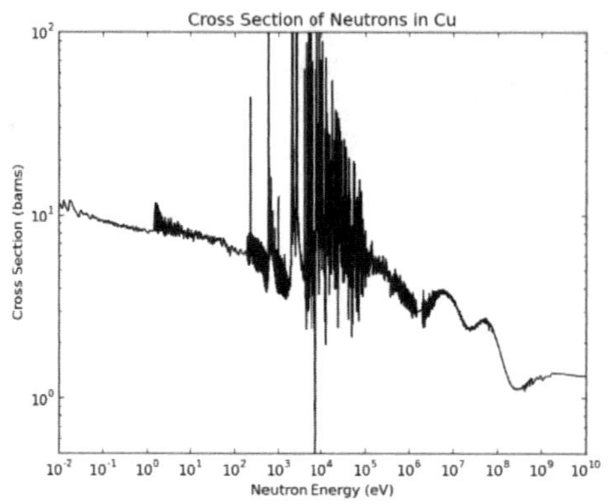

Figure 9. Measured total cross section of neutrons in copper as a function of neutron energy.

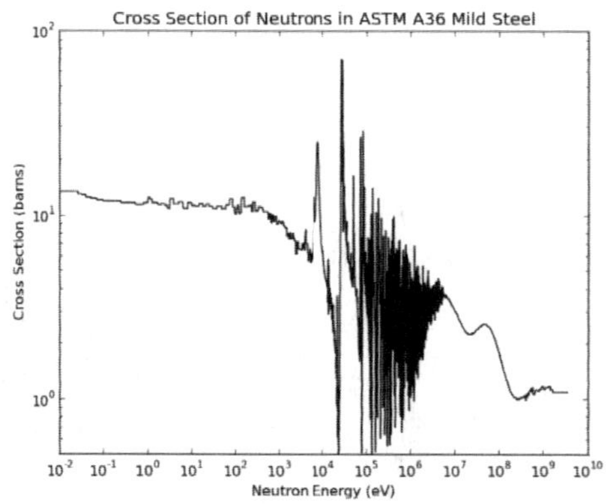

Figure 10. Total cross section of neutrons in steel as a function of neutron energy as a weighted combination of measured data sets for each chemical ingredient.

We have also examined aluminium, to show the behaviour of a structural material with a low atomic number compared to these of other shielding materials. The isotopic composition of aluminium, the default G4_Al material, is the following:
Al27 (Z = 13, N = 27, A = 26.98 g/mole), abundance = 100%

As one can expect, aluminium turns out to be less efficient in stopping high energy particles than iron. The number of particles developed by particle showers in aluminium increases in larger blocks of aluminium (up to 100 cm), and it goes down to initial number (10^6) only when the thickness of the block is about 270 cm. Please see figure 11 where the number of particles as a function of thickness of the block is shown.

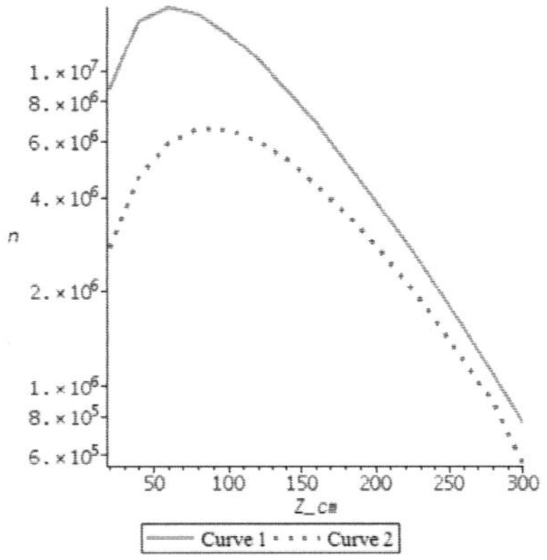

Figure 11. Number of particles detected in the simulation with 10^6 incident neutrons after initial particles are emitted into the blocks of aluminium as a function of thickness of the block. *n* is the number of particles, *Z_cm* is the thickness of the block. Curve 1 (solid) corresponds to all particles (excluding neutrinos), curve 2 (dashed) corresponds to neutrons.

Concrete was also studied for the simulations as very common material, well-known by its applications in different fields, including shielding purposes. The particular material, used for simulations, is regular concrete, which has a large water content that tends to be effective at moderating epithermal neutrons. The components of concrete, the default G4_CONCRETE material, are the following: hydrogen 1%; carbon 0.01%; oxygen 52.9107%; sodium 1.6%; magnesium 0.02%; aluminum 3.3872%; silicon 33.7021%; potassium 13%; calcium 4.4%; iron 1.4%. Density of concrete is 2.3 g/cm^3.

As one can see from figure 12, even if the number of particles, created by particle showers in concrete, increases in relatively small blocks (up to 70 cm), it goes down very slowly and reaches the initial number of neutrons when the thickness of block is about 170 cm. Even with the thickness of the block about 280 cm, the number of neutrons as well as the number of all particles remains significant.

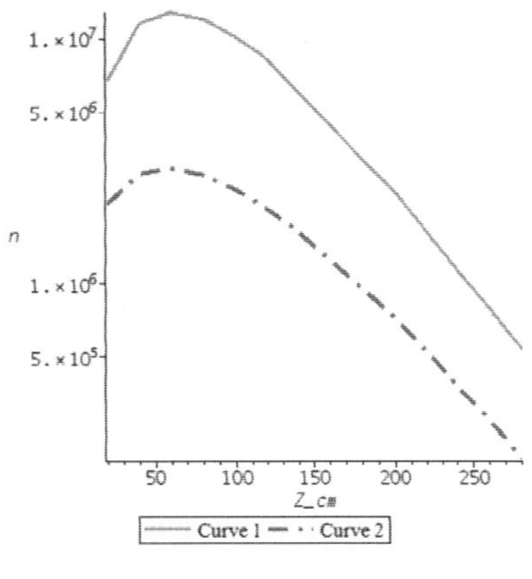

Figure 12. Number of particles detected in the simulation with 10^6 incident neutrons after initial particles are emitted into the blocks of concrete as a function of thickness of the block. *n* is the number of particles, *Z_cm* is the thickness of the block. Curve 1 corresponds to all particles (excluding neutrinos), curve 2 corresponds to neutrons.

Finally, tungsten has been studied for simulation as a dense, high-Z material. The isotopic composition of tungsten, the default G4_W material, is the following:
W180 (Z = 74, N = 180, A = 179.95 g/mole), abundance = 0.12%
W182 (Z = 74, N = 182, A = 181.95 g/mole), abundance = 26.5%
W183 (Z = 74, N = 183, A = 182.95 g/mole), abundance = 14.31%
W184 (Z = 74, N = 184, A = 183.95 g/mole), abundance = 30.64%
W186 (Z = 74, N = 186, A = 185.95 g/mole), abundance = 28.43%

As one can see from figure 13, the number of particles developed by particle showers in tungsten, increases in very small (about 20 cm) blocks. It goes down very quickly, reaches the initial number (10^6) of neutrons with the thickness of the block is about 60 cm. The number of particles goes down to significantly small numbers with the thickness of tungsten block about 180 cm. Tungsten is well known to be very efficient material in stopping and adsorbing particle showers. The main drawback with tungsten is the cost so tungsten can only be used as part of the solution in key areas.

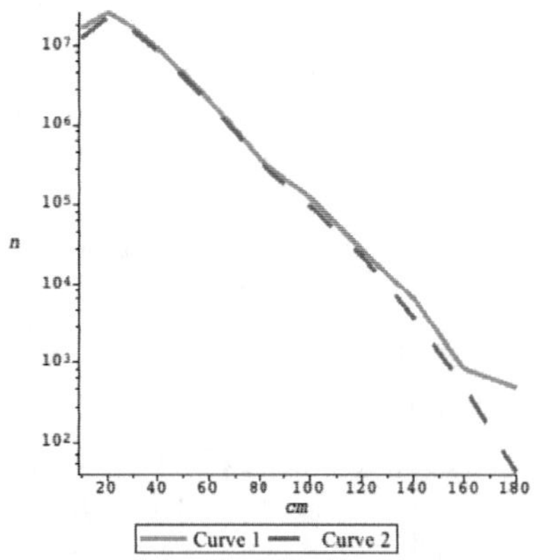

Figure 13. Number of particles detected in the simulation with 10^6 incident neutrons after initial particles are emitted into the blocks of tungsten as a function of thickness of the block. *n* is the number of particles, *Z_cm* is the thickness of the block. Curve 1 (solid) corresponds to all particles, curve 2 (broken) corresponds to neutrons. The artefact at 160-180 cm is caused by the low statistics, due to the effectiveness of tungsten as a shielding material.

Table 1 compares the performance of these different materials in dealing specifically with GeV neutrons.

Current efforts are now investigating laminate shielding solutions, combining different materials effectively to exceed the performance of any single ingredient.

Table 1. The comparison of different materials and combination of materials in terms of particles (and particularly neutrons) coming through them.

Name of the material and thickness	Emerging neutrons detected, % of initial number of neutrons
Iron, 3 m	~ 1%
Copper, 2.6 m	~ 0.06%
Steel, 3 m	~ 0.9%
Aluminium, 3 m	~ 56%
Concrete, 2.6 m	~ 28%
Tungsten, 1.6 m	~ 0.05%

4. Conclusions

Iron-based shielding is commonly used at spallation neutron sources, but has well-known nuclear resonances, producing windows of high transparency between 100 keV and 1 MeV. Structural carbon steel has approximately the same behavior as iron in terms of blocking high energy background, and neither can be thought of as a complete shielding solution without supplementary materials. Copper has been demonstrated at high energy physics labs in multiple roles and has superior properties compared to iron and steel; and copper-based solutions deep inside key shielding areas are seen to be attractive avenue of further research in reaching our background goals.

Tungsten is a well-known high performance shielding material, but can be only a part of the shielding solution because of its cost; we are interested in reducing the total cost of the whole shielding system. Aluminum and regular concrete are used widely in neutron instruments, and readily initiates particle showers at high energy, but do not significantly contribute to containing the subsequent shower.

5. Acknowledgements

The authors wish to thank, in alphabetical order, S Ansell, K Batkov, C Cooper-Jensen, D DiJulio, K Fissum, M Fitzsimmons, S Kenney, O Kirstein, T Krist, D Martin Rodriguez, D Mildner, E Pitcher, T Sato, T Shea, L Tchelidze and L Zanini for useful discussions. Authors based at Oak Ridge acknowledge funding by the Scientific User Facilities Division, Office of Basic Energy Sciences, US Department of Energy.

6. References

[1] Russel G *et al.* 1988 Shielding concerns at a spallation source *Proceedings of the ICANS-X Conference, Los Alamos, NM*

[2] Winn B and Hagen M 2012 Inelastic spurion from the prompt pulse http://neutrons.phy.bnl.gov/HYSPEC/memos_technical_reports.shtm

[3] HYSPEC status reports http://neutrons.ornl.gov/hyspec/documentation/

[4] Gaudio G, Livan M and Wigmans R 2009 *Proceedings of the International School of Physics "Enrico Fermi"* 175 31-77

[5] Völk H and Bernlöhr K 2009 *Exp. Astr.* **25** (1-3) 173-191

[6] Agostinelli S *et al.* 2003 *Nucl. Instum. and Methods in Phys. Res.* A **506** 250-303

[7] Speckmayer P 2010 Impact of the choice of physics list on GEANT4 simulations of hadronic showers in tungsten LCD-Note-2010-002

[8] CMS collaboration 2008 *JINST* S08004

^{3}He spin filter based polarized neutron capability at the NIST Center for Neutron Research

W.C. Chen[1,2], T.R. Gentile[1], R. Erwin[1], S. Watson[1], Q. Ye[1,2], K.L. Krycka[1], B.B. Maranville[1]

[1] National Institute of Standards and Technology, Gaithersburg, Maryland, USA 20899
[2] University of Maryland, College Park, Maryland, USA 20742

E-mail: wcchen@nist.gov

Abstract.
A ^{3}He neutron spin filter (NSF) program for polarized neutron scattering was launched in 2006 as part of the National Institute of Standards and Technology (NIST) Center for Neutron Research (NCNR) Expansion Initiative. The goal of the project was to enhance the NCNR polarized neutron measurement capabilities. Benefitting from more than a decade's development of spin-exchange optical pumping (SEOP) at NIST, we planned to employ SEOP based ^{3}He neutron spin filters for the polarized neutron scattering community. These ^{3}He NSF devices were planned for use on different classes of polarized neutron instrumentation at the NCNR, including triple-axis spectrometers (TAS), small-angle neutron scattering instruments (SANS), reflectometers, and wide-angle polarization analysis. Among them, the BT-7 thermal TAS, NG-3 SANS, and MAGIK reflectometer have already been in the user program for routine polarized beam experiments. Wide-angle polarization analysis on Multi-Axis Crystal Spectrometer (MACS) has been developed for user experiments. We describe briefly the SEOP systems dedicated for polarized beam experiments and polarizing neutron development for each instrument class. We summarize the current status and polarized neutronic performance for each instrument. We present a ^{3}He NSF hardware and software interface to allow for synchronization of ^{3}He polarization inversion (neutron spin flipping) and free-induction decay (FID) nuclear magnetic resonance (NMR) measurements with neutron data collection.
Key words: polarized neutron, SEOP, ^{3}He NSF

1. Introduction

Polarized neutron scattering (PNS) is a powerful tool that probes magnetic structures and morphologies, magnetization density, and magnetic excitations in a wide range of magnetic materials. In 1939, Halpern and Johnson developed the theory of magnetic neutron scattering and showed that the scattered neutron polarization $\boldsymbol{P}_\mathrm{f}$ depends on the orientation of the wave-vector transfer $\boldsymbol{Q}$ with respect to the incident neutron polarization $\boldsymbol{P}_\mathrm{i}$ as given by a classical Halpern-Johnson vector, $\boldsymbol{P}_\mathrm{f} = -(\hat{\boldsymbol{Q}} \bullet \boldsymbol{P}_\mathrm{i})\hat{\boldsymbol{Q}}$, where $\hat{\boldsymbol{Q}} = \boldsymbol{Q}/|\boldsymbol{Q}|$ is the unit vector[1]. More extensive theoretical work on PNS was developed after Halpern and Johnson[2, 3, 4]. In 1969, Moon *et al.* experimentally demonstrated unambiguous separation of magnetic scattering from nuclear scattering, and nuclear coherent scattering from spin-incoherent scattering using uni-axial polarization analysis on a triple axis spectrometer (TAS)[5]. The uni-axial polarization analysis technique was later expanded to the xyz polarization analysis technique for separation of nuclear, magnetic, and spin-incoherent scattering[6]. A more complicated technique, spherical neutron polarimetry, employed the CRYOPAD device[7] to solve intricate structures especially those exhibiting mixed magnetic domains and non-collinear spin structures[8]. Polarized neutron scattering is one of the most challenging neutron instrumentation techniques due to significant intensity reduction from polarized beam production and analysis, necessity of precise spin manipulation, and complicated data reduction and analysis. Supermirror and Heusler crystals have been successfully employed to polarize the incident beam and analyze the scattered beam for decades. Recently many modern neutron instrumentation techniques call for larger beam angular divergences and wavelength bandwidths, and a larger area of detector banks in the interest of large on-sample neutron fluxes and improved detection efficiency. This has made it difficult to apply the standard neutron polarization techniques using supermirrors or Heusler crystals. ^{3}He neutron spin filters (NSF) can (1) polarize a broad wavelength band of neutrons effectively; (2) polarize large area and widely divergent neutron beams; and (3) efficiently flip the neutron polarization (flipping process and efficiency are wavelength-independent) by reversing the ^{3}He nuclear polarization using the adiabatic past passage (AFP) nuclear magnetic resonance (NMR) technique[9]. The development of ^{3}He NSFs has been advanced significantly during the last decade and these NSF devices have been applied or planned to be applied to many polarized neutron instrumentation techniques in most neutron facilities worldwide[10, 11, 12, 13, 14, 15] .

In 2006, the National Institute of Standards and Technology (NIST) Center for Neutron Research (NCNR) launched a multi-year Expansion Initiative to increase the cold neutron measurement capabilities. A ^{3}He NSF program for polarized neutron scattering became part of the Expansion Initiative with the goal of enhancing polarized neutron measurement capabilities. Previously, polarized beam capabilities at the NCNR were limited to polarization analysis on the reflectometer[16], occasionally on both thermal TAS[17] and cold TAS[18], incident polarized beam production on the NG-3 small-angle neutron scattering (SANS) instrument, and neutron spin echo (NSE) instrument[19]. These existing polarized neutron instrumentation techniques utilized supermirror polarizing devices for polarizing the incident beam and analyzing the scattered beam except that Heusler crystals were used on the decommissioned BT-2 thermal TAS. The NIST Physics Laboratory already had a program in research and development of ^{3}He NSFs for neutron science applications using both the spin-exchange optical pumping (SEOP) and metastability-exchange optical pumping (MEOP) methods. Based on substantial progress in SEOP[10], this method was adopted for the NCNR's ^{3}He NSF based polarized neutron program.

This paper presents an overview of polarized neutron development at the NCNR using ^{3}He NSFs over last several years and is organized as follows. Sec. 2 shows the calculated polarized neutronic performance. In Sec. 3 we describe the characteristics and performance of the SEOP system. In Sec. 4, we discuss polarized beam development using ^{3}He NSFs on the BT-7 thermal TAS, NG-3 SANS and the Multi-Axis Crystal Spectrometer (MACS) and their performance. In Sec. 5, we discuss the development of magnetostatic cavity devices used for different instruments. We describe a user-friendly hardware and software interface allowing for synchronization of AFP-based ^{3}He spin inversion and FID (free induction decay) NMR measurements in Sec. 6 and we summarize the paper in Sec. 7.

2. Polarized neutronic performance of a polarized ^{3}He spin filter

A ^{3}He neutron spin filter is a transmission-based neutron-polarizing device. The neutron transmission for each spin state through a polarized ^{3}He cell is given by

$$T^{\pm} = T_{\mathrm{E}}\exp\left[-\mathcal{O}\left(1 \mp P_{\mathrm{He}}\right)\right] \tag{1}$$

where $T^{\pm}$ is the transmission with neutron spin parallel (+) or antiparallel (−) to the ^{3}He spin. T_{E} is the transmission of the ^{3}He glass cell windows. The ^{3}He cells are fabricated from GE-180 glass[20], which has a typical neutron transmission of 0.87 at 5 Å for a typical large spin filter cell. The wavelength dependence of the transmission of glass cell windows has been measured to be relatively flat[21]. $P_{\rm He}$ is the ^{3}He polarization. $\mathcal{O}$ is the opacity of the cell. Because the neutron absorption cross section increases linearly with neutron wavelength, a practical expression for the opacity is $\mathcal{O} = 0.726p\lambda l$, where p is the ^{3}He gas pressure in bar for a cell at a temperature of 295 K, λ is the wavelength in nm, and l is the length of the cell in cm. $\mathcal{O}$ is a dimensionless quantity. The transmission for an unpolarized neutron beam passing through a polarized ^{3}He cell is then given by

$$T_{\rm n} = T_{\rm E}\exp\left(-\mathcal{O}\right)\cosh\left(\mathcal{O}P_{\rm He}\right) = T_0\cosh\left(\mathcal{O}P_{\rm He}\right) \tag{2}$$

where T_0 is the transmission for an unpolarized neutron beam passing through an unpolarized ^{3}He cell ($P_{\rm He}$=0) and is given by $T_0 = T_{\rm E}\exp\left(-\mathcal{O}\right)$. The resulting neutron polarization after passing through a polarized ^{3}He cell is given by

$$P_{\rm n} = \frac{T^+ - T^-}{T^+ + T^-} = \tanh\left(\mathcal{O}P_{\rm He}\right) = \sqrt{1 - \frac{T_0^2}{T_{\rm n}^2}} \tag{3}$$

The flipping ratio F of either the polarizer or analyzer is given by

$$F = \frac{T^+}{T^-} = \exp(2\mathcal{O}P_{\rm He}) \tag{4}$$

Optimization of the trade-off between $P_{\rm n}$ and $T_{\rm n}$ depends on the experiment[22]. A common figure of merit is defined as ${P_{\rm n}}^2T_{\rm n}$ for a polarizer. Fig. 1(a) shows polarized neutronic performance ($P_{\rm n}$, $T_{\rm n}$, T^+, ${P_{\rm n}}^2T_{\rm n}$ and F) as a function of the opacity of a ^{3}He cell at a ^{3}He polarization of 80 %, a typical value that we have achieved recently. A general rule of thumb is $P_{\rm n}$=0.96, $T_{\rm n}$=0.27 for $\mathcal{O}$=2.5 and $P_{\rm He}$=80 %. ${P_{\rm n}}^2T_{\rm n}$ has a broad peak at $\mathcal{O} \cong 2$ in Fig. 1a. In general, the opacity at which ${P_{\rm n}}^2T_{\rm n}$ is maximized, $\mathcal{O}_{\rm p}$, depends on the ^{3}He polarization and is given by

$$\mathcal{O}_{\rm p} = \frac{1}{P_{\rm He}}\tanh^{-1}\left[\frac{\sqrt{8{P_{\rm He}}^2+1}-1}{2P_{\rm He}}\right] = \frac{1}{2P_{\rm He}}\ln\frac{2P_{\rm He}-1+\sqrt{8{P_{\rm He}}^2+1}}{2P_{\rm He}+1-\sqrt{8{P_{\rm He}}^2+1}} \tag{5}$$

Fig. 1(b) shows $\mathcal{O}_{\rm p}$ as a function of $P_{\rm He}$, indicating that $\mathcal{O}_{\rm p}$ is insensitive to the ^{3}He polarization and has a value of ≈ 2 for ^{3}He polarizations lower than 80 %. For ^{3}He polarizations $\geq$80 %, as the ^{3}He polarization increases, the peak location shifts towards a larger $\mathcal{O}_{\rm p}$ value. Fig. 1(b) also shows ${P_{\rm n}}^2T_{\rm n}$ for $\mathcal{O} = \mathcal{O}_{\rm p}$ as a function of the ^{3}He polarization. In most cases, $\mathcal{O}$ values larger than $\mathcal{O}_{\rm p}$ have generally been chosen in order to reach desired neutron polarizations (flipping ratios)[22].

3. NIST SEOP systems for neutron spin filter applications

For all current neutron science applications at the NCNR, we have retrofitted ^{3}He NSF devices to existing instruments. Space constraints make *in-situ* SEOP on the beamline challenging, therefore we have focused on offline operation in which ^{3}He gas is polarized in the lab and transported to the beamline. As soon as the polarized cell is installed on the beam, the ^{3}He polarization starts to decay exponentially with a characteristic relaxation time T_1 assuming that the external field perturbation does not change with time. One of the major tasks for offline operation of ^{3}He NSFs is to achieve as long a T_1 as possible on the beamline. Space constraints and/or stray magnetic fields from the instrument itself or sample environment make T_1's of ^{3}He cells shorter than intrinsic T_1's obtained with a negligible magnetic field gradient. For most of our applications, we refresh polarized ^{3}He cells for each NSF device on the beam daily. It is

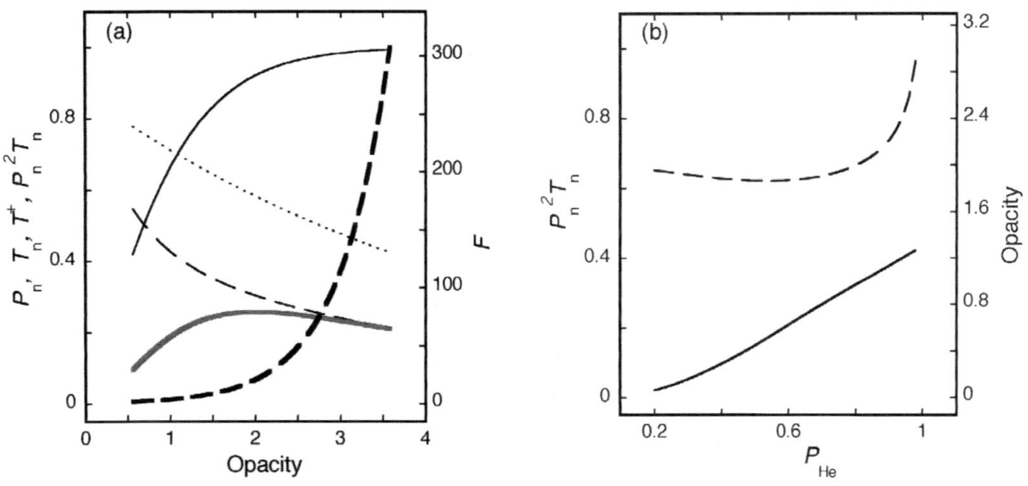

Figure 1. (Color online) (a) Neutron polarization efficiency P_n (solid thin line), transmission T_n (thin dashed line), T^+(dotted line), a figure of merit given by ${P_n}^2T_n$ (solid red thick line), and the flipping ratio F (dashed thick line) as a function of the opacity $\mathcal{O}$ at P_{He}=80 %. (b) the opacity at which ${P_n}^2T_n$ is maximized, $\mathcal{O}_p$, (dashed line) and ${P_n}^2T_n$ (solid line) as a function of P_{He}. The typical empty cell transmission T_E=0.87 is included. Note the flipping ratio is calculated for either the polarizer or analyzer only assuming other neutron polarizing devices are perfect in efficiency as discussed in the text. All quantities in plots are dimensionless.

necessary to have multiple SEOP systems and several cells for each configuration (see the cell database in Ref. [10]). We currently have three well-characterized SEOP systems at NIST used for neutron science applications[10]. Among them, one system has been used for more than a decade for neutron spin filter research and development, and recently has been mainly employed for developing wide-angle cells[23, 24]. Two of them are dedicated to providing polarized ^{3}He gas for polarized neutron experiments at the NCNR[10] as shown in Fig. 2. A fourth system is planned for 2014, allowing us to provide polarized ^{3}He gas to more than one instrument simultaneously.

The general features of the NIST SEOP systems were described in Ref. [10]. Here we briefly describe a new feature of the three systems. Over three years ago, we implemented chirped volume holographic gratings (VHG)[26] for tunable spectral narrowing of high power laser diode bars[10]. VHGs contain planes of varying index of refraction fabricated in photorefractive glass[27, 28], that produce a wavelength-selective reflection. A chirped VHG provides a higher spectral density and better suppression of broadband background as compared to diffraction gratings[31], especially for higher power laser diodes. Recently, we switched to chirped VHGs from diffraction gratings for all high power lasers used in the three systems above. Now, each SEOP system can provide a total power up to 120 W to the optically pumped ^{3}He cell with a spectral linewidth (FWHM) between 0.2 nm and 0.3 nm. This change has resulted in a modest improvement in ^{3}He polarization. We have obtained ^{3}He polarizations of 80 % to 85 % on the neutron beam line for cells nearly 1 liter in volume. A detailed paper on studies of ^{3}He polarization improvement from use of the chirped VHGs will appear in a future publication.

Figure 2. (Color online) One of the NIST SEOP systems showing all the components. The SEOP system includes the NMR system with three pairs of coils (holding field, radio-frequency field coils and pick-up coils) assembled orthogonally each other, NMR control system, high power diode lasers and associated controllers and chiller, and laser optics. The majority of the system is mounted to a movable aluminum 80-20[25] cart.

4. ^{3}He NSF applications at the NCNR

At the NCNR, we have applied ^{3}He NSFs to polarize incident neutrons and analyze scattered neutron beams for SANS, thermal TAS, reflectometers, an interferometer, and wide-angle polarization analysis on the Multi-Axis Crystal Spectrometer. We began providing polarized ^{3}He gas to user experiments in 2008. Fig. 3 summarizes usage of polarized ^{3}He gas for neutron science applications during the last 5 years. On average, the amount of ^{3}He gas polarized is about 150 bar-liter per year for both routine user experiments and ^{3}He NSF tests.

4.1. Polarization analysis on the BT-7 thermal triple axis spectrometer

On the BT-7 thermal TAS, incident neutrons, after passing through a 20 cm by 20 cm double-focusing monochromator of either pyrolytic-graphite(PG)(002) or copper(Cu)(220)[32], are polarized by a ^{3}He NSF device and the scattered neutrons are analyzed by a second ^{3}He NSF device[33]. The maximum beam size at both the ^{3}He polarizer and analyzer is 10 cm in height and 5 cm in width. The ^{3}He polarization of the analyzer was maintained by a magnetically shielded solenoid (MSS)[33, 34], which is attached to the detector arm and located approximately midway between the sample and the PG energy analyzer. For the polarizer, we first employed a MSS to maintain the ^{3}He polarization, but have recently switched to a compact "Magic Box"[35, 36] to facilitate the neutron spin transport[34]. In order to reduce the relaxation rate of the ^{3}He polarization induced by magnetic field gradients[37], we have developed a new magic box for which the mu-metal thickness was doubled from 1.6 mm to 3.2 mm so that it can be operated at a higher field (from 2.3 mT to 3.6 mT) without magnetically saturating the mu-metal. Note that the drive sections of the high field magic box where magnetic coils are located have a total thickness of 6.4 mm. For a typical polarized TAS setup, it is essential to have a capability of longitudinal polarization analysis in both $\boldsymbol{P} \perp \boldsymbol{Q}$ and $\boldsymbol{P}||\boldsymbol{Q}$ configurations[5] at low sample fields. It is also necessary to be able to measure four spin-dependent cross sections in each

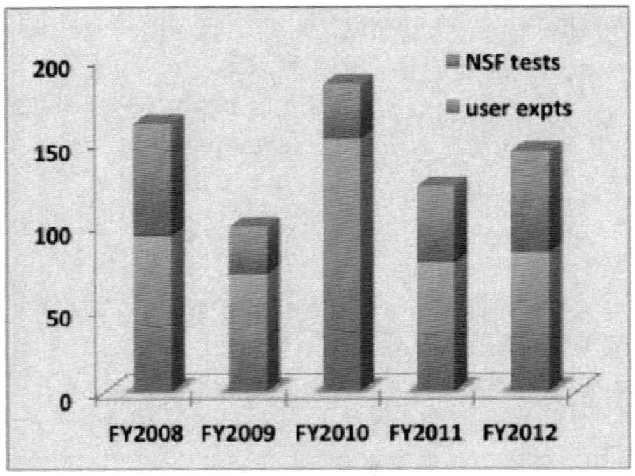

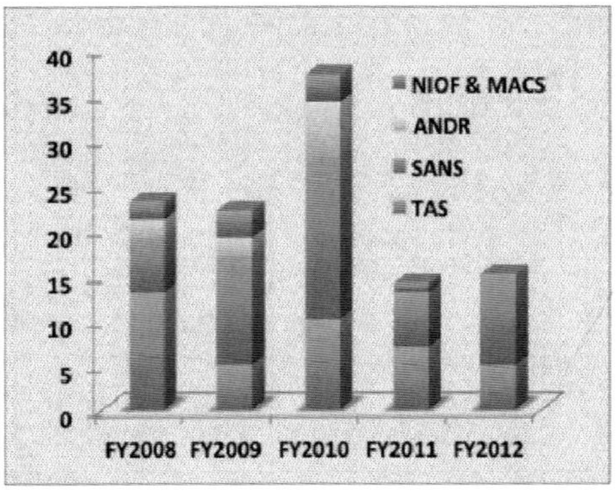

Figure 3. (Color online) Amount of polarized ^{3}He gas produced during the last 5 years (left plot). This includes ^{3}He gas polarized for user experiments and NSF tests. Applications of ^{3}He NSFs to different polarized instruments at the NCNR during the last 5 years (right plot). The instruments that employed ^{3}He NSFs were BT-7 TAS, NG-3 SANS, ANDR reflectometer, MACS, and Neutron Interferometer and Optics Facility (NIOF)[29, 30]. Note the reactor at the NCNR was down from April, 2011 to March, 2012.

configuration. The polarization flipping for both incident and scattered beams was controlled by a precession coil neutron spin flipper at the beginning of polarized beam development on BT-7 in 2006. Because we retrofitted our ^{3}He NSF devices onto the instrument, space was limited for polarized beam implementation. Thus, the neutron spin flipper could only be situated near the sample field coils that provided the neutron polarization axes ($\boldsymbol{P} \perp \boldsymbol{Q}$ and $\boldsymbol{P}||\boldsymbol{Q}$), as mentioned above. For $\boldsymbol{P}||\boldsymbol{Q}$, two issues arose at large scattering angles: (1) decrease of the spin flipper efficiency and (2) non-adiabatic spin transport between the flipper and the sample field. Consequently, the polarized beam setup limited polarization analysis for the $\boldsymbol{P}||\boldsymbol{Q}$ configuration to elastic diffraction experiments at small scattering angles.

Recently we re-designed the polarized beam instrumentation on the BT-7 TAS. The new setup employs newly designed, compact adiabatic spin rotation devices and nearly lossless neutron spin inversion using AFP to invert the ^{3}He polarization[38]. The new apparatus avoids using a precession coil spin flipper and frees up an additional space of $\approx$ 10 cm, which is enough for adiabatic spin transport. The new polarized beam setup on the BT-7 TAS is entirely contained within the ^{3}He NSFs (^{3}He polarizer and analyzer) without the need for external spin flipping devices, resulting in significantly improved polarized neutronic performance. Here, we present a brief summary of the performance. The details of the design and characterization of polarized beam configuration will appear in a future paper.

There are several new notable features for the new polarized beam setup. First, polarized beam calibration is simpler and easier. Using the polarized beam setup provided by the ^{3}He NSFs, it is easy and straightforward to obtain polarization efficiencies since the polarizing efficiencies of the polarizer and analyzer are independently measured from the neutron transmissions or calculated from the ^{3}He polarization[39]. Previously for a conventional polarized beam setup, it was necessary to measure the four spin dependent transmissions using a sample whose spin cross sections are well known to extract polarizing efficiencies of polarizing elements (polarizer, analyzer, each flipper)[40, 41]. The second feature of the new setup is a truly adiabatic spin transport along the entire neutron path. After polarized beam calibration, we characterized the energy dependence and spatial uniformity of the spin transport efficiency. We determined that the overall spin transport efficiency is better than 99 % for neutron energies up to 50 meV for both the $\boldsymbol{P} \perp \boldsymbol{Q}$ and $\boldsymbol{P}||\boldsymbol{Q}$ configurations. We found that the spin transport efficiency is uniform for a wide range of scattering angles, a feature required for the $\boldsymbol{P}||\boldsymbol{Q}$ configuration. The ^{3}He AFP based neutron spin flipping efficiency was nearly 100 % and energy-independent except for a tiny irrecoverable ^{3}He polarization loss per flip. This loss was determined to be $\approx 1 \times 10^{-5}$ and $\approx 3 \times 10^{-5}$ for the polarizer and analyzer, respectively, on BT-7. The nearly lossless ^{3}He polarization per flip is one of the key features as it is desirable to have the loss lower than 1×10^{-4} per flip if flipping happens within several minutes in order to avoid an additional apparent contribution to the effective T_1. For 1×10^{-4} loss, the additional T_1 contribution is $\approx$1000 h for a flip every 0.1 h. Lastly, the ^{3}He polarization was improved from a previous value of 75 % to a current value of 80 % due to use of chirped VHGs described in Sec. 3.

After we fully characterized all polarizing elements, we have tested the new polarized setup to establish the level of performance. We first tested the polarized inelastic performance. This is the most demanding test because spin transport becomes more difficult at higher neutron energies. For this test, we employed the ^{3}He cell Grigio as a polarizer and the cell Sangiovese as an analyzer. Grigio had an opacity of 4.04 at 14.7 meV and its ^{3}He polarization was 80 %. Sangiovese had an opacity of 3.02 at 14.7 meV and its ^{3}He polarization was 78 %. Both cells are cylindrical, have a diameter of $\approx$ 11 cm and a volume of 0.95 liter. We obtained a flipping ratio of 85 for a fully open beam (maximum beam size) with an intensity reduction factor of only 12.0 (both polarizer and analyzer are accounted for) at a neutron energy of 14.7 meV compared to an unpolarized beam configuration; While the measured flipping ratio was reduced to 8.4 at 50 meV, the intensity reduction factor was reduced to 5.9 at 50 meV (overall transmission improved by a factor of 2). The decrease in the flipping ratio and increase in intensity at 50 meV are due to the energy-dependence of the ^{3}He NSF as described in Sec. 2. The calculated flipping ratios were 90 and 8.6 for 14.7 meV and 50 meV, hence the spin transport is nearly perfect at both 14.7 meV and 50 meV.

For the polarized elastic condition, we employed the cell Viognier as polarizer and Syrah as analyzer. Opacities were 3.11 and 2.29 for Viognier and Syrah both at 14.7 meV. ^{3}He polarizations were 80 % for both cells. We obtained a flipping ratio of 30 for a fully open beam with an intensity reduction factor of only 7.7 at a neutron energy of 14.7 meV. Based on the flux measurement[32] in unpolarized mode and a typical transmission of 0.25 for the polarizer, the polarized flux at the sample was calculated to be 0.6×10^7 cm^{-2}s^{-1} at 14.7 meV with the PG(002) monochromator vertically focused and 1.1×10^7 cm^{-2}s^{-1} at 13.7 meV with the monochromator doubly focused, both with a PG filter. Such polarized flux values are comparable to those from instruments IN20 (1.05×10^7 cm^{-2}s^{-1}) and IN22 (0.75×10^7 cm^{-2}s^{-1}) at 14.7 meV at the ILL (Institut Laue-Langevin)[42]. This showed that ^{3}He NSFs in combination with a PG(002) monochromator at ^{3}He polarizations of 80 % well outperformed Heusler crystals. Although the performance of Heusler crystals has improved during the last several years, Kulda *et al.* showed a ^{3}He NSF with a PG monochromator outperformed a Heusler crystal monochromator for wavelengths below 0.18 nm even with a ^{3}He polarization of 48 %[43]. The polarized beam capability with ^{3}He NSFs on the BT-7 thermal TAS at the NCNR has been applied to studies of multiferroics[44, 45, 46], superconductivity[47, 48], and magnetic thin films[49]. All these applications were done with the old polarized BT-7 setup where ^{3}He polarizer and analyzer and precession coil flippers were used.

4.2. Polarization analysis on the NG-3 SANS instrument

We have recently developed SANS polarization analysis on the NG-3 SANS instrument using a ^{3}He NSF[34, 50, 51]. Spin-analyzed SANS is a powerful probe for studies of magnetic morphology with directional sensitivity. The SANS polarization analysis technique developed at the NCNR has recently been employed for hard condensed matter investigations such as magnetic nanoparticles[52, 53, 54], an exchanged-biased system[55], multiferroics[56, 57], and

Journal of Physics: Conference Series **528** (2014) 012014 doi:10.1088/1742-6596/528/1/012014

a giant magnetostrictor[58]. For a typical spin-analyzed SANS setup as shown in Fig. 4, a V-shape Fe/Si supermirror in a transmission geometry is used to polarize the incident neutrons. A precession coil spin flipper is placed immediately after the polarizer. A ^{3}He analyzer is used to spin-analyze the scattered beam from the sample. For the setup, the limited space and a modest stray field from the 1.6 T sample field have prevented us from operating precession coil spin flipper and radio-frequency flippers, positioned immediately after the sample. Therefore, the neutron spin flipping for the scattered beam is done by inverting the ^{3}He polarization using the AFP NMR technique. The ^{3}He polarization of the analyzer is maintained by a compact MSS. The radio-frequency coil required for the ^{3}He polarization inversion was provided by a $\sin\theta$ coil[38], 17 cm in diameter and 20 cm long, to provide an RF magnetic field of 0.02 mT orthogonal to the main holding field without adding material in the beam. The entire ^{3}He analyzer system is under vacuum to minimize air scattering.

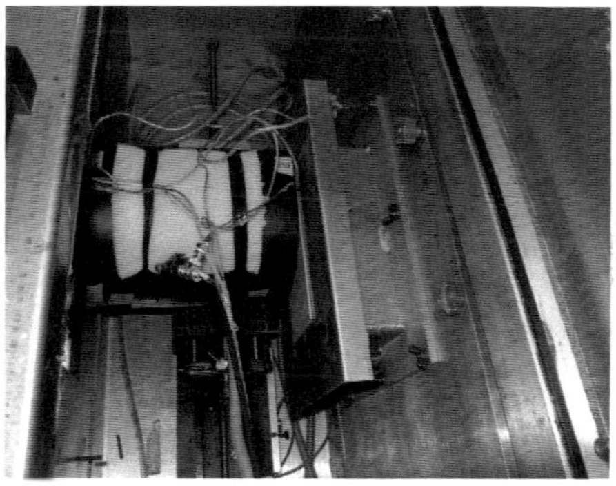

Figure 4. (Color online) A spin-analyzed SANS setup on the NG-3 SANS instrument at the NCNR (the left picture). The neutron beam travels from left to right in the pictures. In the middle of the picture is the 1.6 T electromagnet (black cubic object) with an iron-yoke field return path for reducing the stray field. A supermirror and a spin flipper are located further upstream of the magnet (not shown in the picture). A sample chamber that can be either under vacuum or in air is attached upstream of the large detector tank. Between the magnet and sample chamber, there is an 8-cm gap where an adiabatic spin rotator is located. The MSS maintaining the ^{3}He polarization is located inside the sample chamber and under vacuum during the experiment. The right picture shows a view of the MSS inside the chamber.

In the last few years, a few new features have been developed for improving the SANS polarization analysis capability. The most important improvement is to achieve as large a Q as possible while still maintaining reasonably long relaxation times even at a field of 1.6 T. The length scale of the magnetic features is inversely proportional to Q. For SANS polarization analysis setup combined with a high sample field ($\geq$ 1 T), the achievable maximum Q is smaller than the instrumental Q_{max} defined in the unpolarized setup. The SANS polarization analysis setup has limited the size of magnetic nanoparticles larger than 9 nm for studies of magnetic morphologies at fields lower than saturation[52]. Even for studies of the nanocrystalline alloy at a saturating field, SANS polarization analysis has precluded detection of a magnetic peak at 17 nm[59]. Achieving a larger Q_{max} becomes an essential and challenging task when implementing SANS polarization analysis. This requires either maximizing the diameter of the analyzer cell or minimizing the distance between the sample magnet and the ^{3}He NSF analyzer or both. Fabricating a large diameter ^{3}He cell is practically difficult[10] requiring a larger magnetostatic cavity device, thus demanding more space. A decade of SEOP ^{3}He cell development has led to the ability to make cells with a diameter of 12–13 cm, , currently limited by our GE-180 glass blowing technology. Therefore the most effective method is to decrease the distance between the magnet and analyzer, but this deteriorated the field gradient of the MSS due to a large stray field from the magnet, hence the relaxation time of the ^{3}He polarization decreased. Neutron depolarization from a field direction rotation in the region might also be an issue. During the last few years, we have developed a compact system that can reach $Q_{\rm max}$ of 1 nm^{-1} or higher without offsetting the analyzer cell, while providing the adiabatic spin transport for the overall large, divergent scattered beam and maintaining the relaxation time $\geq$ 100 h.

Here we present a brief summary of the system. The electromagnet we have used provides a field to the sample up to 1.6 T with the field direction with the field direction perpendicular to the beam in the horizontal plane. The magnet was specifically modified with a proper iron-yoke to minimize the stray field at the ^{3}He analyzer. The yoke is 23 cm from the center of the magnet to the edge and the stray field was about 2.5 mT at a position of 8 cm away from the edge of the magnet where the MSS is located. For a MSS, a large hole in each end cap at both ends is necessary for the neutron path. As the hole becomes larger, $Q_{\rm max}$ becomes larger if the cell diameter is not a limiting factor, but the field leakage to the inside solenoid is higher and the field gradient at the polarized ^{3}He cell is worse. Consequently the T_1 of the ^{3}He analyzer cell is shorter. We have developed MSSs of different sizes for a trade-off between $Q_{\rm max}$ and relaxation times as described in Table 1 in Sec. 5. Furthermore, the 8 cm gap between the MSS and magnet required that an adequate, compact spin rotation device be developed to ensure the neutron spin for the entire scattered beam is adiabatically rotated from the transverse horizontal to longitudinal direction within a short distance. We measured four spin-dependent transmissions on the main beam to perform the polarized beam calibration[50, 51] and determined the spin transport was $\geq$ 99%. Mapping the overall instrumental polarization efficiency pixel by pixel on the two-dimensional PSD (position sensitive detector)[60] provided an additional check of the spin transport efficiency for all scattered beams. Using this method, we determined that the spin transport for the entire beam path in our polarized beam setup was $\geq$ 99%. Finally, the ^{3}He polarization loss was determined to be better than 1×10^{-4} per flip even when the magnet operates at 1.6 T. Given a typical polarized SANS measurement time of half an hour, the loss induced a negligible difference in T_1 compared to no ^{3}He flipping.

We routinely operate the polarized SANS setup at two wavelengths, 5 Å and 7.5 Å. We obtained an instrumental flipping ratio of typically 13 at 5 Å and 19 at 7.5 Å if there is no sample depolarization. The instrumental flipping ratio F is given by $F = \frac{(1+P)}{(1-P)}$, where P is the instrumental polarization efficiency. These low flipping ratios are mainly limited by the supermirror polarizer and the spin flipper on the NG-3 SANS instrument. The polarizing efficiency of the single V-shaped Fe/Si supermirror is wavelength dependent, 0.89 and 0.94 for 5 Å and 7.5 Å, respectively. The flipper efficiency is 0.97 (0.985 flipping probability) for both wavelengths. The supermirror polarizer and spin flipper together limit the flipping ratio to be 14 at 5 Å and 21 at 7.5 Å. More recently, we relocated polarized beam experiments from the NG-3 SANS to the NG-7 SANS instrument. At the same time, the NCNR installed a double V-shaped supermirror and an RF flipper for the incident beam to improve the spin flipping efficiency. The new supermirror is expected to have much better polarization efficiency (95 % at 5 Å and 99 % at 7.5 Å). The RF flipper is wavelength-independent and expected to be insensitive to the large wavelength spread present in SANS ($\geq$11 %). It is anticipated to have a flipping efficiency better than 99 %. In these new configurations, the flipping ratios are estimated to be higher than 27 and 55 for 5 Å and 7.5 Å, depending on the final performance of the new supermirror and RF flipper. The transmission of the current polarizer was measured to be 36 %, which for a typical ^{3}He analyzer cell yields an initial intensity loss factor of 5.5 compared to an unpolarized beam.

Another important development was to characterize the spatial variations in polarized neutronic performance[60]. For most of our applications using ^{3}He NSFs, the ends of the blown cylindrical ^{3}He cells are often made rounded for better structural integrity, so the ^{3}He gas thickness may differ for different scattered beams. We have characterized the two-dimensional neutron path variations in opacity, transmission through polarized ^{3}He cells, and neutron polarization generated from a ^{3}He NSF. We found the path length variations in opacity and transmission are about 10 % and 8 %, respectively, from the cell center to the edge for a typical SANS polarization analysis configuration. However, the overall instrumental polarization efficiency is spatially uniform and its variation is about 1 %[60]. Although the spatial variations are not large, it should be corrected for when performing polarization efficiency corrections.

4.3. Wide-angle polarization analysis

Our apparatus for wide-angle neutron polarization analysis[23] consists of a vertical, neutron-compatible, solenoid that contains spin filter polarizer and analyzer cells close to the tail of a sample cryostat. The polarizer cell is contained in a shielded RF solenoid to permit inversion of the ^{3}He polarization via AFP without loss of polarization in the analyzer cell. The wide-angle cells cover an angular range of up to 110 ° on either side of the neutron beam. FID is used to monitor the ^{3}He polarization in the spin filter cells. Using conventional cylindrical cells[10], the apparatus was employed for a study of superconducting $Fe_{1+y}Te_{0.62}Se_{0.38}$[61] on MACS[62]. Recently we reported several upgrades to this apparatus[24], in particular the development of GE180[20] wide-angle cells with relaxation times between 100 h and 400 h, and ^{3}He polarization values, P_{He}, between 0.65 and 0.80. A new measurement in the largest wide-angle cell (denoted Giant, volume of 1.3 liters) yielded P_{He}=0.77 and a relaxation time of 400 h. For this test, the cells were pumped by two 100 W single-bar diode lasers narrowed with chirped VHGs as described in Sec. 3 and an additional 50 W spectrally narrowed laser[63] supplemented the illumination on the ends of the cell.

In April 2013 the upgraded apparatus was tested on MACS (see Fig. 5). Although the MACS apparatus is nominally non-magnetic, we discovered a stray magnetic field from the MACS pre-sample optics that reduced the ^{3}He relaxation times. A mu-metal shield located between the optics and the spin filter apparatus effectively blocked the stray field and improved the field homogeneity at the cells. For the polarizer cell Mourvedre, we obtained a relaxation time T_1=156 h (340 h off-line) and for the analyzer cell Giant we obtained 240 h (400 h off-line). With an improved AFP scheme, the loss per flip for the polarizer was 0.10 % per flip and the associated loss in the analyzer cell was 0.005 % per flip. Four polarizer cells with typical relaxation times of 200 h and and opacity values between 2.2 and 4.8 at 5 Åhave been fabricated. The apparatus will be further tested on MACS in Oct. 2013, and we expect to conduct a neutron scattering experiment soon thereafter.

5. Development of magnetostatic cavity devices

The magnetic field gradient ($|\vec{\nabla} B_{\perp}/B|$) induced ^{3}He polarization relaxation rate is proportional to the square of the gradient and inversely proportional to the partial ^{3}He pressure[37]. On the beam line, the space is often limited for implementation of ^{3}He NSFs. However it is still desired to have a field gradient of 5×10^{-4} cm^{-1} or better so that the field gradient contribution to the relaxation time of the ^{3}He polarization is 600 h or longer at a ^{3}He pressure of 1 bar. For cells used on the beam line, the relaxation times are typically 300 h in a pair of large Helmholtz coils. The relaxation times would reduce to 200 h at a ^{3}He pressure of 1 bar under such a field gradient on the beam line. In addition, when designing magnetostatic cavities it is necessary to connect their magnetic field to the neighboring fields for adiabatic neutron spin transport. Neutron depolarization can occur if a neutron traverses a region in which the magnetic field direction rotates faster than its Larmor precession frequency. Hence, it is necessary to either avoid spin rotation or obtain adiabatic spin transport when designing magnetostatic cavity devices.

Figure 5. (Color online) Photographs of the spin filter apparatus on MACS during the April 2013 test run, without the sample cryostat. The neutron beam exits the monochromator drum (blue, left), passes through the MACS pre-sample optics and into the vertical solenoid. Inside the solenoid (right photo), the beam passes through the metal box containing the polarizer cell and scattered neutrons pass through the analyzer cells. The mu-metal shield discussed in the text can be seen between the pre-sample optics and the solenoid.

Over the last decade, we have developed two types of magnetostatic cavity devices that can provide a large volume, uniform magnetic field for different polarized beam instruments at the NCNR: magnetically shielded solenoids[22, 33, 34, 64, 65] and end-compensated magic boxes[34, 36]. A MSS consists of a solenoid with a cylindrical mu-metal shielding co-axial with the solenoid. Two mu-metal end caps with a hole to pass neutrons are snugly attached to each end of the cylindrical magnetic shielding. Typically our single-layer MSS can tolerate an external stray field up to 2.5 mT and still provide a reasonable field gradient for the ^{3}He cell contained inside for user experiments. A MSS provides a field along the neutron beam line, which often requires an adiabatic $\pi/2$ rotation of neutron spins. Table 1 lists MSSs used on different instruments at the NCNR. For the most compact MSS called Kronos, the field gradient on the NG-3 SANS instrument resulted in a T_1 decrease from 150 h to between 50 h and 90 h, but it yielded a Q increase $\approx$ 40 % higher compared to MSSs called Gemini and Pollux. Both Gemini and Pollux MSS can have T_1's longer than 200 h. To provide a longer T_1 than the MSS Kronos, but maintain the same Q, we developed a double layer MSS called Artemis. The outer shielding is made of lower permeability NETIC[66] mu-metal nested around an inner high permeability CO-NETIC[66] mu-metal shield. The compensation coils are attached to the inner end-caps. We obtained a T_1 of 150 h even with the sample magnet operating at 1.6 T for the MSS Artemis, indicating that the field gradient induced relaxation time is a factor of $\approx$ 4, after correction to the cell T_1 of 400 h in a non-gradient limited field such as a pair of large Helmholtz coils, longer than that in the MSS Kronos, without decreasing the achievable $Q_{\max}$.

A magic box is a coil driven magnetic parallel plate capacitor made of high permeability

CO-NETIC[66] mu-metal. An end-compensated magic box has a short magic box (hence compensated) attached to each end of the box with a small air gap between each section[36]. This minimizes end-effects from a finite magic box and improves the field gradient for a compact box with dimensional constraints. We have demonstrated that the gradient-induced relaxation rate of the ^{3}He polarization for an end-compensated box can be an order of magnitude smaller than that of an uncompensated box[36]. However, experience has shown the box is sensitive to external field gradients, especially a gradient component orthogonal to the holding field[34]. As described in Sec. 4.1, an effective way to minimize the gradient-induced relaxation is to increase the holding field, but without saturating the mu-metal. We constructed a high field magic box made of two layers of mu-metal in the top and bottom magnetic connecting plates and four layers of mu-metal in the drive sections for both the main section and two compensating sections. Table 2 lists two boxes we constructed, one with a single-layer and the other with a double layer, and their performance in the lab and on the beam line. Note that the normalized field gradient on the beam line is sensitive to the external stray field magnitude, and we show the range of the gradient when possible. The lower value corresponds to a stray field of less than 0.1 mT and the higher value corresponds to a stray field of about 0.3 mT. As the holding field of the double-layer box Flattop was increased, the normalized volume-averaged gradient decreased. Increasing the field from 0.9 mT to 3.6 mT improved the relaxation time by more than an order of magnitude.

Table 1. Magnetically shielded solenoids used on different neutron beamlines at the NCNR. D is the outside diameter in cm of the mu-metal shielding, l is the outside length in cm of the shielding, and d_{ec} is the diameter of the end cap hole to pass neutrons. $|\vec{\nabla} B_{\perp}/B|^{\rm i}$ is the volume-averaged field gradient determined from T_1's measured without any stray field for a cylindrical cell 12 cm in diameter and 8 cm long. $|\vec{\nabla} B_{\perp}/B|^{\rm bl}$ is the volume-averaged gradient determined from T_1's measured often with modest stray fields from the sample field and/or adiabatic spin rotation devices on the beam line. On the NG-3 SANS instrument, the stray field is dominated by the 1.6 T electromagnet located very close to the MSS to maintain the ^{3}He polarization of the analyzer. The MSS Artemis has two layers of nested magnetic shielding. The outer shielding is NETIC [66] mu-metal and the inner shielding is CO-NETIC[66] mu-metal. The compensation coils are attached to the inner end-caps. The first (second) number in D, l and d_{ec} refer to parameters for the inner (outer) layer shielding.

MSS name	D cm	l cm	d_{ec} cm	$\lvert\vec{\nabla} B_{\perp}/B\rvert^{\rm i}$ 10^{-4} cm^{-1}	$\lvert\vec{\nabla} B_{\perp}/B\rvert^{\rm bl}$ 10^{-4} cm^{-1}	instrument(s)
Kronos	20.3	25.4	11.4	6.0	11-14	SANS
Gemini	25.4	35.5	11.4	2.0	4.0-7.0	SANS, TAS, ANDR
Pollux	25.4	35.5	11.4	3.5	5.0-7.0	SANS, TAS, ANDR
Artemis	25.4/27.9	33.0/35.5	14.0/14.6	5.0	8.0	SANS

6. Synchronization of automated AFP NMR based ^{3}He spin flipping

As described earlier, one of the advantages of ^{3}He NSFs over conventional solid state neutron polarizing devices is the inversion of the ^{3}He polarization using the AFP NMR technique. Unlike the precession-coil type flipper for which control of the neutron spin state is simply controlled by turning the currents of flipping and cancellation coils off or on, control of the ^{3}He AFP spin flipper is more complicated. First, a National Instrument data acquisition card[67] is used to

Table 2. Magic boxes used on the BT-7 TAS beamline. l, w, and h is the overall length, width and height in cm of the box, B_0 is the field of the magic box for the ^{3}He polarization relaxation measurement. $|\vec{\nabla} B_\perp/B|^{\mathrm{i}}$ is the volume-averaged field gradient determined from T_1's measured without any stray field for a cylindrical cell typically 12 cm in diameter and 8 cm long. $|\vec{\nabla} B_\perp/B|^{\mathrm{bl}}$ is the volume-averaged gradient determined from T_1's measured with modest stray fields from the BT-7 instrument. Cowboy is a single-layer box and Flattop is a double-layer box as discussed in the text.

magic box name	l cm	w cm	h cm	field mT	$\lvert\vec{\nabla} B_\perp/B\rvert^{\mathrm{i}}$ 10^{-4} cm^{-1}	$\lvert\vec{\nabla} B_\perp/B\rvert^{\mathrm{bl}}$ 10^{-4} cm^{-1}
Cowboy	28.4	40	15	0.9	6.3	
Flattop	28.4	40	15	0.9	4.9	up to 100
Flattop	28.4	40	15	2.9	6.1	up to 30
Flattop	28.4	40	15	3.6	6.1	8.5-23

generate a frequency-swept RF signal across the resonance frequency that is also amplitude modulated by a Gaussian function[68]. In addition, we have found that the RF signal should be isolated from the AFP coil with a mechanical relay. Synchronization of the ^{3}He AFP spin flipper with the user experiment has not generally been done in some neutron scattering facilities that use ^{3}He NSFs.

We have developed a system that allows synchronization of not only AFP NMR based ^{3}He polarization inversion but also FID NMR measurements for optional monitoring of a T_1 of the cell. The system consists of a PYTHON program to bridge the COM port to the RS232 interface. The PYTHON program listens to the serial port of the neutron scattering instrument computer. When a text string arrives, the program checks it against an approved list of NMR commands for handling the ^{3}He NSFs, then forwards it over the COM connection to the IGOR Pro software[69] where all NMR commands reside. Further, it forwards back any reply message of NMR commands over the serial port to the instrument computer. The system allows independent development of the NMR software and user-friendly interface on the instrument computer to handle all possible ^{3}He NSF devices (as many as possible). The system allows tracking of the ^{3}He polarization state (hence neutron spin state), number of AFP flips, the time stamp when the ^{3}He polarization is inverted which might be necessary for polarization efficiency correction before data analysis.

7. Conclusions

We have provided an overview of the polarized neutron development using ^{3}He neutron spin filters (NSF) at the NIST Center for Neutron Research (NCNR) during the last several years. We have applied SEOP based ^{3}He NSFs to a wide variety of neutron scattering instruments at the NCNR. Polarized neutronic performance for each instrument class using ^{3}He NSFs has been improved significantly. We have demonstrated the polarized neutronic performance (transmission or intensity, and polarizing efficiency or flipping ratio) is highly predictable when the proper polarized beam calibration is done. We have shown that this polarized beam calibration with ^{3}He NSFs is easy, simple and straight-forward. We have shown ^{3}He NSF based polarizing devices developed at the NCNR are advantageous over conventional neutron polarizers, supermirror and Heusler crystals, in achieving a higher polarized flux and spin-analyzing a wide-angle divergent beam. In some instruments we employ entirely ^{3}He NSFs for the polarized beam setup, while for others we employ a combination of a ^{3}He NSF and a supermirror. We have developed two types of magnetostatic cavities that provide a homogeneous magnetic field environment and/or shielding to the external stray field for use on different beam lines. We have developed the ^{3}He polarization inversion with a nearly lossless polarization feature during inversion using the adiabatic fast passage nuclear magnetic resonance technique. We have developed several spin-exchange optical pumping systems dedicated for polarized beam experiments. We have implemented a user-friendly hardware and software interface that allows easy control of neutron spins during the user experiment. Combined with polarization efficiency correction software and simple polarized beam calibration, polarized beam instrumentation at the NCNR is being established as a convenient tool. Polarized neutron experiments are routinely done on the thermal TAS, SANS, and reflectometer and will soon be established on MACS. Combined with conventional polarized beam techniques such as reflectometry, neutron spin echo, and cold triple axis spectrometry, polarized neutron measurement capabilities have been significantly enhanced at the NCNR.

Acknowledgments

We thank Jeff Anderson, Jack Fuller, Aaron Kirchhoff of the NIST glassblowning and optical shops. We acknowledge Gordon Jones for his great help in ^{3}He neutron spin filter development. We also acknowledge J.A. Borchers, J.W. Lynn, C.F. Majkrzak, C.L. Broholm, J.G. Barker for their scientific guidance of the polarized neutron development. The development and application of neutron spin filters has been supported during the time period of this work in part by the U.S. Dept. of Energy, Basic Energy Sciences. The work utilized facilities supported in part by the National Science Foundation under Agreement No. DMR-0944772.

References

[1] Halpern O, Johnson MH 1939 *Phys. Rev.* bf 55 898
[2] Maleyev SV, Baryakhtar VG, Suris RA 1963 *Sov. Phys. Solid State* **4** 2533
[3] Blume M 1963 *Phys. Rev.* **130** 1670
[4] Marshall W, Lowde RD 1968 *Rep. Prog. Phys.* **31** 705
[5] Moon RM, Riste T, and Koehler WC 1969 *Phys. Rev.* **181** 920
[6] Schärpf O and Capellmann H 1993 *Phys. Stat. Sol. A* **135** 359
[7] Tasset F 1989 *Physica B* **156-157** 627
[8] Brown PJ, Forsyth JB, Tasset F 1993 *Proc. R. Soc. London A* **442** 147
[9] Abragam A 1961 *The Principles of Nuclear Magnetism* Oxford University Press, Oxford, England
[10] Chen WC *et al.* 2011 *Journal of Physics: Conference Series* **294** 012003, and references therein
[11] Andersen KH *et al.* 2005 *Physica B* **356** 103
[12] Babcock E, Ioffe, A 2011 *Journal of Physics: Conference Series* **294** 012005
[13] Jiang CY *et al.* 2013 *Physics Procedia* **42** 191
[14] Beecham CJ *et al.* 2011 *Physica B* **406** 2429
[15] Kira H *et al.* 2011 *Journal of Physics: Conference Series* **294** 012014
[16] http://www.ncnr.nist.gov/instruments/ng1refl/
[17] The decommissioned BT-2 thermal triple-axis spectrometer at the NCNR employed Heusler crystals as both a polarizer and analyzer.
[18] Lee SH, Majkrzak CF 1999 *Physica B* **267-268** 341
[19] http://www.ncnr.nist.gov/instruments/nse/
[20] GE Lighting Component Sales, Bldg. 315D, 1975 Noble Rd., Cleveland, OH 44117. The largest standard GE180 tubing diameter is 16 mm; the 25 mm tubing was procured through a special glass run. Certain trade names and company products are mentioned in the text or identified in an illustration in order to adequately specify the experimental procedure and equipment used. In no case does such identification imply recommendation or endorsement by the National Institute of Standards and Technology, nor does it imply that the products are necessarily the best available for the purpose.
[21] Chupp TE *et al* 2007 *Nucl. Instrum. Meth. A* **574** 500
[22] Gentile TR *et al* 2005 *Physica B* **356** 96
[23] Fu CB *et al* 2011 *Physica B* **406** 2419
[24] Ye Q *et al* 2013 *Physics Procedia* **42** 206
[25] 80/20 Inc., 1701 South East, Columbia City, IN 46725 USA
[26] Optigrate, Inc., 3267 Progress Drive, Orlando, FL 32826
[27] Volodin BL *et al* 2004 *Opt. Lett.* **29** 1891
[28] Moser C and Steckman G 2005 *Photonics Spectra* **36 (6)** 82
[29] Huber MG *et al* 2009 *Nucl. Instrum. Meth. A* **611** 235
[30] Huber MG *et al* 2009 *Phys. Rev. Lett.* **102** 200401; Huber MG *et al* 2009 *Phys. Rev. Lett.* **103** 179903
[31] Chann B, Nelson I, Walker TG 2000 *Opt. Lett.* **25** 1352
[32] Lynn J W *et al* 2012 *J. Res. Natl. Inst. Std. Technol.* **117** 61
[33] Chen WC *et al* 2007 *Physica B* **397** 168
[34] Chen WC *et al* 2009 *Physica B* **404** 2663
[35] Petoukhov AK *et al* 2006 *Nucl. Instrum. Methods. A* **560** 480
[36] McIver JW, Erwin R, Chen WC and Gentile TR 2009 *Rev. Sci. Instrum.* **80** 063905.
[37] Cates GD, Schaefer SR and Happer W 1988 *Phys. Rev. A* **37** 2877
[38] Jones GL *et al* 2006 *Physica B* bf 385-386 1131
[39] Jones GL 2000 *et al Nucl. Instrum. Meth.* **A440** 772
[40] Majkrzak C 1991 *Hanbook of Neutron Scattering* Springer-Verlag, Berlin
[41] Wildes AR 2006 *Neutron News* **17** 17
[42] http://www.ill.eu/instruments-support/instruments-groups/instruments
[43] Kulda J *et al* 1998 *Physica B* **241243** 136
[44] Lee S *et al* 2008 *Phys. Rev. B* **78** 100101(R)
[45] Tian W *et al* 2008 *Phys. Rev. B* **78** 184429
[46] Cabrera I *et al* 2009 *Phys. Rev. Lett.* **103** 087201
[47] Chen Y *et al* 2008 *Phys. Rev. B* **78** 064515
[48] Wang W *et al* 2011 *Phys. Rev. B* **84** 094504
[49] Ratcliff W *et al* 2009 *Adv. Func. Mater.* **21** 1567
[50] Krycka KL *et al* 2009 *Physica B* **404** 2561
[51] Krycka KL, Chen WC, Borchers JA, Maranville BB, Watson SM 2012 *J. Appl. Cryst.* **45(3)** 546
[52] Krycka KL *et al* 2010 *Phys. Rev. Lett.* **104** 207203

[53] Krycka KL *et al* 2010 *J. Appl. Phys.* **107** 09B525
[54] Krycka KL, Jackson J, Borchers JA, Shih J, Briber R, Ivkov C, Grutter C, and Dennis CL 2011 *J. Appl. Phys.* **109** 07B513
[55] Dufour C, Fitzsimmons MR, Borchers JA, Laver M, Krycka KL, Dumesnil K, Watson SM, Chen WC, Won J, Singh S 2011 *Phys. Rev. B* **84** 064420
[56] Ueland BG *et al* 2010 *Phys. Rev. Lett.* **104** 147204
[57] Ramazanoglu M, Laver M, Ratcliff W, Watson SM, Chen WC, Jackson J, Kothapalli J, Seongsu L, Cheong SW, and Kiryukhin V 2011 *Phys. Rev. Lett.* **107** 207206
[58] Laver M, Mudivarthi C, Cullen JR, Flatau AB, Chen WC, Watson SM, and Wuttig M 2010 *Phys. Rev. Lett.* **105** 027202
[59] Michels A, Honecker D, Döbrich F, Dewhurst CD, Suzuki K, and Heinemann A 2010 *Phys. Rev. B* **85** 184417
[60] Chen WC, Erwin R, Watson SM 2013 *Physics Procedia* **42** 163
[61] Thampy V *et al.* 2012 *Phys. Rev. Lett.* **108**, 107002
[62] Rodriguez-Rivera A *et al* 2008 *Meas. Sci. Technol.* **19**, 034023
[63] Quintessence Photonics Corporation, 15632 Roxford Street, Sylmar, California 91342
[64] Chen WC *et al.* 2003 Physica B **335**, 196
[65] Chen WC, Gentile TR, O'Donovan KV, Borchers JA, Majkrzak CF 2004 *Rev. Sci. Instrum.* **75** 3256
[66] Magnetic Shield Corporation, 740 N. Thomas Drive, Bensenville IL 60106. CO-NETIC mu-metal has a higher magnetic permeability than that of NETIC mu-metal.
[67] USB 6251 DAQ card, National Instruments, Austin TX
[68] McKetterick TJ *et al.* 2011 *Physica B* **406** 2436
[69] WaveMetrics Inc. 10200 SW Nimbus, G-7 Portland, OR 97223

^{3}He Neutron Spin Filter cell development program at JCNS

Z Salhi[1], E Babcock[1], P Pistel[2] and A Ioffe[1]

[1] Jülich Centre for Neutron Science, Forschungszentrum Jülich GmbH, Outstation at MLZ, Lichtenbergstrasse 1, 85747 Garching, Germany.
[2] Forschungszentrum Jülich GmbH- Engineering und technologie (ZEA-1) Wilhelm-Johnen-Strasse, 52425 Jülich, Germany.

E-mail: z.salhi@fz-juelich.de

Abstract: In order to produce high-quality ^{3}He Neutron Spin Filters (NSF) with a high polarisation level, it is necessary to achieve a long ^{3}He relaxation time by the reduction of the wall relaxation. This requires one to minimise the amount of impurities at the surface of the glass cells, and to have as few contaminants as possible in the gas filling system. In this report we describe the detailed procedure we employ to produce ^{3}He cells using our newly built filling station. The obtained life times for a number of cells are practically approaching the fundamental limit imposed by the dipole-dipole interaction between ^{3}He atoms.

1. Introduction

High-quality ^{3}He Neutron Spin Filters (NSF) should provide a high polarization and a low absorption of neutron beams; this requires one to achieve high levels of ^{3}He polarization in the ^{3}He cells (containers). The achieved level of ^{3}He polarization is defined, from one hand by the rate of the build-up of ^{3}He polarization and, and from the other by relaxation processes that result in polarization decay. Indeed, a high performance NSF should combine an optimized optical pumping system and ^{3}He cells (containers) providing minimal achievable polarization losses.

Therefore, parallel to the developments of optimal optical pumping methods [1-4] aiming to the increase the maximum levels of polarization, the development of ^{3}He cells (containers) close to ideal, i.e. with minimum the ^{3}He relaxation possible, is necessary. These cells should be optimised to reach a high starting ^{3}He polarisation and to have very low relaxation rate ($\Gamma_1 = 1/T_1$) to maintain the ^{3}He polarisation in the NSF for long time.

Even for perfect cell containers, the decay of ^{3}He polarisation will result in the decrease of neutron polarisation and transmission. Therefore, especially for offline-polarised cells, this decay ideally must be kept to the lowest possible minimum during the experiment. This in turn limits the number of times one has to repolarise the ^{3}He cell or refill it while maximizing neutron performance over any given interval of time. Note that all cells are not identical, therefore determining the exact neutron performance requires each of them to be characterised before or after the usage in the experiment and be monitored over time.

The SEOP method also gives the possibility to polarize the ^{3}He online and in this way balance the ^{3}He spin-relaxation with continual polarization [5]. Since both the SEOP [6] and MEOP [7] method have been shown to give comparable maximum polarizations, optimized online polarization would provide the highest average performance over time as long as the laboratory maximum polarization can be maintained in steady state, or near steady state over time with online polarization. Additionally, when using on-line optical pumping, one can limit how often ^{3}He cell calibrations must be performed as the relative polarization can be monitored via NMR for possible drifts, and the cells ^{3}He pressure is fixed in the typically permanently sealed SEOP cell. Regardless, the smaller the total relaxation rate in a given ^{3}He cell, the greater the level of the achievable steady state ^{3}He polarisation will be, and thus provide the best possible neutron performance, thus good cell preparation is paramount.

In practice, the wall relaxation is an important process; it can dominate the ^{3}He spin relaxation rate or limit eventual performance if care is not taken. Collisions of ^{3}He atoms with ferromagnetic impurities at the cell surface as well as the diffusion of ^{3}He into the glass may cause relaxation because of the precession of the helium magnetic moment about the net local gradient field caused by the magnetic impurities. Additionally for the SEOP process, there are temperature-dependent wall relaxation processes limiting the maximum steady state polarization [8]. If this optical pumping related relaxation can be lowered it would further increase the absolute polarization performance of SEOP polarized cells. Exploring this parameter with our cells will be the focus of future work.

In the following discussion, we will describe the fabrication process of cells made from GE180 glass and our new gas filling system built to clean cells and fill them with highly pure ^{3}He and N_2. Then we provide a summary of the main relaxation characteristics and properties of the cells produced. One highlight of this work is that long wall relaxation time cells are now produced routinely in the lab.

2. ^{3}He relaxation processes in ^{3}He NSF

The ^{3}He polarization in the ^{3}He NSF cell decays to thermal equilibrium with the characteristic time constant T_1. The longitudinal relaxation rate $\Gamma_1 = T_1^{-1}$ results from three main mechanisms for the relaxation of spins:

$$\Gamma_1 = \frac{1}{T_1} = \frac{1}{T_1^{dipole}} + \frac{1}{T_1^{grad}} + \frac{1}{T_1^{wall}}. \tag{1}$$

Here $T_1^{dipole} = 1/807$ (hours) is the dipole-dipole relaxation time due to the interaction between colliding ^{3}He atoms [9]. The magnetic dipole coupling of atoms during a binary collision results in the loss of nuclear polarisation. This relaxation is the fundamental physical process, which ultimately limits the total ^{3}He spin relaxation rate. T_1^{grad} is the relaxation time due to the non-zero gradient of the holding magnetic field and given by:[10-13]

$$\frac{1}{T_1^{grad}} = \frac{7000}{P}\left(\left(\frac{gradB_x}{B}\right)^2 + \left(\frac{gradB_y}{B}\right)^2\right). \tag{2}$$

Where $gradB_x$ and $gradB_y$ are the orthogonal gradients of the static holding field B, and P is the ^{3}He pressure in bar. For example, with a magnetic field gradient $gradB/B \le 10^{-4}\,\mathrm{cm}^{-1}$ the T_1^{grad} is more than 14286h at 1 bar. This value will be used later on this paper as a negligible value for the measurement of the total relaxation time given by Eq.(1) for the practical purposes of this paper. The last term, T_1^{wall}, is the relaxation time due to the interactions of ^{3}He atoms with the cell walls [14-15]. This relaxation is presumably determined by the quality of the glass and possible impurities on its surface and the ^{3}He interactions with them. The glass itself can contain small magnetic impurities like magnetite (the most common impurity in glass), and impurities may be introduced in the cell manufacture and preparation. Then, the diffusion of helium into the glass and to the surface may cause relaxation because of the precession of the helium magnetic moment about the net local gradient field caused by the magnetic impurities [14]. The relaxation properties of ^{3}He on glass may be related to the permeability of the glass, for cells made of Pyrex (borosilicate) glass which has high ^{3}He permeability, such relaxation process can be as fast as 1 hour. Allumosilicate glasses are known to have lower ^{3}He permeability than quartz or Pyrex, that is perhaps why GE180, a boron free allumosilicate glass and the glass of choice for many neutron applications and our ^{3}He cells, gives good performance with respect to ^{3}He relaxation rates. Fortuitously, GE180 being boron-free, has a low neutron absorption is thus also suited to neutron spin filter applications. Discussions of ^{3}He interactions with glasses and coated surfaces are given in refs [16-18], although unfortunately some references did not study GE180, we assume it to have similar properties to the other allumosilicate but boron containing glasses Corning 1723 and 1720 listed in those works.

3. Cell fabrication and filling

We have completed a cell filling station, and cell filling procedure borrowing heavily from the experience and results of NIST [19] and ISIS colleagues [20]. The following sections describe our cell fabrication, preparation, and filling procedure as well as the gas ultra pure gas handling and filling station.

3.1. Cell fabrication and filling.

All our cells are blown from semi-custom large diameter ca. OD=24 mm GE180 glass tubing in the glass workshop of the Forschungszentrum Jülich GmbH. One starts from the glass tubing that is reblown to the desired geometry and annealed at ~785 °C for several hours to remove residual stress in the glass. Then cells undergo safety testing up to a typical pressure of 6 bar.

Each cell is cleaned before being attached to a glass manifold or "string" made from O.D.=12.7 mm Pyrex tubing (Figure1). The cleaning procedure involves several steps: (i) a rinse with soapy water, (ii) several rinses with distilled water and (iii) final rinse with methanol to remove water. The string also contains two ampoules of alkali metals (1g pre-scored ampoules Rb and K-Strem Chemicals 99.9+% purity) [21]. Two small bulbs serve as reserve volumes between the cell and the alkali metal ampoules and glass-encased ferrous slugs for breaking the ampoules while inside the sealed string. The emptied ampoules and slugs are removed or "pulled-off" from the string after intermediate distillation of the alkali-metal into the reserve volume.

The entire string, after attaching the cleaned cell to it and putting in the alkali/metal ampoules and ferrous slugs, is connected to the filling station by inserting the open end of the tube into a UHV Swagelock Ultra-Torr™ connection which serves as the adapter from the glass to the vacuum system of the filling station. Connected to this Ultra-Torr fitting is a short SS bellows which helps insure the glass isn't strained or broken during the baking/filling process, and also allows one to pick up and move the string as needed while working with the cell.

Before filling the cells, the system is leak tested with ^{4}He gas and baked for several days to achieve the residual pressure as low as 10^{-9} mbar. The cells are baked at 400°C and evacuated for at least four days to minimise the impurities on the surface of the glass. The cell string is also baked during this time but at a lower temperature of 200 °C, being careful not to overheat or burn the alkali-metal ampoules. After the pressure in the system reaches equilibrium, the valves between the vacuum system and the gas handling system are closed. The alkali-metal ampoules are then broken under vacuum using a magnet and the ferrous slug which is encased in glass to avoid contamination of the cell with ferromagnetic impurities. The slug is lifted with an external magnet and then dropped onto the tip of the break-seal ampoule with the aide of gravity to open it.

Next, the alkali metal is distilled first to a small bulb (Figure1) in the string using a 530 °C 1.6 kW hot air gun. After this first distillation, the emptied alkali metal ampoules and the portion of the string containing the ferrous slug are removed via sealing-off the connection between the ampoule and the reservoir bulb with a torch. Afterward, the alkali metals are distilled again, this time into the cell until an opaque coating is observed, first with the K and then with a smaller amount of Rb in the desired ratio.

Now the cell can be filled with the required amount of N_2 typically 100 mbar, and ^{3}He through the getters and a liquid nitrogen cold trap in the glass string to ensure a high gas purity. During the filling process the cell pressure is monitored using the high precision MKS Baratron pressure transducer (Figure2) with the accuracy of about 0.05%. For cells with final pressure over atmosphere, the cells are partially submerged in liquid nitrogen until the pressure is low enough (<1 bar) to seal it with the torch in air; this allows for a maximum total cell pressure of about 3.7 bar. Finally, the filled cells are sealed by a torch and pulled-off from the vacuum manifold. Pictures of some of our typical blown cells are shown in Figure 3.

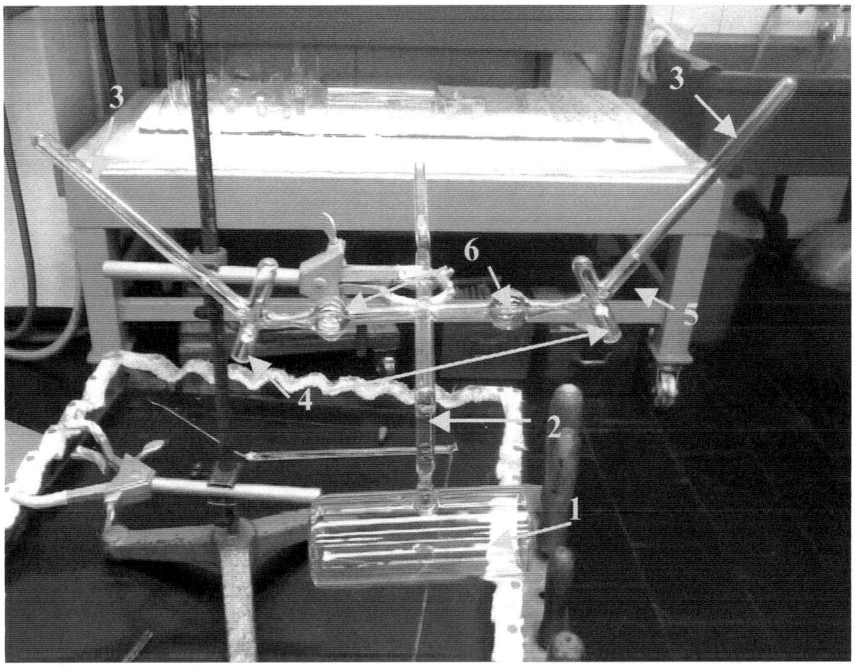

Figure 1. The 5 cm x 15 cm cell named Goldorak (1) attached to a Pyrex string (2), the slanted-vertical pull-off (3) containing alkali metal ampoules, Rb left and K right (4), and the ferrous slug (5) sealed within glass. The small bulbs (6) are the reserve volumes, between these bulbs and the ampoules there is a restriction used to pull-off the emptied ampoules and ferrous slugs before the second distillation of the alkali-metal into the cell.

3.2. Cell filling station. The cell filling station consists of three main sections: a gas handling system, a high vacuum system, and a gas recovery system as shown in (Figure 2). These three components are described in the following sub-sections.

A. Gas handling system. ^{3}He and N_2 gases are provided via 6mm gas tubing, all connections are metal-sealed mostly Swagelock compression fittings. Each gas line has a SAES PS2-CG50 getter and a 0.5 μm particle filter for purifying the gas. The flow-rate is controlled by a pressure reducer placed at the output of each bottle followed by a needle valve.

B. High vacuum system. The high vacuum is obtained using two pumps: a Pfeifer "Adixen" pump model P0385E4 (oil-free), which can pump the system from atmospheric pressure down to 10^{-3} mbar and also acts as the backing pump for a Pfeifer Hicube 80 Pro turbo molecular vacuum pump, which allows to achieve pressures as low as several 10^{-9} mbar.

C. Gas recovery system. A helium tight piston compressor from KNF Neuberger, model PM24721-286.13, which has been specially sealed and modified by the company for use as a ^{3}He compressor [22], is used to collect the gas remaining in the tubes of the gas handling system; the gas is then compressed into a steel gas bottle up to 12 bar. The ^{3}He is then separated from impurities using a Zeolite molecular sieve adsorbent at 77 °K followed by a cold head at 8 °K [23]. The gas recovery system is always kept under a high vacuum (~10^{-9} mbar) to ensure its clearness and that ^{3}He gas will not be contaminated by impurities that may later affect the ^{3}He optical pumping or cell performance.

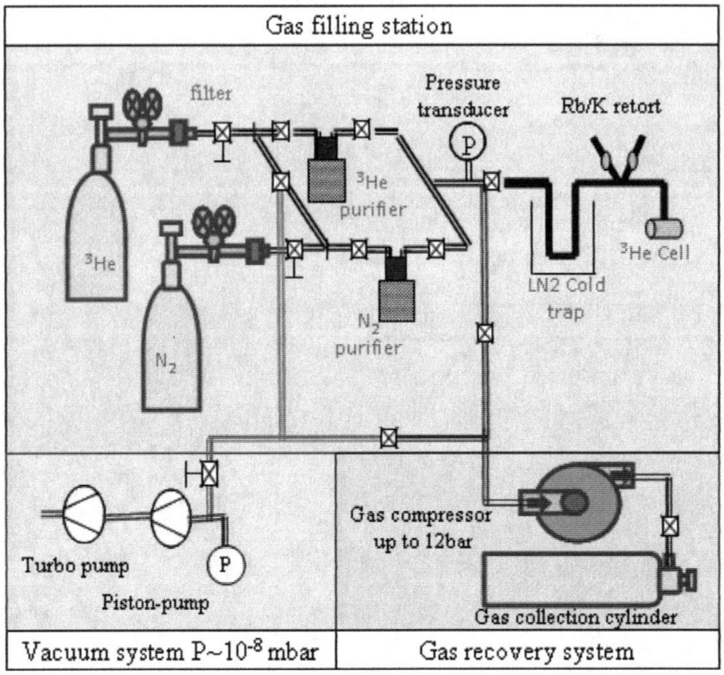

Figure 2. Schematic diagram of the cell preparation system, including the oil-free high vacuum system, the filling station and the recovery system.

4. Cell relaxation time measurement procedure and results.

The ^{3}He cells are polarized using the SEOP technique [1]. For this purpose a cell is placed in a long 2.2 m 6 coil set of 1.4 m diameter. This can provides a magnetic field of up to 18 G (10 G is normal used) at the centre with a designed field homogeneity $gradB / B \leq 10^{-4}\,\mathrm{cm}^{-1}$ in a large volume of about (1x0.5x0.5) m^3. The cell are heated, to temperatures up to 220°C, used to achieve a suitable alkali-metal density using a flowing air oven and are illuminated with circularly polarised light from high power, P>90 W, diode lasers which are spectrally narrowed by chirped volume Bragg gratings (VBG) to achieve a stable linewidth of about 50 GHz at the Rb D_1 optical transition at 794.7 nm (with one particular grating in fixed position, we obtained a linewidth of 35 GHz at 40A or 16.3W output with a linear increase in linewidth of about 1GHz per Ampere to 113 GHz at 120A or 92.5 W output).

The build-up and the decay of the ^{3}He polarisation are monitored using NMR free induction decay (FID): the amplitude of the FID signal is proportional to the ^{3}He polarisation [11, 24]. The measurements are carried out periodically, typically every hour. The exponential decay fit of the FID

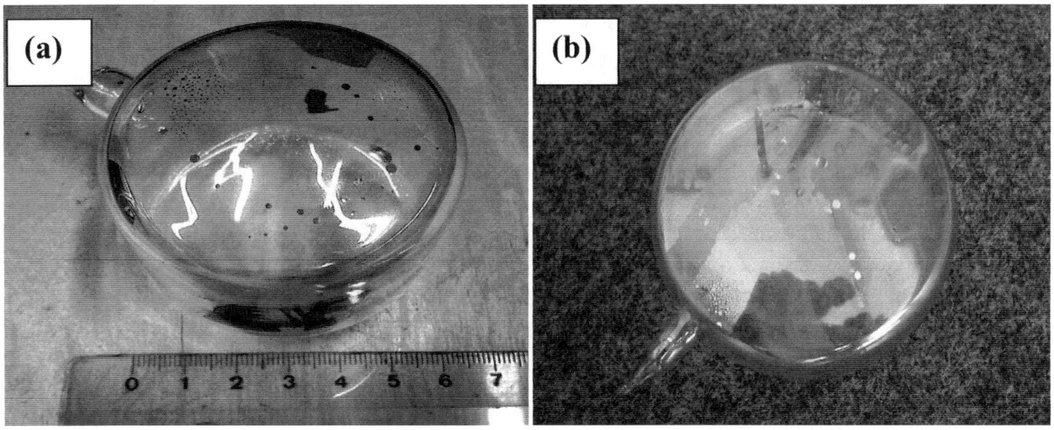

Figure 3. Examples of ^{3}He NSF cells produced using our new filling station: (a) Puck, diameter 6 cm, length 5 cm and (b) Asterix diameter of 12cm, length 5 cm (see Table 1).

amplitude time-dependence gives the value of T_1; for a long T_1 on the order of hundreds of hours, a reliable measurement typically takes several days (see Figure4.). We generally acquire T_1 decay data for several days until the uncertainty in the exponential fit to the data is under 10%.

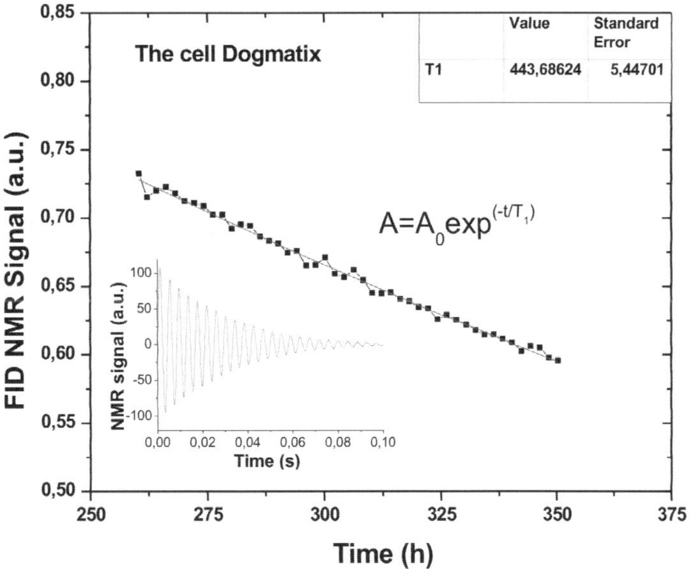

Figure 4. Example of the time dependence of the relative polarisation measured with FID NMR and the exponential fit allowing for the determination of the relaxation time T_1

Parameters of a number of ^{3}He cells made in our lab using the new filling station are presented in Table 2. As one can see, the measured T_1 is regularly approaching the dipole limit for many of the cells, however we must fully characterize the ^{3}He pressures of the sealed cells, this is readily done by measuring the $1/v$ dependent neutron absorption of the cells [25]. Initial test have yielded pressure values 5-10% lower than the fill pressure for Asterix and Willy, further work must be done to characterize the cells pressures in this way and eliminate possible systematic errors. On an interesting side note, even with corrections for the actual ^{3}He pressure, some of these cells may exhibit measured ^{3}He T_1 lifetimes closer to the dipole-dipole limit of than those cited in [26, 27], thus more accurate characterization of them could help to place new experimental limits on the magnitude of axion monopole-dipole coupling strengths as described in those works [26, 27]. However we must also experimentally verify the field gradients in our field coils because from eq. 1, a value at the upper limit of $1\text{x}10^{-4}$ cm^{-1} could decrease our measured T_1 lifetimes by on the order of 5% which should be taken into account for such an analysis.

We quickly note, the first two cells in the table, Maja and Puck have been filled in the beginning of the operation of our filling system while it was still under the cleaning process. The cell Homer, because of the big size and the complexity of the cell geometry (toroid or "doughnut" shape), required an unusually long time period for the glass blowing work and more intrusion with the torch on the glass are is required; therefore it is reasonable to assume more impurities could be embedded into the glass during this process. Nevertheless the obtained relaxation time is very satisfactory for practical purposes.

Cell	Size D x L (cm x cm)	Fill pressure (bar)	T_1 (hours)	T_1^{dipole} (hours)
Maja	5 cm x 5 cm	0.93	550	860
Puck	6 cm x 5 cm	1	500	800
Kurt	6 cm x 5 cm	2.3	350	350
Willy	6 cm x 5 cm	2.3	350	350
Asterix	12 cm x 5 cm	2.7	280	295
Obelix	12 cm x 5 cm	2.1	330	380
Idefix	12 cm x 8 cm	1.7	380	470
Dogmatix	12 cm x 8 cm	1.8	440	445
Homer	22 cm x 8 cm	0.5	1070*	1600
Thekla	8 cm x 5 cm	2.3	290	350
Goldorack	5 cm x 15 cm	2.7	230	295

Table 1. Parameters of different ^{3}He glass cells produced at the JCNS. D and L are the inner diameter and length respectively. The fill pressure is the deduced pressure from the Baratron pressure sensor during cell filling. Many cell pressures have not yet been calibrated via. neutron absorption measurements however initial measurements on Willy and Asterix implied pressures 5-10% lower than the fill pressure. More conclusive measurements will be done at a later time. The uncertainty in the T_1 measurements is ± 10 hours. *If one assumes a gradient of $1\text{x}10^{-4}$ cm^{-1} in our field coils, for Homer, because of the 0.5 bar pressure the T_1 in a theoretical gradient-less field could be up to 200 hours longer.

5. Development of wide-angle GE180 cells suitable for SEOP and MEOP

The use of sealed (or fully blown) ^{3}He cells made of GE180 glass [13] could vastly simplify the equipment required for wide-angle polarization analysis as this would allow the direct polarization of the cell using the SEOP method. Furthermore, fully blown GE180 glass cells may indeed have practical advantages such as pressure resistance and durability compared to other proposed routes such as to form cells from bonded or glued together flat and curved plates of crystal silicon, quartz or even GE180 [28, 29]. Presumably cells made of GE180 could be used as sealed cells to be polarized directly with a SEOP polarizer system as in [29] or to be used as so-called valved cells [28, 30] that are filled with pre-polarized ^{3}He gas from a large scale ^{3}He polarizer employing either the MEOP or SEOP techniques [1-4]. Most likely, silicon or quartz cells are only suitable to be used as valved cells referencing the experience of the NIST, ILL, and ISIS groups which have worked on such cells [29, 30, 28] although we have not pursued these techniques ourselves.

We have successfully manufactured a number of fully blown GE180 wide-angle ^{3}He spin-filter cells. These unique cells have been produced in the glass workshop of the Forschungszentrum Jülich. GE180 is again chosen because of its neutron compatibility and because it is the best experimentally known material for providing long T_1 relaxation times of the ^{3}He gas and maximum polarization using the SEOP method.

A very first all blown "C" or "banana" shaped GE180 cell is shown in the Figure 5a. In Figure 5b a doughnut-shaped blown GE180 cell with a trapezoidal cross section more suitable for a small sample and large area-detector geometry is shown, the trapezoidal cross section shape is also useful to minimise the ^{3}He volume of the cell without affecting the solid angle covered. The C-shaped cells are made by sealing off two points of the doughnut cells. In Figure 6a the picture of a very large (external diameter of 22 cm) doughnut-shaped cell called Homer is shown. This cell was prepared with a Rb and K mixture for hybrid pumping [8] with the procedure described in the previous section. Being intended only for testing of ^{3}He relaxation times of a new Magic-PASTIS coil system [31], it contains 0.5 bar of ^{3}He: the lower ^{3}He pressure makes the cell rather sensitive to the magnetic field gradient relaxation, T_1^{grad} (Eq. (1)), and is thus a good probe for the optimization of the performance of the Magic-PASTIS magnetic cavity. The cell Homer was measured to have T_1 =1074 hours in what we presume to be an "ideal" reference magnetic holding field (see (Eq. (2)). In Figure 6b the very first doughnut-reshaped cell optimized for a typical area detector is shown: its cross section has been flattened starting from cell similar to the one in Figure 6a. To date doughnut and C-shaped cells up to 24 cm in diameter have been constructed. This size is currently limited by the volume of glass that can be maintained at the appropriate temperatures for forming it with a large propane/oxygen burner. We believe even larger cells can be produced by switching to a Hydrogen/oxygen burner that can produce higher temperatures.

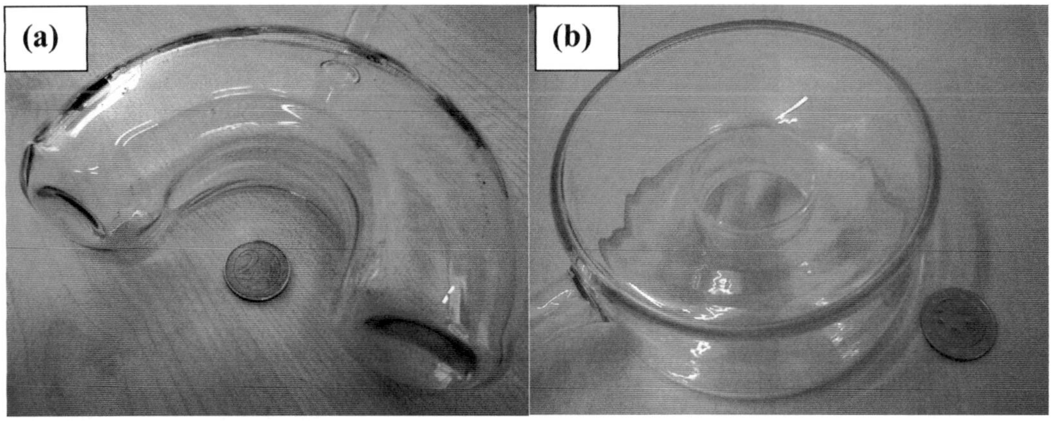

Figure 5. (a) A first C-shaped cell blank with a 6 cm diameter cross-section and 16.5 cm O.D. and (b) a doughnut cell blank of 12 cm O.D. and 6 cm high with more idealized trapezoidal cross-section geometry for the scattered beam.

Figure 6. (a) The first filled doughnut shaped cell called Homer, 22 cm O.D. and 8 cm high with a circular or slightly D-shaped cross section for testing the Magic-PASTIS setup [20]. (b) A very large 22cm O.D. by 9 cm high doughnut cell and 9 cm I.D. inner hole with a trapezoidal cross section for the scattered neutron beam.

6. Conclusions.

The apparatus built to clean and fill SEOP cells operates routinely and we have produced several cells with lifetimes in the excess of several hundreds of hours. This now includes larger diameter cells, up to 12 cm I.D. that can be filled up to the gas pressure of 3.7 bar by submerging them into liquid N_2 while sealing them off. For some cells the obtained lifetime is about 800 hour per inverse bar, which means that we have approached the fundamental limit imposed by the dipole-dipole interaction of ^{3}He atoms. Currently all cells we fill use a K/Rb alkali-metal mixture allowing for higher optical pumping rates and is absolutely necessary to efficiently achieve high ^{3}He polarization in large cells.

7. Acknowledgments.

The authors would like to thank G. D'Orsaneo and A. Schwaitzer (Foschungszentrum Jülich glass workshop) and J. Anderson and A. Kirchhoff (NIST glass workshop) for their help and advice and brainstorming concerning cell fabrication. We are also grateful to T. R. Gentile (NIST Physics Lab), W.C. Chen (NCNR), for their help and useful discussions.

The authors acknowledge funding by BMBF through the collaborative project 05E10CJ1.This research project has been also supported by the European Commission under the 7th Framework Programme thorough the 'Research Infrastructures' action of the 'capacities' Programme NMI3-II Grant number 283883.

References

[1] T.Walker and W. Happer, Spin-exchange optical pumping of noble-gas nuclei. Rev Mod Phys, vol. 69, no. 2, pp. 629–42, 1997.
[2] F. D. Colegrove, L. D. Schearer, and G. K.Walters, "Polarisation of helium-3 gas by optical pumping," Physical Review, vol. 132, pp. 2561–2572, 1963.
[3] J. Brossel, "Optical pumping in weak discharges." Fundamental Appl LASER Phys, pp. 769–90, 1971.
[4] W. Happer, E. Miron, D. Schreiber, and W. A. van Wijngaarden X. Zeng, "Polarization of the nuclear spins of noble-gas atoms by spin exchange with optically pumped alkali-metal atoms," Physical Review A, vol. 29, no. 5, pp. 3092–3110, 1984
[5] E. Babcock, S. Mattauch, A. Ioffe; High level of ^{3}He polarization maintained in an on-beam ^{3}He spin filter using SEOP, Original Research Article:Nuclear Instruments and Methods in Physics Research

Section A: Volume 625, Issue 1, 1 January 2011, Pages 43-46

[6] S.R. Parnell, E.Babcock, K.Nunighoff, M.W.A.Skoda, S.Boag , S.Masalovich, W.C.Chen, R.Georgii, J.M.Wild, C.D.Frost; Study of spin-exchange optically pumped 3He cells with high polarisation and long lifetimes: Nuclear Instruments and Methods in Physics Research A 598 (2009) 774–778

[7] M. Batz, S. Baeßler, W. Heil, E.W. Otten, D. Rudersdorf, J.Schmiedeskamp, Y. Sobolev,and M. Wolf: ^{3}He Spin Filter for Neutrons; Journal of Research of the National Institute of Standards and echnology J. Res. Natl. Inst. Stand. Technol. 110, 293-298 (2005)

[8] E. Babcock, B. Chann, T. G. Walker, W. C. Chen, and T. R. Gentile; Limits to the Polarization for Spin-Exchange Optical Pumping of He3, Phys. Rev. Lett. 96, 083003 – Published 3 March 2006

[9] N. R. Newbury, A. S. Barton, G. D. Cates, W. Happer, and H. Middleton. Gaseous 3He-3He magnetic dipolar spin relaxation. Phys. Rev. A, 48(6), 1993. doi:10.1103

[10] G. D. Cates, S. R. Schaefer, and W. Happer. Relaxation of spins due to field inhomogeneities in gaseous samples at low magnetic fields and low pressure. Phys. Rev. A, 37,1988.

[11] G. D. Cates, D. J. White, T. Chien, S. R. Schaefer, and W. Happer. Spin relaxation due to inhomogeneous static and oscillating magnetic fields. Phys. Rev. A, 38, 1988.

[12] R. Jacob, J. Teter, B. Saam, W. Chen, and T. Gentile, "Low-field orientation dependence of 3He relaxation in spin-exchange cells," Physical Review A,vol. 69, pp. 021401–1–4, 2004.

[13] J. Schmiedeskamp, W. Heil, E.W. Otten, R.K. Kremer, A. Simon, J. Zimmer, "Paramagnetic relaxation of spin polarized He-3 at bare glass surfaces Part I," European Physical Journal D, vol. 38, pp. 427-38, 2006.

[14] R. E. Jacob, S. Morgan, B. Saam, and J. C. Leawoods, "Wall relaxation of 3He in spin-exchange cells," Physical Review Letters, vol. 87, p. 143004, October 2001.

[15] W. Heil, H. Humblot, E. Otten, M. Schafer, R. Sarkau, and M. Leduc, "Very long nuclear relaxation times of spin polarized helium-3 in metal coated cells," Physics Letters A, vol. 201, pp. 337–343, 1995.

[16] W. A. Fitzsimmons, L.L. Tankerslez, and G.K. Walters, "Nature of surface-induced nuclear spi-relaxation of gaseous 3He," Physical Review, vol. 179, pp. 156-65, 1969

[17] A. Deninger, W. Heil, E.W. Otten, M. Wolf, R.K. Kremer, A. Simon, "Paramagnetic relaxation of spin polarized He-3 at coated glass walls Part II," European Physical Journal D, vol. 38, pp. 439-43, 2006

[18] J. Schmiedeskamp, H.J. Elmers, W. Heil, E.W. Otten, Y. Sobolev, W. Kilian, H. Rinneberg, T. Sander-Thommes, F. Seifert, J. Zimmer, "Relaxation of spin polarized He-3 by magnetized ferromagnetic contaminants Part III," European Physical Journal D, vol. 38, pp. 445-54, 2006

[19] W. C. Chen, T. R. Gentile, C. B. Fu, S. Watson, G. L. Jones, J. W. McIver and D. R. Rich, Polarized ^{3}He cell development and application at NIST Journal of Physics: Conference Series 294 (2011) 12003

[20] Stephen Boag: Spin Exchange Optical Pumping of 3He for Neutron Spin Filters, PhD Thesis, University of Nottingham (2008)

[21] Strem chemical- http://www.strem.com/

[22] KNF Neuberger- http://www.knf.de/

[23] Z. Salhi, T. Großmann, M. Gueldner,W. Heil,S. Karpuk, E. W. Otten,D. Rudersdorf, R. Surkau, U. Wolf, Recycling of 3He from lung magnetic resonance imaging. Magnetic Resonance in Medicine Volume 67, Issue 6, pages 1758–1763, June 2012

[24] A. Abragam, Principles of Nuclear Magnetism. London: Oxford University Press, 1966.

[25] T. E. Chupp, J. D. Bowman, W. Chen, K. P. Coulter, M. Dabaghyan,T. Gentile, R. C. Gillis, G. L. Jones, M. Kandes, B. Lauss, S. I. Penttila, D. Rich, M. Sharma, T. B. Smith, A large area polarized 3He neutron spin filter, Nucl. Instr. Meth. A 574-500 (2007).

[26] Yu. N. Pokotilovski, "Limits on short-range spin-dependent forces from spin relaxation," Vol. 686, pp 114-7, 2010.

[27] A.K. Petukhov, G. Pignol, and R. Golub, "Comments on 'Limits on possible new nucleon monopole-dipole interactions from the spin relaxation of 3He gas,'" Physical Review D, vol. 84, p 058501, 2011.

[28] Beecham, C. J.; Boag, S.; Frost, C. D.; et al. He-3 polarization for ISIS TS2 phase I instruments PHYSICA B- Volume: 406 Issue: 12 Pages: 2429-2432 2011

[29] Q. Ye, T.R. Gentile, J. Anderson, C. Broholm, W.C. Chen, Z. DeLand, R.W. Erwin, C.B. Fu, J. Fuller, A. Kirchhoff, J.A. Rodriguez-Rivera, V. Thampy, T.G. Walker, S. Watson, Wide Angle Polarization Analysis with Neutron Spin Filters, Physics Procedia, Volume 42, 2013, Pages 206-212, ISSN 1875-

[30] A.K. Petoukhov, K.H. Andersen, D. Jullien, E. Babcock, J. Chastagnier, R. Chung, H. Humblot, E. Lelievre-Berna, F. Tasset, F. Radu, M. Wolff, H. Zabel, Recent advances in polarised ^{3}He spin filters at the ILL, Physica B 385–386 (2006) 1146–1148
[31] Z. Salhi, E. Babcock, A. Ioffe: ArXiv (2012) aeXiv:1201.5208

Development of portable polarized ^{3}He neutron spin filter and its application to magnetic field imaging at J-PARC

K Sakai[1], T Oku[1], H Hayashida[1], H Kira[2], T Shinohara[1], K Oikawa[1], M Harada[1], K Kakurai[1], K Aizawa[1], M Arai[1], Y Sakaguchi[2] , J Suzuki[2] , T Ino[3] and K Ohoyama[4]

[1]JAEA, Tokai, Ibaraki 319-1195, Japan
[2]Comprehensive Research Organization for Science and Society (CROSS), Tokai, Ibaraki 319-1106, Japan
[3]KEK, Tsukuba, Ibaraki 305-0801, Japan
[4]Institute for Materials Research, Tohoku University, Sendai, Miyagi, 980-8577, Japan

E-mail: kenji.sakai@j-parc.jp

Abstract. In order to expand the measurable neutron energy range up to the energy of epithermal neutrons in polarized neutron experiments, a portable polarized ^{3}He neutron spin filter (NSF) was developed in the Materials and Life science experimental Facility (MLF) at J-PARC. After the generation of ^{3}He polarization through a spin exchange optical pumping method, the NSF was used as a neutron spin polarizer and flipper in a beam line. We constructed a simple neutron spin analysis apparatus with two ^{3}He NSFs and made a preliminary attempt to visualize magnetic fields generated by a coil. The results represented that our apparatus could function for neutron wavelengths between approximately 0.5–5 Å through the use of a neutron time-of-flight method.

1. Introduction

In polarized neutron experiments, researchers have been interested in extending the measurable neutron energy range up to the energy of epithermal neutrons. For example, such an extension in the energy range is useful to investigate domain structures in ferromagnetic materials by measuring the spin rotation and depolarization of polarized neutrons [1, 2]. In addition, the extension would enable studies on verification of a fundamental symmetry and a neutron optical theorem by measuring the correlation terms of a neutron spin around the resonance peaks of neutron nuclear reactions [3, 4, 5]. In order to realize these studies, the development of a neutron spin polarizer and flipper that operates at energies up to the energy of the epithermal neutrons is crucial. A polarized ^{3}He neutron spin filter (NSF) can polarize cold to epithermal neutrons by optimizing the quantity of ^{3}He gas because such NSFs have a large neutron absorption cross section with a strong spin selectivity owing to the presence of a ^{3}He+n (J^π=0$^+$) resonance state at a neutron energy E_n=–518 keV. This NSF is superior as a neutron spin analyzer because the background neutrons scattered from the NSF have little influence. In addition, it can operate as a neutron spin flipper functioning independent of the neutron energy region by controlling the ^{3}He nuclear spins through a nuclear magnetic resonance (NMR) method [4, 6].

A recent improvement in the spin exchange optical pumping (SEOP) technique resulting from the development of high-performance laser systems has enabled the production of ^{3}He gas with a high nuclear spin polarization at pressures of 1–10 atm [7, 8]. This improvement was also very useful to miniaturize the ^{3}He NSF for installing it in a narrow space such as a neutron beam line [9]. A ^{3}He NSF based on an in-situ SEOP system was developed in the Materials and Life science experimental Facility (MLF) at J-PARC [10, 11]. It is capable of sustaining a high stable ^{3}He polarization for a long period of time. However, it takes more than half a day to generate sufficient ^{3}He polarization for the experiments. From the viewpoint of beam availability, we developed a portable ^{3}He NSF, which could be installed in the beam-line after the generation of ^{3}He polarization at a different location. This paper reports the development of this portable ^{3}He NSF and its application to magnetic field imaging at J-PARC.

2. Development of Portable ^{3}He NSF at J-PARC

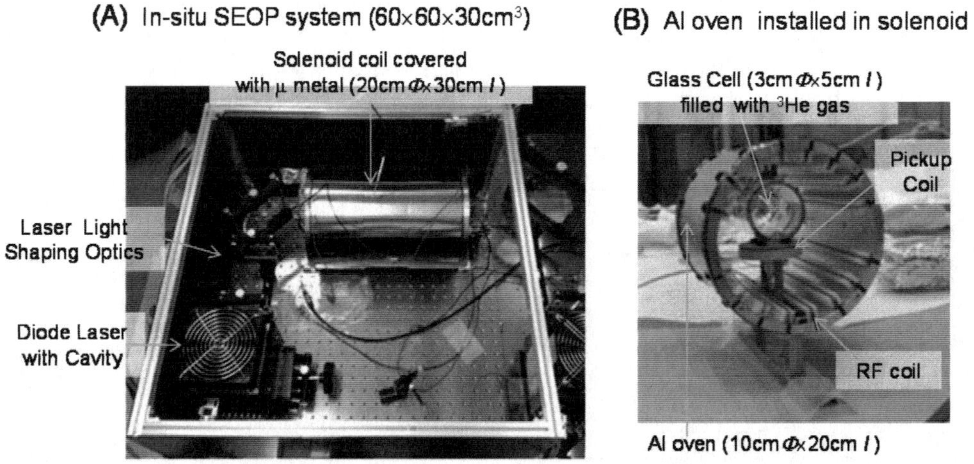

Figure 1. Polarized ^{3}He NSF based on the in-situ SEOP system

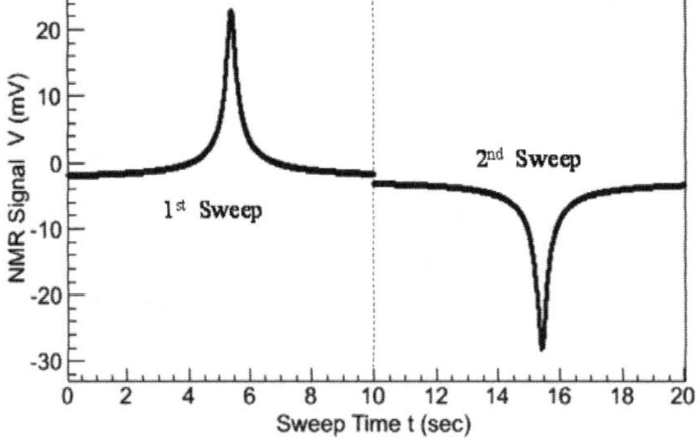

Figure 2. Typical NMR signal V obtained by sweeping the solenoid field twice using a saw wave function

Figures 1 (A) and (B) show the polarized ^{3}He NSF based on an in-situ SEOP system [5, 10, 11]. As shown in figure 1 (A), the ^{3}He NSF consists of a diode laser system (TOYOKO KAGAKU LTDFR-SPL-30) driven at an output power of 40 W with a cavity for shortening the wavelength of a laser light [8, 10], laser light shaping optics, and a solenoid coil of 20 cm in diameter and 30 cm in length for sustaining the ^{3}He polarization. As shown in figure 1 (B), ^{3}He gas along with N_2 gas and a small amount of Rb at 1–3 atm is contained in a cylindrical glass cell of approximately 3 cm in diameter and 5 cm in length [12]. The electron spins of Rb are polarized using the resonant absorption of 795-nm circularly polarized laser light, and the ^{3}He nuclear spins are polarized through spin exchange with the polarized electrons of Rb. The cell is mounted in an oven of 10 cm in diameter and 20 cm in length. The oven consists of aluminum (Al) and is placed at the center of the solenoid. The temperature in the oven is maintained at approximately 450 Kelvin for vaporizing the Rb. These components of the SEOP system are set in a box of 60 × 60 × 30 cm^3 size for shielding them from the laser light. Figure 1 (A) shows the portable solenoid with a magnetic field of 1–2 mTesla generated by a portable battery circuit that is connected to the solenoid along with a DC power supply. It is installed in the beam line following the generation of ^{3}He polarization through the SEOP method.

This ^{3}He NSF is expected to operate for approximately one day according to a typical relaxation time of ~100 h for ^{3}He polarization. In addition, it continuously operates over an extended period of time when it is placed in the same SEOP system installed on the beam line. The ^{3}He NSF is assumed as a convenient neutron polarizer that can be operated immediately after its installation in the beam line without any neutron beam adjustments. However, the ^{3}He polarization should be monitored frequently. As shown in figure 1 (B), a pickup coil and a radio frequency (RF) coil are mounted around the ^{3}He cell for measuring the NMR signal V. In order to apply the ^{3}He NSF as the neutron spin flipper, an adiabatic fast passage (AFP) NMR system is constructed. Such an NMR system is suitable for flipping the ^{3}He nuclear spins in the opposite direction of the solenoid field while maintaining a high ^{3}He polarization. The RF coil shown in figure 1 (B) is developed to operate in the Al oven [13]. Figure 2 shows the V characteristics observed by sweeping the solenoid field twice using a saw wave function by applying a radio frequency of 50 kHz. It indicates that the ^{3}He nuclear spins were flipped because the sign of V at the first sweep is reversed at the second sweep. Although the NMR measurements are useful to monitor the relative time fluctuation of the ^{3}He polarization, it is not very easy to determine the value of ^{3}He polarization directly from the V values. However, we obtained a correlation between the ^{3}He polarization and the V values of the pickup coil by measuring the neutron transmittances of the ^{3}He filter and the NMR signals simultaneously, and this facilitated a continuous monitoring of the ^{3}He polarization values through the NMR signals [5].

3. Application to magnetic field imaging using two portable ^{3}He NSFs

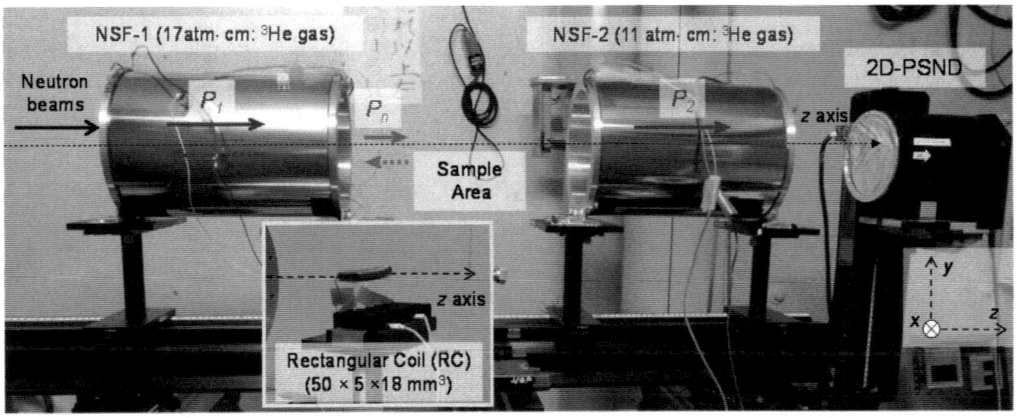

Figure 3. Experimental set up of neutron spin analysis apparatus at BL10

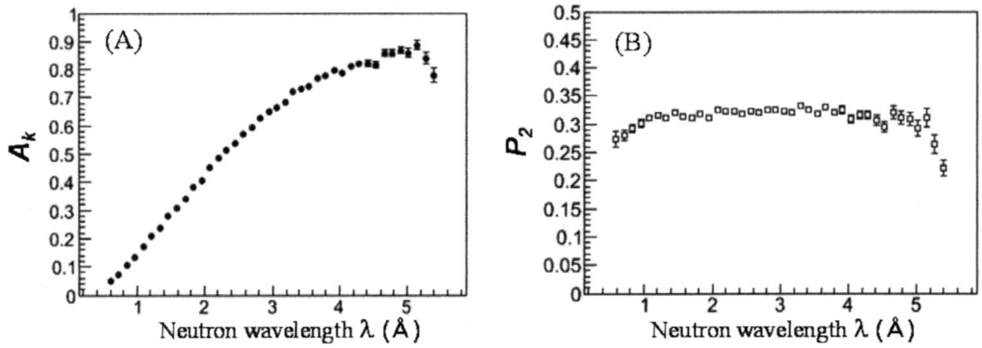

Figure 4. Plots of (A) A_k and (B) P_2 from equation. (2) as functions of λ measured from N-TOF spectrums

A simple neutron spin analysis apparatus with two portable ^{3}He NSFs was constructed at neutron beam line 10 (BL10) of the MLF [14], and a preliminary test was conducted on its feasibility. Figure 3 shows the experimental setup at BL10. Pulsed neutron beams from a spallation neutron source were transmitted along the *z*-axis through the first ^{3}He NSF (NSF-1) used for polarizing neutrons, which were then passed through the sample area to the second ^{3}He NSF (NSF-2) used for analyzing neutron spins. The output neutrons were counted using a two-dimensional position-sensitive neutron detector (2D-PSND) comprising a neutron scintillator and a photomultiplier tube. This apparatus was used to simultaneously analyze neutron spins in cold to epithermal neutrons at a high energy resolution through a neutron time-of-flight (N-TOF) method. A neutron spin polarization P_n was sustained around the sample area along the *z*-axis by the leakage fields from the solenoids of both NSF-1 and NSF-2. The ^{3}He cell containing the ^{3}He gas of 17 atm·cm was placed in NSF-1 and that of 11 atm·cm was placed in NSF-2. The ^{3}He nuclear spin of NSF-1 was flipped in the same and opposite direction of the field along to the *z*-axis through the AFP-NMR method. The relative values of ^{3}He polarizations P_1 and P_2 of NSF-1 and NSF-2 were monitored by observing the NMR signals V_1 and V_2. In addition, a correlation between P_1 and V_1 of the pickup coil in NSF-1 was observed previously.

The neutron polarization P_n of the transmitted beams after they were passed through NSF-1 is expressed as

$$P_n = \tanh(N_1 \sigma l_1 P_1) \qquad (1),$$

where N_1 and l_1 denote the number density and the thickness of ^{3}He gas in NSF-1 and σ denotes a neutron absorption cross section of ^{3}He nucleus. P_1 was deduced from the V_1 values monitored during the experiments. On the other hand, the neutron spin analyzing ability of the apparatus can be evaluated from the neutron spin asymmetry A_k, which is defined as

$$A_k = \frac{T_+ - T_-}{T_+ + T_-} = P_n \tanh(N_2 \sigma l_2 P_2) \qquad (2),$$

where N_2 and l_2 denote the number density and the thickness of ^{3}He gas in NSF-2 and T_+ (T_-) denotes the neutron transmittance on NSF-2 with its neutron spin ***s*** parallel (antiparallel) to the ^{3}He nuclear spin of NSF-2. Figure 4 (A) shows the values of A_k as a function of neutron wavelength λ with the measurement of T_+ and T_-. For evaluating the feasibility of our apparatus, we attempted to derive P_2 by substituting the values of A_k in figure 4 (A) and the $P_1 \approx 0.45$ derived from the V_1 in Eq. (1) and (2). Figure 4 (B) shows P_2 as a function of λ measured from the N-TOF spectrum. The P_2 values were expected to remain constant, independent of the neutron wavelength. In fact, the P_2 values in figure 4

(B) are nearly constant for λ between approximately 1 and 5 Å, although they decrease rapidly for $\lambda >$ 5 Å owing to the influence of the background, which decreases neutron transmittance. In addition, although the values presented in figure 4 (B) are lower for λ between 0.5 and 1 Å owing to the influence of the small A_k and the low detection efficiency of the 2D-PSND, they could be sufficiently distinguished through statistical errors. This was possible because the high-intensity pulsed neutron source at J-PARC has a high neutron count rate for a small λ region. Figure 4 represents that the neutron spin analysis apparatus at J-PARC could function for λ = 0.5–5 Å through the use of the N-TOF method.

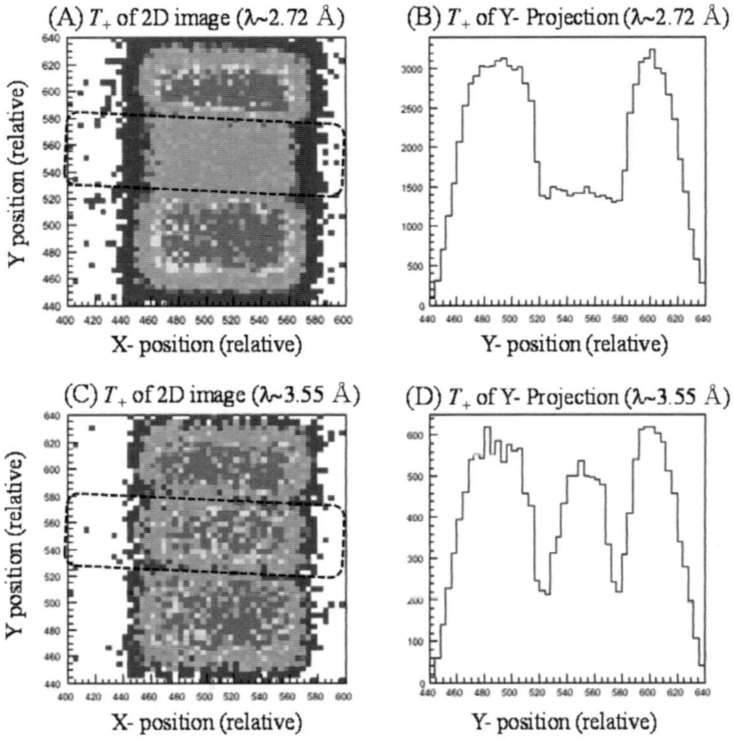

Figure 5. (A) and (C) show the 2D neutron images of T_+ through the RC and (B) and (D) show their corresponding y-projection components for λ = 2.72±0.27 and 3.55±0.27 Å, respectively.

An attempt was made to image magnetic fields using the apparatus. As shown in figure 3, a rectangular coil (RC) of 50 × 5 × 18 mm^3 size was mounted in the sample area in order to apply its magnetic field $\boldsymbol{H}_R$ in the x-direction. The neutron spin asymmetry A_k' occurring owing to the presence of the RC is defined as

$$A'_k = \frac{T_+ - T_-}{T_+ + T_-} = P_n \tanh(N_2 \sigma l_2 P_2) \sin(\gamma_n H_R t) \qquad (3),$$

where γ_n and H_R denote the neutron gyromagnetic ratio and the absolute value of $\boldsymbol{H}_R$, respectively. The variable t denotes the neutron passage time along the z-axis through the RC of length l_R and is proportional to λ according to the relation of $t = m\lambda l_R h^{-1}$, where m and h denote the neutron mass and the Plank's constant, respectively. In equation (3), the term $\gamma_n H_R t$ corresponds to the neutron spin rotation angle θ around the x-axis, also known as the Larmor precession angle. This spin rotation is caused by the presence of $\boldsymbol{H}_R$. T_+ (T_-) in equation (3) denotes the neutron transmittance with the neutron spin parallel (antiparallel) to the ^{3}He nuclear spin of NSF-1. Figures 5 (A) and (C) show the neutron 2D images measured around a beam cross section of about 10 × 15 mm^2. Figures 5 (B) and (D) show the *y*-projection components corresponding to T_+ in figures 5 (A) and (C), respectively. The areas enclosed by the dashed lines in figures 5 (A) and (C) represent the inside of the RC where $\boldsymbol{H}_R$ is applied along the *x*-axis. The difference between the T_+ values around the RC in figures 5 (B) and (D) corresponds to the difference between the spin rotation angles θ for $\lambda \approx 2.7$ and 3.6 Å. Figures 6 (A), (B), and (C) present the values of T_+, T_-, and A_k' as functions of λ measured around the inside of the RC, where T_+ and T_- are normalized by the N-TOF spectrum in the absence of the RC. In figure 6 (C), the obtained values of A_k' form a periodic curve as predicted in equation (3). The peak values of A_k' around $\lambda \approx 1.8$ and 3.6 Å correspond to the spin rotation angles $\theta \approx 2\pi$ and 4π radian, and it derives that neutron spins rotate one time whenever λ increase by 1.8 Å. By substituting $\theta \approx 2\pi$ and $\lambda \approx 1.8$ Å in the relation $\theta = \gamma_n H_R m\lambda\, l_R h^{-1}$, the value of $H_R \approx 4.2$ mTesla was obtained, which is consistent with that directly measured by a magnetic sensor.

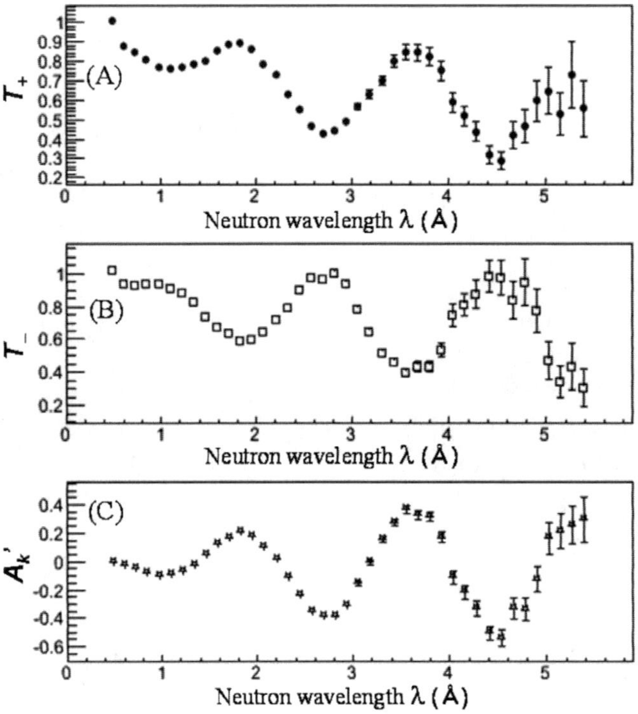

Figure 6. (A), (B) and (C) represent the T_+, T_- and A_k' around the inside of the RC as functions of λ

4. Summary

In order to expand the measurable neutron energy range up to the energy of the epithermal neutrons in polarized neutron experiments, we developed a portable ^{3}He NSF on the basis of the SEOP and AFP-NMR methods. Further, a neutron spin analysis apparatus with two portable ^{3}He NSFs was constructed on BL10 in MLF at J-PARC, and its feasibility was tested. An attempt was made to image the magnetic fields generated by an RC. The preliminary result on the feasibility of the apparatus represents that it could function for λ = 0.5–5 Å through the use of the high-intensity pulsed neutron source beams at J-PARC. Although a ^{3}He polarization of > 0.6 is usually generated in the ^{3}He cell by using our SEOP system [15], the P_1 and P_2 values of NSF-1 and NSF-2, respectively, could not be maintained at such high levels in the preliminary experiment. After the generation of ^{3}He polarization in NSF-1 and NSF-2 in the adjusting room housing the SEOP system, the portable devices were mounted on the narrow space in BL10, which is surrounded by radiation-shielding blocks. In the experiments, the polarizations of the NSFs were sustained within the relaxation rate of ^{3}He until they were mounted in BL10, where an unexpected ^{3}He depolarization occurred. In addition, the depolarization depended on the positions of the NSFs along the beam line. Such a depolarization may have been caused by the inhomogeneity of the magnetic fields around the beam line, which in turn may have been caused by the materials that were used as versatile tools placed around the beam line. The depolarization may affect the field condition around the beam line. After the experiments were conducted, only NSF-1 was placed along the beam line in order to investigate the cause of depolarization by measuring the NMR signals. It was confirmed that the depolarization decreased when many materials as possible were removed from around the beam line. However, the reduction in depolarization was not sufficient in comparison with the relaxation rate. Further planning to improve the magnetic shield of the neutron spin analysis apparatus is underway. If the P_1 and P_2 values are maintained at > 0.6, it is possible to expand the measurable λ range because of an increase in the neutron spin analysis ability that can be evaluated from A_k around a small λ region as well as an increase in the neutron transmittance around a large λ region. Further improvements to the apparatus are underway in order to realize its practical application.

Acknowledgments
We are grateful to the staff of MLF at J-PARC for their help and useful suggestions. This research was partly supported by the Grants-in-Aid for Scientific Research (No.22604005) and the Quantum Beam Fundamental Development Program, MEXT.

References

[1] Alfimenkov V P *et al,* 1997 *J. Appl. Phys.* **81** (3) 1358
[2] Shinohara T *et al,* 2011 *Nuclear Instr. and methods* A **651** 121
[3] Bunakov V E *et al,* 1998 *Phys. Lett.* B **429** 7
[4] Haseyama T *et al,* 2002 *Phys. Lett.* B **534**/1-4 39
[5] Sakai K *et al,* 2012 *Journal of Physics: Conf. Ser.* **340** 012037
[6] Chen W *et al,* "Enhancing polarized neutron measurement capabilities using ^{3}He spin filters" in these proceedings
[7] Babcock E *et al,* 2005 *Applied Optics* **44**(5) 3098
[8] Schnitzler C *et al,* 2007 *In Proc. SPIE* **6456** 645612
[9] Ino T *et al,* 2009 *Physica* B **404** 2667
[10] Oku T *et al,* "Developments and application of an in-situ SEOP ^{3}He neutron spin filter at J-PARC/MLF" in these proceedings
[11] Kira H *et al,* 2011 *Physica* B **406** 2433
[12] Ino T *et al,* 2005 *Physica* B **356** 109
[13] Ino T *et al,* 2012 *Journal of Physics: Conf. Ser.* **340** 012006
[14] Oikawa K *et al,* 2008 *Nuclear Instr. and methods* A **589** 310
[15] Hayashida H *et al,* "Developments and demonstration of in-situ SEOP ^{3}He filter system for neutron spin analyzer on polarized neutron reflectmeter SHARAKU at J-PARC" in these proceedings

Recent polarizing supermirror projects at the ILL

T Bigault, G Delphin, A Vittoz, V Gaignon and P Courtois

Institut Laue Langevin, 6 rue Jules Horowitz, BP156, Grenoble Cedex 9,France.

E-mail: bigault@ill.fr

Abstract. We present a status of recent projects involving the in-house production of neutron multilayer optics, mainly polarizing supermirrors, at the ILL. Our main "mass production" project is for the wide-solid angle analyzing benders for the future instrument WASP (Wide Angle Spin Echo). The current status of this project based on Co/Ti supermirrors, which spans several years, will be presented. Some parameters of polarizing supermirrors for cavity polarizers, mainly based on Fe/Si supermirrors and produced in the past few years for various ILL instruments, are also reported. Some supermirror samples produced in order to study depolarization effects are also mentioned.

1. Introduction

The multilayers were fabricated by magnetron sputtering, on the two large area coating machines run by the ILL multilayer group. Reactive sputtering is used in some cases in order to adjust scattering length densities and interface structures [1]. The first chapter is a progress status of the supermirror production for the WASP analyzers. The second one presents supermirrors produced for several cavity devices, with more focus on the D33 polarizers. The last one is a short report on the type of multilayers produced for the depolarization study.

2. Status of the WASP project

The production of supermirror analyzers for the future ILL spin-echo instrument WASP is in progress. The analyzer design, based on Schärpf-type C-shaped benders and described in [2], is inspired by the design of the D7 analyzers upgrade [3]. We remind a few parameters from the design: the supermirrors are Co/Ti with m=2.8, double-side coated on borated float glass. The coating process and typical supermirror performance are the same as in [3]. The substrate dimensions, 254x141x0.2 mm^3, were chosen to fit the instrument geometry and to maximize the number of substrates coated per process run. The mirrors are then mounted into cassettes which hold them with a bending radius of about 7 m. Each cassette, consisting of 37 mirrors, corresponds to a horizontal angle of analysis of 1° from the sample. The full angular analyzer coverage of the designed instrument would be 240°, which corresponds to 1.25 m^2 of detector coverage at 3 m from the sample, or 636 m^2 of supermirror to be coated. The production started in 2008, with an annual rate steadily increasing from then on; today enough mirrors have been produced to cover more than 50° (Figure 1). The instrument commissioning is foreseen at the end of 2015, with a target of 90° of analyzer coverage. There will be a deviation from the extrapolation in 2013 because the control system of the coating machine was renewed. This upgrade is expected to improve the process reliability in the following years.

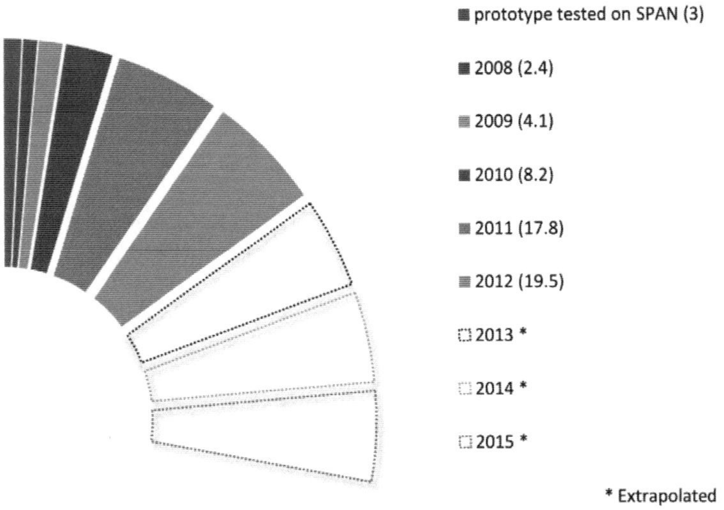

Figure 1: Supermirror analyzers' production status for WASP represented in terms of detector angular coverage. The number of cassettes produced (or degrees covered) is indicated in parentheses. The extrapolated production for 2013, 2014 and 2015 is the average over the last 3 years.

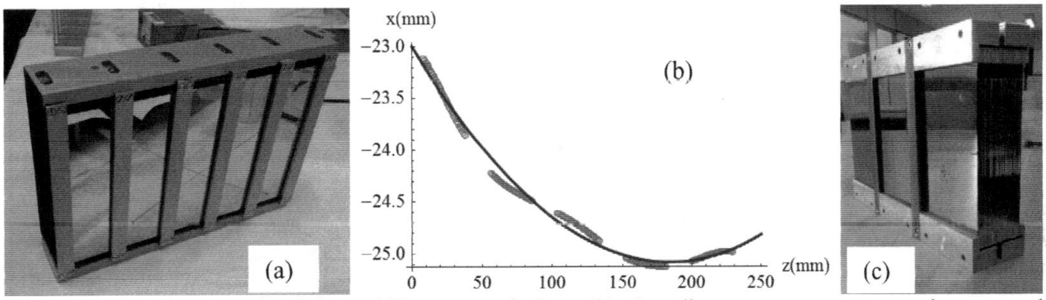

Figure 2 : (a) Photograph of the "old" cassette design. (b) Coordinate measurement on the external surface of the side mirror of a cassette of the type shown in (a) (red circles), and best circle fit (blue solid curve) corresponding to a radius of 8.1m. The z-axis is along the neutron beam path. (c) Photograph of the cassette design under development.

Some measurements were performed with the first mirrors produced, on the SPAN instrument at the BERII reactor in Berlin. The mirrors were mounted into three cassettes, which were used as an analyzer in front of a 2D position sensitive detector. A velocity selector was used to characterize the cassettes over the wavelength range to be used by the WASP instrument. Two π-flippers were used in order to perform polarization corrections. Across the wavelength range covered by the design (0.4 – 1.2nm), the performance of the analyzer block could be estimated from the measurements by correcting from instrumental effects. This gives an analyzing efficiency with a maximum of about 95% at 0.5nm with possibly a slight decrease for shorter and longer wavelengths, and a roughly constant transmission of about 51% of polarized beam. When taking into account the detection area effectively covered by the analyzers, this is within the design specifications and close to the analyzer performance predicted by Monte Carlo simulations. More details concerning the measurement procedure, data analysis and comparison with simulations can be found in [2].

Nevertheless, dimensional metrology revealed that the mechanical design of the cassettes, also inspired from the D7 project, presents some bending non-homogeneity along the mirrors (Figure 2-b).

The best-fit circle has a radius of 8.1m, instead of 7m expected from the design. Still, this radius is low enough to prevent direct line of sight through the 1mm channels between the mirrors, which would allow transmitting some neutrons without analyzing them. From the difference between the measured coordinates and the best-fit circle, an RMS slope error of 4mrad was calculated. This imperfection should affect the transmission performance of the cassettes. This could for instance contribute to the small difference between simulated and measured cassette performance observed in [2].

The design is being revised and a new prototype is being tested (Figure **2**-c), in order to reach more uniform bending. The essential change is in the type of mechanical spacers that are inserted between the mirrors before pressing them, in order to define the neutron channels. Vertical aluminum wires placed in the neutron path and perpendicular to it were replaced by horizontal shims placed out of the neutron path and parallel to it. The stack of mirrors and spacers is pressed through continuous contacts along the mirrors edges, instead of using discrete contact points in the old design. A small part of the mirrors height is lost for placing the shims, but this is compensated by the suppression of the absorption by aluminum wires. The characterization of the shape errors resulting from this new mechanical design is in progress.

3. Cavity devices

From 2011, several supermirror coatings were produced in order to fabricate transmission polarizers in cavities (see e.g. [4] for the operating principle) or wavelength filters in cavities.

3.1. Design parameters

The cavities were designed by taking into account the operating wavelength range of each instrument, the m-value of the supermirrors that could be achieved with satisfying quality, the characteristics of the incident beam, and the available space on the instrument. Resulting parameters for these various devices are presented in Table 1. In the case of D33, the polariser design was given by the calculations performed in [5], together with wavelength filters which we also produced. There are two interchangeable Fe/Si polarisers in order to cover the whole wavelength range of the instrument. Similarly, two interchangeable wavelength filters based on non-magnetic Ni/Ti are used to cut the longer wavelengths, with two different cut-off values. Other parameters like the characteristics of the coating on the guide faces, not detailed here, were deduced from [5].

3.2. Fabrication

The devices consist in guide sections with grooves in opposite faces in order to maintain polarising mirrors at a given angle from the incident beam. The guide sections, including the internal guide non-depolarizing coatings when needed, were produced by guide manufacturing companies. We present here the supermirror coatings on silicon substrates that were produced and inserted into the assembled guide sections. In every case, the substrates were single crystal double-side polished silicon substrates, 0.55 mm in thickness, and identical supermirrors were coated on both sides. The supermirrors required for the devices in Table 1 were coated in several coating runs. As an area of 800x190mm^2 can be coated at once, several substrates were coated simultaneously in each run together with a test glass.

Table 1. Parameters of cavity polarizers or wavelength filters, fabricated for various ILL instruments.

Instrument and device type	Geometry	Incidence angle (°)	Supermirror m-value	Operating wavelength range[(1)] (nm)	Supermirror materials	Device overall Length (mm)
IN11 polariser [(2)]	V-shape	0.908	3.2	0.29 – 1.43	Fe/Si	2000
D22 polariser [(3)]	V-shape	1.04	3.2	0.33 – 1.64	Fe/Si	1242
D33 polariser	V-shape	0.72	3.6	0.20 – 1.13	Fe/Si	1200
D33 polariser	single blade	1.4	3.6	0.39 – 2.20	Fe/Si	1200
D33 wavelength filter	V-shape	1.32	1.2	0 – 1.11	Ni/Ti[(4)]	842
D33 wavelength filter	single blade	2.2	1.2	0 – 1.85	Ni/Ti[(4)]	842

[(1)] Assuming a parallel beam incident on mirror, without divergence. For the wavelength filters, wavelengths greater than the indicated positive cutoff are rejected.
[(2)] Replacement.
[(3)] Upgrade.
[(4)] Non-depolarizing supermirror.

3.3. Characterization: the example of mirrors for the D33 polarisers

At least one test glass from each run was measured by polarized neutron reflectometry with our test instrument T3. As an example, Figure 3(a) shows an average over several of these measurements, weighted by the proportion of mirrors from each coating run actually mounted in the D33 V-shaped polarizer. Figure 3(b) shows polarized neutron transmission measured on a double-side coated mirror that was mounted afterwards in the D33 V-shaped polarizer. In both cases the only correction to the data is background subtraction, as the polarizer and flipper efficiency are not known. Thus the measured polarization ("P" in Figure 3) has to be taken as a lower limit.

In order to estimate the cavity polarizer performance with realistic mirror characteristics, the transmission data presented in Figure 3(b) were converted into transmission for both spin states as a function of wavelength. The incident angle is set to the design values of the D33 V-shaped polarizer and single-blade polarizer, and the attenuation in the substrate is included. The results are presented on Figure 4 for both devices, in terms of polarization efficiency, total transmission of a non-polarized beam, and usual figure-of-merit P^2T. Again, the polarization has to be taken as a lower limit, due to the previous remark on T3 measurement and to the fact that the applied magnetic field is about 100mT on D33, instead of 60mT on T3. The curves on Figure 4 only characterize the properties of the double-side coated mirrors in transmission at given incidence angles for a perfectly collimated beam. Other effects, like incoming beam divergence and reflections on the guide faces, can be taken into account by performing a ray-tracing simulation with the full geometry, using these data as an input. This work is in progress, by using the approach described in [5], and the results will then be compared with the performances of the complete devices on D33. The measurement of these performances on the instrument is also in progress.

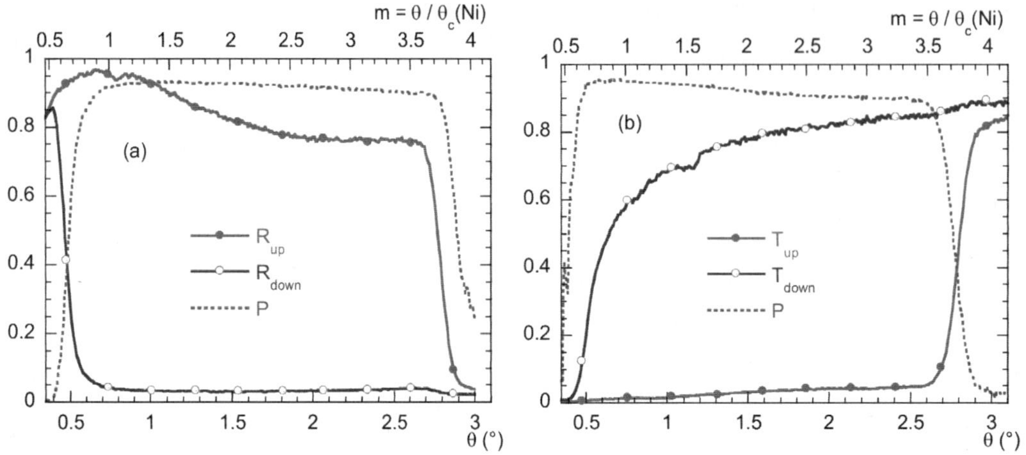

Figure 3 : (a) Polarised neutron reflectivity measurements on single-side coatings, averaged over the coating runs made for the D33 V-shaped polariser. (b) Polarised neutron transmission measurement of a double-side coated D33 mirror. The wavelength is 0.75nm, and a field of 60mT was applied along the mirrors hard magnetization axis. See text for details.

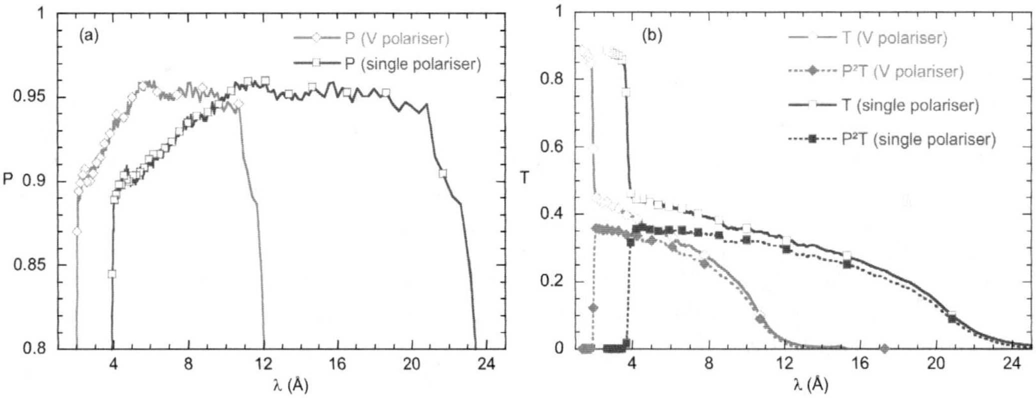

Figure 4: Polarisation efficiency P (a), total transmission of a non-polarised beam T and figure-of-merit P^2T (b) deduced from measured transmission of Figure 3(b), in the case of the D33 V-polariser and single polariser, with double-side coated mirrors at respective incidence angles of 1.4° and 0.72°. This would be in the case of a single transmission through the mirror of a perfectly collimated beam.

4. Multilayers for the depolarization study

The new generation of neutron beta decay experiments necessitates a very high polarization of large, cold neutron beams [6]. This puts high requirements on the performance of polarizing supermirrors, which becomes limited by depolarization. Due to this effect, the polarization of a perfectly polarized neutron beam can be decreased when it is reflected by or transmitted through a supermirror coating. A systematic experimental study of parameters influencing such depolarizing effects is in progress. The measurement setup is a test bench developed on PF1B at the ILL, based on ^{3}He opaque spin filter cells, which can measure depolarization of the neutron beam down to 10^{-4} accuracy. For this study, a series of samples was produced to investigate the effect of several coating parameters. It was based on Co/Ti and Fe/Si material pairs, using the same deposition conditions as for usual supermirror production. The number of individual layers was bounded by 1 and 541 in the case of Co/Ti, or by 1 and 903 in the case of Fe/Si. Single layer coatings were either Co or Fe with thickness ranging from 50 to 200 nm. In most cases the layer thickness sequence was given by a standard supermirror design.

Three samples were produced in order to isolate the effect of the magnetic layer thickness: one was coated according to the thickness sequence of an m=2.1 Fe/Si supermirror, and two others were coated with a constant Fe thickness of 2 and 10 nm, respectively.

Details about the method and the results of the first experiment can be found in [7], where mostly Co/Ti supermirrors were studied. The main conclusions of this work are that the depolarization increases with the m-value, and that it can be reduced by increasing the magnetizing field, up to about 1 T. At 0.1 T, the depolarization can be of the order of 5.10^{-3} or more. The remaining samples, including Fe/Si coatings, "non-supermirror" coatings and supermirrors from other sources, were measured in a second beamtime [8]. Among other results, the measurements performed on the multilayers with a constant Fe thickness indicate a more important depolarization when the Fe layers are thinner and much less magnetic field dependence than for supermirrors. This suggests that the magnetic layer thickness is a critical parameter for depolarization by multilayers.

This method gives interesting insights into the mechanisms of magnetization in supermirrors, which are complex multilayer structures, in an applied magnetic field. This is of interest both for fundamental understanding and for applications in neutron optics. For instance, this study reveals some depolarization mechanisms occurring up to applied magnetic field of 1 T in some cases, much higher than field intensities commonly used to "saturate" supermirror polarizers. The present method then allows quantifying small depolarization levels that probably reveals incomplete saturation, which would be challenging to measure by other techniques like SQUID magnetometry or polarized neutron off-specular scattering, as explained in [9]. An extensive report on the results from the neutron depolarization study can be found in [9], where depolarization dependence on many parameters is reported: applied magnetic field, m-value, incidence angle, wavelength, reflection/transmission geometry, layer material, etc.

Acknowledgements
The authors would like to thank the people involved in these various projects for fruitful discussions and interaction, their feedback on supermirror performance and their interest in supermirror development, and among them K. H. Andersen, C. Dewhurst, P. Fouquet, B. Farago, P. Falus, D. Bazzoli, C. Klauser and T. Soldner.

References

[1] Hoghoj P, Anderson I, Siebrecht R et al. 1999 Neutron polarizing Fe/Si mirrors at ILL *Phys.* B **267-268**, 355-9

[2] Fouquet P, Farago B, Andersen K H et al. 2009 Design and experimental tests of a novel neutron spin analyzer for wide angle spin echo spectrometers *Rev. Sc. Instr.* **80**, 095105

[3] Stewart JR, Deen PP, Andersen KH, Schober H, Barthelemy JF, et al. 2009 Disordered materials studied using neutron polarization analysis on the multi-detector spectrometer, D7 *Journ. Appl.Cryst.* **42**, 69-84

[4] Mezei F, 1988 Very High Reflectivity Supermirrors And Their Applications, *SPIE Proc.* **983**, 10

Krist T, Lartigue C and Mezei F 1992 Transmission geometry supermirror neutron polarizer device *Phys.* B **180-181**, 1005

[5] Dewhurst C D 2012 Modelling of wavelength cut-off filters and polarising mirrors in a neutron guide. *Nucl. Instr. Meth.* A **683,** 16-23

[6] Dubbers D, Abele H, Baeßler S et al. 2008 A clean, bright, and versatile source of neutron decay products *Nucl. Instr. and Meth.* A **596**, 238-247

[7] Klauser C, Bigault T, Rebrova N et al. 2013 Ultra-Sensitive Depolarization Study of Polarizing CoTi Supermirrors with the Opaque Test Bench *Phys. Procedia* **42**, 99-105

[8] Klauser C et al., in preparation.

[9] Klauser C., 2013 High precision neutron polarization for PERC, *Ph.D. thesis, Vienna University of Technology.*

Magic-PASTIS for XYZ polarization analysis using SEOP polarized ^{3}He gas

E Babcock[1], Z Salhi[1], P Pistel[2], G Simeoni[3], A Ioffe[1]

[1] Jülich Centre for Neutron Science at MLZ, Lichtenbergstr. 1, 85747 Garching Germany, [2]Forschungszentrum Jülich - Engineering und Technologie (ZEA-1) Wilhelm-Johnen-Strßae 52425 Jülich Germany, [3]Forschungs-neutronenquelle Heinz Maier-Leibnitz, Lichtenbergstr. 1, Garching Germany.

E-mail: `e.babcock@fz-juelich.de, z.salhi@fz-juelich.de`

Abstract. We present progress towards a complete system for neutron polarization analysis on a time-of-flight (TOF) neutron spectrometer with a large area/angle detector array. Finite element calculations have been used to model the field gradients of a newly proposed PASTIS coil set, which uses a wide-angle banana shaped ^{3}He Neutron Spin Filter cell (NSF) to cover a large neutron scattering solid angle. The final goal of this insert is to enable X-Y-Z polarization analysis to be installed on the future hot/thermal time-of flight spectrometers, although the method is also applicable to thermal/cold spectrometers as well. The components of this system, such as the magnetic field coils and design are applicable to neutron spectroscopy with wide angle detector arrays in general, and the ^{3}He wide angle cell developments for polarized inelastic neutron scattering are independent from the methods used to polarize the gas as well.

1. Introduction

Neutron polarization analysis, PA, is a powerful tool for separation of nuclear spin-incoherent background, analysis of complex magnetic structures and the study of magnetic excitations [1]. Several wide angle spectrometers with polarization analysis exist or are under construction [2, 3, 4, 5]. The PA can be performed in a variety of ways depending on the instrument's parameters and energy range using either supermirror analyzer arrays [2], or ^{3}He neutron spin filters as described in this paper and in the references [5, 6, 7, 8]. The performance of the super mirror-based systems, is normally limited by the analyzer height and integration over the height of the detectors, as well as the maximum analyzable energy which is normally about 20 meV or less [9].

For ^{3}He based systems, a sufficiently low magnetic field gradient must be maintained over the volume of the neutron spin filter cell in order to obtain ^{3}He polarization lifetimes of a practical and useful level (see section 2). Further for any method to obtain information on the $Q_\perp$ and $Q_\parallel$ of the momentum transfer vector Q over a large solid angle detector, the sample field must be changeable between the three orthogonal X-Y-Z directions [10]. In the case of the ^{3}He NSF, since the cell occupies the same field volume as the sample, the large volume highly uniform ^{3}He holding field used to maintain long ^{3}He NSF lifetimes, must also be able to be changed between the X,Y and Z directions. We present a design study with finite element magnetic field (FEM) calculations of possible X-Y-Z field configurations suitable for maintaining the polarization of the ^{3}He and is adapted to the TOPAS instrument geometry. Two clear options exist, a magnetized mu-metal geometry, similar to reference [7], or a resistive coil set similar to reference [6], however in our proposed designs, certain key differences exist which build on the experiences from prior devices. First we provide the reader with a review of the sources of ^{3}He relaxation, and how they should be optimized for effective neutron spin filter use.

2. ^{3}He relaxation time

Here to ease the discussion for the reader we provide a short review of ^{3}He nuclear spin relaxation in a closed volume. The ^{3}He polarization in NSF cell decays to thermal equilibrium with a characteristic time constant T_1. The longitudinal relaxation rate, $\Gamma_{He} = T_1^{-1}$ consists of three main mechanisms for relaxations of spins and is characterized by:

$$\Gamma_1 = \frac{1}{T_1} = \frac{1}{T_1^{dipole}} + \frac{1}{T_1^{gradient}} + \frac{1}{T_1^{wall}} \tag{1}$$

T_1^{dipole} is the dipole-dipole relaxation time due to interaction between colliding ^{3}He atoms [11]. The magnetic dipole coupling of atoms during a binary collision results in the loss of nuclear polarization. This rate is well known and described by:

$$\frac{1}{T_1^{dipole}} = \frac{p}{817}, \tag{2}$$

where p is given the pressure in bar and 817 is the dipole-dipole relaxation rate constant in hours. This dipole relaxation effect is practically relevant at the one to several-bar pressures used for neutron spin filter cells and forms the upper limit to the maximum on beam storage times as the other two terms can theoretically approach 0 in ideal cases. For example at p=2.7 bar the maximum possible storage time constant is $T_1^{dipole} \simeq 300$ h.

In the presence of magnetic field gradient, a diffusing atom will transition through field gradients orthogonal to the B_0 magnetic holding field [12]. If the motion through these field gradients creates an effective field change with a rate close to or exceeding the Larmour frequency of the atoms, then spin flips can be inducted causing a loss of magnetization. $T_1^{gradient}$ is the relaxation due to diffusive motion through orthogonal field gradients and given by:

$$\frac{1}{T_1^{gradient}} = \frac{7000}{p}\left(\left(\frac{\Delta B_x}{B_0}\right)^2 + \left(\frac{\Delta B_y}{B_0}\right)^2\right), \tag{3}$$

where ΔB_x and ΔB_y are the orthogonal gradients of the static holding field B_0. For example, to maintain reasonable neutron spin filter performance over time while refreshing the polarized ^{3}He gas or cell less than once once per day[8] for a cell of $p = 2.7$ bar,

$$\left(\left(\frac{\Delta B_x}{B_0}\right)^2 + \left(\frac{\Delta B_y}{B_0}\right)^2\right)^{1/2} \leq 10^{-3}\text{cm}^{-1} \rightarrow T_1^{gradient} \geq 100\text{h}, \tag{4}$$

here the 2-2.7 bar pressure is chosen to reduce relaxation due to field gradients, while still allowing for a long T_1^{dipole}.

The final term T_1^{wall} in equation 1 is the relaxation time due to the interactions with the cell wall and is the least understood ^{3}He relaxation mechanism. This relaxation depends purely on the characteristics of the container which can contain magnetic impurities like magnetite (the most common impurity in glasses). The diffusion of helium into the glass may also cause relaxation by precession of the helium moment about the net local gradient field caused by the magnetic impurities. Many references have been written as to possible origins and characterizations of this relaxation and how it can be changed through application of external fields [13, 14, 15].

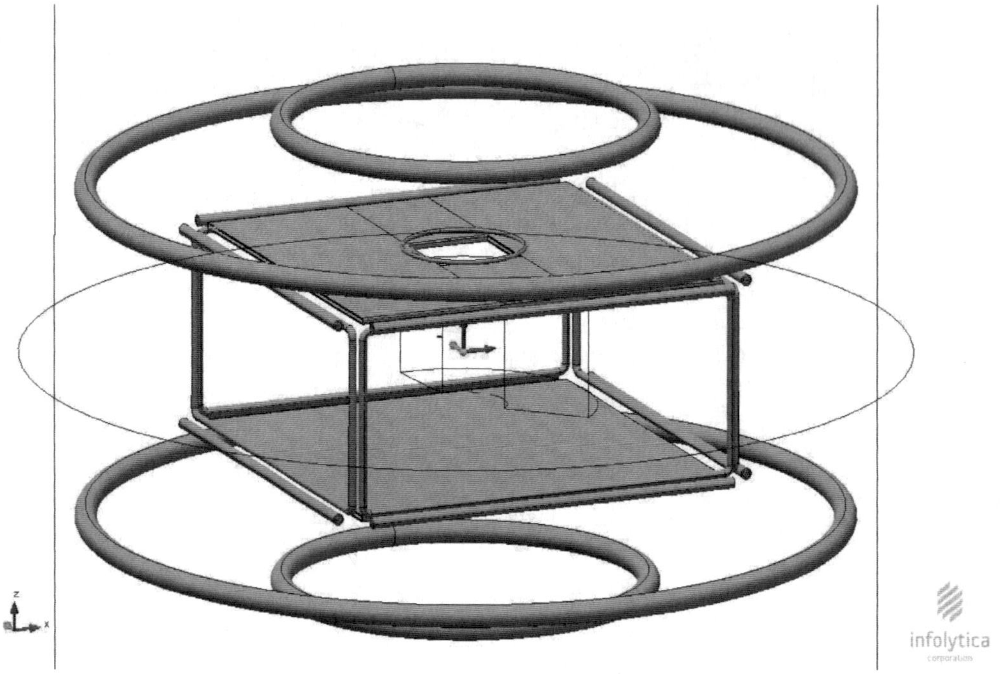

Figure 1. The model written in 3D finite element magnetic modelling software [16], showing the geometry of our concept. The square plates are 35 cm by 35 cm and the largest circular coils are D = 70 cm. (Color online)

3. Magnetic system design

Several steps have been made in the design of wide angle polarization analysis using polarized ^{3}He for analysis of $Q_\parallel$ and $Q_\perp$ magnetic scattering, i.e. X-Y-Z polarization analysis [10]. Previously a wide angle polarization analysis prototype called PASTIS was developed at the Institut Laue Langevin, where a set comprised of compensated X, Y and Z coils were used to create a homogeneous magnetic field in three orthogonal directions in a compact format suitable to known neutron instrumentation with acceptable ^{3}He relation times [6]. The idea has been further explored in the ILL and ISIS with a more exotic magnetized μ-metal configuration and a further minimization of dead areas with the original compensated coils geometry respectively [5, 7].

Our so called Magic-PASTIS design is based on an open μ-metal geometry. The design used for finite element calculations using commercially available software [16] is shown in figure 1. By introducing μ-metal sheets we aim to produce an improved homogeneous magnetic field in a large region while reducing the blind areas due to the coil frames of the more conventional compensated square/circular coils of the original, and optimized resistive coil PASTIS systems [5, 6]. Those systems for example have eight vertical dead areas (in the horizontal plane) of approximately $\geq 5^\circ$ at $\pm 18^\circ, \pm 71^\circ$ and $\pm 109^\circ$ whereas the design presented here has a $> 40^\circ$ opening in the vertical plane and only four small 3° dead angles, one every 90°, in the horizontal plane. For either system these dead angles can be rotated arbitrarily with respect to the incident beam. The center of the configuration is the sample, placing the ^{3}He cell significantly off center, but this is largely defined by the typical sample environment geometry of any real neutron spectrometer that is designed to accept sample magnets and cryostat, which also place the sample at the center.

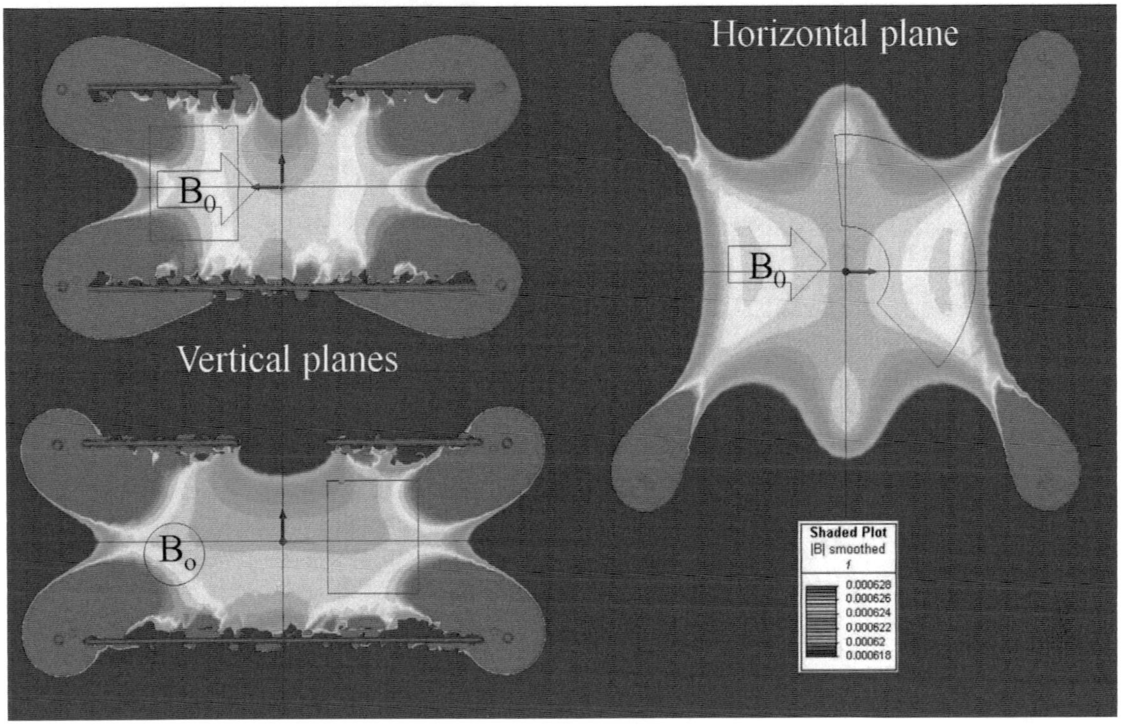

Figure 2. A color-shaded plot of the calculated Magnetic field distribution for the three planes of the x or y direction from FEM simulations. Each color gradient is a relative change of 8.33×10^{-4}. The size of the square plates is 35 cm x 35 cm.

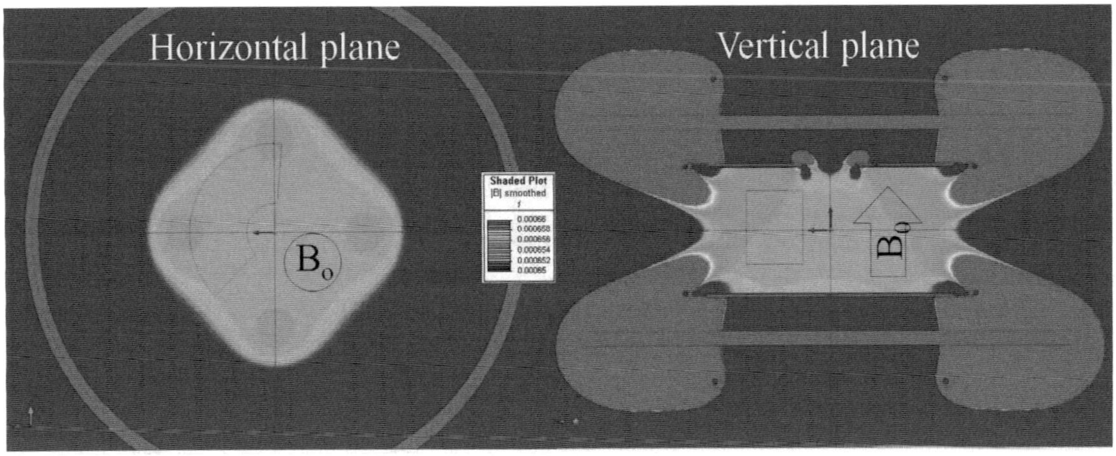

Figure 3. A color-shaded plot of the calculated Magnetic field distribution for the two planes of the z direction field from FEM simulations. Each color gradient is a relative change of 7.57×10^{-4}. The diameter or the main circular coils is 70 cm.

Figure 4. The current status of the magic PASTIS system. The system is now being optimized for the values of currents and fine positioning of coils. (Color online)

As discussed in the Section 2, in order to keep the depolarization of the polarized ^{3}He sufficiently low it is necessary that the field gradient is below about 10^{-3}cm^{-1} over the volume of a large neutron spin filter cell, see Equation 4. This level would allow the cells to remain on the instrument long enough to make experiments feasible by reducing the time spent refilling, calibrating and transporting polarized cells/gas to the beamline to once per day or less.

Figure 2 shows finite element calculations of the created magnetic field in the X or Y direction, and figure 3 is the same for the Z direction. It is clear that the inner part is highly homogeneous, including also up to the radius of the ^{3}He cell. As seen in figure 1, the coil setup fulfils the requirement of a large solid-angle opening with minimal dead area, allowing the polarization analysis over our large solid angle detectors. We have constructed a prototype of this new setup which will now be optimized experimentally for current values and optimal positioning of the coils starting with the ideal geometry from the FEM model. A picture of the completed prototype can be seen in figure 4. In parallel, the construction of the incident beam polarizer based on the design of the MARIA reflectometer in-situ SEOP analyzer is continuing [17]. Once TOPAS is brought into operation it will provide polarization analysis covering a wide angular range from day 1.

4. Development of wide angle GE180 cells for SEOP or MEOP polarized gas

The use of sealed ^{3}He cells made of GE180 glass could vastly simplify the equipment required for wide angle PA as they would allow direct polarization of the cell using the SEOP method [18]. Furthermore, an all blown glass construction using GE180 may indeed have practical advantages such as pressure resistance and durability compared to other proposed routes, such as attempts from other groups to form cells from bonded, or glued-together, flat and curved plates of crystal silicon, quartz, or even GE180 [5, 18]. Presumably such cells if made of GE180 could be used as sealed cells and be polarized directly by a SEOP polarizer system as in [18] or be used as valved

Figure 5. A first doughnut shaped cell for magic PASTIS testing called Homer. This cell is filled with 0.5 bar of ^{3}He to make it sensitive to magnetic field gradients. The T_1 lifetime of this cell is 1074 ± 20 hours. (Color online)

cells which are filled with pre-polarized ^{3}He gas from a large scale ^{3}He polarizer employing either the MEOP or SEOP technique [5, 7, 19].

Consequentially, the route towards suitable wide angle ^{3}He spin-filter cells we have been pursuing is that of an all-blown GE180 construction. Such cells are unique firsts, and results of the work in the Jülich glass workshop who have been developing the techniques. GE180 is chosen because of its suitability for good ^{3}He T_1 relaxation times, neutron compatibility, and the fact it is the best know material for use in SEOP polarization of the ^{3}He gas [20]. First doughnut shaped cells are formed by making a hole through the center of a large cell.

In figure 5 we show a picture of a very large, D =22 cm doughnut-shaped cell called Homer which was prepared with a Rb/K hybrid mixture [21], 0.07 bar N_2 and 0.5 bar of ^{3}He. This cell was made for tests of ^{3}He relaxation times in the new magic-PASTIS coil system from figure 4. As was shown in equation 3, the low ^{3}He pressure will make the cell sensitive to magnetic field gradient relaxation and a good probe to optimize the performance of the magic PASTIS magnetic cavity. Aided by the low ^{3}He pressure, good glass handling and cell preparation techniques, Homer was measured to have a 1074 hour T_1 in what we presume to be an nearly "ideal" magnetic holing field which has a relative field gradient better than 1×10^{-4} cm^{-1}. C-shaped cells optimized for spectroscopy applications are then made by sealing off sections of the doughnuts to obtain the desired angular acceptance.

5. Inelastic scattering test from GE180

The inelastic neutron scattering properties of GE180 glass was not known. Therefore we characterized the inelastic neutron scattering properties on TOFTOF [22], to determine if such cells can eventually be used for spectroscopy applications and/or under what conditions. Shown in figures 6 and 7 are the results of this test.

A sample of the glass was prepared in the same way as a ^{3}He spin filter cell is prepared [23]. The inelastic neutron spectrum for these samples was then measured of a broad range of resolutions and incident neutron wavelengths. The broad dynamic range, high incident energy measurement using 1.6 Å neutrons was performed at 70 °K and 298 °K where no difference

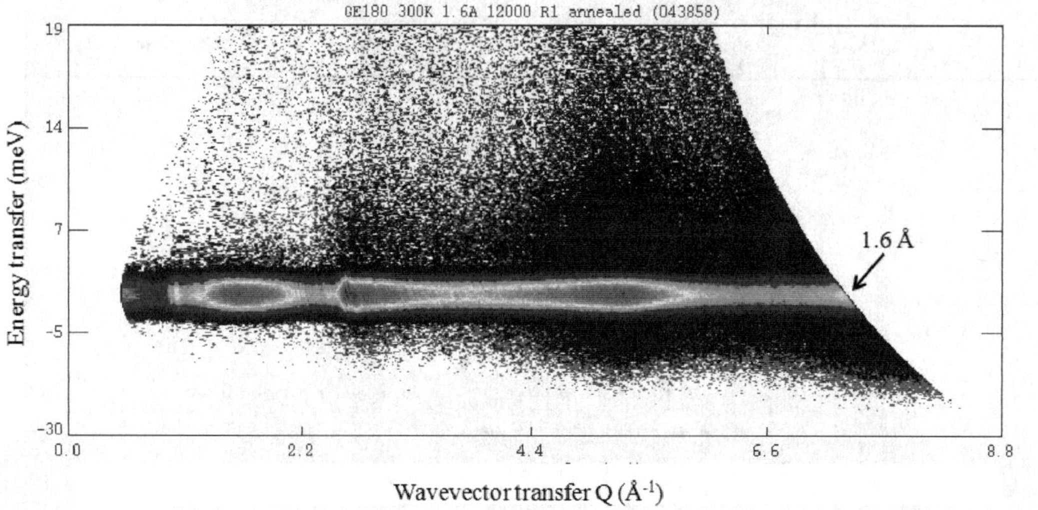

Figure 6. Wide dynamic range plot of E (meV) vs. Q (Å^{-1}) taking with an incident neutron wavelength of 1.6 Åat 298 °K. One can see a diffraction pattern which largely resembles the main component of the glass, $Si0_2$. Since this diffraction is at relatively high Q, it could be largely reduce with coarse collimation after the ^{3}He cell. Furthermore, this diffraction is nearly purely elastic in nature. (Color online)

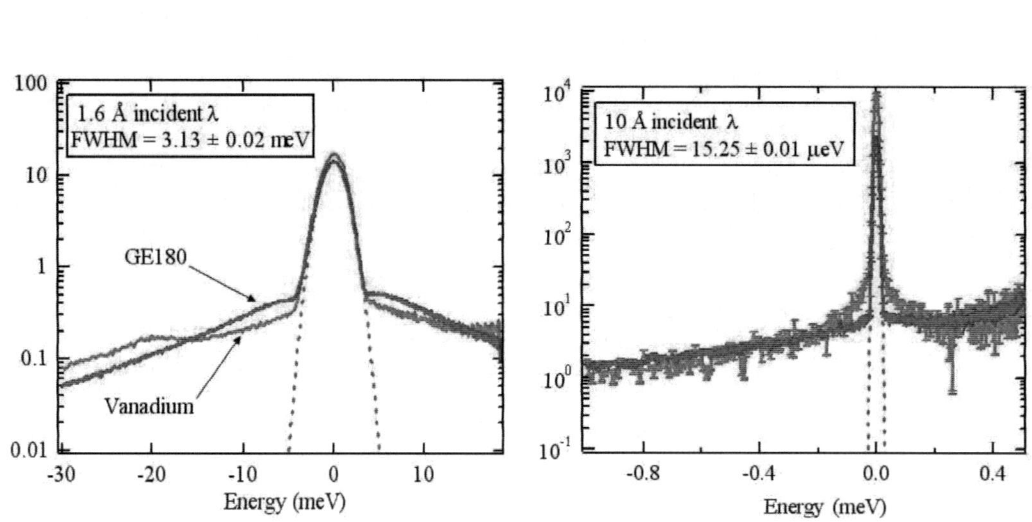

Figure 7. An example spectrum from GE180 (red) on TOFTOF spectrometer at MLZ. Left is a broad dynamic range/ low resolution scan which shows little inelastic contributions at large energy transfer. Right is a high resolution scan which shows that to the highest available resolution of TOFTOF, using GE180 one recovers the pure instrumental resolution. For comparison on both graphs a Vanadium sample (green) is also ploted, in the high enegy resolution graph to the right, GE180 is the lower curve. Neither data set has been scaled. (Color online)

in the spectra could be observed. Furthermore no appreciable structure or excitations from the glass were observed as can be seen in the the 1.6 Å spectrum at 298 °K shown in figure 6. By looking at the energy slices of this data for the two extreme incident energies, corresponding to 1.6 Å (left), and 10 Å (right) in figure 7, one sees that the GE180 scattering is largely elastic, and the pure instrument resolution of TOFTOF down to the minimum value of 15 μeV is readily recoverable as seen by the comparison to a vanadium sample. Future tests will compare the absolute levels of scattering from other possible wide angle ^{3}He cell materials, namely pure quartz and single crystal silicon to that of GE180.

6. Conclusions

Work is steadily progressing towards the realization of complete Magic-PASTIS wide angle polarization analysis system for neutron spectroscopy. The Magic-PASTIS magnetic cavity will be tested and optimized and the GE180 cells will be prototyped and prepared. Once the critical mass of components is available we will proceed with neutron sample measurements and data collection and analysis techniques.

6.1. Acknowledgments

We acknowledge support from the BMBF through the collaborative project 05E10CJ1 of the ESS design update work package K3 for which Dr. Z. Salhi is employed as a Post Doc. We also point out the ever useful conversation and exchange of ideas with colleagues working in spectroscopy, Dr. P. Deen,(ESS AB, Lund Sweden), Dr. W. Schweiker (Forschungszentrum Jülich GmbH) and Dr. J.R. Stewart (ISIS, STFC, Rutherford Appleton Lab, Chilton U.K.)as well as colleagues working with ^{3}He polarization techniques and application to wide angle PA, Dr. T.R. Gentile and Dr. W.C. Chen (NIST, Gaitersburg MD USA), and Dr. S. Boag (ISIS, STFC, Rutherford Appleton Lab, Chilton U.K.) and finally J. Anderson and A. Kirschhoff (NIST glass workshop, Gaithersburg MD USA).

References

[1] Moon RM, Riste T, Koehler WC 1969 *Phys. Rev.* **181** 920-&
[2] Stewart JR, Deen PP, Andersen KH *et al.* 2009 *J. of Appl. Crystal.* **42** 69-84
[3] Schweika W, Boni P 2001 *Physica B* **297** 155
[4] Voigt J, Babcock E, Brueckel Th *J. of Phys. Conf. Series* **211** 012032
[5] Beecham CJ, Boag S, Frost CD *et al.* 2011 *Physica B- Condensed Matter* **406** 2429-32
[6] Stewart JR, Andersen KH, Babcock E, *et al.* 2005 *Physica B-condensed matter* **385-86** 1142-5
[7] Andersen KH, Jullien D, Petoukhov AK, Mouveau P, Bordenave F, Thomas F, and Babcock E, 2008 *Physica B-condensed matter* **404** 2652-4
[8] Babcock E, and Ioffe A 2011 *J. of Phys. Conf. Series* **294** 012005
[9] Babcock E, Salhi Z, Ioffe A 2013 *J. of the Phys. Soc. of Japan* **82***suppliment* SA30
[10] Scharpf O and Capellmann H 1993 *Physica Status Solidi A - App. Res.* **135** 359-79
[11] Newbury NR, Barton AS, and Cates GD, *et al.* 1993 *Phys. Rev. A* **48** 4411-20
[12] Cates GD, Schaefer SR, and Happer W, 1988 *Phys. Rev. A* **37** 2877-85
[13] Jacob RE, Teter J, Saam B, *et al.* 2004 *Phys Rev. A* **69** 021401
[14] Schmiedeskamp J, Heil W, and Otten EW *et al.* 2006 *European Phys. J. D* **38** 427-38
[15] Deninger A, Heil W, Otten EW, *et al.* 2006 *European Phys. J.* **38** 439-43
[16] MagNet 2D/3D Electromagnetic Field Simulation Software, Infolytica Co. Montral, Qubec H2X 4B3 Canada, http://www.infolytica.com/en/products/magnet/
[17] Babcock E, Mattauch S and Ioffe A 2011 *Nuclear Ints. & Meth. in Phys. Research Sec. A - Accelerators Spectrometers Detectors and Associated Equiptment* **625** 43-6
[18] Fu CB, Gentile TR, Jones GL, Chen WC, Erwin R, Watson S, Broholm C, Rodriguez-Rivera JA, Scherschligt J, 2011 *Physica B- Condensed Matter* **406** 2419-23
[19] Xemed LLC Durham NH 03824 USA, https://www.xemed.com/products/magnilium/
[20] Rich DR, Gentile TR, Smith TB 2002 *et al. Appl. Phys. Lett.* **80** 2210-2
[21] Babcock E, Nelson I, Kadlecek S, Driehuys B, Anderson LW, Hersman FW, and Walker TG 2003 *Phys. Rev. Lett.* **91** 123003

[22] http://www.mlz-garching.de/toftof
[23] Salhi Z, Babcock E, Ioffe A 2014 *IOP Conference Series* this issue

Lifetime behaviour and polarization stability in ^{3}He neutron spin filter cells

S.Boag[1], C.Y.Jiang[2], X.Tong[2] and S.R.Parnell[3]

1 ISIS Neutron and Muon Facility, Rutherford Appleton Laboratory, Chilton, Didcot, Oxfordshire, OX11 0QX, UK

2 Instrument and Source Design Division, Oak Ridge National Laboratory, Oak Ridge, Tennesse, USA

3 Centre for the Exploration of Energy and Matter, Indiana University, Bloomington, Indiana, USA

E-mail: stephen.boag@stfc.ac.uk

Abstract. The successful use of ^{3}He cells in polarized neutron applications demands that once polarized the gas has both a long relaxation time and high polarization. In spin-exchange optical pumping the cell characteristics determine the maximum polarization as well as the lifetime.

We report on our experience of neutron spin filter cells, showing results on the orientation dependence of such cells. We observe a previously observed lifetime hysterisis effect in ^{3}He cell relaxation lifetime and include new measurements of the angular dependence of the relaxation lifetime, that are characteristic of a dipolar effect which diminishes upon repeated pumping/cooling cycles. We present demonstrations of magnetization effects in ^{3}He and ^{129}Xe cells and show the applicability of using sol-gel coatings on reducing the magnetization effect in ^{129}Xe cells.

We report recent results of the stability of the polarization in new SEOP cells, which was evident for 2 states, one with high energy spin state, low T_1 orientation and the second with low energy spin state, high T_1 orientation. These results are suggestive of an abnormal relaxation and have only been observed for cells with short (T_1 <10 Hours) lifetimes.

1. Introduction

For spin-exchange optical pumping (SEOP) [1] the maximum achievable polarization is limited by a balancing of the optical pumping rate and relaxation rate of the noble gas, as well as the individual "X-factor" [2] for a ^{3}He cell. Hence it is important to minimise the noble gas relaxation for applications requiring high (75%+) polarization.

One such application is that of neutron spin filters [3, 4, 5], where it is essential to achieve high ^{3}He polarization and maintain this for a long period of time. This is to achieve the highest attainable neutron polarization/analysing power and limit time dependent corrections for the changing neutron polarization. Wall relaxation can play a significant part in maintaining the gas polarization. Many extensive investigations into wall relaxation rates of ^{3}He cells have produced key breakthroughs allowing vast improvements in the lifetimes (T_1) of cells [6, 7, 8, 9]. However, there is still a great deal of variation in the achieved relaxation rates of cells and many often exhibit magnetization and magnetic field orientation dependence effects [10]. The aim of this proceedings is to add to the existing observations of these effects.

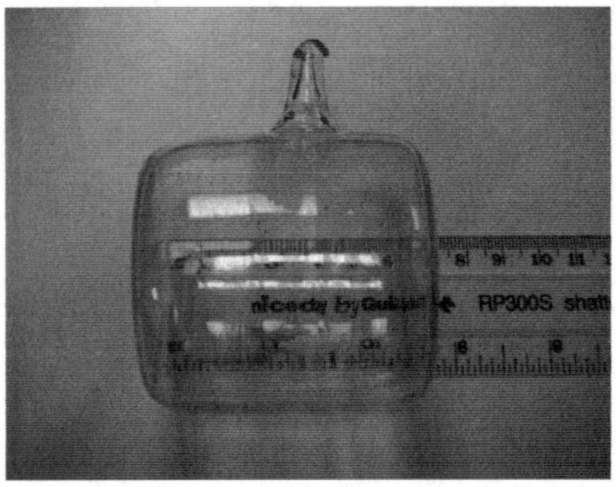

Figure 1. A typical ^{3}He neutron spin filter cell.

In the case of continuously pumped ^{3}He neutron spin filters [11, 12, 13], provided that the optical pumping rate is much greater than the relaxation rate ($\Gamma_{He} = 1/T_1$) of the cell, the cell should reach and maintain a steady-state polarization. However, we report on recent measurements of the ^{3}He polarization stability of continuously pumped cells with high (T_1 <10 hours) levels of relaxation.

2. ^{3}He Neutron spin filter cells

SEOP ^{3}He cells are generally produced from reblown GE180 glass tubing, an alumosilicate glass that has been shown to yield lifetimes of 100's of hours that are dominated by dipole-dipole interactions only [9, 14]. Furthermore, the lack of Boron and other such materials with large neutron cross sections means that there is only a small amount of absorption and scattering from the neutron transmission through the cell itself and this is largely independent of energy. A typical neutron spin filter cell (figure 1) is cylindrical in shape, which provides a homogeneous area and known path length for the neutrons to pass through.

In order to make SEOP cells that achieve high gas polarization and have long gas relaxation times it is important to reduce the wall relaxation in the cell by having as few contaminants as possible in the system. To achieve this the cell production rig is based around a turbo molecular pumped vacuum system, obtaining a base vacuum of approx 1×10^{-7} mbar. Furthermore the ^{3}He and ^{129}Xe, N_2 and alkali metals used to fill the cell are of very high purity. Full descriptions of the cell production procedure employed for the cells used in this work can be found here [15, 16, 17]. Upon production the cells are initially polarized in laboratory based SEOP systems and their room temperature lifetime (T_1) measured using Free Induction Decay (FID) NMR [18].

3. Observations of orientation dependence in neutron spin filter cells

The following section contains results showing the so called "orientation dependence" that has been observed previously in ^{3}He SEOP cells used for medical imaging studies [10], as well as ^{3}He neutron spin filters [16, 19, 20]. The effect is exhibited as a difference obtained for the T_1 values, dependent upon the orientation of the cell within the magnetic holding field. In table 1 the low (T_{1A}) and high (T_{1B}) values for T_1 are measured before and after a 180° rotation of the cell within the magnetic field, either by physical rotation of the cell, or by changing the polarity of the magnetic field. The difference in between the values in the two orientations varies from cell to cell and the effect is seemingly independent of the alkali metal element used and has been seen in cells coated with Rb, K and hybrid Rb-K mixtures [16, 21].

For the cells studied in this work the following behaviour was observed, firstly a new cell (i.e. no previous exposure to the magnetic field or laser) is polarized as normal and then a T_1, which we refer to as T_{1A} is measured. The cell is then reversed by 180° and a second relaxation time,

Table 1. High and low orientation T_1's for a selection of cells produced at ISIS and ORNL.

Cell name	Fill location	Alkali metal	^{3}He Presure (bar)	T_{1B} (High)	T_{1A} (Low)
Llewellyn	ISIS	Rb	0.71	124	65
Sammy	ISIS	Rb	0.65	96	56
Bob	ISIS	Rb	0.65	110	20
Cooper	ISIS	Rb	2.32	236	129
Billy	ISIS	Rb	2.08	119	50
Dorothy	ISIS	Rb	0.8	463	443
Buttercup	ISIS	K	1.5	94	58
Blue Devil	ORNL	RB-K	1.33	151	36
Chang	ORNL	RB-K	2.01	230	150
Messi	ORNL	RB-K	1.61	151	67
Smokey	ORNL	RB-K	1.98	85	75
Xavi	ORNL	RB-K	0.84	90	60

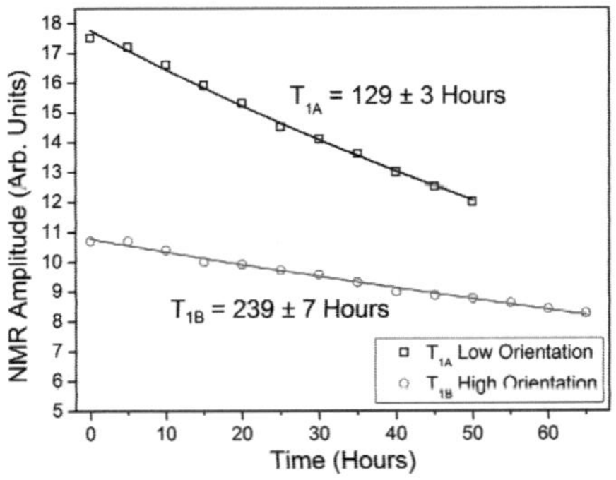

Figure 2. First relaxation data for cell Cooper, including T_1 values in low and high orientation. The cell was pumped in the low orientation, the relaxation time was measured, then turned through 180° and measured again.

T_{1B} is measured. In this case it was always observed that $T_{1B} > T_{1A}$ which is shown clearly in figure 2.

3.1. T_1 hysteresis

Cells will usually show a further T_1 hysteresis effect, whereby the high/low orientation T_1's will begin to decrease/increase and converge upon a value between the original two T_1's. We investigated the cell Rb14, called Bob, further as with an initial high and low T_1 of 110 and 20 hours respectively this cell showed a large orientation dependence.

As illustrated in figure 3 the cell was firstly polarized and the T_1 measured in the low orientation (T_{1A-A}), then repolarized in the low orientation, turned through 180° and measured in the high orientation (T_{1A-B}). It is then repolarized and measured in the high orientation (T_{1B-B}) and then finally it is repolarized in the high orientation, turned through 180° and measured back in the low orientation (T_{1B-A}). The results show that as the cell is cycled through the orientations the T_1's begin to converge, in this case to a value at roughly the mean of the relaxation rates.

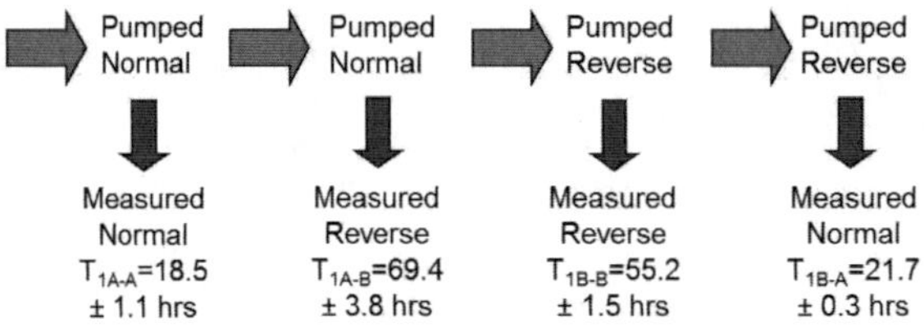

Figure 3. Orientation rotation and hysteresis cycle for cell Rb14 called Bob

3.2. Spherical cell investigation

The orientation and T_1 hysteresis behaviour can be seen in greater detail from an investigation of the effect on T_1 of a spherical cell, 5cm⌀ containing 2 bar ^{3}He, measured through a 360° cycle. The spherical cell was used to remove any geometrical effects, either through any holding field inhomogeneities or the resultant magnetic dipolar field that would arise from polarized ^{3}He in a cylinder.

The cell was initially polarized in the low T_1 orientation, then the T_1 measured for approximately 50 hours before turning the cell through 45° and measuring the new T_1. It is then repeatedly turned through 45° and each subsequent new T_1 measured, the cell was not repolarized between orientations. Once a complete set (i.e. 360°) of measurements had been obtained the cell was then repolarized in the high orientation and the measurements repeated.

Plots of the ^{3}He relaxation rate as a function of angle can be seen in figure 4. In both plots the 0° position refers to the low T_1 orientation. From figure 4(a) it is clear that the maximum lifetime is achieved when the cell is orientated at 180° from the initial polarizing position, regardless of which orientation the cell was pumped in. Furthermore, the difference between the high and low orientations decreased after the cell was repolarized. In this case the longest T_1 stayed at approximately the same value, whilst the lower values increased.

A polar plot of the data (figure 4(b)) reveals the behaviour to be like that of a dipole. It is clear that the high and low orientations are the dominant directions, with the T_1 scaling between these values as the cell is rotated. The suggestion of a hysteresis effect can also be seen clearly in the polar plot, as the cell is rotated through the magnetic field and then repumped. The effect appears to decrease as the T_1 for each orientation improves.

3.3. Cell magnetisation

The presence of magnetic impurities within the cell walls is further hinted at by the magnetisation effects seen by many groups [10, 22]. It was found that exposure to a high magnetic field (>1000 G) could cause a vast reduction in the T_1 of a cell, although this could often be recovered by degaussing the cell.

To investigate the effect further we took the cell Fred, which no longer showed any orientation behaviour and placed it between the poles of a 1500 G electromagnet. This caused a dramatic reduction in the T_1 of the cell from 40 to <1 hours, making it difficult to achieve any polarization within the cell. However, by using a degaussing procedure we were able to partially recover the T_1 of the cell, although with a reintroduction of an orientation dependence to 13 and 25 hours. To degauss the cell it was placed in the electromagnet and the polarity of the field was periodically reversed whilst the field strength was gradually decreased.

A further demonstration of the magnetisation effect occurred when a series of ^{129}Xe cells were fabricated for MRI studies in Pyrex. They were polarized using an external cavity diode laser described in reference [23]. Polarimetry at high field was performed on a 1.5 T GE Signa HDx scanner using a home built birdcage coil of similar dimensions and construction to that

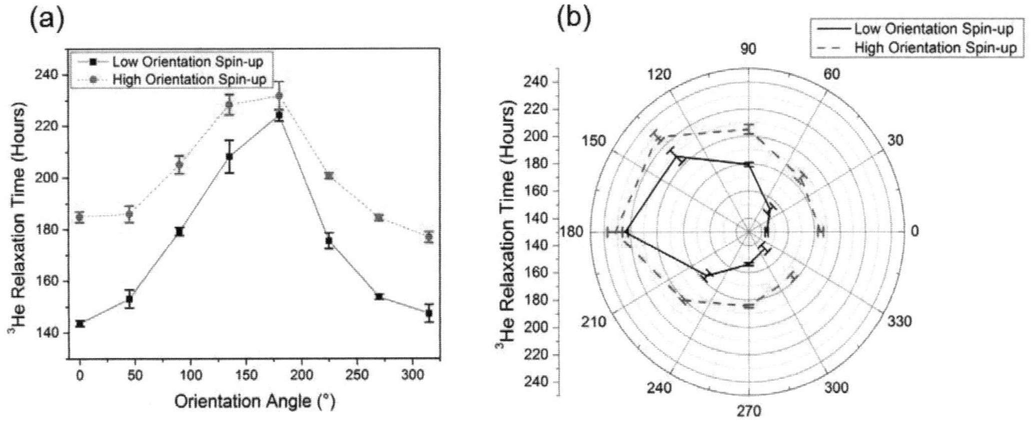

Figure 4. Line (a) and polar (b) plots showing the effect on T_1 of rotating the orientation angle of the spherical cell "Laura." Low and high orientation denotes the orientation in which the cell was polarised for each cycle.

used previously for ^{3}He [24]. Using this volume coil allows for uniform excitation of all the spins in the cell, care was taken to measure and minimise the flip angle and correct the magnetisation to account for losses in magnetisation from the RF to get an accurate determination of T_1. Low field NMR was performed using our home built NMR system [18]

At low field ($\approx$ 20 G) a $T_1 \approx$ 50 minutes [25] was found for sealed and valved uncoated cells. For bare Pyrex cells in the 1.5 T field the ^{129}Xe T_1 measured was $\approx$ 2 minutes. Further attempts to repolarise this cell resulted in negligible polarization and only a low voltage FID was observed from the cells. This is consistent with earlier investigations [20], which showed that magnetisation of alkali coated cells occurs after exposure to high magnetic field. To remove this problem a series of Pyrex cells were coated with a commercially available sol gel coating (similar to that used in [26, 27]). The coated cells were measured at 1.5 T and ^{129}Xe T_1 of 156 $\pm$ 10 minutes and 183 $\pm$ 10 minutes were recorded for two orientations 180° apart. In this case the coating appeared to work as a solution to the strong magnetisation effect. However, once this had been removed as a dominant relaxation effect, they appear to now exhibit orientation dependance.

4. Polarization decay during spin-up process

A further effect, observed at the Oak Ridge National Laboratory, occurred during the spin-up process for several new cells with a short (< 10 hours) lifetime. The cells have exhibited an unusual effect during ^{3}He polarization build-up, whereby the polarization appears to reach saturation, but then begins to slowly decay. Figure 5 shows an example of this effect, measured in laboratory conditions using FID NMR.

Similar effects have been seen previously at the Los Alamos Neutron Science Center and the ILL [28, 29], however these cases occured during in-situ pumping experiments in the presence of high neutron flux. The investigations revealed two effects: firstly, an immediate increase in alkali metal relaxation as the neutron flux was introduced; and a slow decrease in the optical transmission of the cell due to the build up of a white film on its inside walls (likely an Rb compound formed due to neutron interactions in the cell).

For the results shown in figure 5 a hybrid K/Rb spin filter cell containing 1.9 bar of ^{3}He was polarized initially in the high energy (alkali metal optically pumped to (J=1/2, m_j=1/2))

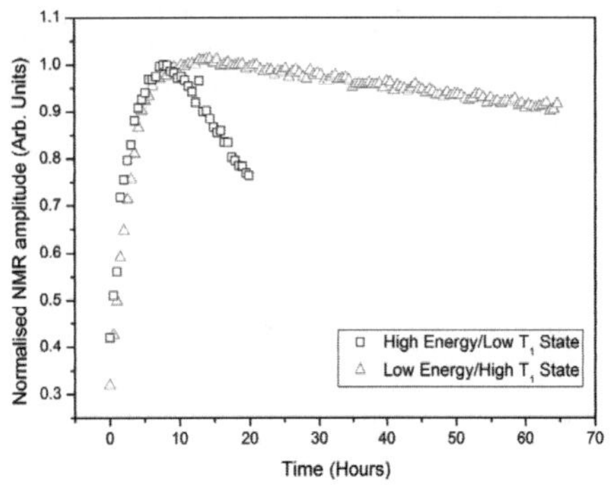

Figure 5. Slow decay effect seen during ^{3}He polarization build up. The cell was initially polarised in the high energy state.

state at 200°C and the decay was observed. A T_1 was then measured, giving 3.6 hours. The cell orientation was then rotated by 180° and the laser helicity reversed, thereby polarizing the cell in the low energy (alkali metal optically pumped to (J=1/2, m_j=-1/2)) state. The T_1 for this second orientation was then measured, yielding 5.1 hours. The plots in figure 5 have been normalised to account for a change in amplitude observed between each spin-up state. This was likely to have been caused by movement of the NMR coil during the cell reversal and means that no reliable comparison of the amplitudes of the two spin-ups can be made. The normalised plots reveal that in this case the effect is more prevalent in the initial polarization state, i.e. high energy and low T_1 orientation. Fits to the early, i.e. pre-saturation section of each spin-up curve reveal a similar rate (≈ 0.4 hours^{-1}). However, fits to the post-saturation sections reveal a decay rate of 0.038 hours^{-1} for the high energy, low T_1 state and 0.006 hours^{-1} for the low energy, high T_1 state.

There are a number of potential causes for this observed effect. Before speculating over lifetime effects from the ^{3}He itself, it is important to remove any potential instrumentation or measurement effects. For example drifts in the cell temperature or laser wavelength/power are likely to cause changes in the alkali metal polarization or density, which could lead to a change in the overall ^{3}He polarization over time. More extensive investigations into the spin-up process, such as measuring laser absorption or the alkali metal polarization, relaxation or density [29] over time, are required to remove this as a possibility.

Another potential cause is radiation damping [30], an effect that can occur during the FID process. After the initial pulse that "tips" the ^{3}He into the transverse plane, the precessing spins can induce a current in the NMR coil, creating a "counter" field that causes the magnetisation to rotate back into the longitudinal field direction. This will cause a change in the measured FID transverse relaxation time (T_2^*) that will scale with the cell polarization. The effect will also be different for each polarization state; in the low energy state T_2^* will decrease, whereas for the high energy state T_2^* will increase, sometimes showing a non-exponential form that will be unreliable for FID polarimetry [31]. Further analysis of the individual FID's was not possible for the data shown in figure 5, although investigations of T_2^* over time for each state would reveal any radiation damping effects. Furthermore in this case the position of the NMR coil relative to the cell was changed between measurements, which would also alter any radiation damping effects. Ultimately though, absolute polarization measurements, such as neutron transmission [5] or EPR, are required to reveal whether the observed decrease is a real decrease in ^{3}He polarization, or an artifact of the NMR measurements.

5. Discussion

The orientation effect is clearly affecting the maximum achievable T_1 in ^{3}He cells. The observations in this work show a history based effect, where high and low T_1 orientations are set in on the initial spin-up orientation, followed by a gradual decrease in the effect with subsequent spin-ups. Furthermore, as indicated by the spherical cell investigation the effect is dipolar and will change as a function of angle.

The existence of strong magnetisation and hysteresis effects suggests that there are ferromagnetic impurities present within the cell. GE quotes low impurity levels of 0.03% for Fe_2O_3, however the magnetisation effect has also been seen in cells produced from "iron free" glass [32]. Further investigations from Schmiedeskamp *et al.* identified dust like magnetite (Fe_3O_4) particles located close to, or at the inner glass surface as a potential source for magnetisation. It is unknown how these impurities may have entered the glass, although it is likely to be during the cell manufacturing process.

However, the existence of magnetic impurities alone does not fully explain the orientation effect, as the magnetic field strength used for SEOP is usually too low (in our case 1mT) for significant magnetization to occur. The overall mechanisms behind the orientation effect are clearly complex. Jacob *et al.* [19] speculated that orientation dependence could be dependent on interactions of the alkali metal with impurities within the cell (either from the cell walls, or introduced by the alkali itself). In this work, the addition of a sol gel coating appeared to remove, or reduce the strong magnetisation effect for pyrex ^{129}Xe cells, although it left the cells with an orientation dependance. This appears to agree with the predictions of Jacob *et al.*, as the introduction of a coating will alter the ineractions of the polarized gas with the cell walls, but not the alkali metal. However, although not widely reported, the orientation effect has been seen in uncoated cells [33], suggesting that this may not always be the case. It is worth noting however that an uncoated cell will usually be attached to a source of polarized gas using a valve, and that magnetic effects have in some cases been traced to valves [20].

The polarization stability results shown in section 4 demonstrate an undesirable effect that may occur when using continuously pumped in-situ ^{3}He polarization devices. Although further investigations are required, if the effect is still present after instrumentation and measurement based factors have been discounted or removed then the effect could potentially be another demonstration of unusual lifetime behaviour in ^{3}He cells.

Although the orientation effect clearly has a contribution to the ^{3}He relaxation rates within the cell, it is still possible to produce cells where the T_1 is hundreds of hours, allowing high (>75%) ^{3}He polarizations, thereby creating cells suitable for use in neutron experiments. However, care needs to be taken to not magnetise cells, which could affect the proximity of the cell to high magnetic field sample environment equipment, or even the field strength of the main holding field. The magnetisation effect has large implications in MRI and NMR applications due to the proximity of the high magnetic fields required. In all cases it is clear that either cell demagnetisation is needed, or that the processes or materials that contribute to the orientation dependence, or magnetization of, ^{3}He cells must be isolated and removed from the cell making process.

6. Acknowledgments

Research at the Oak Ridge National Laboratory's Spallation Neutron Source was sponsored by the Scientific User Facilities Division, Office of Basic Energy Sciences and the U.S. Department of Energy. Steven Parnell acknowledges funding through a National Science Foundation Grant DMR-0956741. The authors would like to thank Juan Parra-Robles for assistance with the Sol-Gel coating, Graham Norquay for the ^{129}Xe T_1 measurements and Jim Wild for useful discussions.

7. References

[1] Walker T G and Happer W 1997 *Rev. Mod. Phys.* **69** 629–642
[2] Babcock E, Chann B, Walker T G, Chen W C and Gentile T R 2006 *Phys. Rev. Lett.* **96** 083003
[3] Heil W, Anderson K, Hoffmann D, Humblot H, Kulda J, Lelievre-Berna E, Scharpf O and Tasset F 1998 *Physica B* **241-243** 56–63
[4] Gentile T R, Babcock E, Borchers J A, Chen W C, Hussey D, Jones G L, Lee W T, Majkzrak C F, Donovan K V, Snow W M, Tong X, te Velthuis S G E, Walker T G and Yan H 2005 *Physica B: Condensed Matter* **356** 96 – 102
[5] Boag S, Parnell S R, Frost C D, Andersen K H and Babcock E 2007 *Physica B* **397** 179–181
[6] Fitzsimmons W A, Tankersley L L and Walters G K 1969 *Phys. Rev.* **179** 156–165
[7] Timsit R S, Daniels J M and May A D 1971 *Canadian Journal of Physics* **49** 560–575
[8] Heil W, Humblot H, Otten E, Schafer M, Sarkau R and Leduc M 1995 *Physics Letters A* **201** 337–343
[9] Rich D R, Gentile T R, Smith T B, Thompson A K and Jones G L 2002 *Appl. Phys. Lett.* **80** 2210
[10] Jacob R E, Morgan S W, Saam B and Leawoods J C 2001 *Phys. Rev. Lett* **87** 143004
[11] Boag S, Babcock E, Andersen K H, Becker M, Charlton T R, Chen W C, Dalgliesh R M, Elmore S D, Frost C D, Gentile T R, Anton R L, Parnell S R, Petoukhov A K, Skoda M W A and Soldner T 2009 *Physica B: Condensed Matter* **404** 2659 – 2662
[12] Babcock E and Ioffe A 2011 *Physica B: Condensed Matter* **406** 2448 – 2452
[13] Tong X, Jiang C Y, Lauter V, Ambaye H, Brown D, Crow L, Gentile T R, Goyette R, Lee W T, Parizzi A and Robertson J L 2012 *Review of Scientific Instruments* **83** 075101
[14] Parnell S R, Babcock E, Nünighoff K, Skoda M W A, Boag S, Masalovich S, Chen W C, Georgii R, Wild J M and Frost C D 2009 *Nuclear Instruments and Methods in Physics Research Section A* **598** 774–778
[15] Chang L J, Lee C H, Su H C, Ioffe A, Bussmann K, Parnell S R and Boag S 2009 *Chinese Journal of Physics* **47** 159–165
[16] Chen W C, Gentile T R, Fu C B, Watson S, Jones G L, McIver J and Rich D R 2011 *Journal of Physics: Conference Series* **294** 012003
[17] Jiang C Y, Tong X, Brown D R, Lee W T, Ambaye H, Craig J W, Crow L, Culbertson H, Goyette R, Graves-Brook M K, Hagen M E, Kadron B, Lauter V, McCollum L W, Robertson J L, Winn B and Vandegrift A E 2013 *Physics Procedia* **42** 191 – 199
[18] Parnell S R, Woolley E B, Boag S and Frost C D 2008 *Meas. Sci. Technol* **19** 045601
[19] Jacob R, Teter J, Saam B, Chen W C and Gentile T R 2004 *Phys. Rev. A* **69** 021401(R)
[20] Hutanu V, Rupp A, Klenke J, Heil W and Schmiedeskamp J 2007 *J. Phys. D: Appl. Phys.* **40** 4405
[21] Babcock E, Nelson I, Kadlecek S, Driehuys B, Anderson L W, Hersman F W and Walker T G 2003 *Phys. Rev. Lett.* **91** 123003
[22] Hutanu V and Rupp A 2005 *Physica B* **356** 91–95
[23] Parnell S R, Deppe M H, Parra-Robles J and Wild J M 2010 *Journal of Applied Physics* **108** 064908
[24] Teh K, de Zanche N and Wild J M 2007 *Journal of Magnetic Resonance* **185** 164–172
[25] Norquay G, Parnell S R, Xu X, Parra-Robles J and Wild J M 2013 *Journal of Applied Physics* **113** 044908
[26] Hsu M F, Cates G D, Kominis I, Aksay I A and Dabbs D M 2000 *Appl. Phys. Lett.* **77** 2069–2071
[27] Ye Q, Laskaris G, Chen W, Gao H, Zheng W, Zong X, Averett T, Cates G D and Tobias W A 2010 *The European Physical Journal A* **44** 56–61
[28] Sharma M, Babcock E, Andersen K H, Barron-Palos L, Becker M, Boag S, Chen W C, Chupp T E, Danagoulian A, Gentile T R, Klein A, Penttila S, Petoukhov A, Soldner T, Tardiff E R, Walker T G and Wilburn W S 2008 *Phys. Rev. Letters* **101** 083002
[29] Babcock E, Boag S, Becker M, Chen W C, Chupp T E, Gentile T R, Jones G L, Petukhov A K, Soldner T and Walker T G 2009 *Physical Review A: Atomic, Molecular, and Optical Physics* **80** 033414
[30] Bloembergen N and Pound R V 1954 *Physical Review* **95** 8–12
[31] Parnell S R, Boag S, McKetterick T J and Wild J M 2011 *Journal of Physics: Conference Series* **294** 012010
[32] Schmiedeskamp J, Heil W, Otten E W, Kremer R K, Simon A and Zimmer J 2006 *The European Physical Journal D - Atomic, Molecular, Optical and Plasma Physics* **38** 427–438
[33] Mada M O 2009 *Application of Hyperpolarised Helium-3 in Lung Functional Magnetic Resonance Imaging* Ph.D. thesis The University of Nottingham

Development and demonstration of in-situ SEOP ^{3}He spin filter system for neutron spin analyzer on the SHARAKU polarized neutron reflectometer at J-PARC

H Hayashida[1], T Oku[1], H Kira[2], K Sakai[1], M Takeda[1,3], Y Sakaguchi[2], T Ino[4], T Shinohara[1], K Ohoyama[5], J Suzuki[2], K Kakurai[3], M Mizusawa[2], N Miyata[2], D Yamazaki[1], R Maruyama[1], K Soyama[1] and M Arai[1]

[1]J-PARC Center, Japan Atomic Energy Agency, Tokai, Ibaraki 319-1195, Japan
[2]Comprehensive Research Organization for Science and Society, Tokai, Ibaraki 319-1106, Japan
[3]Quantum Beam Science Directorate, Japan Atomic Energy Agency, Tokai, Ibaraki 319-1195, Japan
[4]High Energy Accelerator Research Organization (KEK), Tsukuba, Ibaraki 305-0801, Japan
[5]Institute for Materials Research, Tohoku University, Sendai, Miyagi 980-8577, Japan

E-mail: hayashida.hirotoshi@jaea.go.jp

Abstract. A new neutron reflectometer, SHARAKU, with a vertical sample-plane geometry was installed at beam line 17 at J-PARC Materials and Life Science Facility. Although a polarizing supermirror was previously installed as a neutron spin analyzer on SHARAKU, a ^{3}He spin filter is advantageous because it can cover a large solid angle. An in-situ SEOP ^{3}He spin filter system using a new compact laser unit has been developed for the analyzer. In this paper, we report a successful off-specular measurement with the new compact in-situ SEOP analyzer at SHARAKU.

1. Introduction

Polarized neutron-scattering methods are powerful tools to study magnetic materials, and these methods require some polarizing device for the neutron beam. ^{3}He neutron spin filters (NSFs) are advantageous over magnetic supermirrors because they cover a large solid angle and produce polarized shorter neutron beam wavelengths. Hence, ^{3}He NSFs become popular devices for polarized neutron scattering experiments in recent years. We have also been engaged with the development of ^{3}He NSFs using the spin exchange optical pumping (SEOP) technique. The SEOP technique enables us to use a ^{3}He NSF with in-situ pumping of ^{3}He gas, which is a huge advantage for some experiments because the in-situ pumping technique allows us maintain a stable ^{3}He gas polarization over long measurement times. In fact, the development of the in-situ SEOP ^{3}He NSF systems in Japan has achieved a 73 % polarization of ^{3}He gas [1,2].

A new neutron reflectometer, SHARAKU, with vertical sample-plane geometry was installed at the beam line 17 (BL17) at the J-PARC Materials and Life Science experimental Facility (MLF) [3]. Magnetic properties in thin magnetic films are one of the main targets for SHARAKU which provides measurement capabilities of the specular and off-specular reflection. Polarizing supermirrors of reflection type (Fe/Si) have been installed as a neutron spin analyzer for SHARAKU. Some of the polarizing supermirrors are stacked vertically in order to cover a wide area (100 mm horizontal, 50 mm vertical), which analyze a polarized neutron beam with wavelengths longer than 0.2 nm [3,4]. However, the stacked supermirrors are not exactly aligned at equal angles. Therefore the analyzer gives an effect to off-specular patterns. The ^{3}He NSF is able to overcome this problem. Moreover, the ^{3}He NSF enables a large solid angle coverage and polarization of neutron beams with shorter wavelengths by controlling the pressure length of the ^{3}He gas [5-12]. An in-situ ^{3}He NSF has been developed for the analyzer on SHARAKU. A new compact laser unit has been installed inside the in-situ SEOP ^{3}He NSF to simplify the laser beam adjustment. Off-specular measurement using the in-situ SEOP ^{3}He NSF with the new compact laser unit was performed at SHARAKU. A Fe/Cr multilayered thin film with a giant magnetoresistance effect was used as a sample [13]. The experimental setup and performance of the in-situ SEOP ^{3}He NSF is reported in Section 2, and the results for the off-specular pattern measurements are describe in Section 3.

2. Experimental setup and performance of the in-situ SEOP ^{3}He NSF

Figures 1(a), (b) and (c) show a schematic diagram, picture of the experimental setup, and picture of the inside of a box of in-situ ^{3}He SEOP NSF, respectively. A V-shape transmission polarizer, which consists of Fe/Si supermirrors and a Drabkin spin flipper, has been installed in front of the sample stage [14]. The polarizer can be moved in horizontally, and we can choose polarized or unpolarized modes for experiments. SHARAKU has nine slits in its beam path. The eighth and ninth slits are shown in Figs. 1(a) and (b). There is a large area, about 1500 mm in length and 500 mm in width, between eighth and ninth slits, and the in-situ ^{3}He SEOP NSF is set to a wide area. A guide coil is set downstream of the eighth slit to avoid neutron spin depolarization. A two-dimensional position-sensitive detector is set downstream of the ninth slit. Figure 1(c) shows the inside of the box of the in-situ ^{3}He NSF and the compact laser unit; the quartz mirror and solenoid coil are also shown. The ^{3}He cell is set inside of the solenoid coil. The size of the ^{3}He cell is 35 mm in diameter and 55 mm in length. The pressure length-product of the cell is 11 bar cm. The cell was fabricated using GE180, and Rb is doped inside the cell. The quality factor, $Q = P_n^2 T_r$,where P_n is the polarization of the neutron beam and T_r is the transmission intensity of neutron beam, shows similar performance to a ^{3}He cell because the P_n and T_r are in a trade-off relationship. The Q factor curve of the ^{3}He cell with a pressure length of 11 bar cm exhibits a wavelength peak at 0.23 nm, which is evaluated with 70% P_{3He}. Because a neutron intensity peak of the time-of-flight (TOF) spectrum for SHARAKU is 0.25 nm, which is nearly identical to the peak of the Q factor, the ^{3}He cell with pressure length of 11 bar cm is selected for this experiment. The relaxation time of the ^{3}He cell was 130 h, which was estimated from a measurement of the NMR signal of ^{3}He. Because the diameter of the cell is 35 mm, only a square area of 25 mm by 25 mm can be covered, which is not sufficient for SHARAKU. Therefore we must develop a larger ^{3}He cell, at least 150 mm in diameter, for the SHARAKU analyzer.

A diode laser element, volume holographic grating (VHG) element, and quarter-wavelength plate are contained inside the laser unit, whose dimensions are 135 mm (L), 135mm (W), and 100 mm (H). The laser unit is a joint-developed instrument by the authors and Toyoko Kagaku Co., Ltd. The laser unit is very compact, and its portability is drastically improved. The laser output power is 30 W with a peak wavelength of 794.7 nm. The unit is cooled with an air-cooled fan, and the temperatures of the diode laser element as well as the VHG element are controlled by a Peltier element, which enables us to control the peak of the output laser beam.

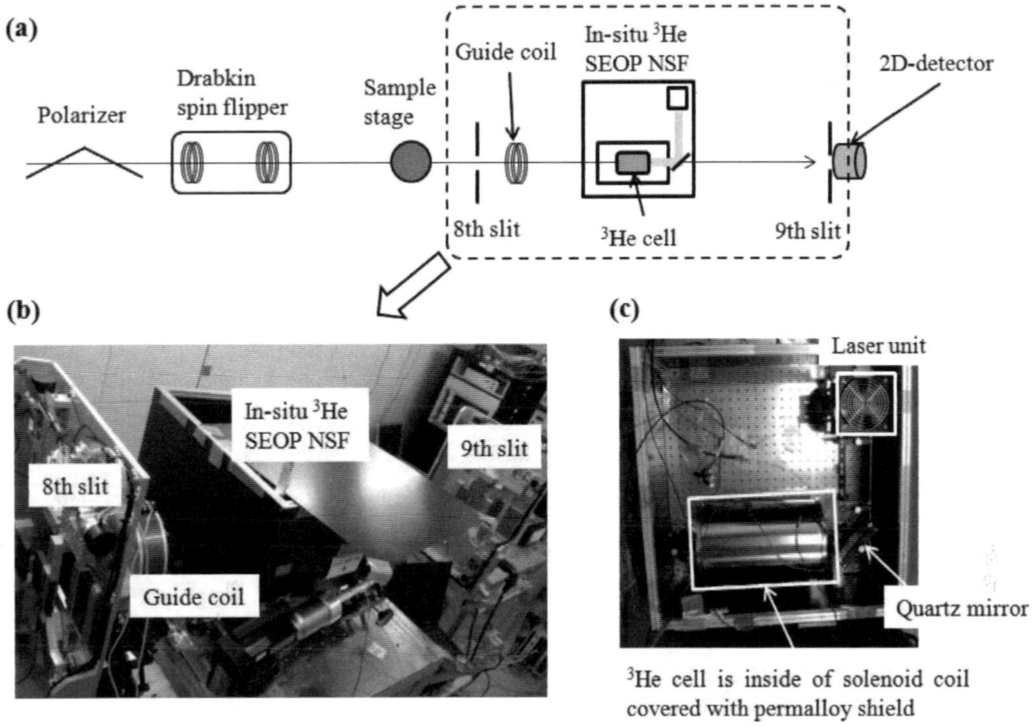

Figure 1. (a) Schematic diagram of the experimental setup at SHARAKU. (b) Picture of the setup of the in-situ ^{3}He SEOP NSF. (c) Picture showing the inside of the BOX of the in-situ ^{3}He SEOP NSF.

A performance test on the in-situ SEOP ^{3}He NSF was performed without a sample. The wavelength dependence of the neutron beam transmission intensity through the ^{3}He cell was measured. The polarization of ^{3}He gas can be obtained from the neutron beam transmission intensity of the ^{3}He cell filled with unpolarized ^{3}He gas T_0, and those with polarized ^{3}He gas T as follows [15].

$$\begin{aligned} T &= T_E e^{-\sigma\rho t} \cosh P_{3He}\, \sigma\rho t, \\ T_0 &= T_E e^{-\sigma\rho t}, \\ P_{3He} &= \frac{1}{\sigma\rho t} \operatorname{acosh} \frac{T}{T_0} \\ \sigma\rho t &= 7.97 \times 10^{-2}\, p\, t\, \lambda \end{aligned} \quad (1)$$

where P_{3He} is the polarization of ^{3}He, σ is the neutron absorption cross section for the unpolarized ^{3}He, ρ is the number density of ^{3}He determined, t is the length of the ^{3}He cell, and T_E is the transmission of an empty cell. p is the ^{3}He pressure at 0℃, and λ is the neutron wavelength. The wavelength dependences of T and T_0 were measured with the TOF neutron beam at BL17 and the polarization of ^{3}He P_{3He} was obtained, as shown in Fig. 2. The result shows the polarization of ^{3}He to be 68 %. Moreover the in-situ SEOP system remained stable over 5 days and the ^{3}He gas polarization of 68 % was also stable over the duration of the experiment.

Figure 3 shows the wavelength dependence of the overall instrumental asymmetry A. The A was obtained from I_{ON} and I_{OFF} as follows

$$A = \frac{I_{ON} - I_{OFF}}{I_{ON} + I_{OFF}} \quad (2)$$

where I_{ON} is the TOF spectrum of the neutron intensity with the Drabkin spin flipper ON and I_{OFF} is TOF spectrum with the flipper OFF. The ^{3}He NSF was used as an analyzer. The instrumental asymmetry was more than 95 % for wavelengths longer than 0.35 nm and was stable over the off-specular measurements, which are described in the next section. Polarization efficiencies of the polarizer and Drabkin spin flipper were included in the instrumental asymmetry A shown in Fig. 3.

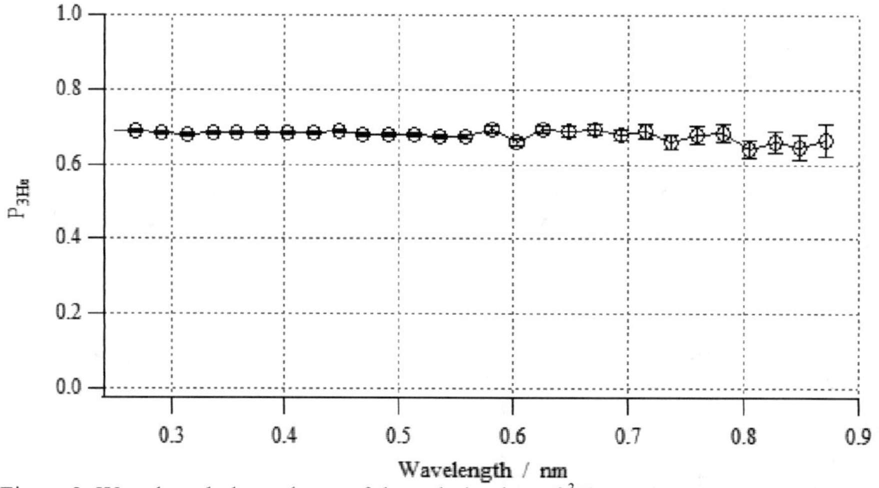

Figure 2. Wavelength dependence of the polarization of ^{3}He.

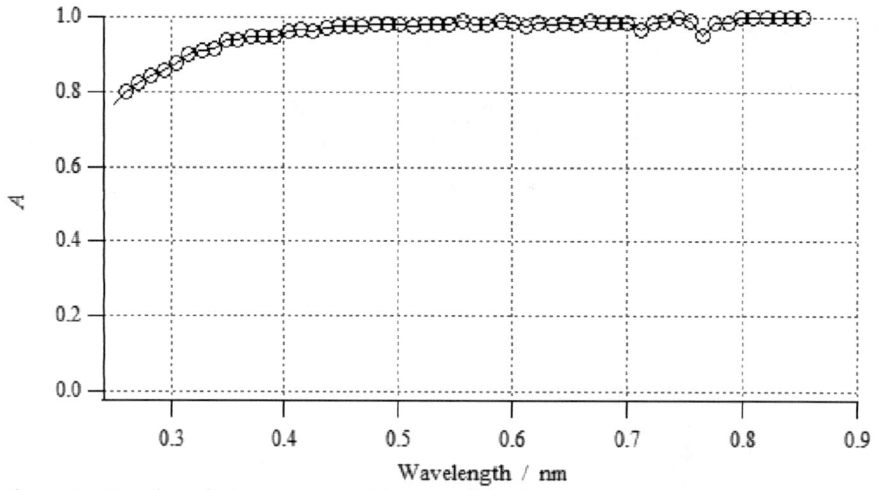

Figure 3. Wavelength dependence of the overall instrumental asymmetry.

3. Results of the off-specular measurement

A demonstration of off-specular measurement using the in-situ SEOP ^{3}He NSF was performed at SHARAKU. The Fe/Cr multilayered thin film with the giant magnetoresistance effect was used as a sample. A 10 kOe maximum magnetic field can be applied to the sample. For low magnetic field conditions, the Fe layers create antiferromagnetic correlations that produce a specular peak and off-specular reflection. On the other hand, in a strong magnetic field, antiferromagnetic correlations are suppressed and both the specular peak and off-specular reflection disappear [13]. Because the polarizing supermirror on the SHARAKU is a transmission type, neutron spins that are antiparallel to a guide field can be produced. On the ^{3}He NSF analyzer, neutron spins that are parallel to a guide field can be transmitted. Hence, specular and off-specular reflections containing a spin flip, (-, +), can be measured using the OFF mode of the Drabkin flipper. On the other hand, those reflections with a non-spin flip, (+, +), can be measured using the ON mode of the Drabkin flipper. In this experiment, measurements containing a spin flip (-, +) and non-spin flip (+, +) were performed with a 200 Oe (low magnetic field condition) and 10 kOe (strong magnetic field condition) magnetic field, respectively.

Figure 4 shows the results of this measurement indicated on an Q_x-Q_z map. Q_x and Q_z show the momentum transfers of the in-plane and depth structures on the Fe/Cr multilayered sample, respectively. Polarization efficiency correction was performed for all results shown in Fig. 4. Specular reflection appears at $Q_x = 0$, while off-specular reflection appears at $Q_x \neq 0$. Figures 4(a) and (b) show the results of the Q_x-Q_z map with an applied magnetic field of 200 Oe. Figure 4(a) is the map with a spin flip (-, +), and (b) is the map with a non-spin flip (+, +). Off-specular reflections around $Q_z = 0.8$ nm^{-1} are clearly observable for both the spin flip (-, +) and non-spin flip (+, +) conditions, which indicates the existence of antiferromagnetic correlations. In the spin flip condition (-, +), the off-specular reflection is much clearer than that of the non-spin flip condition (+, +). These features are consistent with the features reported in Ref. [3].

Figure 4(c) and (d) are the results of the Q_x-Q_z map with an applied magnetic field of 10 kOe. Figure 4(c) is the map with a spin flip (-, +), and Fig. 4(d) is the map with a non-spin flip (+, +). Off-specular reflections around $Q_z = 0.8$ nm^{-1} disappeared for both the spin flip (-, +) and non-spin flip (+, +) conditions, which indicates that antiferromagnetic correlations are suppressed. These features are also consistent with the features reported in Ref. [3].

For our demonstration of the off-specular measurement at SHARAKU using an in-situ SEOP ^{3}He NSF, consistent results with the features reported in Ref. [13] were successfully observed. However, the size of the ^{3}He cell was not large enough to cover the entire scattering area (100 mm horizontal and 50 mm vertical) at SHARAKU. High polarizations of ^{3}He gas provide high transmission neutron beam intensities and high neutron beam polarization at short wavelengths of 0.2 nm. Hence, we plan to continue the development of a larger ^{3}He cell and SEOP system with a high polarization of ^{3}He gas.

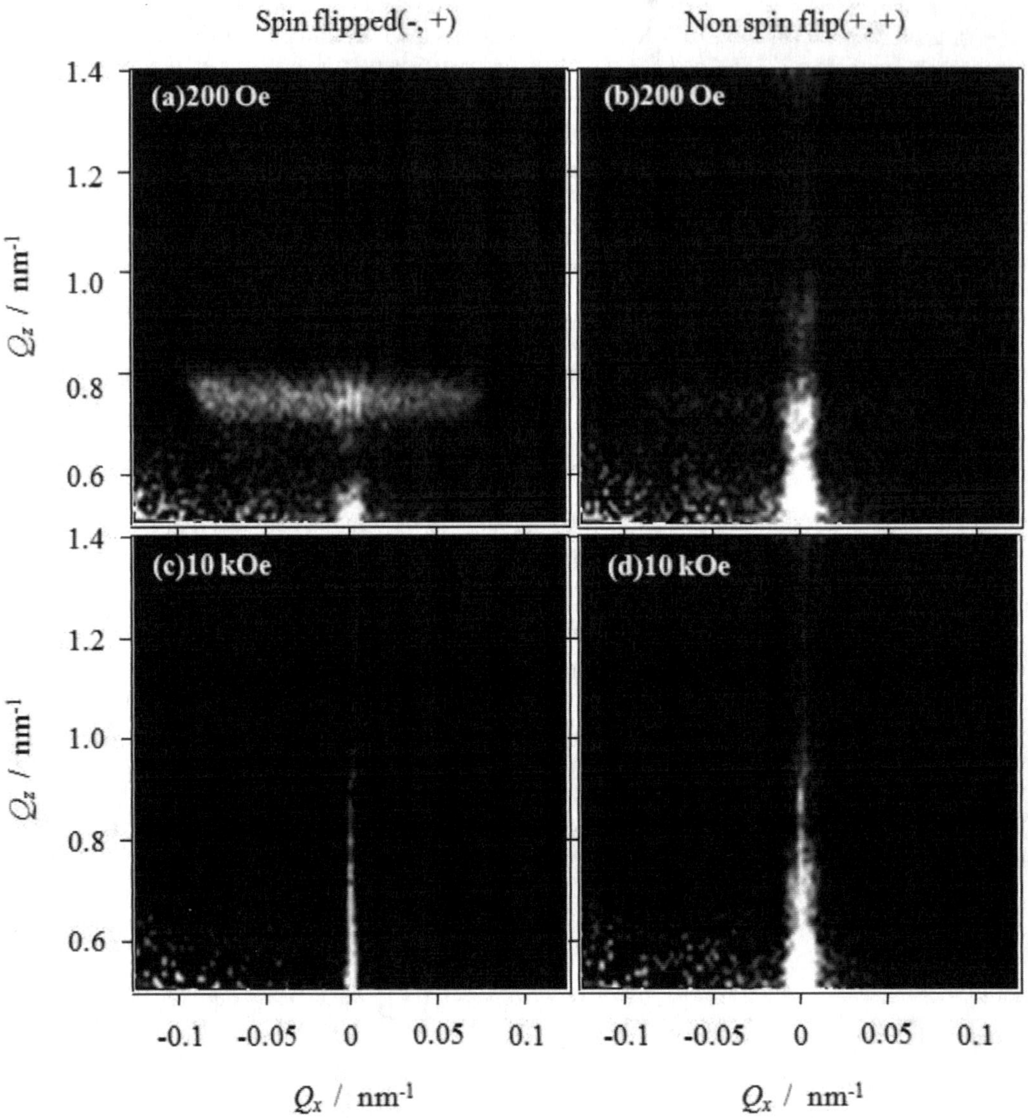

Figure 4. Off-specular measurement results indicated in the Q_x-Q_z map. (a) and (b) are the results with an applied magnetic field of 200 Oe. (a) is the map with a spin flip (-, +) and (b) is the map with a non-spin flip (+, +). (c) and (d) are the results with an applied magnetic field of 10 kOe. (c) is the map with a spin flip (-, +) and (d) is the map with a non-spin flip (+, +).

4. Summary

A new compact laser unit for the SEOP ^{3}He NSF was installed, and a performance test of the in-situ ^{3}He SEOP NSF was performed at SHARAKU. In addition we were able to achieve the following: the polarization of ^{3}He reached 68 %, polarization of neutron spin was more than 95 % for wavelengths longer than 0.35 nm, and the polarization remained stable for 5 days, which shows a good performance for practical use. Moreover, the portability of the in-situ SEOP ^{3}He NSF was drastically improved.

The first test of the compact in-situ SEOP ^{3}He NSF for the measurement of an off-specular reflection was performed at SHARAKU. The Fe/Cr multi-layered thin film with a giant magnetoresistance effect was used as a sample. For the low magnetic field condition with a weak guide field of 200 Oe, the off-specular reflection was clearly observed, which indicates the existence of antiferromagnetic correlations. For the strong magnetic field condition with an applied field of 10 kOe, the antiferromagnetic correlations were suppressed and the off-specular reflection disappeared. These results are consistent with those reported in Ref. [13].

A demonstration of in-situ SEOP ^{3}He NSF with the new compact laser unit and off-specular measurement using the in-situ SEOP ^{3}He NSF was successfully performed at SHARAKU.

Acknowledgments
This work was supported by the Quantum Beam Fundamentals Development Program, MEXT.

References

[1] Kira H, Sakaguchi Y, Oku T, Suzuki J, Nakamura M, Arai M, Endoh Y, Chang J L, Kakurai K, Arimoto Y, Ino T, Shimizu M H, Kamiyama T, Ohoyama K, Hiraka H, Tsutsumi K and Yamada K 2011 *J. Phys.: Conference Series **294** 012014.*
[2] Ino T, et al, 2009 *Physica B **404** 2667.*
[3] Takeda M, et al, 2012 *Chinese J. Phys. **50** 161.*
[4] Hayashida H, Takeda M, Yamazaki D, Maruyama R, Soyama K, Kubota M, Mizusawa T, Yoshida N and Sakaguchi Y 2013 *Physics Procedia **42** 130-135.*
[5] Jones G.L., et al, 2006 *Physica B **385-386** 1131.*
[6] Chupp T.E., et al, 2007 *Nucl Instrm. Meth. A **574** 500.*
[7] Boag S, et al, 2009 *Physica B **404** 2659.*
[8] Tong X, et al, 2012 *Rev. Sci. Instrm. **83** 075101.*
[9] Nickel B, et al, 2001 *Rev. Sci. Instrm. **72** 163.*
[10] Radu F, et al, 2003 *Physica B **335** 6367.*
[11] Chen W.C, 2004 *Rev. Sci. Instrm. **75** 3256.*
[12] Anderson K.H, et al, 2006 *Physica B **385-386** 1134.*
[13] Takeda M, Endoh Y, Kamijo Y and Mizuki J 1995 *Physica B **213&214** 248.*
[14] Becker J, Ebert M, Grossmann T, Heil W, Hofmann D, Humbolt H, Leduc M, Otten W E, Rlhe D, Siemensmeyer K, Surkau R, Steiner M, Tasset F and Trautmann N 1997 *Physica B **234-236** 1078.*
[15] Culter K.P, et al, 1990 *Nucl. Instrm. Meth. A **288** 463.*

A new type of wide-angle supermirror analyzer of neutron polarization

V G Syromyatnikov [1,2], V A Ulyanov [1], V Lauter [3], V M Pusenkov [1], H Ambaye[3], R Goyette [3], M Hoffmann [3], A P Bulkin [1], I N Kuznetsov [1] and E N Medvedev [1]

[1] Neutron Research Department, Petersburg Nuclear Physics Institute, Orlova Roscha, Gatchina, St. Petersburg, 188300, Russia
[2] Physics Department, St. Petersburg State University, Ulyanovskaya 1, Petrodvorets, St. Petersburg, 198504, Russia
[3] Neutron Sciences Directorate, Oak Ridge National Laboratory, P.O. Box 2008, Oak Ridge, TN 37831, USA

E-mail: svg@pnpi.spb.ru

Abstract. We describe here a new type of wide-angle supermirror-based multichannel analyzer configured in the fan orientation. The increased channel width allows for reflections only from one of the channel's walls, so that the overlap of beams propagating through neighboring channels is avoided. However the straight beam, which is unavoidably propagating through the channels with the increased width, is blocked by an absorbing mask at the entrance of the analyzer. The neutron transmission of such analyzer is 22% higher and the number of the supermirrors needed to cover the same beam cross section is 16% less in comparison with a conventional fan analyzer. Results of the calculations and first tests of the analyzer at the Magnetism Reflectometer at Oak Ridge National Laboratory, USA, are presented.

1. Introduction

In the past few years, two-coordinate position sensitive detectors (2d-PSD) with the sensitive area greater than 400 cm^2 have been widely used at polarized neutron instruments. Therefore, there is an urgent need in wide-angle polarization analyzers allowing for polarization analysis (PA) over the whole large area of the detected beam. The use of wide-aperture fan type multichannel supermirror analyzer (MSA) (see Fig. 1) is one of the possible choices that suites well the requirements of neutron reflectometry.

Two wide-angle fan type supermirror analyzers have been manufactured in recent years in the PNPI. The first one, designed for the use with a 1-dimensional (linear) PSD at the TOF reflectometer REMUR (JINR, Dubna) [1], consists of 94 channels and has the following characteristics: the angular acceptance of scattered beam is about 1.5 degrees, the cross-section of the entrance window is (40×120) mm^2 (w × h) and is optimized for the neutron wavelength band of $\lambda = 1.5 - 10\,\text{Å}$. The analyzing (polarizing) efficiency averaged over the wavelength band is about 0.95 - 0.96.

Another wide-angle fan type MSA designed for the neutron reflectometer NeRo (HZG, Germany) [2] has been used for a monochromatic beam of cold neutrons with a wavelength of 4.35Å.

This analyzer has been installed in front of the 2d-PSD with the sensitive area of 250 × 250 mm^2. It consists of 90 channels, has an entrance window of (90×160) mm^2 (w × h) and provides an anglular acceptance of about 6.1 deg. The analyzing (polarizing) efficiency of such analyzer is about 0.98 at above the given wavelength.

In the present paper we describe a wide-angle MSA designed for the Magnetism Reflectometer at the SNS (ORNL, USA) [3, 4] that operates in a wavelength range from 2.5 - 8Å.

2. Generic fan supermirror analyzer

The design of the analyzer for the Magnetism Reflectometer at the SNS is based on two MSAs mentioned in Ch.1, whose generic layout is depicted in Fig. 1.

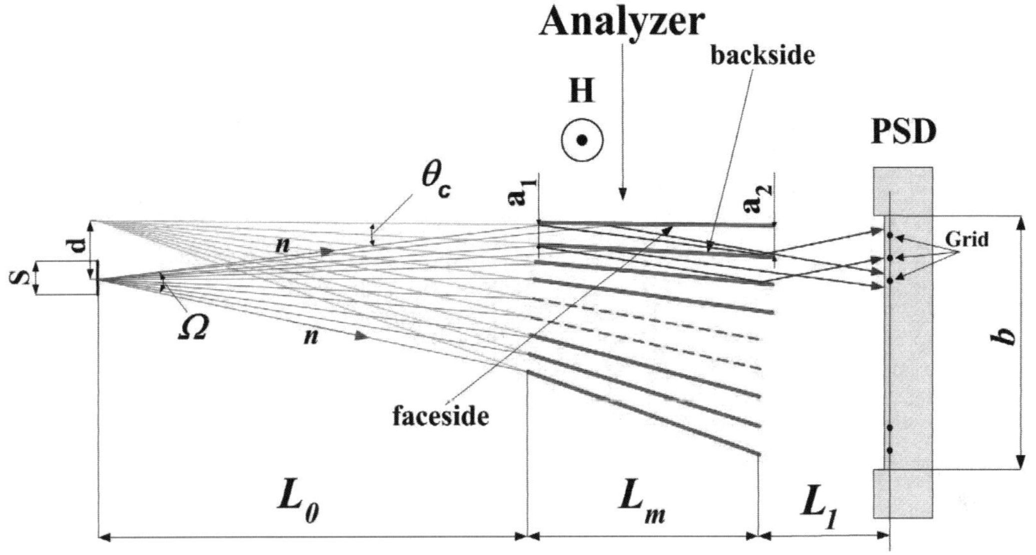

Figure 1. Sketch of the layout of a generic fan type MSA.

The entrance of the analyzer is at the distance L_0 from the center of the sample and covers a beam of the width *s*. Each channel of the analyzer consists of two supermirrors of projected length L_m labeled as "*faceside*" (FS) and "*backside*" (BS). They are separated by wedge-shaped spacers, widening from the entrance to the exit of the analyzer and are installed above and below the supermirrors along the channels. The supermirror coatings for FS and BS are m_1 and m_2, respectively. The channel widths at the entrance is a_1, the width at the exit is a_2 and the constant channel height is *h*. The neighboring channels of the analyzer are turned relative to each other by the angle $\Delta\theta = (a_1 - a_2)/L_m$. The exit of the analyzer is placed at a distance L_1 from the surface of a 2d-PSD with a width of *b*, so that the illuminated area of the detector is *(b* x *h)*. The analyzer is placed in the external magnetic field *H* parallel to the plane of the supermirrors; the magnitude of the magnetic field should be close to the saturation field of the supermirrors. The turquois solid lines, which are the prolongation of the channel walls, are crossing in one point, the analyzer focus, at the distance L_0 from the analyzer entrance. The focus is shifted by distance *d* from the sample axis. The maximal glancing angle θ_{max} under which the neutron beam impinges on the walls of the channel shouldn't exceed θ_c, which is the critical angle of the supermirror coating for neutrons with the minimal used wavelength. Ω is the acceptance angle of the analyzer. Further on, we will call such analyzer “a conventional fan MSA”: The parameters of such an conventional fan MSA optimized for an analyzer-detector distance of 110 mm are calculated according the algorithm described in Ref. [2] and given in Table 1.

Journal of Physics: Conference Series **528** (2014) 012021 doi:10.1088/1742-6596/528/1/012021

Table 1. Parameters of the conventional fan MSA optimized for analyzer-detector distance of 110mm.

Length of supermirrors	L_m = 100 mm
Height of supermirrors	h_0 = 150 mm
Height of analyzer channel	h = 120 mm
Strength of magnetizing field	H = 650 G
Number of supermirrors in analyzer	N = 185
Polarizing Fe/Si supermirror coating: FS	m_1 = 4.0
BS	m_2 = 2.5
Distance from sample to entrance of analyzer	L_0 = 2220 mm
Maximal glancing angle in the channel	θ_c = 12.6 mrad
Thickness of glass substrate	d_0 = 0.4 mm
Width of the beam blocked by sample	s = 2 mm
Channel width at the entrance	a_1 = 0.942 mm
Channel width at the exit	a_2 = 1.0 mm
Displacement of the analyzer focus relative to sample axis	d_0 = 6.4 mm
Distance from analyzer exit to detector	L_1 = 110 mm,
Sensitive width of 2d-PSD	b = 220 mm
Acceptance angle for the beam reflected from sample	Ω = 6.36°
Entrance cross-section of analyzer	(120 x 247) mm^2

Q-dependences (Q is the momentum transfer) of the reflection coefficients of supermirrors for the spin-up component of the neutron beam are shown in Fig. 2. Using them one can calculate [5] the transmission function of a single channel of the analyzer (Fig. 3).

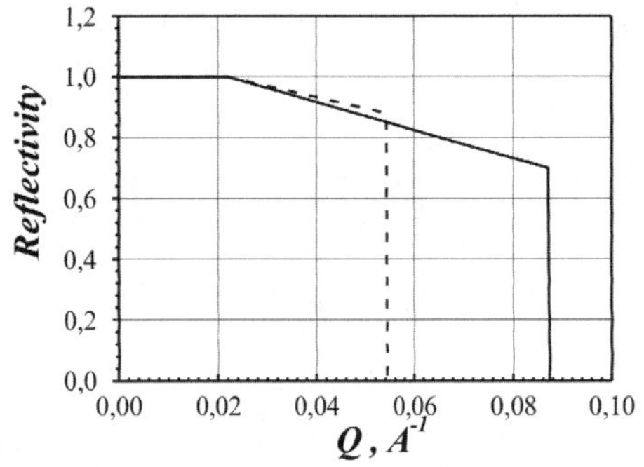

Figure 2. Reflection coefficients for the spin-up neutrons for supermirrors with m_1 = 4 at the FS wall (solid line) and m_2 = 2.5 at the BS wall (dash line) as functions of Q.

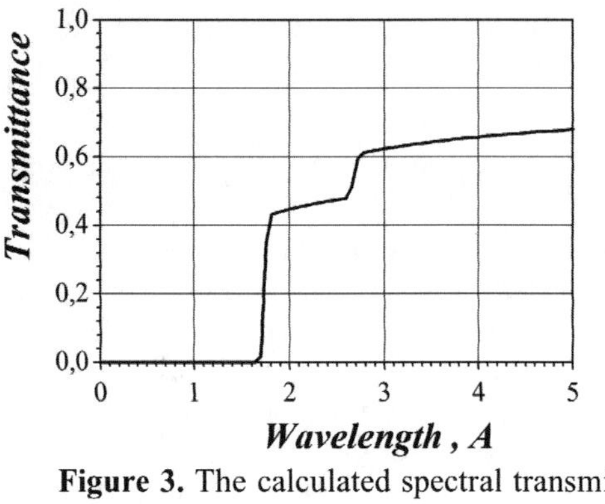

Figure 3. The calculated spectral transmission for the spin-up neutrons for a single channel of the conventional fan analyzer with m_1 = 4 at the FS wall and m_2 = 2.5 at the BS wall.

The calculated spatial distribution of the beam transmitted through a single channel of the conventional fan MSA at the detector surface separated by 110 mm is shown in Fig. 4. One can see that neutrons with the wavelengths λ=2Å and 3Å are transmitted to one peak of 0.85 mm width. However, there is also a second peak for λ=3Å, so that the full width of the reflected beam becomes 3.07 mm. Thus, there is a double reflection (from the BS wall) for neutrons with a wavelength larger than 3Å because of the small (about 1mm) channel width at the analyzer exit that results in the peak at 3mm (Fig.4). This leads to the undesirable overlap of beams propagating through the neighboring channels. The simplest solution of this problem would be to place the detector sensitive area very close to the exit of the analyzer. However, practically this distance could only be reduced to 66 mm that is not sufficient to avoid the overlap.

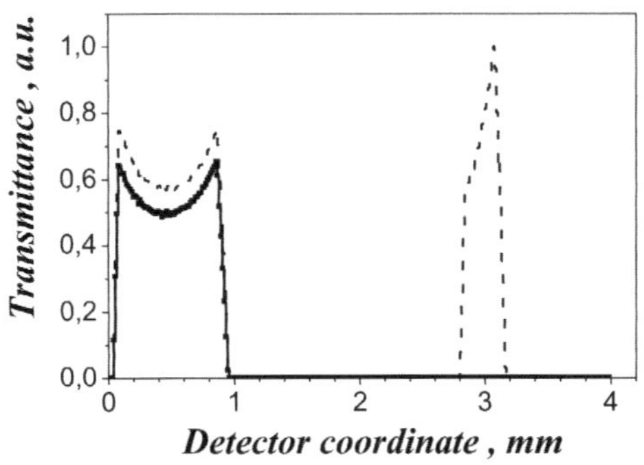

Figure 4. Calculated intensity distributions for one channel of the conventional fan analyzer. The solid line corresponds to spin-up neutrons λ=2Å, the dash line corresponds to spin-up neutrons with the wavelength λ=3Å.

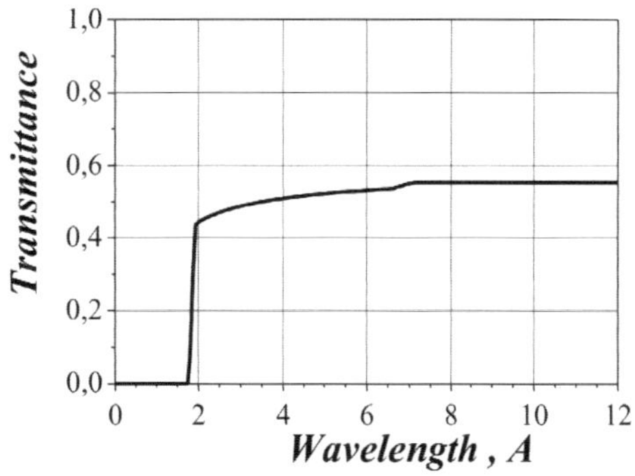

Figure 5. The spectral transmission of a single channel of the conventional fan analyzer (m_1 = 3.7 (FS) and m_2 = 0 (BS)).

Trying to solve the situation with the reflection from the BS wall, we painted it with an absorbing layer with m_2 = 0 (e.g. a gadolinium oxide layer). This allows for suppression of the parasitic reflection from the BS wall (peak at 3mm), however the width of the transmitted beam for both wavelengths is decreased to 0.8 mm. The transmission of spin-up neutrons through one channel of the conventional MSA as a function of the neutron wavelength is presented in Fig. 5. The maximum transmission of 0.55 is achieved for neutrons with wavelengths above 7.1Å.

3. The fan analyzer with an absorbing mask at the channel entrance

As it was shown in Ch.2, the neutron reflection from the BS of the narrow analyzer channel results in undesirable broadening of the outgoing beam (see also Fig. 6a), however this effect can be suppressed by painting the backside of the channel with an absorbing layer. In this chapter we will describe another approach that allows us to obtain even better results.

The idea is to increase the width of the analyzer channels, so that any double reflection will be avoided and simultaneously to block the part of the neutron beam, which can propagate through the channel without touching its walls and therefore will not be analyzed (see Fig. 6b). This can be achieved by using a neutron absorbing mask at the entrance of the analyzer. This mask is made of stripes of a thin polyester film coated with gadolinium oxide. The thickness of the mask is calculated using simple geometrical considerations and is equal to $a_1^* - a_1$ where a_1^* is the width of the channel at the entrance and a_1 is the effective channel width with the accounting for the mask thickness:

$$a^*_1 = L_{m2}\cdot(\theta_c - s/L_0)/(1+L_{m2}/L_0); \qquad a_1 = L_m\cdot(\theta_c - d_0/L_0)/(1+L_m/L_0),$$

The parameters of such improved analyzer and adopted to the requirements of the Magnetism Reflectometer at the SNS, with a wavelength range of (2.5-8)Å and an entrance cross-section of 250×120 mm^2 (w × h), are given in Table 2. One can see, that the thickness of the mask should be not less than 0.12 mm to block the straight beam. The calculated spatial distribution in the beam transmitted through a single channel for s=2 mm is presented in Fig. 7: the straight beam is fully blocked and the width of the reflected beam is about 1.3 mm, which is less than a_2+d_0=1.65 mm, the period of the channels at the exit of the analyzer. It means that the overlap of neutron beams passing through neighboring channels is avoided. The calculated spectral transmission through one channel for spin-up neutrons is presented in Fig. 8; the SM coatings of walls of the analyzer are $m_1 = 3.7$ (FS) and $m_2 = 0$ (BS).

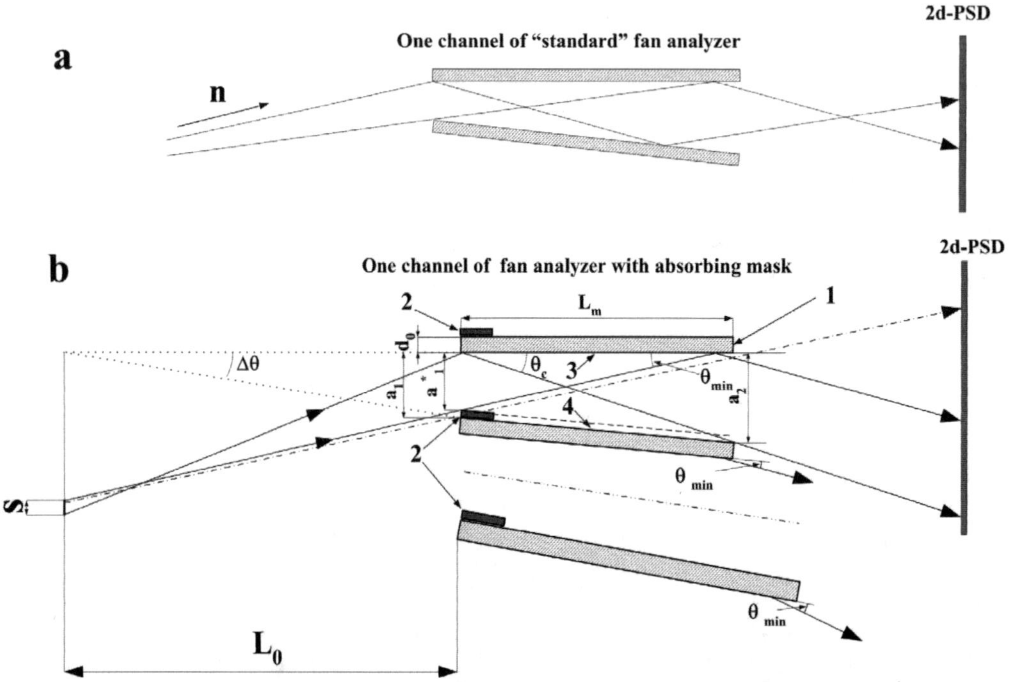

Figure 6. Layout of a conventional analyzer (a) and the analyzer with the neutron absorbing mask at the entrance (b). 1 – glass substrate with thickness d_0; 2 – absorbing mask; 3 – FS wall; 4 – BS wall.

Table 2. The optimized parameters of the new variant of fan analyzer with absorbing mask.

Length of supermirrors	$L_m = 100$ mm
Height of supermirrors	h_0=150 mm
Height of the analyzer channel	$h = 120$ mm
Strength of a magnetizing field	$H = 650$ G
Number of supermirrors in the analyzer	$N = 159$
Polarizing *Fe/Si* supermirror coating: FS	$m_1 = 3.7$
BS	$m_2 = 0$
Distance from a sample to an entrance of the analyzer	$L_0 = 2260$ mm
Maximal glancing angle in the channel	$\theta_c = 12.5$ mrad
Thickness of a glass substrate of supermirror	$d_0 = 0.4$ mm
Width of the beam blocked by sample	$s = 2$ mm
Channel width at entrance with spacer	$a_1 = 1.180$ mm
Actual width of channel at entrance including mask	$a^*_1 = 1.059$ mm
Channel width at the exit	$a_2 = 1.25$ mm
Distance from the analyzer exit to the detector	$L_1 = 66$ mm
Sensitive width of 2d-PSD	$b = 220$ mm
Acceptance angle for the beam reflected from a sample	$\Omega = 6.36°$
Entrance cross-section of the analyzer	(120 x 250) mm^2

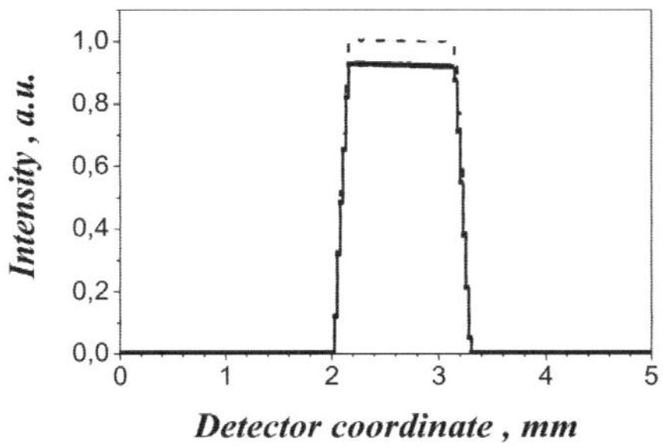

Figure 7. The calculated spatial distribution of spin-up neutron intensity passing through the analyzer with a mask at its entrance. The solid line corresponds to neutron intensity for λ=2Å, the dash line for λ=3Å.

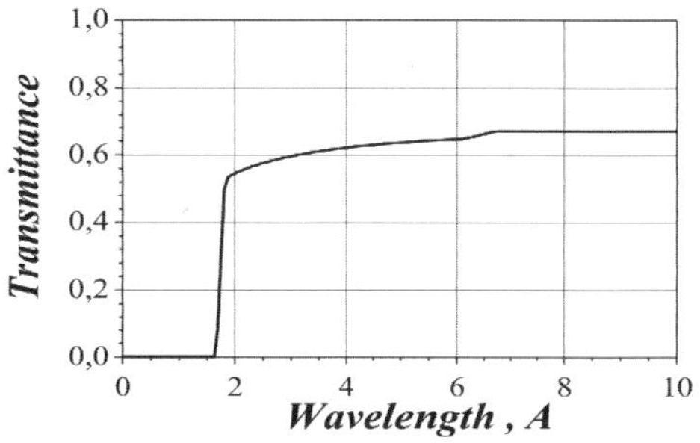

Figure 8. The calculated spectral transmission through one channel of analyzer with the mask at its entrance for spin-up neutrons.

As one can see from Fig. 8, the maximal value of the transmission for this analyzer is 0.67 for λ>6.7 Å, i.e. about 22 % higher than for a conventional fan analyzer (Fig. 5) over the whole used wavelengths range. It should be noted that due to the increased width of the channel's exit for this version of the analyzer one needs 16% less supermirrors than for a conventional analyzer covering the neutron beam of the same cross-section.

Principally, because of the use of an absorbing mask, no coating of the BS wall is necessary. However, the deposition of a supermirror coating only on one side of a thin glass substrate results in mechanical tensions and in the deformation of the flat substrate [6]. To compensate for this effect, one can cover the BS wall with a layer of the gadolinium oxide, as in the conventional fan MSA.

4. Results of test experiments

The analyzer improved by absorbing masks has been designed in the PNPI for the Magnetism Reflectometer at the SNS on the basis of the above presented design using 159 polarizing neutron Fe/Si supermirrors made by SwissNeutronics [7]. The magnetic system of the analyzer [8] has been built on the basis of permanent rare-earth NdFeB magnets. In this way a magnetic field of up to 65mT (at the middle of the magnetic system) are achieved over the a height of 120mm of the super mirror channels and with a high uniformity of the magnetic field over the supermirror's area.

The test of the analyzer has been carried out at the polarized neutron beam in a wavelength band from 2.2 – 4.7 Å. The detector surface has been separated from the analyzer exit face by 66 mm. The polarization efficiency of each 10th analyzer channels has been measured using a highly efficient radio-frequency adiabatic spin-flipper. The wavelength dependency of the polarizing efficiency for the central (0) and the two side channels (40). (-50) are presented in Fig. 9, showing a very high level of polarizing efficiency: (40): 0.988±0.002, (0): 0.987±0.002 (-50): 0.989±0.002, respectively. The spectral transmission of spin-up neutrons for the central channel is shown in Fig. 10: as it is seen the transmission is rather close to the calculated values (cf. Fig. 8). The results of the measurements on other the analyzer channels will be presented elsewhere [9].

5. Conclusions and acknowledgements

We present a new version of wide-angle high efficiency supermirror fan polarization analyzer with a neutron absorbing masks at the entrance designed for the operation with a large size 2d-PSD. The first analyzer has been designed and manufactured in the PNPI for the use at the Magnetism Reflectometer (SNS, Oak Ridge National Laboratory) that operates with a wide neutron wavelength

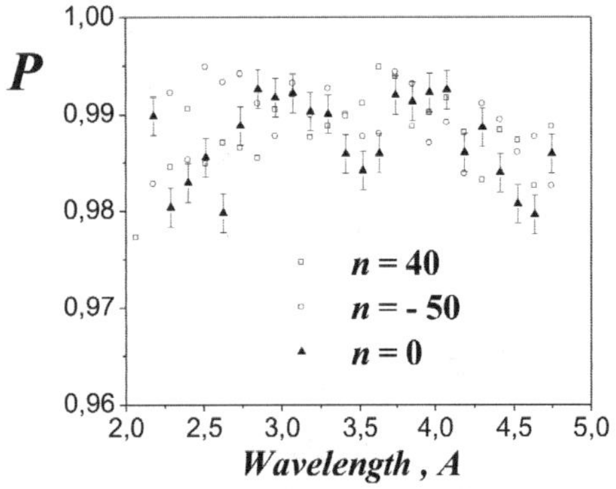

Figure 9. Spectral polarizing efficiency for three analyzer channels (n = 40, 0, -50).

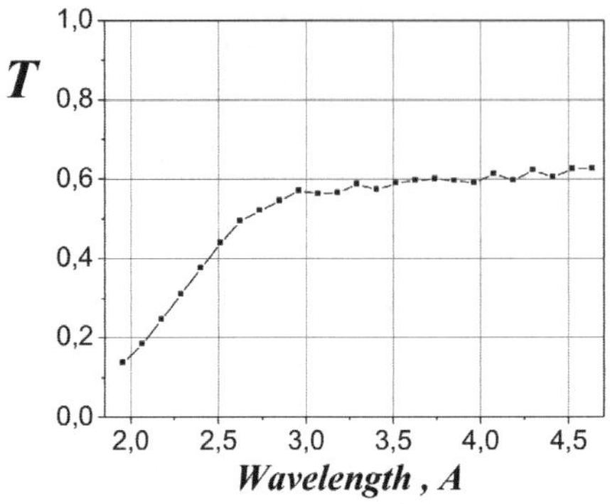

Figure 10. Wavelength dependence of transmission through the central analyzer channel for spin-up neutrons.

band of (2.2 – 4.7) Å. In comparison with the conventional type of fan analyzer, the presented analyzer requires 16% less supermirrors and demonstrates significant improvements in the neutron beam transmission that is 22 % higher in the whole range of used wavelengths.

Test measurements on 10 selected channels have demonstrated a very high analyzing power and high luminosity of the analyzer: the averaged level of polarizing efficiency over the wavelengths band (2.2 – 4.7) Å is about 0.99 and the transmission for spin-up neutrons is close to the calculated value and approaches 60% for wavelengths above 3.7 Å (has been only tested for the central channel).

This work was partly supported by RFBR grant No. 12-02-12066-ofi_m. The work performed at SNS at ORNL is sponsored by U.S. Department of Energy, Office of Energy Efficiency and Renewable Energy. Authors also thank S.I. Kalinin for help in the designing of the analyzer, A.G. Gilev for calculations of the analyzer magnetic system, N.K. Pleshanov for his interest in this work.

References

[1] Ul'yanov V A, Nikitenko Yu V, Pusenkov V M, Kozhevnikov S V, Jernenkov K N, Pleshanov N K, Peskov B G, Petrenko A V, Proglyado V V, Syromyatnikov V G, Schebetov A F, *Preprint JINR* E13-2004-161, Dubna;
Nikitenko Yu V, Ul'yanov V A, Pusenkov V M, Kozhevnikov S V, Jernenkov K N, Pleshanov N K, Peskov B G, Petrenko A V, Proglyado V V, Syromyatnikov V G, Schebetov A F 2006 *Nuclear Instruments and Methods A* **564** 395
[2] Syromyatnikov V G, Schebetov A F, Lott D, Bulkin A P, Pleshanov N K, Pusenkov V M 2011 *Nuclear Instruments and Methods A* **634** 126
[3] Lauter V, Ambaye H, Goyette R, Lee W H, Parizzi A and Klose F 2009 *Physica B* **404** 2543
[4] http://neutrons.ornl.gov/mr/
[5] Pusenkov M, Schebetov A. Gibcus H, Gommers R, F. Labohm L, de Haan V, van Well A. 2002 *Nuclear Instruments and Methods A* 492 105.
[6] Kumar M S, Boni P, Tixier S, Clemens D 1998 *Physica B* **241-242** 95
[7] http://www.swissneutronics.ch
[8] Gilev A G, Ulyanov V A, Syromyatnikov V G, Lauter V, Kuznetsov I N, Medvedev E N 2013 *Preprint PNPI* N2916 15 p (in Russian)
[9] Lauter V, Syromyatnikov V G, Ul'yanov V A et al., to be published

Journal of Physics: Conference Series **528** (2014) 012022 doi:10.1088/1742-6596/528/1/012022

Neutron spin filtering with dynamically polarized protons using photo-excited triplet states

T.R. Eichhorn[1,2], **N. Niketic**[1], **B. van den Brandt**[1], **P. Hautle**[1], **W.Th. Wenckebach**[1,4]

[1] Paul Scherrer Institute, CH - 5232 Villigen PSI, Switzerland
[2] Laboratory of Functional and Metabolic Imaging, Ecole Polytechnique Fédérale de Lausanne (EPFL), CH-1015 Lausanne, Switzerland

E-mail: patrick.hautle@psi.ch

Abstract. The use of polarized protons as neutron spin filter is an attractive alternative to the well established neutron polarization techniques. The spin-dependent neutron scattering cross section of protons is usefully large up to the sub-Mev region. Employing optically excited triplet states for the dynamic nuclear polarization (DNP) of the protons, low temperatures and strong magnetic fields are not required and the apparatus can be simplified.

The triplet DNP method can be used to build a reliably working neutron spin filter that is operated in 0.3 T and about 100 K. The high proton polarization of 0.5 obtained is presently still limited by the cooling of the sample. The corresponding analyzing power of $A \sim 0.5$, obtained with the 5 mm thick sample, can be further increased using a longer sample. Interesting possibilities for a triplet spin filter are opened by the use of neutron optics elements that allow to adapt the beam to the filter cross section.

1. Introduction

Dynamic nuclear polarization (DNP) [1], a method traditionally used to create polarized targets with which the role of spin in nuclear and particle interactions is investigated, has also opened new possibilities in neutron science where the strong spin dependence of the neutron scattering on protons is exploited [2, 3, 4]. Proton polarization values close to unity have been achieved in some cases employing the classical scheme of dynamic nuclear polarization (DNP). One polarizes electron spins by cooling them down to low temperature (~1 K) and applying a strong magnetic field (2.5 - 5 T) and then transfers their high polarization to the nuclear spins by means of a microwave field. These stringent conditions can be relieved with a more recent and very promising DNP method that uses optically excited triplet states as the source of the electron polarization [5]. Here the triplet electron spin is polarized as a result of the selection rules of the intersystem crossing from the excited singlet to the triplet state. Then neither low temperatures nor high fields are required and the apparatus can be simplified, which is very attractive for applications. Recently we have demonstrated that the triplet method can be used to build a reliably working spin filter for neutrons operating at 0.3 T at a temperature of 100 K [6]. This note gives a short introduction to the principles of triplet DNP and reports on recent progress of the "triplet spin filter" project.

[4] academic guest

2. Neutron spin filtering with polarized protons

Shapiro and coworkers demonstrated that dynamically polarized protons offer an attractive possibility to realize a broad band neutron spin filter, as the spin-dependent neutron-proton cross section is usefully large up to the sub-MeV region [7]. However, an actual implementation of a polarized proton spin filter has been so far restricted to a few special cases, e.g., for high precision neutron polarimetry [8, 9] or to polarize neutrons of thermal [11] and epithermal energy where it is the only method currently available [12]. This is probably due to the technical requirements of a classical DNP system, which however can be significantly relaxed when employing the triplet DNP technique to realize a spin filter.

The working principle of a polarized proton spin filter is based on the fact that the singlet cross section for neutron-proton scattering is much higher than the triplet cross section [13]. Hence, neutrons polarized anti-parallel to the protons will thus be much stronger scattered than those polarized parallel. This is the case for coherent and incoherent scattering. It is customary to define a spin-dependent effective cross section

$$\sigma_{\pm} = \sigma_0 \pm \sigma_p P \; , \tag{1}$$

where the + and the − stand for the two eigenstates of the neutron spin with respect to the target's polarization axis. The spin dependent part is the product of the 'polarization cross section' σ_p and the nuclear polarization P. The term σ_0 denotes the neutron spin-independent cross section.

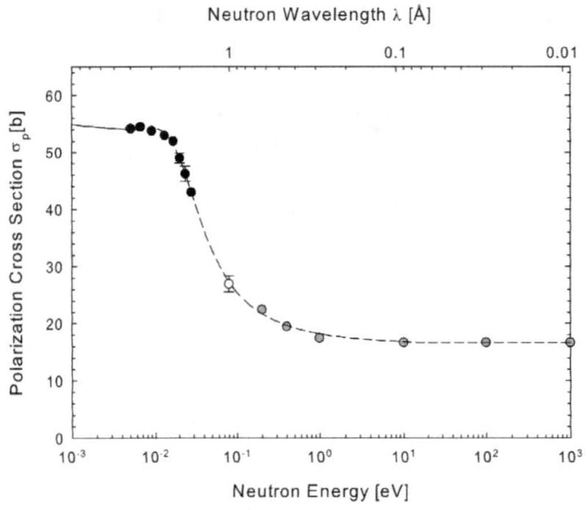

Figure 1. Typical trend of the polarization cross section σ_p as function of the neutron energy (bottom) and wavelength (top), respectively. The data points are a compilation of literature values for different materials: propanediole: black dots [14], ethylene glycol: open dot [11], LMN: gray dots [7]. The dashed line is to guide the eye, and the solid line gives the theoretical value for completely incoherent scattering and absorption.

After passing through the spin filter with a proton density N and a thickness d, the two spin components of the neutron beam are attenuated by a factor $\exp(-\sigma_{\pm} N d)$. Thus, an initially unpolarized neutron beam will have a polarization after the filter of

$$A = \tanh(\sigma_p P N d) \tag{2}$$

which is called the filter analyzing power. The thicker the filter, the higher the filter analyzing power for given $(\sigma_p N)$, and the lower the transmission T of the neutron beam

$$T = \exp(-\sigma_0 N d)\cosh(\sigma_p P N d)\ . \tag{3}$$

Note that for a given analyzing power A, the total transmission T increases with the filter polarization P.

The polarization cross section σ_p for a specific material is normally determined experimentally as structural effects observed for slow neutrons can only be treated well theoretically if the sample structure is sufficiently well known. Fig. 1 gives the general trend for the value of σ_p as a function of the neutron energy. It is qualitatively correct for most hydrogenous materials. For long wavelengths above about 5 Å, the cross section is determined by incoherent scattering and absorption on bound nuclei. With decreasing wavelength the neutrons start to resolve the chemical structure of the sample, giving rise to coherent scattering, most prominent between 2 and 5 Å. For even shorter wavelengths inelastic scattering starts to have an effect and the cross section approaches for energies larger than inter-atomic bonds ($E > 1\,\text{eV}$) the one given by the scattering of isolated free protons [15].

3. Polarizing protons via photo-excited triplet states

A proven system for triplet DNP is a naphthalene molecular host crystal doped with a small concentration of pentacene guest molecules (Fig. 2). The latter can be optically excited into a triplet state where the triplet electron spin is strongly aligned. Using a short optical pulse the

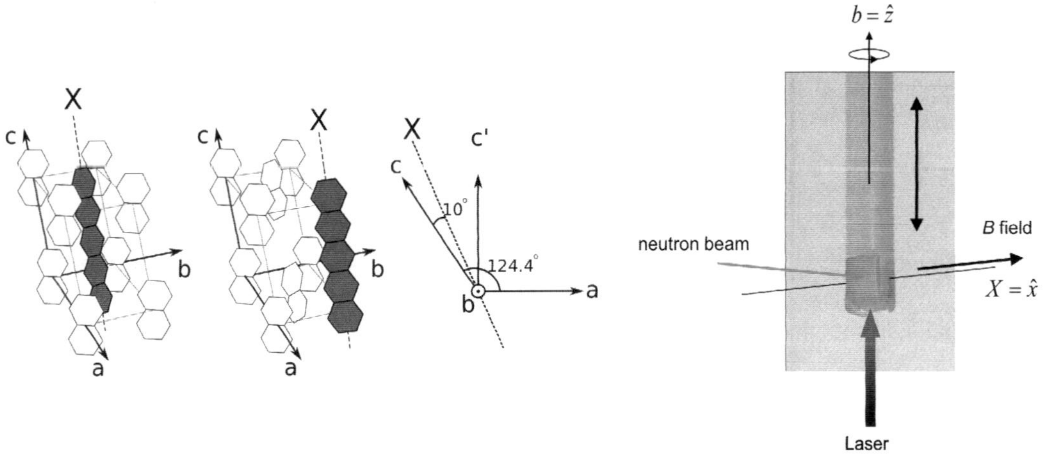

Figure 2. Left: The crystal structure of pure naphthalene is monoclinic with a unit cell with axes $a \sim 8.1$ Å, $b \sim 5.9$ Å, $c \sim 8.6$ Å and $\beta = 124.4°$ [16]. The a-b plane is the cleavage plane. Two naphthalene molecules can be replaced at two sites by one pentacene molecule whose X-axis lies in the a-c plane at an angle of 10° to the c-axis. Right: The crystal is cut such that its bottom/top corresponds to the crystal a-c plane and is mounted on a holder made of polychlorotrifluorethylene (PCTFE). By rotation around the vertical b-axis the orientation $B \parallel X$ can be set.

molecule is excited from the singlet ground state S_0 to the lowest excited singlet state S_1 (Fig. 3). From there the system either decays back to ground state under the emission of fluorescence, or decays via intersystem crossing (ISC) to the lowest triplet state T_1. While the life time of the lowest excited singlet state S_1 is only about 20 ns, the life time of the lowest triplet state T_1 is of the order 50 μs, allowing its use for DNP. The triplet spin levels are split by fine coupling and can be further split with an externally applied magnetic field, typically oriented along the X-axis of the pentacene molecule (see Fig. 2). With $B \parallel X$ two ESR lines are observed at ~ 0.3 T — separated by about 1.5 GHz — corresponding to the transitions -1 to 0 and 0 to 1. In triplet DNP only one of the two ESR transitions is used and the two energy levels involved may be regarded as a highly polarized effective spin-$\frac{1}{2}$ system. For $B \parallel X$ the selection rules for the ISC process strongly favor the occupation of the triplet level T_0 resulting in a high "effective" electron spin polarization of ~ 0.9 [17]. Furthermore the electron spin lattice relaxation time is longer than the life time of the triplet state. Hence, the triplet electron spin polarization is not only independent of temperature and strength of the applied field, but also almost completely maintained during the life time of the triplet state.

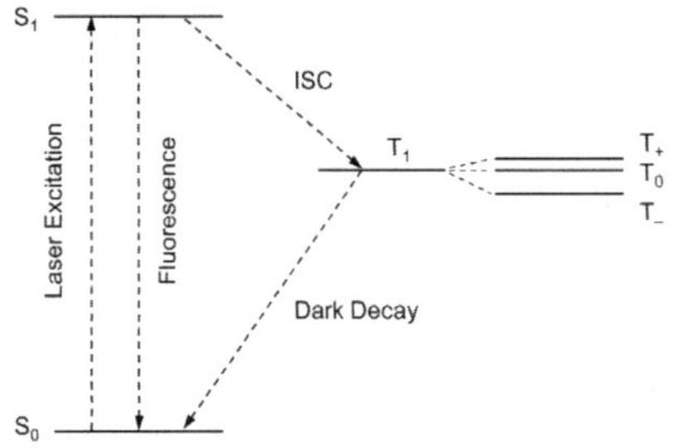

Figure 3. Simplified energy level scheme of the pentacene molecule, showing the photo physical processes during illumination with laser light. The triplet spin levels are split by fine coupling and can be further split with an externally applied magnetic field, typically oriented along the X-axis of the pentacene molecule. They are denoted as T_+, T_0 and T_-.

The Integrated Solid Effect (ISE) scheme is proven to be the most efficient method to transfer the triplet electron polarization to the nuclear spins [5]. In this method the externally applied magnetic field is adiabatically swept over the inhomogeneously broadened ESR line (typically in 15 μs) under microwave irradiation, which allows all electron spin packets to constructively participate in the polarization transfer. For more details we refer to [18, 19]. As in classical DNP, the nuclear spin polarization thus created is localized around the triplet guest molecules but spreads out further over the host molecules via spin diffusion. By repeating the cycle many times the proton spin polarization is accumulated and approaches the maximum value given by the electron polarization. Since the proton spin lattice relaxation time in naphthalene is already very long at liquid nitrogen temperature, $\sim$ 24 h in a field of 0.3 T, compared to a typical polarization build-up time of $\sim$ 1 - 2 h, triplet DNP experiments need moderate cryogenic means only. High proton polarizations of $P > 0.30$ had been reached in fields of only 0.3 T at temperatures of 77 and 100 K [20, 21].

4. Spin filter apparatus

A detailed description of the apparatus is given in [22] and we only shortly overview its main components in the following. The large high quality spin filter crystals are grown with a self-seeding vertical Bridgman technique [23] from zone-refined protonated naphthalene and pentacene-h_{14} or pentacene-d_{14} purified by sublimation. The pentacene concentration is determined with optical transmission spectroscopy and is typically between 2 and 4×10^{-5} mol pentacene per mol naphthalene. From the single crystal samples of approximately $5 \times 5 \times 5\,\mathrm{mm}^3$ are cut such that its bottom/top corresponds to the crystal a-c plane and are mounted on a

PCTFE holder (see Fig. 2). The sample is introduced in a helium flow cryostat, cooled to typically 100 K and the pentacene molecules are excited into the triplet state with a Yb:YAG disk laser (Jenlas disk IR50) feeding an LBO crystal to generate the second harmonic at 515 nm, which proved to work excellent for DNP with pentacene-d_{14} [24]. The light is transported via a multi mode fiber to an optical stage at the bottom of the cryostat which collimates the unpolarized light to a beam of 7 mm waist shining along the b-axis of the sample crystal. In typical experiments we use pulses of 400 ns length with a repetition rate of 400 Hz up to 4 kHz and an energy of approximately 1 mJ per pulse at the entrance window of the cryostat. A static magnetic field of about 0.3 T is applied using a small electro magnet and DNP is performed with an X-band pulsed ESR system operating at 9.3 GHz that generates the ISE passages synchronized to the laser. The latter also enables us to observe the ESR signal of the triplet states and orient the static magnetic field along the X-axis of the pentacene molecules. The proton polarization is measured both by NMR, where the enhanced signal is compared to the signal measured in thermal equilibrium, and neutron transmission.

5. Performance of the spin filter

The neutron spin filter apparatus has been operated and characterized at the BOA beam line of the continuous spallation neutron source SINQ at the Paul Scherrer Institute in Switzerland [25]. In a first experiment the wavelength dependent polarization cross section $\sigma_p(\lambda)$ for naphthalene has been determined over the spectrum of the white cold beam to high precision with a moderate proton polarization only [6]. The knowledge of the polarization cross section then provides a method complementary to NMR, to characterize the samples DNP performance, i.e. the filter polarization is readily determined with a high absolute precision (better than 3 %) via a simple flipping ratio measurement. Moreover, the neutron beam can be used to measure the proton spin polarization as a function of position in the naphthalene sample, which was found to be homogeneous, even at low laser power.

The filter performance could be considerably improved employing a more efficient disk laser system in combination with pentacene-d_{14} doped naphthalene crystals. A proton spin polarization of $P > 0.40$ is now achieved in two hours and can furthermore be increased to $P > 0.50$ by reducing the heat input by lowering the repetition rate of the pulsed DNP scheme [24]. This corresponds to an analyzing power of $A \sim 0.5$ for the 5 mm thick filter.

First experience was gained with focusing elements that adapt the large beam diameter to the small filter sample. A focus de-focus scheme was set up where a parabolic mirror focuses the incoming beam of 25×25 mm horizontally and vertically on the sample and a second mirror of identical focal length restores the original beam properties [26].

6. Outlook and Conclusions

With a filter polarization of $P \sim 0.5$ already considerable analyzing powers can be achieved as is illustrated in Fig. 4 for different sample thicknesses. Taking into account the figure of merit $Q = A^2 \times T$, a thickness of 7.5 mm is ideal for cold neutrons and filter polarizations up to 0.8. For shorter wavelengths the lower cross sections suggest an optimum filter thickness of about 10 mm.

We expect that the maximum filter polarization can be further increased by a better cooling of the sample and by optimizing the DNP process. An improved cryogenic system under construction addresses the former aspect. Furthermore a detailed study of the ISE scheme used for the dynamic polarization of the protons has just been completed [18, 19] and should provide the necessary information to further optimize the parameters of the DNP process. In a next step the scaling of the filter size will be addressed which requires a redesign of the microwave cavity structure. Here a cross section of around 1 cm^2 will be envisaged.

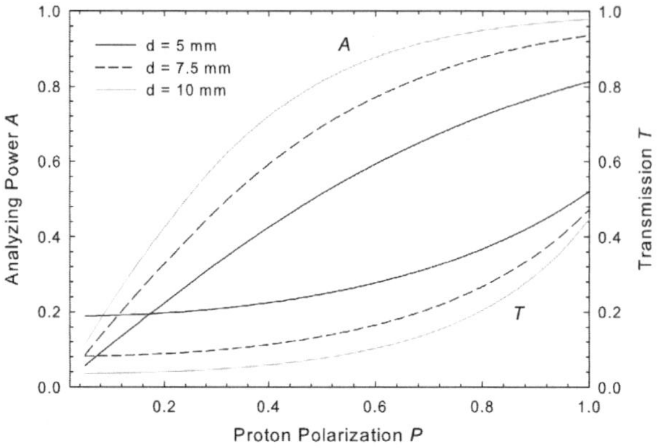

Figure 4. Analyzing power (left axis) and transmission (right axis) for different thicknesses of the naphthalene filter for the cold white neutron beam at BOA, which is peaked at $\lambda \sim 3\,\text{Å}$.

A key point of the project will be the study of the filters' integration with neutron optical devices. Different neutron optics concepts for standard scattering schemes will be evaluated by Monte Carlo simulations and the most promising cases experimentally investigated. Here the use of the spin filter as primary beam polarizer for a very broad range of neutron energies will be the first scheme to study.

7. Acknowledgements
This work was supported by the Swiss National Science Foundation grants 200020_144424 and 200021_143297.

References
[1] Abragam A and Goldman M 1978 *Rep. Prog. Phys.* **41** 395
[2] Hayter J B, Jenkin G T, White J W 1974 *Phys. Rev. Lett.* **33** 696
[3] Stuhrmann H B 2004 *Rep. Prog. Phys.* **67** 1073
[4] van den Brandt B, Glättli H, Grillo I, Hautle P, Jouve H, Kohlbrecher J, Konter J A, Leymarie E, Mango S, May R P, Michels A, Stuhrmann H B and Zimmer O 2006 *Eur. Phys. J.* B **49** 157-65
[5] Henstra A, Lin T-S, Schmidt J, Wenckebach W Th 1990 *Chem. Phys. Lett.* **165** 6-10
[6] Haag M, Eichhorn T R, van den Brandt B, Hautle P, Wenckebach W Th 2012 *Nucl. Instrum. Methods Phys. Res.* A **87** 91-97
[7] Lushchikov V I, Taran Yu V, Shapiro F L 1970 *Sov. J. Nucl. Phys.* **10** 669
[8] Zimmer O, Müller T M, Hautle P, Heil W, Humblot H 1999 *Phys. Lett.* B **455** 62-8
[9] Aswal V K, van den Brandt B, Hautle P, Kohlbrecher J, Konter J A, Michels A, Piegsa F M, Stahn J, Van Petegem S, Zimmer O 2008 *Nucl. Instrum. Methods Phys. Res.* A **586** 86
[10] P. Hautle, W. Heil, D. Hofmann, H. Humblot, T.M. Müller, O. Zimmer, Proc. 9th Int. Workshop on Polarised Sources and Targets. Nashville 2001, p. 274, World Scientific, Singapore, 2002.
[11] Hiramatsu S, Isagawa S, Ishimoto S, Masaike A, Morimoto K, Funahashi S, Hamaguchi Y, Minakawa N, Yamaguchi Y 1978 *J. Phys. Soc. Jpn.* **45** 949-53
[12] Delheij P P J, Bowman J D, Frankle C M, Haase D G, Langston T, Mortensen R, Penttila S, Postma H, Seestrom S J, Yen Yi-Fen 195 *Nucl. Instrum. Methods Phys. Res.* A **356** 120-1
[13] Glättli H and Goldman M 1987 *Methods of Experimental Physics* vol 23, part C (Academic Press, New York) pp 241-86
[14] Hautle P, Heil W, Hofmann D, Humblot H, Müller T M, Zimmer O 2002 *Proc. 9th Int. Workshop on Polarised Sources and Targets, Nashville 2001* (World Scientific, Singapore) p 274
[15] Turchin V F 1965 *Slow neutrons* (Israel Program for Scientific Translations)
[16] Brock C P and Dunitz J D 1982 *Acta Cryst.* B **38** 2218-28

[17] van Strien A J and Schmidt J 1980 *Chem. Phys. Lett.* **70** 513-7
[18] Henstra A, Wenckebach W Th 2013 *Mol. Phys.* DOI: 10.1080/00268976.2013.861936
[19] Eichhorn T R, van den Brandt B, Hautle P, Henstra A, Wenckebach W Th 2013 *Mol. Phys.* DOI: 10.1080/00268976.2013.863405
[20] Iinuma M, Takahashi Y, Shaké I, Oda M, Masaike A, Yabuzaki T 2000 *Phys. Rev. Lett.* **84** 171-4
[21] Wakui T, Hatano M, Sakai H, Uesaka T, Tamii A 2005 *Nucl. Instrum. Methods Phys. Res.* A **550** 521-34
[22] Eichhorn T R, Haag M, van den Brandt B, Hautle P, Wenckebach W Th, Jannin S, Van der Klink J J, Comment A 2013 *J. Magn. Reson.* **234** 58-66
[23] Selvakumar S, Sivaji K, Arulchakkaravarthi A, Balamurugan N, Sankar S, Ramasam P 2005 *J. Crystal Growth* **282** 370-5
[24] Eichhorn T R, Haag M, van den Brandt B, Hautle P, Wenckebach W Th 2013 *Chem. Phys. Lett.* **555** 296-9
[25] http://www.psi.ch/sinq/boa/
[26] Rantsiou E et al. 2014 *these proceedings*

Neutron spin manipulation optics: basic principles and possible applications

N K Pleshanov[1]
Neutron Research Department, Petersburg Nuclear Physics Institute,
Orlova Roscha, Gatchina, St. Petersburg, 188300, Russia.

E-mail: pnk@pnpi.spb.ru

Abstract. A number of basic elements for neutron spin manipulation optics (NSMO) based on Larmor and non-Larmor (quantum) precessions under reflection are considered. It is concluded that transition to 3D in neutron polarization optics may bring additional instrumental possibilities. New neutron optical devices will include spin turners (particularly, $\pi/2$-turners and π-turners, or flippers), spin precessors and antiprecessors, 3D-polarizers, 3D-analyzers, 3D-rotators, spin manipulators, hyper-polarizers. The innovative neutron optics is directly applicable to developing 3D polarization and polarimetry techniques, such as reflectometry with 3D-polarimetry, Neutron optical Spin Echo (NoSE), including compact NoSE and TOF NoSE schemes. A hyper-polarizer is a device which not only separates neutrons with the opposite spins, but also flips the 'wrong' spins. Thus, hyper-polarizers can double the intensity of polarized neutron beams, although a gain in the intensity can be achieved only with the increase either in the angular divergence or in the width of the beam, in full accordance with the Liouville theorem. The tasks to be solved for implementation of the NSMO concepts are discussed.

1. Introduction

Quantum mechanically, the neutron spin evolution is the result of superposition of the coherent spin states. Wherever the moduli of the opposite spin components do not change in time and space, the spin evolution may be described as precession. The quasi-classical Larmor precession picture and the quantum approach can be reconciled, when the spin-dependent scattering is negligible. Particularly, in a homogeneous magnetic medium both quasi-classical and quantum approaches lead to precession with the exactly Larmor frequency, even in the essentially non-classical case when the energies of the opposite spin components are different [1]. It is only when the neutron state is a superposition of non-orthogonal coherent states with different energies, the spin behavior in a homogeneous magnetic medium may have no classical counterpart [1,2].

In this paper we assume that polarized neutrons are prepared by stationary devices and the energies of the opposite spin components are equal. When the neutron with a mass m and a magnetic moment μ passes a distance l with the velocity $(v_+ + v_-)/2$ (t is the respective travel time) in a homogeneous magnetic medium, its spin turns (precesses) by an angle

$$\varphi_{\rm L} = (k_- - k_+)l = (k_- - k_+)\frac{\hbar}{m}\frac{(k_+ + k_-)}{2}t = \omega_{\rm L} t, \qquad (1)$$

[1] To whom any correspondence should be addressed.

where $\omega_L = 2|\mu|B/\hbar$ is the Larmor frequency, $k_\pm = \sqrt{k^2 - \kappa_\pm^2}$ are the wave numbers, corresponding to the velocities $v_\pm$ of the neutron in the states with the spin up (+) and down (–) the magnetic induction vector **B** of the medium, $\kappa_\pm = \hbar^{-1}\sqrt{2m(V_n \pm V_m)}$ are the critical wave numbers of the medium with a nuclear potential V_n and the magnetic potential $V_m = |\mu|B$ for the (±) spin states; $k = \hbar^{-1}\sqrt{2mE}$, E is the neutron energy. Since $\mu<0$, the neutron spin precession about ***B*** is anticlockwise. In most cases $k >> |\kappa_\pm|$, and one obtains

$$\varphi_L \cong l(\kappa_+^2 - \kappa_-^2)/k = ClB\lambda, \quad C = 8\pi m|\mu|/h, \tag{2}$$

the well known law of proportionality ($C \cong 4.63 \cdot 10^5$ T^{-1}m^{-1}nm^{-1}) of the precession angle to neutron path l [m], field magnitude B [T] and neutron wavelength λ [nm].

The orientation of the spin may be changed as a result of reflection from a layered structure with magnetic moments not collinear to the incident neutron spin. Thus, one can design magnetic structures that provide the required change of the spin direction.

For specular reflection of neutrons from a layered structure the Schrödinger equation is reduced to one-dimensional equation with the energy (θ is the glancing angle)

$$E_\perp = E\sin^2\theta. \tag{3}$$

Its plane wave solutions for the incident and reflected neutrons are, respectively,

$$\Psi_0(z) = \begin{pmatrix} |\psi_0^+(z)|\cdot\exp[i\alpha_+(z)] \\ |\psi_0^-(z)|\cdot\exp[i\alpha_-(z)] \end{pmatrix}_Z, \quad \Psi_r(z) = \begin{pmatrix} |\psi_r^+(z)|\cdot\exp[i\beta_+(z)] \\ |\psi_r^-(z)|\cdot\exp[i\beta_-(z)] \end{pmatrix}_Z \tag{4}$$

(z is the surface normal coordinate, Z is the quantization axis).

Further we consider magnetically collinear layered structures and suppose the quantization axis Z to be parallel to their magnetization. If the spin-up and spin-down reflectivities differ ($R_+ \neq R_-$), the angle between the spin and Z changes. Therefore, it is only when $R_+ = R_-$, one may speak about precession of the spin under reflection. Then the orientation of the reflected neutron spin at the reflecting surface ($z = 0$) can be found by rotating the incident neutron spin about Z by an angle

$$\varphi = (\delta_- - \delta_+)_{z=0} = (\alpha_+ - \beta_+)_{z=0} - (\alpha_- - \beta_-)_{z=0}, \tag{5}$$

where $\delta_\pm$ are the shifts of the phases of the plane waves reflected in the (±) spin states relative to the phases of the incident waves in the same (±) spin state. The precession angle φ is defined so that it is positive for an anticlockwise (like Larmor) turn of the neutron spin. In order to observe the precession, the spins of the incident neutrons should be inclined to the direction of the magnetic fields in the reflecting layers. However, no one can guarantee now that it will be a Larmor precession.

Of interest for practical applications are the cases when the reflectivities from magnetic structures are close to 1. Then one may speak about neutron spin manipulation optics (NSMO). The possibility of building neutron spin turners on the basis of total reflection from magnetically anisotropic thin films has been mentioned in Ref. [3]. In the present paper the concepts of NSMO on the basis of magnetically collinear layered structures are developed. The magnetic moments are assumed to be in-plane and parallel to one another, although they may be inclined to sufficiently low external fields. It is justified in the first approximation, e.g., for numerous magnetic layers in remanent supermirrors [4-6] with the easy axis induced by sputtering conditions. Such coatings have been applied in remanent polarizers and analyzers [7]. Since low external fields do not affect the results noticeably, in further analysis the external fields are assumed to be absent. Numerical calculations are carried out with the *generalized matrix* method [8,9] (reinvented later under the name "*supermatrix* method" [10]).

2. Quantum precession of the neutron spin under total reflection

The precession condition is satisfied when neutrons in both spin states are totally reflected ($R_\pm = 1$). Consider precession under total reflection in the simplest cases.

2.1. Larmor precession coating (LPC)

First of all, consider total reflection of neutrons from a non-magnetic substrate coated with a magnetic film of thickness *d*, reflection from the upper boundary of which is negligible (figure 1). Taking account of the fact that the neutron traverses the magnetic film twice, before and after reflection from the substrate, the precession angle in terms of the wave vector transfer $q = 4\pi\sin(\theta)/\lambda$ (λ is the neutron wavelength, θ is the glancing angle) is

$$\varphi_{\mathrm{L}} = qd\left[\sqrt{1-(q_-/q)^2} - \sqrt{1-(q_+/q)^2}\right] \quad (q \le q_{\mathrm{s}}), \tag{6}$$

where $q_\pm = 2\kappa_\pm$ are the critical wave numbers of the magnetic film for neutrons with the opposite spins. The condition $q \le q_{\mathrm{s}} = \hbar^{-1}\sqrt{2mV_{\mathrm{s}}}$ corresponds to the inequality $E_\perp = E\sin^2\theta \le V_{\mathrm{s}}$, i.e. to the condition of total reflection from the substrate with a potential V_{s}.

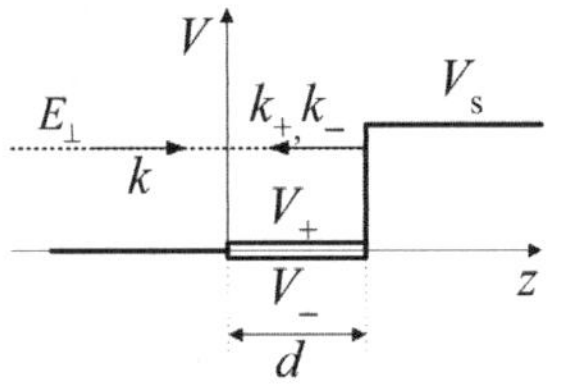

Figure 1. Total reflection of the neutron from a non-magnetic substrate in the presence of a weakly reflecting magnetic film (see details in the text).

Reflection from the upper boundary is negligible, when $q >> |q_\pm|$. Then

$$\varphi_{\mathrm{L}} \cong \frac{d(q_+^2 - q_-^2)}{2q} = 8\pi CB\frac{d}{q} = CB\lambda\frac{2d}{\sin\theta}. \tag{7}$$

This formula coincides with equation (2) for Larmor precession, since the length of the neutron path in the magnetic film is $l = 2d/\sin(\theta)$.

Such a magnetic coating that weakly reflects neutrons and induces Larmor precession will be called further as "Larmor precession coating" (LPC). Note that the precession angle is then proportional to the neutron path, *B* and λ, i.e. it decreases with *q* as q^{-1}.

Generally, definition (5) does not lead to Larmor precessions with the well-known relations. Consider two special cases of non-Larmor spin precessions, which are of interest from both conceptual and practical points of view.

2.2. Total reflection from the boundary with a magnetic medium

Reflection from the boundary with a magnetic medium has been considered in Ref. [11]. The simple formula derived [3] for the phase shift of the neutron wave under total reflection can be used to obtain the precession angle for the total reflection (TR) of both spin components from the boundary with a magnetic medium (figure 2-a):

$$\varphi_{\mathrm{TR}} = 2[\arccos(q/q_+) - \arccos(q/q_-)] \quad (q \le q_\pm). \tag{8}$$

The inequality $q \le q_\pm$ is equivalent to $E_\perp \le V_\pm$.

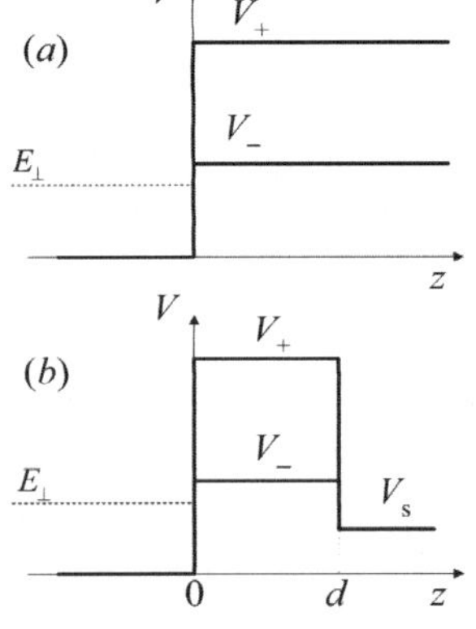

Figure 2. Total reflection of the neutron from the boundary with (a) a magnetic medium and (b) a thick magnetic film.

The precession is about the vector **B** in magnetic medium, anticlockwise and increasing with q from 0 at $q = 0$ to a maximum $2\arccos(q_- / q_+)$ at $q = q_-$, so the precession angle can reach $\pi/2$ (when $V_+ \geq 2V_-$), but never can reach π. Note that the decrease of the precession angle with the wavelength corresponds with the fact that the neutron wave with a shorter wavelength penetrates deeper into the non-classical under-barrier region (with the field).

In practice, the precession is easier to observe [3,12] for reflection from a sufficiently thick magnetic film on a smooth substrate (figure 2-b). E.g., for an iron film thicker than 100 nm, the precession angle can be calculated with equation (8).

2.3. *Total reflection with spin splitting*

Even more peculiar precession is induced by spatial separation of the opposite spin components ('spin splitting'), which has been studied [13] and is exploited in NSE spectrometers [14] at J-PARC (Tokai, Japan).

When the spin-up and spin-down neutron components are totally reflected from different boundaries of a magnetic layer (the spin-down neutron reflection from its upper boundary is supposed to be negligible) with a thickness d and potential eigenvalues $V_\pm$ (figure 3), the precession angle is

$$\varphi_S = qd\sqrt{1-(q_- / q)^2} + (\varphi_{TR}^- - \varphi_{TR}^+) \qquad (E_\perp \leq V_+, V_s), \tag{9}$$

i.e., it is defined by the phase difference related to different lengths of paths of neutron waves representing the states with the opposite spins as well as to the phase shifts $\varphi_{TR}^\pm$ under their total reflection from the respective boundaries:

$$\varphi_{TR}^+ = -2\arccos\sqrt{E_\perp / V_+}\,, \tag{10a}$$

$$\varphi_{TR}^- = -2\arccos\sqrt{(E_\perp - V_-)/(V_s - V_-)}\,. \tag{10b}$$

Account of the change in the kinetic energy and the relative height of the barrier for the spin-down neutron in the medium before the barrier (V_s) is taken in equation (10b).

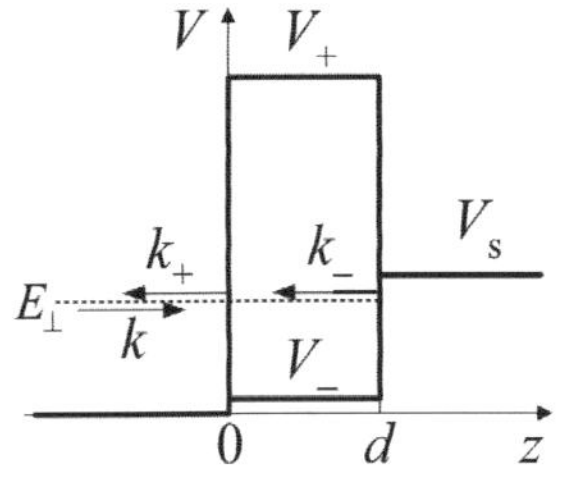

Figure 3. Total reflection with the spin splitting: the spin-up and spin-down neutron components undergo total reflection at the upper and lower interfaces of the magnetic film with a thickness d on a non-magnetic substrate with a potential V_s.

The precession is non-Larmor, anticlockwise for most parameters, but it can also be made clockwise ('antiprecession'), e.g. when $V_- = 0$ and $qd < \varphi_{TR}^+ - \varphi_{TR}^-$. However, when the total reflection conditions for the spin-up and spin-down components are identical ($V_+ = V_s$ and $V_- = 0$), the precession angle is positive and exactly proportional to q:

$$\varphi_S = qd = 4\pi d\sin(\theta) / \lambda\,. \tag{11}$$

As in the case of total reflection from the boundary with a magnetic medium, the non-classical behavior of the spin is related to purely quantum features of the interaction, in which the probabilities to find one and the same particle in the states with the opposite spins in a region with magnetic fields may essentially differ. Such interaction is impossible for a classical particle, which can be found at any time in a certain point and with a certain angular momentum.

3. Neutron spin manipulation optics: basic elements

Quantum aspects of the neutron interaction with magnetic layers open new possibilities for spin manipulations with neutron optical spin selectors, spin turners (π/2- and π-turners, by way of example) and spin processors. In order to analyze the potentialities of neutron spin manipulation optics, in further calculations the magnetic layers are assumed to be homogeneously magnetized along the anisotropy axis, even when the external fields are switched off (perfect remanence).

3.1. Neutron optical spin selectors

All known neutron polarizers, including those on the basis of polarizing coatings, are spin selectors. Remanent polarizers can be magnetized in either of opposite directions thus eliminating the need of spin flippers [15,4] or can be rotated about their surface normals for polarized beam preparation and vector polarization analysis [16].

3.2. Neutron optical spin turners

Consider the possibility of building neutron spin turners on the basis of magnetically anisotropic thin films. The calculations are carried out for reflectors turning the spin by the angles $\pi/2$ and π, which are the most interesting for the practice. The task of rotation of the spin by any arbitrary angle is solved in the same manner.

The spin turner may be characterized by the orientation of the rotation axis and by the spin rotation angle γ. The efficiency of a spin turner can be defined as the portion of neutrons reflected with the spin in the desired direction $\boldsymbol{u}$, which is obtained by turning the incident beam polarization by the angle γ about the rotation axis, i.e.,

$$\varepsilon(\gamma) = (1 + P_u)/2, \quad (12)$$

where P_u is the projection of the reflected beam polarization onto $\boldsymbol{u}$. Note that this is a generalization of the conventional definition of the efficiency of flippers. Further calculations are carried out on the assumption that the rotation axis is perpendicular to the incident neutron spin.

Curve 1 in figure 4-*a* is the precession angle φ calculated as a function of q in the region of total reflection of neutrons from a perfect boundary with magnetized bulk Fe. In practice, it is rather difficult to magnetize a bulk Fe to saturation and polish its surface to roughness about 1 nm. These problems are easily solved by using an Fe layer on a smooth substrate. For an Fe layer thicker 100 nm, the q-range of (almost) total reflection remains practically the same and the spin precession is described by curve 1. The use of a glass substrate with a sufficiently high potential ($V_s > V_-$) ensures total reflection even from an Fe layer as thin as 20 nm (curve 2). An Fe film, 100 nm or thicker, would be required with a silicon substrate ($V_s < V_-$). An oxide layer formed in air on the Fe layer surface sets up a barrier for neutrons and decreases the precession angle (curve 3).

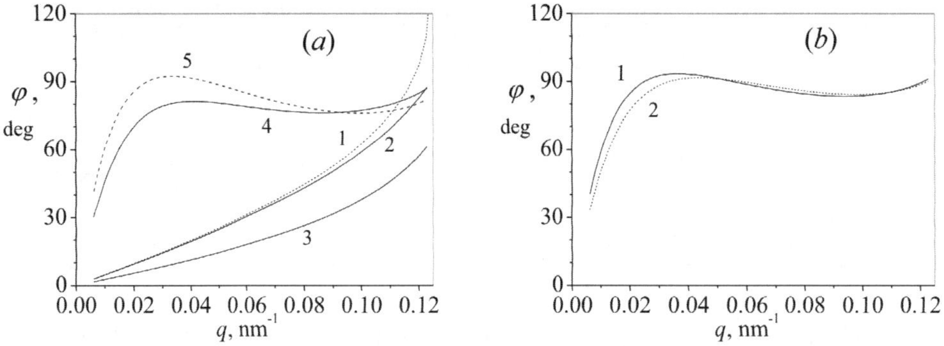

Figure 4. The precession angle φ calculated as a function of q in the total reflection region ($R_{\pm}= 1$) (*a*) from (1) magnetised bulk Fe ($V_n = 211$ neV, $V_m = 132$ neV); (2) Fe layer (20 nm); (3) Fe layer (20 nm) with a surface Fe_3O_4 (3 nm) oxide layer; (4) Fe layer coated with a Ti (15.5 nm) layer; (5) Fe layer (20 nm) coated with a Ti layer (20.5 nm) oxidized (TiO_2 3 nm) in air.
(*b*) from an Fe layer (20 nm) coated with (1) TiH layer (14 nm, -65 neV), (2) TiH layer (16.5 nm, -60 neV) oxidized to TiO_2 (3 nm).
The layered structures are on the glass substrate ($V_s = 97$ neV).

As the precession angle essentially depends on q, spin turners based on total reflection from a single boundary with magnetic medium can work only with strongly collimated and monochromatic beams. So the ideas to relax these restrictions are required.

First, a layer with a negative potential V_b and an optimized thickness b (figure 5) can be used to form a medium before the total reflection boundary (*biased total reflection*). The phase shifts under biased total reflection from the barrier are defined by the kinetic energy $E_\perp - V_b$ and the barrier height $V_\pm - V_b$ in the medium before the barrier. As a consequence, the range of precession angles in the total reflection region decreases (cf. (10-a) and (10-b)). The precession is also affected by interference of the waves reflected from the boundaries of the additional layer and, optimizing its thickness b, one can additionally smoothen the dependence $\varphi(q)$ at the values of q, at which reflection from the upper boundary is still insignificant. Curve 4 in figure 4-*a* is for an Fe layer coated with a Ti layer; curve 5 represents the effect of its oxidation. The mean precession angle in the working q-range is somewhat below 90°, but it can be increased by hydrogenating titanium (figure 4-*b*).

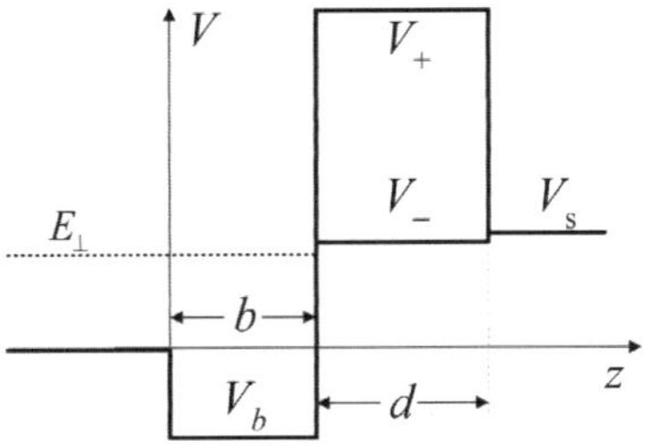

Figure 5. Total reflection of the neutron from the boundary with a thick magnetic film on the substrate.

The second idea boils down to balancing the non-Larmor precession (increasing with q) under reflection from the Fe layer with the Larmor precession (decreasing with q) in LPC on top of the Fe layer. One can use LPC made as a (multi)layer with a low (average) nuclear potential and comparatively low (average) magnetization to minimize reflection from its upper boundary. Curve 1 in figure 6 is calculated for LPC on the basis of a uniform layer with parameters optimized for a $\pi/2$-turner and gives an example of the smoothing of the dependence $\varphi(q)$. The unfavorable effect of oxidation on the efficiency (curve 2) may be reduced to a considerable degree by depositing a thin Ti layer, which forms, due to oxidation in air, an *antireflection bilayer* Ti/TiO_2 [17,18] for neutrons (curve 3). The possibility of using a multilayer LPC, in which magnetic and non-magnetic thin layers are alternated (Co/Ti, Co/TiZr, CoFe/TiZr, etc.), is also confirmed by calculations. One must bear in mind this possibility in other examples of using LPC given below.

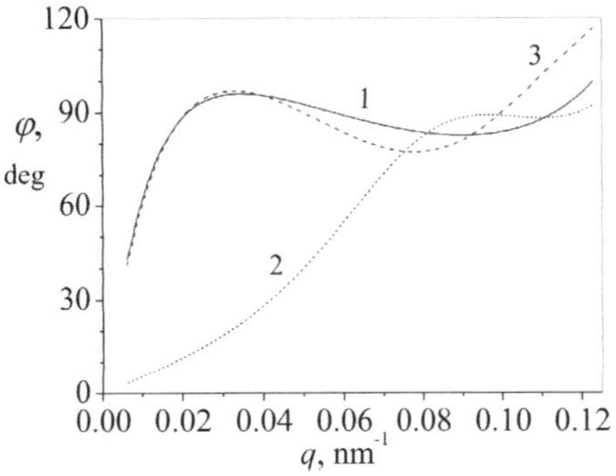

Figure 6. The precession angle φ calculated as a function of q in the region of total reflection from an Fe film (20 nm) coated with (1) LPC (35 nm, V_n = -10 neV, V_m = 5.5 neV); (2) LPC (40 nm, V_n = 0 neV, V_m = 3 neV) and a surface oxide (3 nm, 70 neV); (3) LPC (40 nm, V_n = 0 neV, V_m = 3 neV) and the bilayer Ti(5 nm)/TiO_2(3 nm). The layers are on the glass substrate (V_s = 97 neV).

The efficiencies of the $\pi/2$-turners considered above are compared in figure 7. In all cases the effect of oxidation was taken into account. The thickness of the iron film is only 20 nm, because the glass substrate ensures total reflection of spin-down neutrons (V_s is even slightly above the potential of iron,

V_{-}). An Fe film with a thickness no less than 100 nm would be required with a silicon substrate (V_s < V_{-}). The mean efficiencies of all π/2-turners in the working q-range from 0.02 to 0.123 nm^{-1} are above 0.99, so they can be of practical importance.

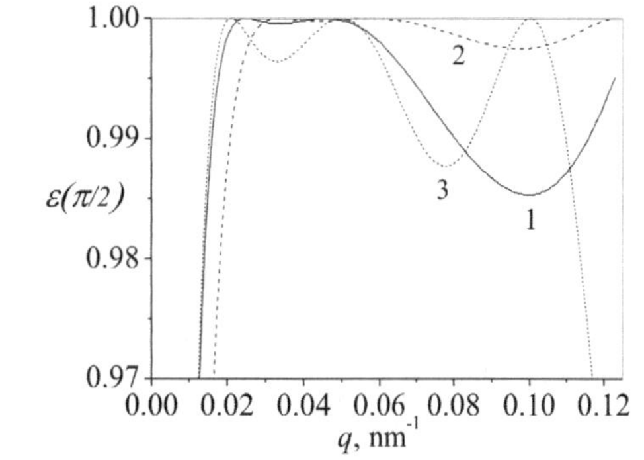

Figure 7. The efficiencies $\varepsilon(\pi/2)$ calculated as a function of q in the region of total reflection for the π/2-turners on the basis of an Fe film (20 nm) coated with (1) Ti (20.5 nm) / TiO_2 (3 nm); (2) TiH (16.5 nm, -60 neV) / TiO_2 (3 nm); (3) LPC (40 nm, V_n = 0 neV, V_m = 3 neV) / Ti(5 nm)/TiO_2(3 nm). The efficiencies averaged in the q-range from 0.02 to 0.123 nm^{-1} are: (1) 0.9936, (2) 0.9987 and (3) 0.9923.

Performance of the spin turners may be improved by further optimization of the subsidiary layers and by combining the two approaches to smoothing the dependence $\varphi(q)$. However, the condition $E_{\perp} = E\sin^2\theta \le V_{-} = V_n - V_m$ for the total reflection of neutrons from the boundary with a medium with nuclear V_n and magnetic V_m potentials implies that the working glancing angles are very small. This is the chief drawback of such spin turners, so new ideas are required.

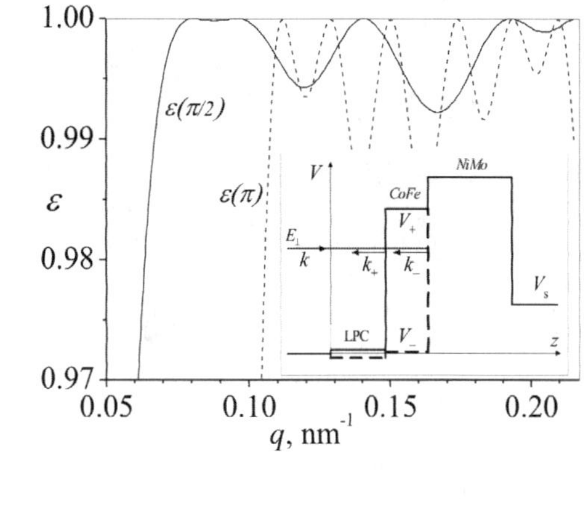

Figure 8. The efficiencies ε calculated as a function of q in the region of total reflection from spin-splitting structures and LPC. The solid curve is the efficiency of a π/2-turner NiMo(100 nm) / CoFe (5 nm) / LPC (57 nm, V_n = -10 neV, V_m = 10 neV) / Ti(5 nm)/TiO_2(3 nm); the dotted curve is the efficiency of a π-turner NiMo(100 nm) / CoFe (10 nm) / LPC (130 nm, V_n = -10 neV, V_m = 10 neV) / Ti(5 nm)/TiO_2(3 nm). The mean efficiency of the π/2-turner in the q-range from 0.075 to 0.215 nm^{-1} is 0.9976; the mean efficiency of the π-turner in the q-range from 0.11 to 0.215 nm^{-1} is 0.9948.

Inset: Neutrons in the spin-up and spin-down states experience total reflection from the LPC/CoFe and CoFe/NiMo (non-magnetic) boundaries, respectively. The spin-down neutron potential V_{-} of the magnetic layers is shown by the dashed lines. The Ti/TiO_2 bilayer is not shown.

Not only π/2-turners, but also π-turners can be built on the basis of spin-splitting structures. As before, we can use LPC to balance the non-Larmor precession. Since the spin-splitting structure may have a higher potential for total reflection of spin-down neutrons, the glancing angles for the spin turners can be increased. The mean efficiencies of such π/2- and π-turners with optimized layers and an antireflective Ti/TiO_2 bilayer also exceed 0.99 in the working q-ranges (figure 8).

What is more important, a magnetic LPC on top of a non-magnetic periodic multilayer may by itself induce the Larmor precession of the spins of the Bragg-reflected neutrons by any desired angle.

Due to small relative widths of the Bragg peaks, spin turners with record glancing angles and record efficiencies can be built. For a non-magnetic multilayer and a weakly reflecting LPC, the precession condition ($R_+ \cong R_-$) is well satisfied. At the Bragg peak (large q), even layers of Co, Fe, or CoFe are weakly reflecting and can be used as LPC. In the examples given in figure 9 the averaged-over-the-peak efficiencies of the π/2- and π-turners with the proper thicknesses of LPC exceed 0.999. When the period of the multilayer decreases, the Bragg peak width becomes lesser and the efficiency of the spin turner increases.

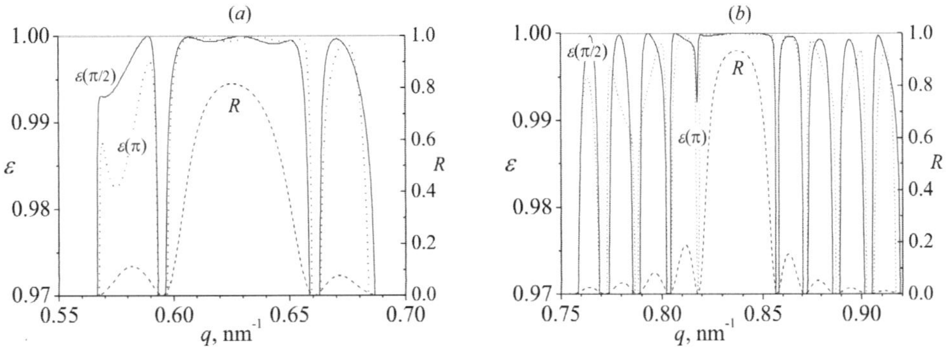

Figure 9. The reflectivities R and the efficiencies ε of the spin turners in the vicinity of the Bragg peak for non-magnetic multilayers (*a*) [NiMo(5.32nm)/Ti(4.94nm)]x20 and (*b*) [NiMo(3.88nm)/Ti(3.73nm)]x50. The multilayers are coated with a Co layer of thickness (*a*) 46 nm and (*b*) 92 nm (π/2- turners), and (*a*) 62 nm and (*b*) 123 nm (π-turners).

Earlier Zabel [19] has mentioned the possibility of using an antiferromagnetic (AF) superlattice as a flipper. When its magnetization is perpendicular to spins of incident neutrons, they are flipped within the Bragg magnetic reflection peak of 1/2 order. Yet, it follows from calculations that the π–layers on non-magnetic multilayers are more efficient than AF superlattices; in addition, the efficiency of the AF superlattices as flippers decreases when their period diminishes. Still another advantage of the π–layers is a comparative ease of their production. Besides, spins of Bragg-reflected neutrons can be rotated by an arbitrary angle depending on the thickness of the magnetic layer (LPC).

The fact that rotation of neutron spins may be combined with the beam monochromatization is advantageous in some applications, but in most cases it will reduce the instrument luminosity. Therefore, consider the possibilities of producing spin turners on the basis supermirrors, which can reflect neutrons in the widest q-range (supermirrors with m=7 have been produced [20]).

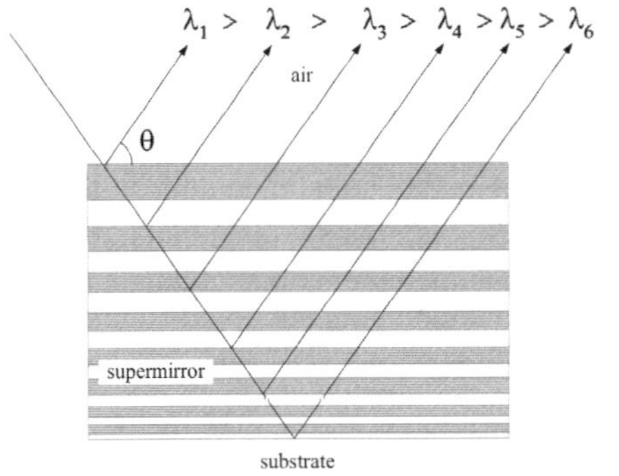

Figure 10. Reflected neutrons with lesser wavelengths traverse greater distances in the supermirror.

First of all, note that the path of neutrons reflected with larger q is longer, i.e. they traverse more layers, in the supermirror. It is illustrated in figure 10 for one glancing angle and different wavelengths. Reflection of neutrons with a given q is coherently enhanced at a group of bilayers with close-to-the-Bragg-condition thicknesses. Other bilayers are weakly reflecting and, if magnetic, induce Larmor precession. These features in neutron interaction can be used to design supermirror-based spin turners: neutrons reflected with larger q traverse more layers in the supermirror, and at the same time more magnetic layers are required to turn their spins by a given angle.

So, one approach is inclusion of magnetic layers into a non-magnetic supermirror. Thus, to build a $\pi/2$-turner, Ti in 4 out of 300 NiMo/Ti bilayers was substituted by Co (in bilayers with numbers 1, 15, 77, 128 as counted from thicker ones), and a Co layer (7 nm) on top of the (m=3) supermirror was added (figure 11). 7 substitutions were made to build a π-turner. The efficiencies are moderate, but the method can certainly be developed by optimizing the parameters of the included magnetic layers.

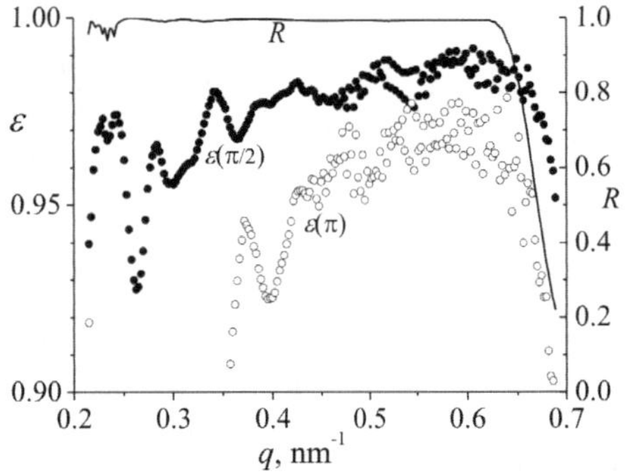

Figure 11. The reflectivities R and the efficiencies ε of spin turners on the basis of the NiMo/Ti (m=3) (300 bilayers) supermirror, in which some of the Ti layers are substituted by Co layers with the same thicknesses. An additional Co layer with a thickness 7 (14) nm is deposited on top of the $\pi/2$ (π)-turner. The points are obtained by convolution with the resolution $\Delta q/q = 0.03$.

Another approach is to use layers (uniform or composed of thinner layers of different materials) with low average magnetization. Calculations of the efficiencies of NiMo/X (m=3) $\pi/2$- and π-supermirrors built with specially tuned magnetization of layers X are represented in figure 12. The material of the magnetic layers with required properties (see the neutron potentials in the caption) is designated as X; its elaboration is a task still to be solved. In principle, instead of Ti layers, NiMo layers could be substituted with magnetic layers, e.g., on the basis of Ni.

The efficiencies are good (Fig. 12), but only at larger q. Thicker bilayers function poorly. Removing 20 thicker bilayers, we can obtain very good efficiencies of $\pi/2$- and π-turners at large q (figure 13-*a*). Still better efficiencies (exceeding 0.999) of the $\pi/2$- and π-filters are obtained without 50 and 100 thicker bilayers (figure 13-*b*,*c*).

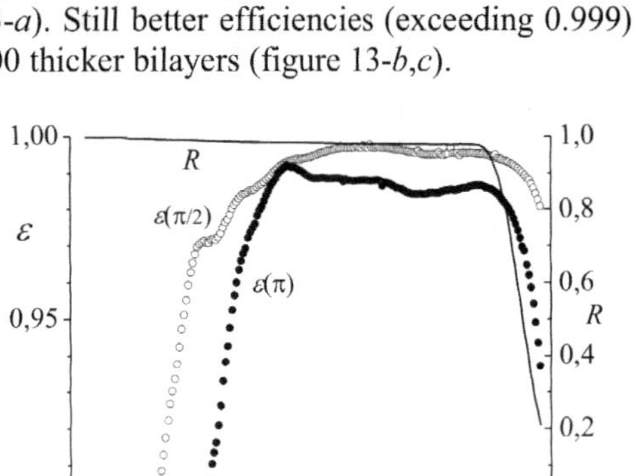

Figure 12. The reflectivities R and the efficiencies ε of spin turners on the basis of the NiMo/X (m=3) (300 bilayers) supermirror obtained from NiMo/Ti (m=3) sequence by substitution of Ti by X with the potentials $V_n = 0$, $V_m = 1.5$ neV ($\pi/2$-turner) and $V_m = 3$ neV (π-turner). An additional Co layer with a thickness 26.5 (53) nm is deposited on top of the $\pi/2$ (π)-turner. The points are obtained by convolution with the resolution $\Delta q/q = 0.03$.

To design supermirror-based spin turners with high efficiency in the entire q-range, more sophisticated tuning of the parameters of magnetic layers is required. Particularly, the use of total reflection with the spin splitting essentially improves the efficiency of spin turners at low q.

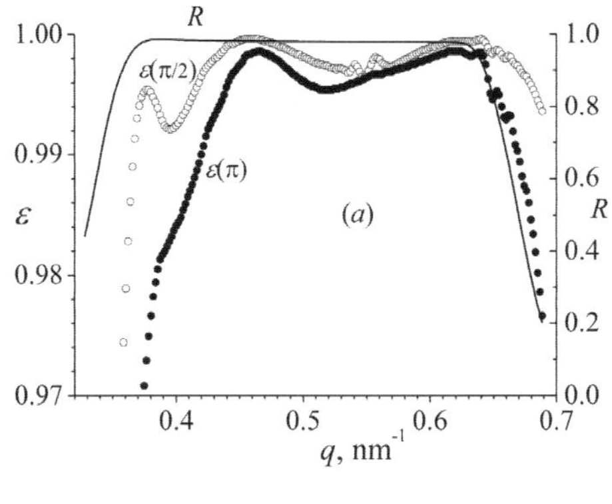

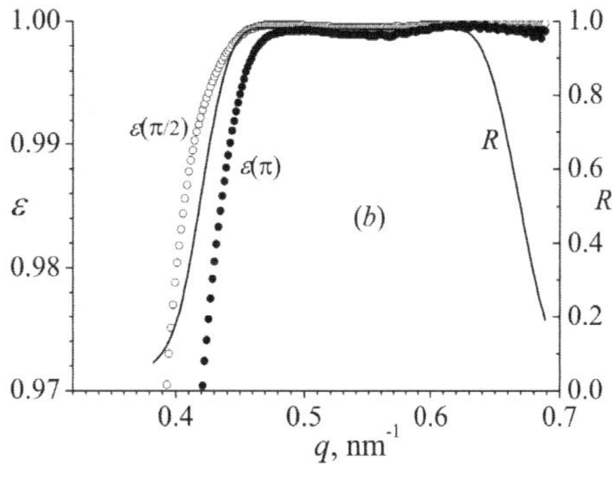

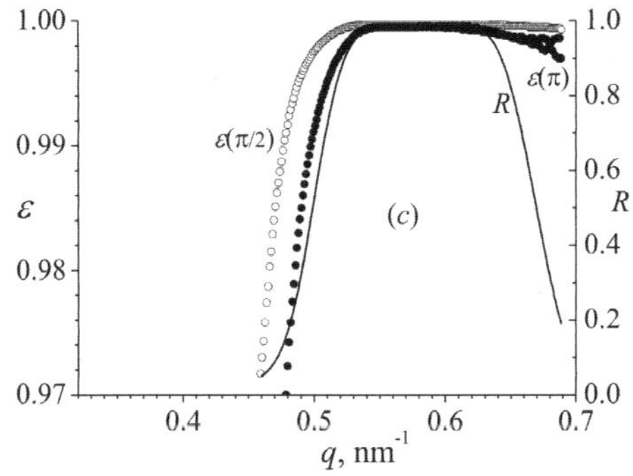

Figure 13. The reflectivities R and the efficiencies ε of spin turners on the basis of the filters obtained from NiMo/X (m=3) supermirror by removing (a) 20, (b) 50 and (c) 100 thicker bilayers. In calculations the potentials of the X layers are $V_n = 0$, $V_m = 1.5$ neV (π/2-turner) and $V_m = 3$ neV (π-turner). A top Co layer is additionally deposited with a thickness (a) 29.5 (59) nm, (b) 33 (66) nm, (c) 38 (76) nm for the π/2 (π)-turners. The points are obtained by convolution with the resolution $\Delta q/q = 0.03$.

3.3. Neutron optical spin processors

Neutron optical spin processors are spin turners designed to achieve large precession angles (analogue of precession coils used in the NSE arms). They may be characterized by the precession rate $|d\varphi/dq|$ that defines the NSE length to be obtained when such precessors are placed in the NSE arms:

$$\rho_{NSE} = |d\varphi/dq|. \tag{13}$$

In a quantum precessor developed in J-PARC (Tokai) a magnetic multilayer is stacked on top of a non-magnetic gap layer, which is above a non-magnetic multilayer, so that the spin-up and spin-down components reflect with a given q from different multilayers [21]. Multilayers on the basis of permalloy $Fe_{55}Ni_{45}$ are saturated in fields as low as 5 Oe. Multilayer quantum precessors have been applied in multilayer NSE [14] and multilayer interferometry [22].

We shall consider new possibilities related to using supermirrors. Two types of spin precessors can be envisaged: (1) Larmor precessors on the basis of LPC backed by a non-magnetic supermirror and (2) quantum precessors made as a stack of two supermirrors, each reflecting one of the spin components. We suppose that the condition $R_+ \cong R_-$ is fulfilled in both cases.

In the Larmor precessors a thick LPC, i.e. magnetic coating weakly reflecting neutrons in the working q-range, is deposited above a non-magnetic coating. Larmor precession of spins is induced when neutrons pass LPC before and after being reflected from the coating underneath. As an example, calculations of $|d\varphi/dq|$ for a Co layer with a thickness 1 μm on top of a non-magnetic NiMo/Ti (m=3) supermirror are represented in figure 14. Curve 1 is calculated according to the formula

$$\left|\frac{d\varphi_1}{dq}\right| \doteq 8\pi C \frac{Bd}{q^2} \tag{14}$$

obtained from an approximate formula (7) for Larmor precessions under reflection. Account of a difference in the paths of neutrons with the opposite spins in the Co layer due to spin-dependent refraction has been taken in calculation of curve 2. The refraction correction becomes conspicuous for thick LPC and at small q. The exact numerical calculation of $|d\varphi/dq|$ (points) agrees with curve 2. In a NSE experiment with monochromatic neutrons, the ρ_{NSE}-scan can be made by changing the glancing angles. In TOF measurements the spatial correlation functions can be obtained simultaneously for a range of values of ρ_{NSE}. Note that the values of $|d\varphi/dq|$ in the working q-range are proportional to the LPC thickness, so the range of ρ_{NSE} can be optimized.

The scatter of points (Fig. 14) is related to differences in reflection of spin-up and spin-down neutron waves in the presence of the magnetic layer. Due to interference of the waves reflected at different interfaces and a large thickness of the magnetic layer, the resultant phase shifts are sensitive even to slight differences in the spin-up and spin-down neutron wave amplitudes. The scatter diminishes when the magnetization of LPC is reduced (with the proportional increase of its thickness). Thus, development of Larmor processors suggests the use of materials with sufficiently low magnetization and sufficiently high remanence.

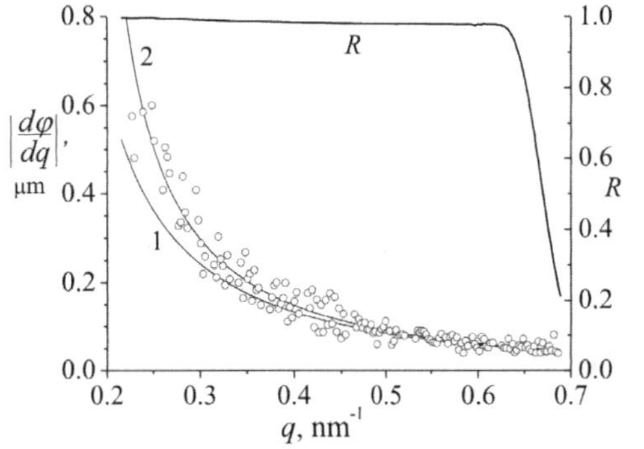

Figure 14. The precession rate $|d\varphi/dq|$ and the reflectivity R calculated for the Larmor precessor on the basis of a Co layer (1 μm) backed by the supermirror NiMo/Ti (m=3, 300 bilayers): (1) according to Larmor precession formula, (2) corrected for refraction, (o) exact numerical calculations.

Point out additional possibilities related to quantum precessors on the basis of supermirrors. First, by analogy with multilayers, consider using non-magnetic NiMo/TiZr supermirror (SM1) beneath and magnetic CoFe/TiZr supermirror (SM2) on top. The materials can be chosen so that $V_+(\text{CoFe}) = V(\text{NiMo})$ and $V_-(\text{CoFe}) = V(\text{TiZr}) = 0$. With the same sequence of thicknesses the equivalent layers in different supermirrors, reflecting the spin-up and spin-down neutrons with the same q, are always at a distance equal to the total thickness D of a supermirror. So $\varphi = Dq$, and $|d\varphi/dq| = D$ for any q, which is in agreement with numerical calculations (figure 15-*a*). The scatter of points is related to the discrete nature of the reflector now containing numerous magnetic layers with perfect interfaces.

Using a non-magnetic gap layer of thickness G between the supermirrors, one can change the precession rate to $|d\varphi/dq| = D+G$, or partly compensate a difference in the optical paths in SM1 and SM2, if any. Further we assume that $G = 0$.

The reflectivity R_+ of a real supermirror with rough interfaces may noticeably differ from 1, and spin-up neutrons will partly traverse through the magnetic supermirror and reflect from the non-magnetic supermirror. Calculations show that, as a consequence, not only the condition $R_+ \cong R_-$ is violated, but also the dependence $\varphi(q)$ is perturbed.

Perturbations due to unwanted reflections can be minimized when both supermirrors in the stack are magnetic, each reflecting one of the spin components. E.g., a CuTi/CoFe supermirror with $V(\text{CuTi}) = V_+(\text{CoFe})$ reflects only spin-down neutrons. It may be either beneath or on top of the CoFe/TiZr supermirror. It is in the first case that the spin splitting leads to precession with a minimum scatter of points (figure 15-*b*). (Multilayer precessors can also be improved in the same manner.) In the second case, the precession is in the opposite direction (antiprecession). Owing to refraction of the spin-up component in the CuTi/CoFe supermirror (the mean potential for this component is not equal to 0), the antiprecession rate depends on q (Fig. 15-*b*).

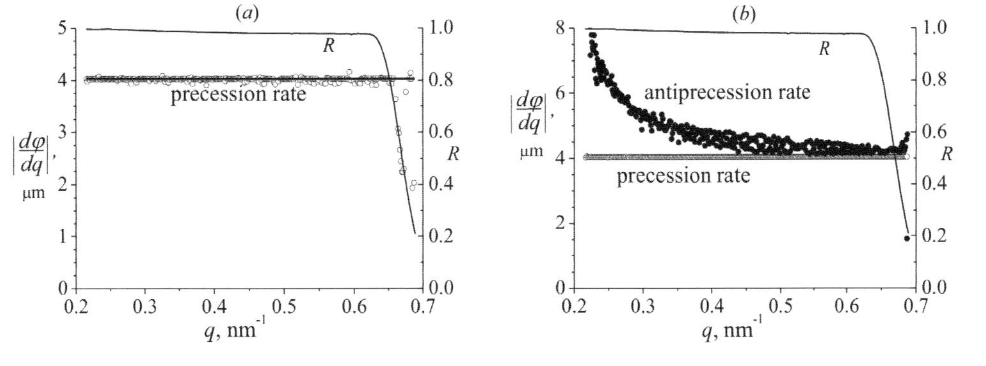

Figure 15. The precession rate |dφ/dq| and the reflectivity R calculated for the quantum precessor *(a)* on the basis of a CoFe/TiZr supermirror (m = 3, 300 bilayers) backed by a NiMo/Ti supermirror with the same sequence of thicknesses. The points (o) are exact numerical calculations, the solid line is the total thickness D= 4.035 μm of a supermirror;
(b) on the basis of a stack of CoFe/TiZr and CuTi/CoFe supermirrors with the same sequence (m = 3), the CuTi/CoFe supermirror being either beneath (o) or on top (•).

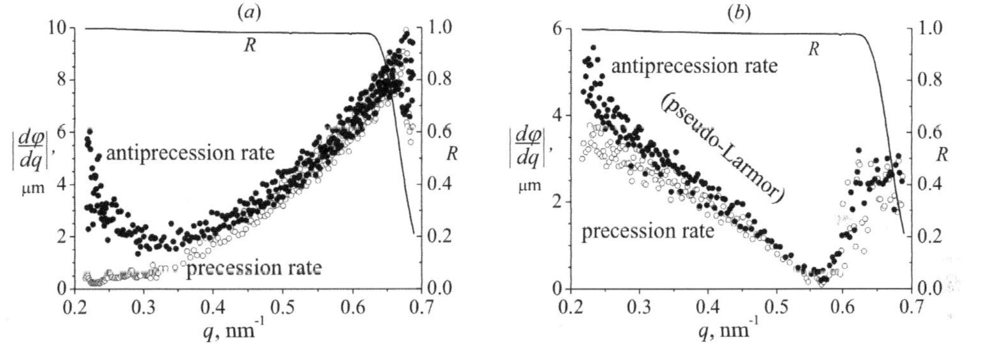

Figure 16. The precession rate |dφ/dq| and the reflectivity R calculated for the quantum precessor on the basis of CoFe/TiZr (m = 3) и CuTi/CoFe (m = 3) supermirrors with inverted sequence of thicknesses of (a) the upper and (b) the lower supermirror, when the CuTi/CoFe supermirror is beneath (o) and on top (•).

Spin precessors with a constant rate |dφ/dq| can be used in NSE experiments with beams of large divergence or/and wide wavelength band. A set of quantum precessors with different D would be required to measure the spatial correlation function. In principle, one may use external fields to scan near discrete values of ρ_{NSE}.

Inverting the sequence of thicknesses (thinner bilayers above thicker ones) in one of the supermirrors, one can introduce dependence of |dφ/dq| on q. The distance between the equivalent layers increases (decreases) with q, when the sequence of the upper (lower) supermirror is inverted. Correspondingly, precession and antiprecession rates will be increasing (decreasing) with q (Fig. 16). The phases of the neutron waves reflected from initial and inverted sequences of the equivalent layers are different, even when the reflectivities are the same. It explains the non-monotonic behavior of |dφ/dq| as a function of q. Precession and antiprecession with |dφ/dq| that decreases with q may be called pseudo-Larmor. The dependences of |dφ/dq| on q may be corrected by means of gap layers.

4. Innovative neutron optics instrumentation

NSMO basic elements have been considered. It has been concluded that some elements can work with beams of neutrons with wide wavelength band and large divergence. Combining the NSMO elements, one can design innovative neutron optics instrumentation. Rotation of NSMO elements about their reflecting surface normals may bring additional instrumental possibilities. Magnetic screens can be used to eliminate external fields for the efficient work of NSMO devices. Since the elements are based on very thin coatings, the polarized neutron devices can be made quite compact.

4.1. Neutron 3D polarizers

Using a spin-selector followed by a $\pi/2$-turner, we can build a 3D polarizer producing beams with polarization in any desired direction from an initially unpolarized beam. A scheme of the 3D polarizer is represented in figure 17. Note that the design with the reflecting surface on the opposite side is also possible.

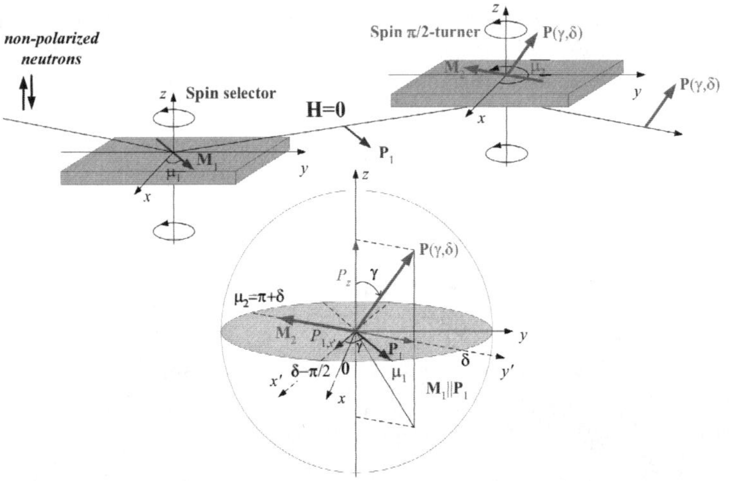

Figure 17. Scheme of a (broad band) 3D polarizer and its theory. The spin selector and the $\pi/2$-turner can be rotated about their reflecting surface normals. For the sake of visibility, the magnetization $\mathbf{M}_2$ of the $\pi/2$-turner coating is drawn on the back side of the substrate.

The reflecting surfaces of the spin-selector and the $\pi/2$-turner are assumed to be parallel to each other and coincide with the equatorial plane of the unitary sphere (figure 17). The z-axis is along the normal to the reflecting surfaces, so the x-axis and the y-axis are in the equatorial plane. Define the desired polarization vector **P** by the polar angle δ of its projection onto the equatorial plane and by its tilt γ relative to the z-axis. The angle δ is counted from the x-axis to the y-axis.

Designate the direction of the projection of the vector **P** onto the equatorial plane на as y' and introduce the x'-axis in this plane. By definition, the direction of y' is defined by the angle δ. The polarization vector $\mathbf{P}_1$ after the spin selector is in the equatorial plane. Rotation of $\mathbf{P}_1$ about the y'-axis by $\pi/2$ leaves the component $P_{1,y'}$ unchanged, and aligns the component $P_{s,x'}$ along the z-axis. Provided that $P_{1,x'} = P_z = \cos\gamma$, we obtain the vector **P** with the preset angles δ and γ. In the coordinates (x', y') the vector $\mathbf{P}_1 = (\cos\gamma, \sin\gamma)$, so in the coordinates (x, y) its direction is defined by the polar angle

$$\mu_1 = \gamma-(\pi/2-\delta) \tag{15}$$

(see also figure 17). Since $\mathbf{P}_1 \parallel \mathbf{M}_1$, μ_1 is the angle for the magnetization of the spin selector. Taking account of the sign of precession, the magnetization of the $\pi/2$-turner should be opposite to the y'-axis (figure 17), i.e.

$$\mu_2 = \pi+\delta\,. \tag{16}$$

Therefore, the 3D-polarizer can form a neutron beam with any direction of the polarization $\mathbf{P}(\delta,\gamma)$, provided that the spin selector and the $\pi/2$-turner can be rotated about their surface normals.

3D-polarizers for preparation of monochromatic beams can be made more compact (figure 18), when a substrate transparent for neutrons (silicon, quartz, etc.) with LPC is placed immediately above the polarizing multilayer. An unpolarized beam, after traversing the transparent substrate and LPC, is monocromatized and polarized in the multilayer; being reflected from the multilayer, the polarized beam passes through LPC with magnetization and thickness optimized for turning spins by $\pi/2$ (note that the polarized neutrons pass LPC only once). With independent rotations of the substrates about their surface normals, any preset direction of the polarization $\mathbf{P}(\delta,\gamma)$ of the monochromatic beam is achieved with the magnetizations of the polarizing multilayer and the $\pi/2$-LPC at angles (15) and (16), respectively.

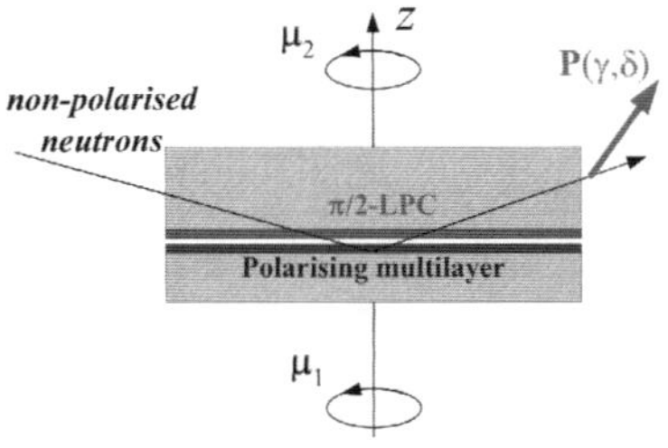

Figure 18. Scheme of a compact 3D polarizer for producing monochromatic beams with polarization in any desired direction. The two substrates can be rotated independently to align magnetizations of the polarizing multilayer and the $\pi/2$-LPC spin turner in the directions required.

4.2. Neutron spin 3D rotator

A π-turner followed by a $\pi/2$-turner can work as a 3D rotator (figure 19), rotating spins from any direction $\mathbf{P}_0(\delta_0,\pi/2)$ in the reflecting plane to any desired direction $\mathbf{P}(\delta,\gamma)$. The theory of the 3D rotator is similar to that of the 3D polarizer. First, we can rotate $\mathbf{P}_0$ in the equatorial plane from δ_0 to the angle $\delta_1 = \gamma-(\pi/2-\delta)$ by directing the magnetization of the π-turner at the angle

$$\mu_1 = (\delta_0 + \delta_1)/2 = [(\delta_0 + \delta)+(\gamma - \pi/2)]/2. \quad (17)$$

With the magnetization of the $\pi/2$-turner at the angle μ_2 given by (16), we obtain the reflected beam with the desired polarization $\mathbf{P}(\delta,\gamma)$. A more detailed analysis shows that there are four pairs of μ_1 and μ_2 that produce the same result. The magnetizations at $\mu_1 = [(\delta_0 + \delta)-(\gamma - \pi/2)]/2$ and $\mu_2 = \delta$ would also transfer $\mathbf{P}_0(\delta_0,\pi/2)$ to $\mathbf{P}(\delta,\gamma)$. The other pairs are obtained by substituting $\mu_1 + \pi$ for μ_1 in the first two combinations.

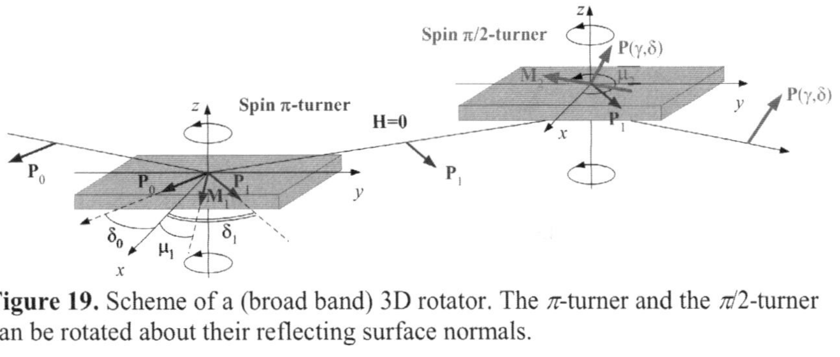

Figure 19. Scheme of a (broad band) 3D rotator. The π-turner and the $\pi/2$-turner can be rotated about their reflecting surface normals.

4.3. Neutron spin manipulator.

Adding a $\pi/2$-turner before the π-turner in the scheme in figure 18, one can rotate spins from any direction to any direction (figure 20). The additional $\pi/2$-turner is needed to rotate spins from an initial direction $\mathbf{P}_0(\delta_0,\gamma_0)$ into the reflecting plane. Therefore, its magnetization should be at the angle δ_0 (or $\delta_0 + \pi$), i.e. along the projection of $\mathbf{P}_0$ onto the reflecting plane. Then the polarization vector of neutrons reflected from the first $\pi/2$-turner will be in the reflecting plane at an angle δ_1. And the task is reduced to the one considered in section 4.2.

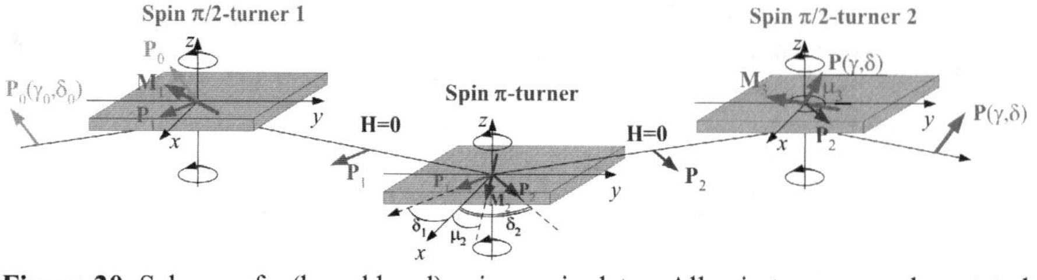

Figure 20. Scheme of a (broad band) spin manipulator. All spin turners can be rotated about their reflecting surface normals.

4.4. Neutron 3D analyzers

A new possibility of measuring three components of the polarization vector is also to be mentioned. First, two polarization components in the reflecting plane of a spin selector are measured by rotating it about its surface normal and finding intensities for four orientations of its magnetization. Then, introducing a $\pi/2$-turner before the spin selector (inverting the scheme of the 3D polarizer in figure 17), one can measure the polarization component perpendicular to the reflecting plane of the spin selector. It suffices to measure intensities for opposite magnetizations of either $\pi/2$-turner or spin selector. For monochromatic beams, a $\pi/2$-LPC spin turner can be placed above the spin selector, as shown in figure 18. An even more distinct solution would be a polarizer coated by a $\pi/2$-LPC, with their magnetizations being perpendicular.

4.5. Neutron hyper-polarizers

Common polarizers are conceived as spin selectors transmitting neutrons with the desired spin (50%) and eliminating neutrons with the 'wrong' spin (50%). Define a hyper-polarizer as a device which not only separates neutrons with the opposite spins, but also flips the wrong spins, thus polarizing more than 50% (up to 100%) of neutrons. A gain in polarized neutron intensity can be achieved only with the increase either in the angular divergence (hyper-polarizers of type I) or in the width (hyper-polarizers of type II) of the beam, so that the total phase volume doubled by the spin subspace is conserved (the Liouville theorem).

4.5.1. Hyper-polarizers of type I. Two variants of the scheme of a hyper-polarizer of type I are represented in figure 21. The polarizing filter on the basis of a periodic multilayer (or truncated supermirror) reflects only spin-up neutrons in the range from $q^+_{\min} = 4\pi \sin\theta^+_{\min} / \lambda_{\max}$ to $q^+_{\max} = 4\pi \sin\theta^+_{\max} / \lambda_{\min}$. The extreme angles and wavelengths in the formulas are defined by mean glancing angle θ_+ and beam divergence $\Delta\theta$, as well as by mean wavelength and wavelength band width. The spin-down neutrons traverse the polarizing filter and reflect from a spin-flipping π-coating (flipper). The reflecting surface of the flipper is turned relative to the filter surface by an angle α. The angle α is chosen so that neutrons with flipped spins traverse the filter without being reflected:

$$\alpha > \theta^+_{\max} \frac{\lambda_{\max}}{\lambda_{\min}} - \theta^+_{\min}. \tag{18}$$

The reflection q-range of the flipper should include the interval from $q^-_{\min} = 4\pi \sin(\theta^+_{\min} + \alpha) / \lambda_{\max}$ to $q^-_{\max} = 4\pi \sin(\theta^+_{\max} + \alpha) / \lambda_{\min}$. Condition (18) is obtained on the assumption that all glancing angles are small; more generally, the flipper coating and the angle α must provide that $q^-_{\min} > q^+_{\max}$.

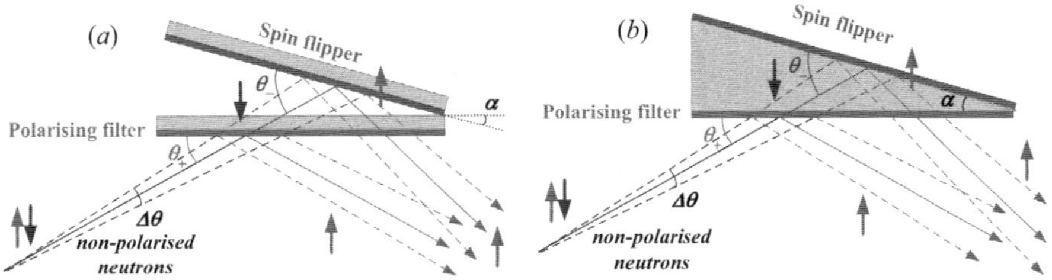

Figure 21. Schemes of a hyper-polarizer of type I: (*a*) angle design and (*b*) wedge design.

The spin-up neutron reflectivity of the polarizing filter should be close to 1, since spin-up neutrons passing through the filter are reflected from the flipper with the "wrong" spin. E.g., for the polarizing filter one could use a thick CoFe film totally reflecting spin-up neutrons at the glancing angles below the critical angle and transmitting spin-down neutrons. Being reflected from a multilayer flipper, neutrons should traverse the CoFe film at glancing angles exceeding the critical angle.

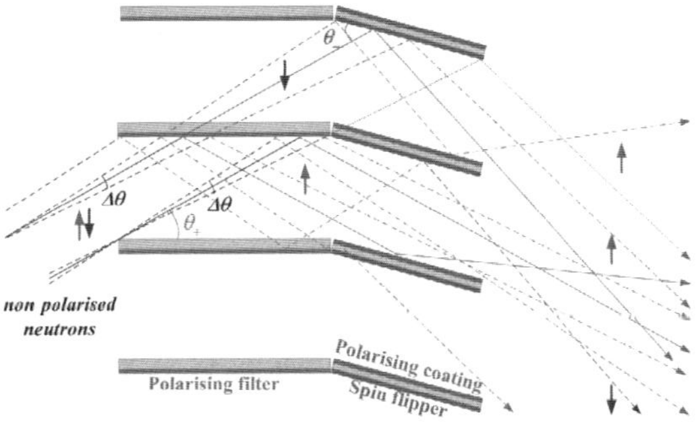

Figure 22. Multichannel design of a hyper-polarizer of type I.

Other conditions for hyper-polarization of neutron beams are also possible, the main requirement being that neutrons reflected from the flipper either traverse through ($\theta_- > \theta_+$) or miss (it is possible when schemes in figure 21 are inverted, i.e. $\theta_- < \theta_+$) the polarizing filter. To intercept a wide beam, one can build a hyper-polarizer from several hyper-polarizing angles (wedges) or use a multichannel design (figure 22). Hyper-polarizers of type I are fit for hyper-polarization of collimated monochromatic beams.

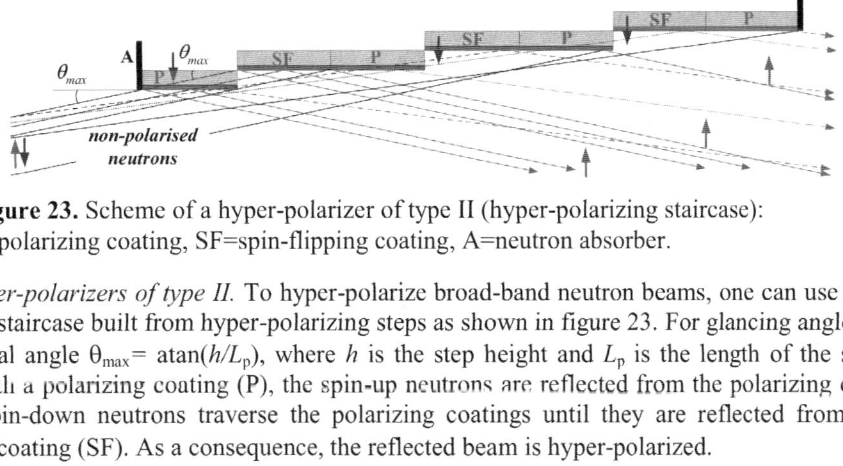

Figure 23. Scheme of a hyper-polarizer of type II (hyper-polarizing staircase): P=polarizing coating, SF=spin-flipping coating, A=neutron absorber.

4.5.2. Hyper-polarizers of type II. To hyper-polarize broad-band neutron beams, one can use a hyper-polarizing staircase built from hyper-polarizing steps as shown in figure 23. For glancing angles below the diagonal angle $\theta_{max} = \text{atan}(h/L_p)$, where h is the step height and L_p is the length of the substrate (wafer) with a polarizing coating (P), the spin-up neutrons are reflected from the polarizing coatings, and the spin-down neutrons traverse the polarizing coatings until they are reflected from a spin-flipping π-coating (SF). As a consequence, the reflected beam is hyper-polarized.

4.5.3. Additional possibilities for hyper-polarization. Mention two additional possibilities. The first is the scheme of a hyper-polarizer which has been suggested earlier [16] and allows of hyper-polarizing a broad-band neutron beam without spin-flip reflection. The spin-up neutrons are reflected from a polarizing coating on top of a layered structure with a gradually rotating magnetization, in which spins of transmitted neutrons reverse adiabatically. The scheme became realizable after a publication by Kraan and Rekveldt [23] on the possibility of using a single domain wall in a permalloy film as a neutron spin flipper.

Secondly, an unpolarized beam can be split into sub-beams with the opposite spins not only by reflection, but also by spin-dependent refraction, e.g. between two sharp magnetic poles (Stern-Gerlach splitting). Reflecting one of the sub-beams from a neutron optical π-turner, one can achieve hyper-polarization. Moreover, moving the π-turner from one sub-beam to the other is equivalent to using a flipper.

4.6. Double-reflection spin turners

All rays reflected consequently from two mirrors with reflecting surfaces at an angle β to each other (periscope geometry in Fig. 24) are deviated by the same angle:

$$2(\beta_1 + \beta_2) = 2\beta, \qquad (19)$$

where β_1 and β_2 are the glancing angles of reflection from mirrors 1 and 2, accordingly. It means that layered structures that produce spin precession by angles proportional to q, i.e. to the glancing angle θ ($\sin\theta \cong \theta$ for thermal and cold neutrons), can be used to build spin turners for monochromatic beams. According to (19), spins of all monochromatic neutrons consequently reflected from two identical spin turners based on spin-splitting coatings will be rotated by one and the same angle

$$\varphi = 4\pi\beta D/\lambda, \qquad (20)$$

which depends on neutron wavelength λ, spin-splitting distance D and angle β. In particular, for a certain wavelength it can be made $\pi/2$ and π (figure 24).

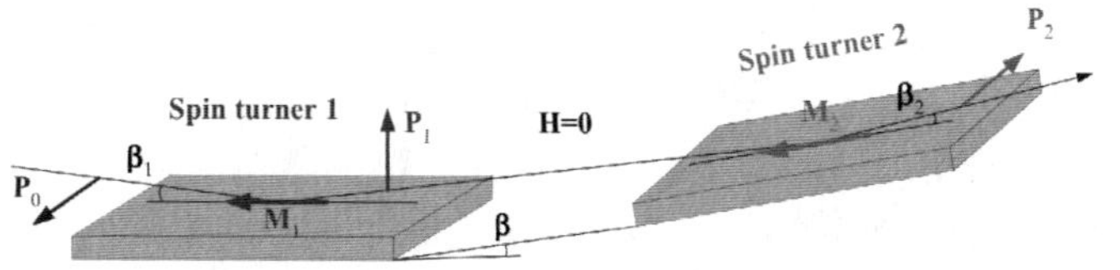

Figure 24. Scheme of a double-reflection spin flipper. Two identical $\pi/2$-turners are based on spin-splitting coatings.

4.7. Multichannel spin turners.

Multichannel design of spin turners is possible, provided that the number of reflections from the spin-turning coatings in each channel can be fixed (say, reduced to 1, when one of the channel walls is weakly reflecting and absorbs neutrons). Combining a multichannel $\pi/2$-turner with a multichannel analyzer, one could measure at least two neutron polarization vector components.

5. Discussion

The existing polarized neutron optics is built on the basis of polarizing coatings that work as spin selectors. A variety of elements for neutron spin manipulation optics (NSMO) based both on Larmor and non-Larmor precession under reflection has been considered. It can be concluded that transition to 3D in neutron polarization optics brings additional instrumental possibilities. New neutron optical devices may include spin turners (particularly, $\pi/2$-turners and π-turners, or flippers), spin precessors and anti-precessors, 3D-polarizers, 3D-analyzers, 3D-rotators, spin manipulators, hyper-polarizers. Possible applications of the innovative neutron optics range from 3D polarization and polarization analysis to new neutron optics instrumentation. Hyper-polarizers can double the intensity of polarized neutron beams. In accordance with the Liouville theorem, a gain in intensity can be achieved only with the increase either in the angular divergence (hyper-polarizers of type I) or in the width (hyper-polarizers of type II) of the beam. However, hyper-polarizers of type I can be focused onto the sample. It is easier to transport a neutron beam with a lower cross section and then polarize it with a hyper-polarizer of type I or II. Besides, neutron beams at the outlets of the existing neutron guides can be hyper-polarized to achieve a gain in the polarized neutron intensity. Another strategy is to place a hyper-polarizer in front of the conventional polarizer to achieve a gain in both intensity and polarization of the beam. The possibility to build a hyper-polarizer by sputtering a polarizing coating onto a permalloy film, the thickness of which is optimum for formation of single domain walls, is also worth studying. Hyper-polarization of a Stern-Gerlach-split beam with a π-turner might also be of interest in some applications.

NSMO is directly applicable to developing such techniques as reflectometry with 3D-polarimetry for both monochromatic and white neutron beams, Neutron optical Spin Echo (NoSE), including compact NoSE and TOF NoSE schemes, based only on neutron optical devices ($\pi/2$-turners, π-turners and precessors). Development of polarimetry with VCN and even UCN on the basis of the neutron optical spin turners is a challenge that can now be addressed. UCN polarimetry is an equivalent of ellipsometry, but with extreme sensitivity to surface magnetism. New UCN sources and progress with NSMO may allow of developing this technique.

NSMO can be made broad-band, which is of special interest for applications at impulse reactors and spallation sources. As the working parts of NSMO are very thin coatings, the devices can be made quite compact and used, e.g., for handling micro-beams [24,25].

The potential of the innovative optics is still to be revealed by studies of spin precession phenomena with real structures. Coercivity and remanence of coatings on the basis of magnetic layers are of decisive importance. Surface oxidation is also to be addressed as an important issue for the work of the spin turners. The magnetic layers for spin turners are too thin to produce significant stray fields. However, stray fields from numerous magnetized layers of remanent supermirrors in the absence of the guide field may detectably rotate the neutron polarization vector [12]. Efficient ways of compensating stray fields have been proposed and substantiated by numerical calculations in Ref. [12]. They may be needed and can be developed for supermirror-based spin selectors and spin precessors. Numerous problems may be envisaged in developing spin precessors based on Larmor and non-Larmor spin precessions under reflection from magnetic layered structures. Step by step approach is required to find appropriate solutions.

Acknowledgements
The work was partly supported by RFBR grant No. 12-02-12066-ofi_m.

References

[1] Pleshanov N K 2001 *Physica* **B 304** 193
[2] Pleshanov N K 2000 *Phys. Rev.* **B 62** 2994
[3] Pleshanov N K 1994 *Physica* **B 198** 70
[4] Böni P, Clemens D, Kumar M Senthil, Pappas C 1999 *Physica* **B 267-268** 320
[5] Pleshanov N K, Bodnarchuk V, Gähler R, Korneev D A, Menelle A, Metelev S V, Pusenkov V M, Schebetov A F, Ul'yanov V A 2001 *Physica* **B 297** 126
[6] Kumar M Senthil, Böni P 2002 *J. Appl. Phys.* **91** 3750
[7] Pappas C, Kali G, Krist T, Böni P, Mezei F 2000 *Physica* **B 283** 365
[8] Pleshanov N K 1996 *Z. Phys.* **B 100** 423
[9] Pleshanov N K, Pusenkov V M 1996 *Z. Phys.* **B 100** 507
[10] Rühm A, Toperverg B P, Dosch H 1999 *Phys. Rev.* **B 60** 16073
[11] Pleshanov N K 1994 *Z. Phys.* **B 94** 233
[12] Pleshanov N K, Axel'rod L A, Zabenkin V N, Syromyatnikov V G, Ul'yanov V A 2008 *J. Surface Investigation* **2** 846
[13] Ebisawa T, Tasaki S, Kawai T, Hino M, Achiwa N, Otake Y, Funahashi H, Yamazaki D, Akiyoshi T 1998 *Phys. Rev.* **A 57** 4720

[14] Tasaki S, Ebisawa T, Maruyama R, Achiwa N, Kawai T, Kawabata Y, Hino M, Yamazaki D 2003 *Physica* **B 335** 234
[15] Korneev D A 1992 *Neutron Optical Devices and Applications*: *Proc. SPIE* vol 1738, ed C F Majkrzak, J L Wood p 477
[16] Pleshanov N K 2001 *Physica* **B 297** 131
[17] Pleshanov N K 2010 Nucl. Instrum. Methods **A 613** 15
[18] Matveev V A, Pleshanov N K, Bulkin A P, Syromyatnikov V G 2012 *J. of Physics: Conf. Ser.* **340** 012086
[19] Zabel H 1994 *Physica* **B 198** 156
[20] http://www.swissneutronics.ch
[21] Funahashi H, Ebisawa T, Haseyama T, Hino M, Masaike A, Otake Y, Tabaru T, Tasaki S 1996 *Phys. Rev.* **A 54** 649
[22] Hino M, Achiwa N, Tasaki S, Ebisawa T, Kawai T, Yamazaki D 2000 *Phys. Rev.* **A 61** 136071
[23] Kraan W H and Rekveldt M T 2008 *Nucl. Instrum. Methods* **A 596** 422
[24] Kozhevnikov S V, Rühm A, Ott F, Pleshanov N K, Major J 2011 *Physica* **B 406** 2463
[25] Rühm A, Kozhevnikov S V, Ott F, Radu F, Major J 2013 *Nucl. Instrum. Methods* **A 708** 83

Neutron spin manipulation devices using YBCO films

T.Wang[1], F.Li[1], S.R.Parnell[1], W.A.Hamilton[2], H.Kaiser[3], A.L.Washington[1], D.V.Baxter[1] and R.Pynn[1,2]
[1]Indiana University, Centre for the Exploration of Energy and Matter, Bloomington, IN, USA
[2]Neutron Science Directorate, Oak Ridge National Laboratory, Oak Ridge, TN,USA
[3]University of Missouri Research Reactor Center, Columbia, MO, USA

Abstract. The Meissner effect in a thin-film superconductor can be used to create a sharp boundary between regions of different magnetic field and hence can be used as a component of neutron spin manipulation devices. We have developed two cryogenic neutron spin manipulation devices using single-crystal, high-T_c, YBCO films, which can be cooled without using liquid cryogens and eliminate small angle scattering associated with polycrystalline films.

The devices are a spin flipper and a spin precession device both of which use 350-nm-thick YBCO films covered with gold on a 0.5 mm thick sapphire substrate. The spin flipper consists of one such film mounted on an oxygen-free copper frame and connected to a closed-cycle He refrigerator. The flipper is capable of working with a maximum neutron beam size of 42 × 42 mm^2 and can be used with both vertical and horizontal guide fields.

The spin precession device was constructed by mounting two of the YBCO films parallel to one another with an H-magnet between them. By changing the current through the H - magnet, the precession of the neutron polarisation between the films can be controlled. Tests at the Low Energy Neutron Source (LENS) show that this device is capable of generating controlled spin precession for a neutron beam up to 20 × 20 mm^2 in cross section.

1. Introduction

Precise manipulation of the neutron spin is vital in a number of types of neutron scattering experiments [1]. These can be broadly arranged into three categories: (1) magnetic structure investigation either in elastic (for example [2, 3]) or inelastic scattering (for example [4]), (2) incoherent background subtraction for hydrogenous materials [5] and (3) Larmor labelling techniques involving either scattering angle [6, 7] or energy encoding [8]. For all of these applications highly efficient devices are required to manipulate the neutron spin. These rely upon a combination of non-adiabatic transitions and/or precession of the neutron spin in a magnetic field.

The Meissner effect in a superconducting film can be used to provide sharp well defined field boundaries and the use of superconducting tape allows high currents and hence high magnetic fields to be achieved. The Meissner effect in low-temperature niobium foil has been used in a number of devices pioneered by the ILL in France; a white beam spin flipper (Cryoflipper) [9, 10] and a vector polarimeter (Cryopad) [11]. In the former (Cryoflipper) the superconductor is used to provide an abrupt non-adiabatic transition between two separated anti-parallel fields. In Cryopad, the orientation of the neutron polarisation can be placed in any direction parallel to a Nb Meissner screen by the use of rotatable guide fields. This polarisation may be further rotated by a precise magnetic field that is contained between two Nb Meissner screens. Precise control of the precession is only possible by the use of abrupt transitions in the magnetic field provided by the superconductor which separates the guide and precession magnetic fields. Using Cryopad, information not accessible by other means can be obtained about the spatial arrangement of the magnetisation vector within a sample. As described by Brown [2], the control of incident beam polarisation and the directional analysis of the scattered neutron polarisation allows the full polarisation tensor to be determined.

In Cryopad the use of superconducting screens effectively creates defined regions for spin precession. In another device called MuPAD [12], designed to achieve the same goals, superconducting screens are not used. Rather the penetration of external fields is prevented by mu-metal shields and a combination of two precession fields orthogonal to each other is used to control the neutron polarisation vector.

While Cryopad performs very well, it requires considerable operational infrastructure and it is not designed for use at pulsed neutron sources. Furthermore, it is more complex than is needed for applications with small neutron scattering angles, such as small angle neutron scattering (SANS) and reflectometry. To address these concerns, we decided to design a simple, compact device that uses high temperature superconductors as Meissner screens. A key component of such a device is the precession region. This paper describes the current status of our design of this component.

2. YBCO material details and small angle scattering

Single crystals of Yttrium Barium Copper Oxide (YBCO) grown on sapphire are now commercially available (Ceraco GmbH, Germany) and can be cooled below their superconducting transition using a closed cycle refrigerator (CCR) with a base temperature of $\approx$ 10K. This simplifies the device, lowers the maintenance requirements and cost of operation. As we have previously shown [13] these films are highly transparent to neutrons and are not expected to produce significant small angle neutron scattering (SANS), in contrast to the case of polycrystalline Nb where SANS can occur from grain boundaries.

The penetration of magnetic flux quanta into the superconducting sheet is a significant concern for the operation of a cryoflipper. The critical field of high-T_c materials is anisotropic and various values have been reported in the literature [14, 15, 16], all in the range of $\approx$100-200 G for films and single crystals of YBCO. The strength of magnetic guide fields is much less than the magnitude of the critical magnetic field, so we do not expect flux penetration problems.

The films used have a high critical temperature (87 K). Each one consists of a $78\times100\times0.5$ mm^3 sapphire substrate with a 40 nm CeO_2 buffer layer onto which is grown a 350 nm thick single-crystal YBCO film, capped with 100 nm of gold.

In order to check the assertion that the films produce negligible SANS, a series of tests were performed using unpolarised neutrons at the LENS [17, 18] SANS instrument. In all cases, no change in scattering was observed from the instrumental background [13]. From these measurements a wavelength independent transmission of 98.4 $\pm$ 0.1 % was calculated over a neutron wavelength region of 4 Å to 14 Å. The cutoff at 4 Å is due to the Be filter on the LENS SANS instrumentt.

3. Cryogenic system

Both devices presented in this proceedings use the same cryogenic device and electromagnetic guide fields. The later are mounted on either side of a short, rectangular cryostat tail, which contains the YBCO film. The cryostat tail and electromagnets are 200 mm long along the neutron beam. The guide fields can be used in either a horizontal or vertical configuration, both of which produce magnetic fields that are approximately parallel to the YBCO film in the absence of the film. The soft iron pole pieces of each guide field are 52 mm apart. Sapphire windows on either side of the cryostat tail allow for high neutron transmission whilst minimising parasitic scattering and background effects. The housing incorporates a standard commercial

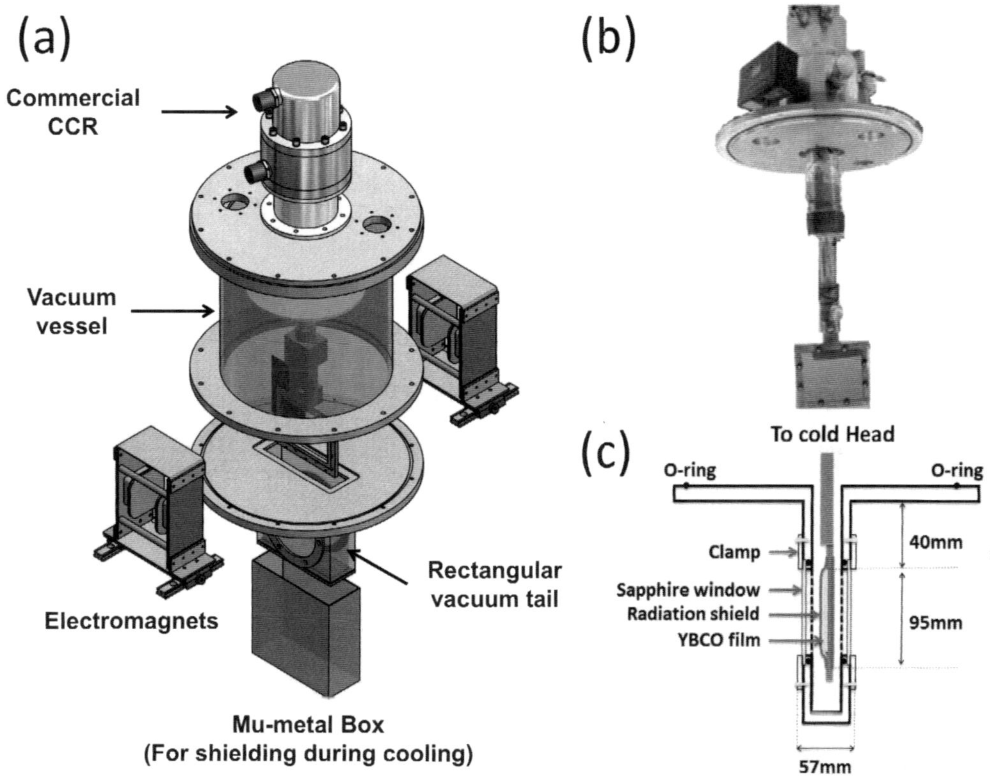

Figure 1. (a) Exploded view schematic of the cryoflipper components showing the electromagnets in the horizontal orientation. (b) Picture detail of the CCR with the YBCO film in an oxygen free copper mount. (c) Schematic of the cryostat tail, showing YBCO film placement and arrangement of the aluminium radiation shields.

closed cycle refrigerator (CCR) (Sumitomo CH-204, USA). The vacuum vessel is fabricated from aluminium to eliminate potential distortions of the magnetic field. Figure 1 shows a schematic and a photograph of the device.

4. Cryogenic spin flipper

The design of the cryoflipper and electromagnets is presented in [13] whilst the efficiency for both pulsed and continuous sources is presented in [19]. The efficiency of the device was determined using a two flipper measurement which allows for decoupling of the individual flipper efficiencies (for example [20]). In all cases the beam size was defined by Cd slits placed close to the cryoflipper.

Flipping efficiencies measured at LENS are shown in Figure 2 for an electromagnet current of 3A with a 42×42 mm^2 beam. This shows a uniform efficiency over a broad neutron wavelength range demonstrating that the device is suitable for white neutron beams. Furthermore the average efficiency is $99.33 \pm 0.25\%$. With optimisation of the electromagnet current the flipper efficiency increases to a maximum of $99.5 \pm 0.3\%$ for a current of 5A. A number of different beam sizes were tested from 10×10 mm^2 to 42×42 mm^2 and no statistically relevant difference

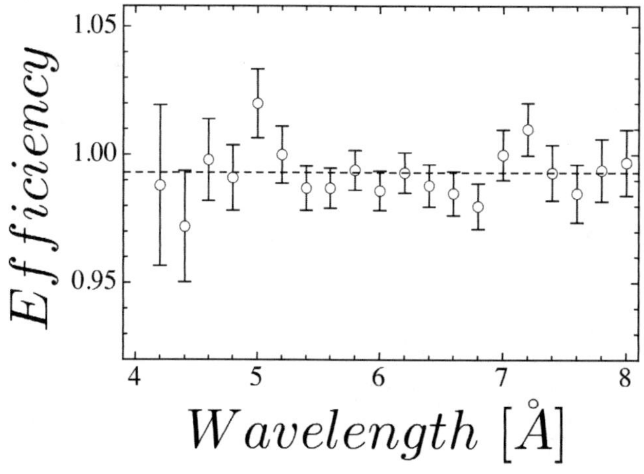

Figure 2. Wavelength dependence of the extracted cryoflipper efficiency for a $42 \times 42mm^2$ neutron beam obtained on the SESAME instrument at LENS. The dashed line is the average over the wavelength range shown (99.3± 0.3%).

was observed.

Measurements were performed with the magnetic field from electromagnets in both the horizontal and vertical directions. As the film is rectangular this places the field along the long or short axis of the film respectively. Mounting the electromagnets such that the field is pointing along the short axis of the film (78 mm) produces a slightly lower efficiency ($\approx$ 1% lower for a 41×41 mm^2 beamsize) than when the field is along the long (100 mm) axis, indicating that the film size is the principal limiting effect on the device efficiency. It appears likely that larger guide fields and a larger YBCO film could be used to flip the polarisation of even larger cross-section neutron beams if required. The results are reproducible within error from cycle to cycle, including dismantling and reassembly of the device [19].

5. Cryogenic spin precession device

Using the same vacuum vessel and electromagnets a spin precession device was also constructed. This consists of two YBCO films, as discussed in section 2. The YBCO films are parallel to one another, separated by 15mm with an H-magnet between them (Shown in Figure 4(a)). Thin cryogenic mu-metal is used as a magnetic flux return path and ensures that the region between the two YBCO films is magnetically isolated. The pole pieces are wound using YBCO superconducting tape (Superpower, USA). By changing the current through the tape, the precession of neutron polarisation between the parallel YBCO films can be controlled.

In order to optimise the dimensions of the device and gain a better understanding, the whole assembly of YBCO, cryogenic mu-metal and superconducting tape (Superpower, USA) was modelled using commercial finite-element software (Magnet©, Infolytica, Canada). Shown in figure 3 (a) is the field profile in the y-z plane (the same plane as the YBCO films) mid-way between the two films. The contours show a highly uniform region between the pole pieces, however the uniformity decreses with increasing beam size. This is more apparent in 3 (b) which shows the modulus of the magnetic field integrated through the device. The simulation shows that the field is well contained by the YBCO and mu-metal pieces. As the field integral along the x-axis varies slightly with position across the area probed by the neutron beam, the total precession angle of the polarisation will depend on position within the beam. The simulation indicates that for an average precession angle of π the precession angle will have a standard deviation of 2.7 ° for a 20×20 mm^2 beam size and a 4A current at a neutron wavelength of 3.90 Å. For 15×15 mm^2 and 10×10 mm^2 beam size the precession angle deviates by 1.6 ° and 0.7 ° respectively.

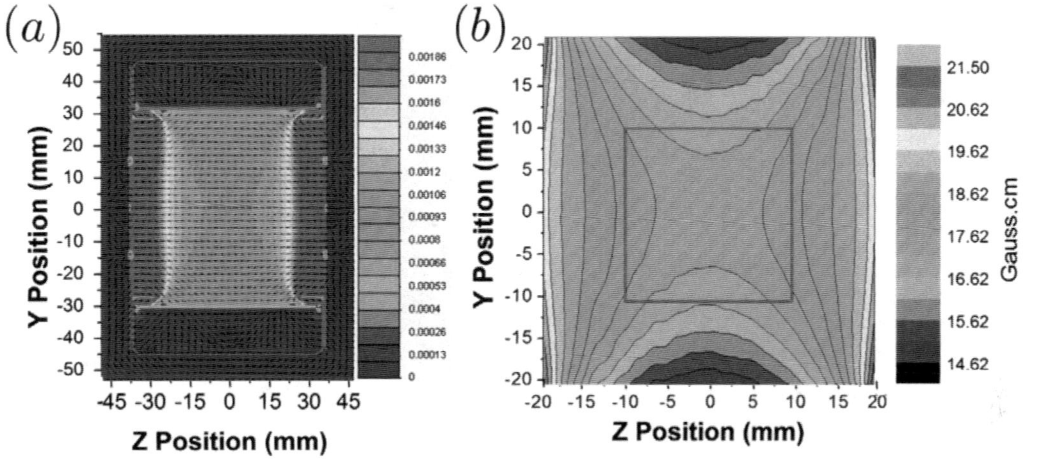

Figure 3. Simulation result using MagNet© software. (a) is a contour and vector plot of the field in the mid-plane (x=0) of the device. (b) shows a magnetic field integral map of the device at 4 A. The field integral within the central region defined by the red box is 17.4 ± 0.28 G cm. Note z is the vertical direction, x the direction of the neutron beam with y orthogonal to the other two axes.

Simulations were also performed for a rectangular solenoid wound from one layer of the superconducting tape. This option was not chosen for two reasons: the field integral is lower compared to that achieved using the separated pole pieces in the H-magnet configuration and, secondly, most of the superconducting tapes available commercially now contain a certain amount of gadolinium, which greatly reduces the transmission. Neutron transmission measurements were performed on a number of different superconducting tapes, but only one of them was found to have high transmission. However this tape was on a steel backing which cannot be bent with a small radius of curvature.

An exploded view of the final device design is shown in Figure 4 (a) along with the coordinate axes used in this contribution. The whole device is only 15 mm in length along the beam and the separation between the pole pieces, shown in figure 4 (b) is 41 mm. The solenoid has two layers of 5 mm wide superconducting tape with four turns on one layer and five on the other.

The prototype device was measured on the SESAME beam line at LENS. We employed a continuously pumped nuclear spin polarised ^{3}He analyser [21], instead of the supermirror device used to test the flipper. The device was cooled to below T_c of the YBCO films inside a mu-metal box to prevent flux trapping. The procedure for the experiment was to measure the zero-current polarisation ($P_{I=0}$) first and to use this to normalise the polarisation measured with current on (P_I). This omits the polarisation losses from other sources but measures only the dephasing loss from the precession. Flipping was performed by reversing the incident electromagnet, using the first YBCO film as a cryoflipper, as described earlier.

The measurement of $P_I/P_{I=0}$ is shown in Figure 5 for a 20×20 mm^2 beam for both positive and negative precession fields. Results for positive and negative field agree, indicating that there was very little magnetic flux trapped between the YBCO films. With a 4A current, the

(a)

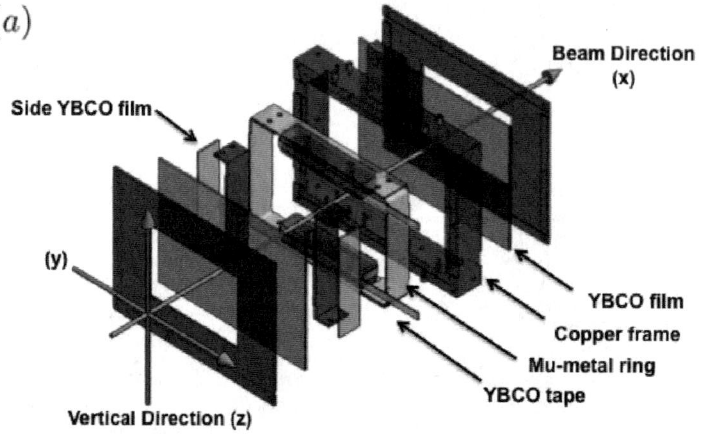

(b)

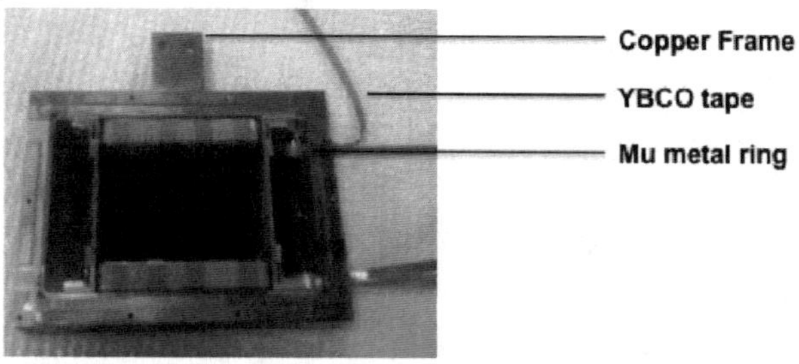

Figure 4. (a) shows an exploded view of the precession device. Also indicated is the coordinate frame used in this communication, where x is the direction of the incident neutron beam. The bottom picture (b) shows the physical precession device with the front YBCO film removed. Note the mu-metal ring and superconducting tape. (Colour version available online)

polarisation returns almost to its full magnitude after a π rotation from which we conclude that the field integral does not vary significantly over the area of the neutron beam. The experiment shows that the mu-metal and YBCO combination provides a good zero field chamber (with current in the device off) implying that these materials may be utilised to provide a zero field region, similar to that of Cryopad and MuPAD. Measurements were performed with a range of beam sizes and the behaviour of the ratio of the polarisation with current on/current off ($P_I/P_{I=0}$) is well described by;

$$P_I/P_{I=0} = (1 - a\lambda)Cos(\sigma\lambda) \tag{1}$$

where the constant a is the depolarisation per Å caused by the variation of the field integral over the neutron beam and σ is a constant of proportionality between precession angle and the current in the precession coils. Fitting the data to this function gives good agreement as shown in Figure 5. At 4A the π flip occurs at 3.87 $\pm$ 0.01 Å which is close to the value obtained from simulation (3.90 Å). The measured depolarisation was 1.6%, 0.4% and 0.1% for beam sizes of 20×20 mm^2, 15×15 mm^2 and 10×10 mm^2 respectively. Simulation results gave 0.37%, 0.10% and 0.02% for these beam sizes, somewhat less than we measure. However, because our neutron source produces a pulse width of 600 μs there is also a spread in values of neutron wavelength for any given time of flight. Correcting for this leads to a predicted depolarisation of 0.65 %,

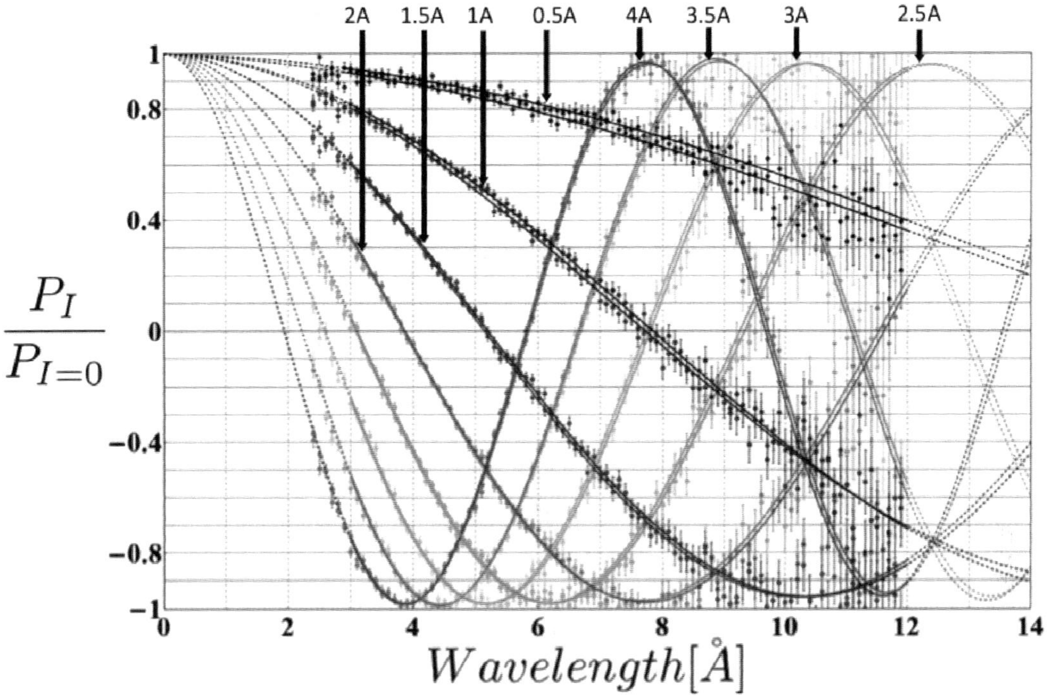

Figure 5. Measured polarisation with different currents (P_I) through coils divided by polarisation with no current ($P_{I=0}$) for a 20 × 20 mm^2 beam size. Points are experimental data and solid lines are fitted curves. Current used is labelled on the top with each current having both positive and negative values as shown by the same colour. Note data is shown from 2.2-12Å, whilst the fits were performed over the regions indicated by the solid lines. These lines when dotted are extrapolated from the fitted region.

0.29 % and 0.19 % for beam sizes of 20×20 mm^2, 15×15 mm^2 and 10×10 mm^2 respectively. This is in good agreement for the 10×10 and 15×15 mm^2 beam sizes although we measure a significantly higher depolarisation for the 20×20 mm^2 than is predicted. This may be a result of slight offsets in the positioning of the device at the larger beam size, although we have not investigated this point in detail.

The limiting of this device to 4A was due to heating at the superconductor/copper contacts interface. Changing these contacts allows for significantly higher currents to be achieved and will be reported in a later publication. Also, changing the current as a function of time will allow us to achieve the same precession angle for a range of neutron wavelengths.

6. Conclusion

The simple devices shown in this proceedings demonstrate the applicability of high T_c YBCO superconductors to polarised neutron instrumentation. The cryoflipper has a high efficiency and and easy to setup for a white beam. The device is robust, compact and requires very limited operational overhead. The precession device shown here could be mounted close to the sample. Combining two of these precession regions, one before a sample and the other after it would allow a compact vector polarimeter to be constructed. Combined with rotatable guide fields and a zero field sample chamber, a full spherical neutron polarimeter can be constructed. The total length of the device would be ≈ 0.4 m including the external guide fields.

Further uses of this technology include Wollaston prisms for spin echo scattering angle encoding applications [6, 22, 7] and work is underway to develop such a device.

7. Acknowledgments

This project is supported by National Science Foundation grant DMR-0956741. We would like to acknowledge Eddy Lelièvere-Berna (ILL, France) for useful discussions on the radiation shield construction, Jack Doskow (Indiana University Bloomington, USA) for producing the drawings and Xin Li (Indiana University Bloomington, USA, now Oak Ridge National Laboratory, USA) for assistance with the SANS measurements.

Construction of LENS was supported by the National Science Foundation grants DMR-0220560 and DMR-0320627, the 21st Century Science and Technology fund of Indiana, Indiana University, and the Department of Defence.

8. References

[1] Williams W G 1988 *Polarized Neutrons (Oxford Studies in Neutron Scattering and Condensed Matter), Oxford, Oxford University Press*

[2] Brown P J 2001 *Physica B: Condensed Matter* **297** 198 – 203

[3] Moon R M, Riste T and Koehler W C 1969 *Physical Review* **181** 920

[4] Regnault L P, Tasset F, Lorenzo J E, Roberts T, Dhalenne G and Revcolevschi A 1999 *Physica B: Condensed Matter* **267** 227–235

[5] Babcock E, Salhi Z, Appavou M S, Feoktystov A, Pipich V, Radulescu A, Ossovyi V, Staringer S and Ioffe A 2013 *Physics Procedia* **42** 154–162

[6] Pynn R 1980 *Neutron Spin Echo, edited by F. Mezei, Lecture Notes in Physics, Vol. 128, pp. 159–177. Heidelberg: Springer*

[7] Rekveldt M T, Plomp J, Bouwman W G, Kraan W H, Grigoriev S and Blaauw M 2005 *Review of Scientific Instruments* **76** 033901

[8] Mezei F 1972 *Zeitscrihft Fur Physik* **255** 146

[9] Tasset F, Gardner P and Saïdane K B 1982 *Int. Conf. The Neutron and its Applications, 13–17 Sept, Cambridge (UK) (Conf. Handbook 8.Q.6), p. 124.*

[10] Fitzsimmons M R, Lütt M, Kinder H and Prusseit W 1998 *Nuclear Instruments and Methods A* **411** 401–416

[11] Tasset F 1989 *Physica B: Condensed Matter* **156** 627–630

[12] Janoschek M, Roessli B, Keller L, Gvasaliya S, Conder K and Pomjakushina E 2005 *Journal of Physics Condensed Matter* **17** L425–L430

[13] Parnell S, Kaiser H, Washington A, Li F, Wang T, Baxter D and Pynn R 2013 *Physics Procedia* **42** 125 – 129

[14] Lamura G , Aurino M , Andreone A and Villegier J C 2009 *Journal of Applied Physics* **106** 053903

[15] Tochihara S, Yasuoka H, Mazaki H, Osada M and Kakihana M 1999 *Journal of Applied Physics* **85** 8299–8306

[16] Bohmer C, Brandstatter G and Weber H W 1997 *Supercond. Sci. Technology* **10** A1

[17] Baxter D V, Cameron J M, Derenchuk V P, Lavelle C M, Leuschner M B, Lone M A, Meyer H O, Rinckel T and Snow W M 2005 *Nuclear Instruments and Methods B* **241** 209–212

[18] Lavelle C M, Baxter D V, Bogdanov A , Derenchuk V P, Kaiser H , Leuschner M B, Lone M A, Lozowski W , Nann H , Przewoski B v, Rernmes N , Rinckel T , Shin Y , Snow W M and Sokol P E 2008 *Nuclear Instruments and Methods A* **587** 324–341

[19] Parnell S R, Washington A L, Kaiser H, Li F, Wang T, Hamilton W A, Baxter D V and Pynn R 2013 *Nuclear Instruments and Methods in Physics Research Section A: Accelerators, Spectrometers, Detectors and Associated Equipment* **722** 20–23

[20] Kraan W H, Grigoriev S V, Rekveldt M T, Fredrikze H, de Vroege C F and Plomp J 2003 *Nuclear Instruments and Methods A* **510** 334–345

[21] Boag S, Babcock E, Andersen K, Becker M, Charlton T, Chen W, Dalgliesh R, Elmore S, Frost C, Gentile T, Anton R L, Parnell S, Petoukhov A, Skoda M and Soldner T 2009 *Physica B: Condensed Matter* **404** 2659 – 2662

[22] Pynn R, Fitzsimmons M R, Lee W T, Shah V R, Washington A L, Stonaha P and Littrell K 2008 *Journal of Applied Crystallography* **41** 897–905

International Workshop on Neutron Optics and Detectors (NOP&D 2013) IOP Publishing
Journal of Physics: Conference Series **528** (2014) 012025 doi:10.1088/1742-6596/528/1/012025

Grazing Incidence Neutron Spin Echo Spectroscopy: Instrumentation Aspects and Scientific Opportunities

O. Holderer[1], H. Frielinghaus[1], S. Wellert[2], F. Lipfert[1], M. Monkenbusch[3], R. von Klitzing[2], D. Richter[1,3]

[1]Jülich Centre for Neutron Science (JCNS) at Heinz Maier-Leibnitz Zentrum (MLZ), Forschungszentrum Jülich, Lichtenbergstr. 1, 85747 Garching, Germany

[2]Institut für Chemie, Technische Universität Berlin, Berlin, Germany

[3]Jülich Centre for Neutron Science (JCNS) and Institute for Complex Systems, Forschungszentrum Jülich, D-52425 Jülich, Germany,

E-mail: o.holderer@fz-juelich.de

Abstract. Grazing Incidence Neutron Spin Echo Spectroscopy (GINSES) opens new possibilities for observing the thermally driven dynamics of macromolecules close to a rigid interface. The information about the dynamics can be retrieved as a function of scattering depth of the evanescent neutron wave, on the length scale in the range of some 10-100 nm. Using a classical neutron spin echo spectrometer with a laterally collimated beam, dynamics can be measured in grazing incidence geometry. We show examples of how the interface modifies the dynamics of microemulsions, membranes and microgels. Instrumental details and possible improvements for this technique will be presented. The key issue is the low intensity for dynamics measurements with an evanescent neutron wave. Conceptual questions how a specialised instrument could improve the experimental technique will be discussed.

1. Introduction

Small angle neutron scattering under grazing incidence conditions (GISANS) allows to study structural features close to a rigid interface [1][2]. The neutron beam is directed to an interface between typically a silicon block and the sample below the critical angle. An evanescent neutron wave is then travelling parallel to the surface with a penetration depth of 10 - 100 nm perpendicular to the surface. The scattered neutrons from the evanescent wave are recorded to get structural information lateral and perpendicular to the rigid interface on a nanometer length scale. Recently, this technique has been also applied to study the dynamics of mesoscopic objects close to the rigid interface [4][5][8] with Neutron Spin Echo (NSE) spectroscopy. NSE spectroscopy has the highest energy resolution of all spectroscopic techniques [6][7] in neutron scattering and allows to study the dynamics of macromolecules such as thermal fluctuations of surfactant membranes [9][10] or internal motion of proteins [3]. NSE spectroscopy under grazing incidence conditions (GINSES) is sensitive to thermal fluctuations close to the interface and by varying the penetration depth of the evanescent wave the dynamics as a function of distance to the interface can be measured. The penetration depth is of the order of 10-100 nm. Figure 1 shows a bicontinuous microemulsion measured in GISANS geometry. The interface imposes a lamellar phase for the first membrane layers, giving rise to a Bragg peak in the detector image, while the bulk microemulsion at larger distances to the interface contributes with a

Debye-Scherrer like scattering contribution [4]. The typical Q-value for a GINSES experiments is indicated with the red circle in the GISANS plot (on the Q_y=0 axis). The large sample cell and the slit geometry of the beam results in a poor resolution for $Q_y \neq 0$, all measurements so far were restricted to the Q_y=0 axis. Sensitivity for fluctuations perpendicular to the interface is obtained, e.g. the membrane fluctuations of microemulsions.

2. GINSES at an existing (standard) NSE spectrometer

A series of experiments under grazing incidence has been carried out at the J-NSE spectrometer at the FRM II [7]. An entrance aperture in front of the first precession coil has limited the beam laterally to a width of 2 mm, the sample aperture has been adapted to the footprint of the sample cell (also about 2 mm). The sample cell consists of a silicon block (1 or 2 cm thick) where the beam enters at the front side and is then totally reflected at the inner surface to the sample. Different types of sample cells have been used (from GISANS or reflectometry experiments) with a length of 8 or 15 cm. The sample cell can be temperature controlled with a water thermostat. The scattered beam passes through the silicon block and follows then the standard flight path of the NSE spectrometer to the detector. The only difference to a standard experiment is therefore the use of the strong lateral collimation with slit apertures and the appropriate sample cell. The collimation and the small sample volume illuminated with the evanescent wave leads to an intensity loss of the order of $1/10^4$. A very high instrumental stability (stable currents, no thermal effects at the coils) and a low magnetic and radiation background of the surrounding is mandatory for such GINSES experiments.

The GINSES technique has been applied to microemulsions [4][5][12] and microgels [11] near interfaces. As an example, the dynamics of a lamellar microemulsion, consisting of d-decane and D_2O to equal amounts, and 32% of the surfactant $C_{10}E_4$4, and the influence of the scattering vector is presented (Figure 2). Due to the low scattering intensity in most experiments the whole 2D-detector has been evaluated to capture the largest possible solid angle. When the statistics is good enough, the detector can be divided into regions of different Q. Here, three Q-values have been extracted from measurements on lamellar microemulsions by dividing the 2D detector into three slices of different Q. Besides the lamellar phase in GISANS geometry, also a bulk lamellar has been measured for comparison. The sample is placed in a cell made especially for lamellar microemulsions, having 60 quartz plates over the 30 mm sample width. The microemulsion orients along the quartz plates with the normal of the planes parallel to Q. The SANS data (top left in Figure 2) showed distinct peaks of the lamellar ordering. NSE measurements with the center of the detector set to Q=1.1 nm^{-1} have been performed just in front of the second peak. The relaxation time in such microemulsions typically slows down significantly at the position of the correlation peaks, with $1/\tau \propto 1/S(Q)$ (often referred to as "de Gennes narrowing"). The slowing down of the dynamics when approaching the correlation peak at Q=1.5 nm^{-1} with the relaxation time increasing from ~140 ns for Q=1.0 nm^{-1} to 180 ns for Q=1.3 nm^{-1} (top right in Figure 2) reflects the $1/\tau \propto 1/S(Q)$ behaviour of the relaxation time. The situation is different for measurements of the same sample under grazing incidence (bottom left and right in Figure 2). There the flat interface is only imposed on one side of the sample, which obviously reduces the range of the ordering. The relaxation time decreases continuously with increasing Q. The difference in the relaxation time between grazing incidence conditions and incident angles above the critical angle is not significant. The highest Q has also the fastest dynamics in GINSES-measurements on this sample (in bicontinuous microemulsions a Q^3-dependence would be expected for $1/\tau$[9], here the Q-range is too limited to quantify the exact Q-dependence). The difference in absolute relaxation time is also at least partly attributed to the much better ordering of the microemulsion in the lamellar cell.

A second group of experiments focused on the internal dynamics in a thin layer of adsorbed charged microgel particles [11]. Although the intensity is extremely low, the intermediate

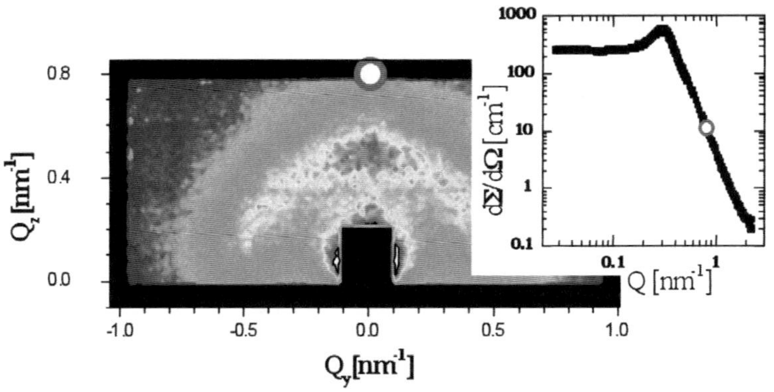

Figure 1. GISANS detector image of a bicontinuous microemulsion near a Si interface. It shows a superposition of the peak from the lamellar phase at the interface and the Debye-Scherrer like ring from the bulk microemulsion. The inset shows a scattering curve of a standard SANS experiment of a bicontinuous microemulsion with the characteristic correlation peak. Most GINSES experiments have been carried out at Q=0.8 nm^{-1} as a compromise between probing local dynamics at high Q and high intensity at low Q.

scattering function S(Q,t) could be measured for rather low Q-values of Q=0.6 nm^{-1}. The dynamics of these particles was strongly dependent on the charge of the microgels. The charge state was tuned by protonation of the carboxylic acid groups in the network. For being in the easily accessible dynamic range, even neutral and slightly charged particles were studied in the first set of experiments. Compared to bulk phase diffusion the dynamics seems to be strongly slowed down in this system. Moreover, first data suggest a depth dependence of the dynamics. More details can be found elsewhere [11].

3. New concepts

3.1. Instrument

The flight path after the scattering event has to be undisturbed and the solid angle covered by a 2d-detector of about 30x30 cm^2 has to be used to gather the maximum possible intensity. The usual setup as for a normal NSE instrument is required, including the correction coils for reducing the field inhomogeneity of the off-axis flight paths in the solenoid. The two precession regions of a spin echo spectrometer (before the scattering event and after the scattering event) have to be symmetrical. Therefore, the magnetic layout of a GINSE instrument must be basically the same as for a standard NSE instrument.

Possible gains can be achieved in front of the precession paths by preparing a beam already in the desired shape (laterally narrow, but high). The flight path of typically 4 m between the end of the neutron guide and the sample in a NSE spectrometer acts as a collimation distance, the lateral divergence of the beam has to be small enough such that the neutrons hit the sample. Fig. 3 shows a possible neutron optics design which improves the intensity at the sample. McStas simulations were made where the beam (initially 30 mm in height and 45 mm in width in this example, the scale on the figure is different for x- and y- direction) is split into 3 pieces, the outer two pieces are then guided on top and below the middle part respectively. In a first section, the outer pieces are directed upwards and downwards respectively, until they are one height of the neutron guide displaced, then the three parts of the beam are reflected by a polarizing mirror into the new direction, each polarizer section displaced by half of the polarizer length. In this way, all parts of the beam are finally on top of each other, with the same divergence as in

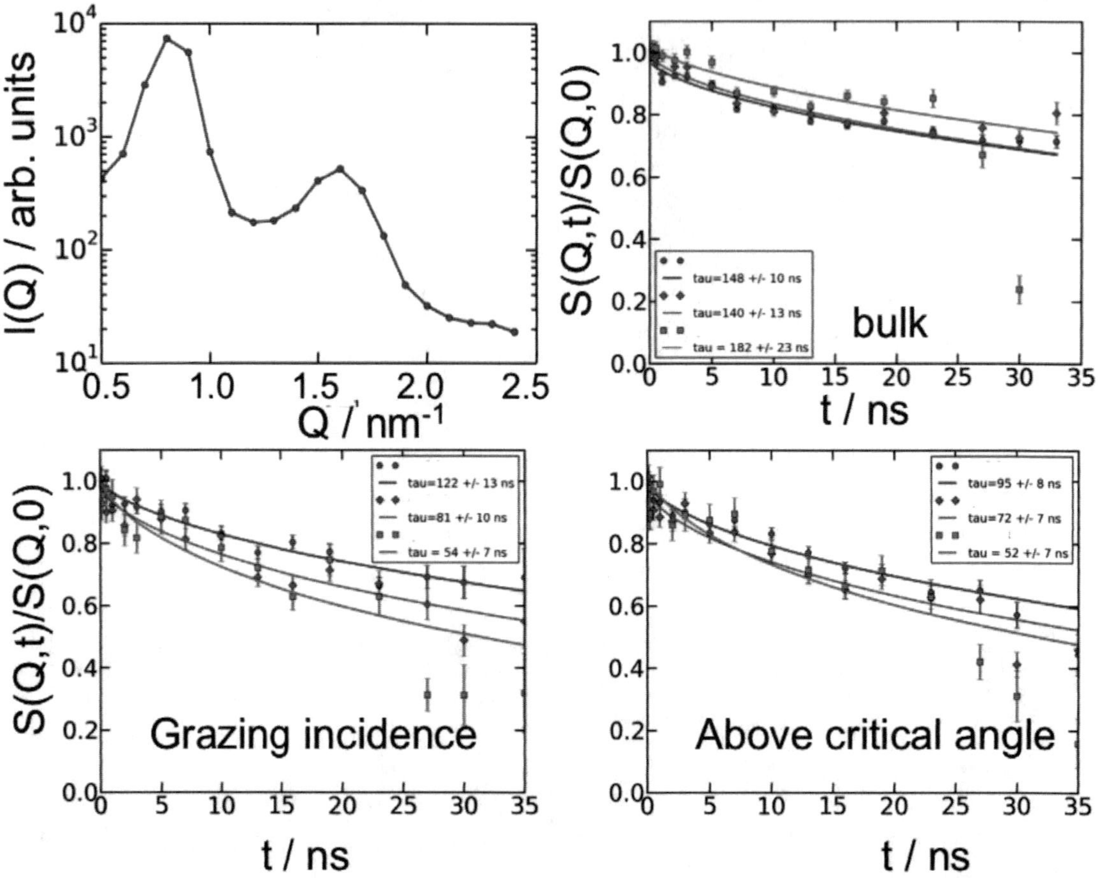

Figure 2. Top left: Small angle scattering of a lamellar microemulsion (bulk sample), oriented in a cell with 60 parallel quartz plates. The dynamics at Q=1.0 - 1.3 nm^{-1} shows the slowing down when approaching the next scattering peak (top right; the plots showing S(Q,t) show always the three Q-values 1, 1.1, 1.3 nm^{-1} in this order). With the GINSES sample cell the slowing down is not observed, neither under grazing incidence conditions, nor above the critical angle (bottom left and right respectively).

the beginning. The ideal gain of such a configuration would be a factor of 3, the first McStas-trials lead to a factor of 2 in gained intensity. The gain has been determined from the McStas simulations with a monitor having the size of the sample footprint (3mm wide, 50mm heigh) at the sample location (approx. 4 m behind the end of the neutron guide). Only neutrons with a sufficiently small divergence reach therefore the sample monitor. The divergence has been checked with a lambda-divergence monitor, which registers the divergence for the different wavelengths (Fig. 3, right). In front of the beam splitting section and at the end of the neutron guide, the divergence was approximately $\pm 0.15°$. This setup requires at least 3 reflections, including one polarizing reflection. A setup for beam shaping with only two reflections might be possible with two inclined mirror planes for each part of the neutron beam.

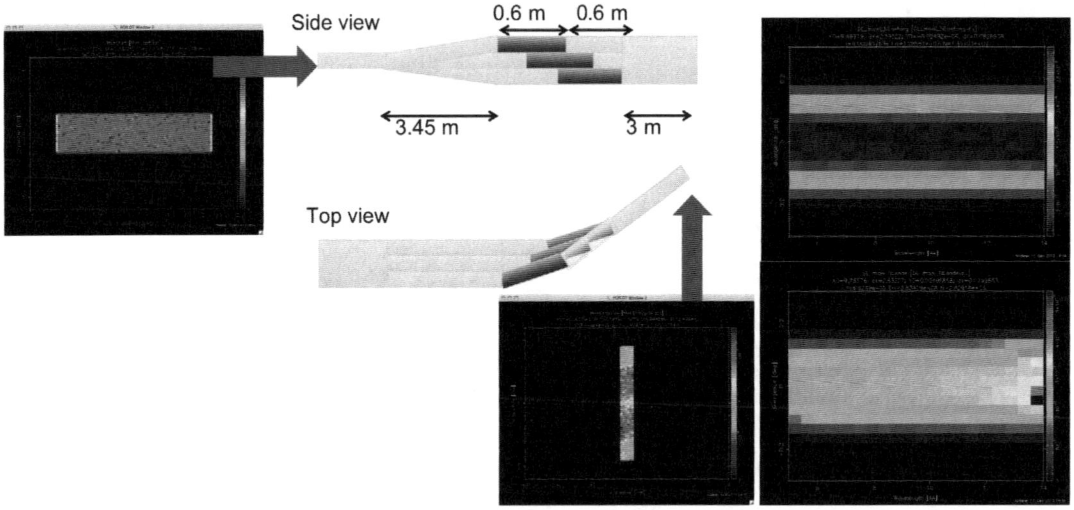

Figure 3. right: Possible realization of a neutron guide system with optimized geometry. The divergence is preserved and not increased with this setup, giving the advantage over a simple trumpet- and antitrumped design. left: plots of a neutron divergence monitor: divergence vs wavelength in the range of 5-14 Å, at the beginning of the beam splitter and at the end of the neutron guide system. In both cases the divergence is approximately $\pm 0.15°$.

3.2. Sample cell and environment

A gain in intensity (possibly by a factor of 2-3) is possible with an optimized cell design. The beam height (which is limited to about 40 mm at the J-NSE spectrometer) could be further increased along with a higher sample cell. The longest cells we have used so far are with 150 mm at the limit, where the transmission of the Si block is still acceptable. A prism in front of the sample cell allows to get a very well defined scattering depth, even for large wavelength bands [12]. Although first tested at a pulsed source at the SNS-NSE for the larger wavelength band available there [13], this option is especially suited also for instruments at continuous sources. The well defined scattering depth for all wavelengths at one setting allows to change the wavelength while staying at the same scattering depth. One can then easily extend the measured timerange with some extra Fourier times at longer wavelengths (a procedure which is commonly used for standard measurements). Secondly, the extreme angles (incident angle of "zero" or the critical angle) can be approached with a much higher precision when the scattering depth is better defined.

An important topic are auxiliary techniques at the sample position. It is desirable to have a quality control of the sample right at the start as well as for the duration of the rather long experiment (typically one to several days for one sample). Different kinds of monitoring and measuring equipment at the sample position can be envisaged. Diffusing wave spectroscopy might be an option for online monitoring of concentrated and turbid samples. One interesting technique in this context is evanescent wave dynamic light scattering (and variations thereof), which can provide dynamic information on larger length- and timescales than GINSES. Besides the extension of the dynamic range it can also monitor the quality of the sample.

4. Conclusion

We presented examples of neutron spin echo investigations near a solid surface with the possibility of varying the penetration of the evanescent wave and hence the probing distance.

Thermal fluctuations of microemulsions and microgel particles in the vicinity of the interface have been investigated, and modifications of the thermal motion by the solid surface could be detected. The low intensity of the technique due to the strong lateral collimation of the beam requires a very stable instrument with a low and constant background. An optimized neutron beam transport and sample environment would allow to increase the intensity at the sample by a factor of about 5 compared to the situation at current NSE instruments which would allow to study a wider class of samples. Auxiliary monitoring and measuring techniques at the sample position like e.g. evanescent wave dynamic light scattering could help to gain additional information simultaneously and to monitor the sample quality throughout the rather long experiment.

[1] P. Müller-Buschbaum, J.S. Gutmann, R. Cubitt, M. Stamm, *Colloid Polym Sci* **277**, 1193-1199 (1999)
[2] M. Kerscher, P. Busch, S. Mattauch, H. Frielinghaus, D. Richter, M. Belushkin, G. Gompper, *Phys. Rev. E* **83**, 030401 (2011)
[3] R. Biehl, B. Hoffmann, M. Monkenbusch, P. Falus, S. Preost, R. Merkel, D. Richter, *Phys. Rev. Lett.* **101**, 138102 (2008)
[4] H. Frielinghaus, M. Kerscher, O. Holderer, M. Monkenbusch, D. Richter, Phys. Rev. E 85, 041408 (2012);
[5] H. Frielinghaus, O. Holderer, F. Lipfert, M. Kerscher, S. Mattauch und D. Richter, EPJ Web of Conferences, 33:03005 (2012)
[6] M. Monkenbusch, R. Schätzler, D. Richter, *Nucl. Instr. and Meth. In Phys. Res. A* **399**, 301 (1997)
[7] O. Holderer, M. Monkenbusch, R. Schätzler, H. Kleines, W. Westerhausen, D. Richter, *Meas. Sci. Technol.* **19**, 034022 (2008)
[8] M. Walz, S. Gerth, P. Falus, M. Klimczak, T.H. Metzger, A. Magerl, *J. Phys. Cond. Mat.* **23**, 324102 (2011)
[9] O. Holderer, H. Frielinghaus, D. Byelov, M. Monkenbusch, J. Allgaier, D. Richter, *J. Chem. Phys.* **122**, 094908 (2005)
[10] S. Wellert, M. Karg, O. Holderer, A. Richardt, T. Hellweg, *Phys.Chem.Chem.Phys.* **13**, 3092 (2011)
[11] S. Wellert et al., in preparation
[12] H. Frielinghaus, O. Holderer, F. Lipfert, M. Monkenbusch, N. Arend, D. Richter, *Nucl. Instr. and Meth. In Phys. Res. A* **686**, 71-74 (2012)
[13] M. Ohl et al., *Nucl. Instr. and Meth. In Phys. Res. A* **696**, 85 (2012)
[14] M. A. Plum, S. D. B. Vianna, A. Unger, R. F. Roskamp, H.-J. Butt, B. Menges and W. Steffen, Soft Matter 7, 1501 (2011)
[15] G. Fytas, S. H. Anastasiadis, R. Seghrouchni, D. Vlassopoulos, J. Li, B. J. Factor, W. Theobald and C. Toprakcioglu, Science 274, 2041 (1996)

The Neutron Beam Expansion Program at the Bragg Institute

F Klose[1], P Constantine, SJ Kennedy and RA Robinson
Australian Nuclear Science and Technology Organisation
B87, Locked Bag 2001, Kirrawee DC NSW 2232, Australia

E-mail: Frank.Klose@ansto.gov.au

Abstract. The Bragg Institute is operating the neutron scattering science facilities at the Australian research reactor OPAL. The first set of seven neutron scattering instruments was provided as part of the OPAL construction project which was completed in 2007. During the period 2008 - 2013, the instrument suite was significantly expanded by a further seven instruments. In addition to this, major investments were made to establish a world-class infrastructure for supporting these instruments, including new sample environments, ^{3}He polarisers/analysers, additional neutron guides and a Be filter option for chemical spectroscopy.

1. Introduction

Over the past few years, the Bragg Institute went through a large instrumentation expansion phase, doubling the number of instruments from 7 to 14 and adding or upgrading support infrastructure such as neutron guides and sample environments. The funding for the program came from the Australian government through ANSTO internal funding for the TOF inelastic spectrometer PELICAN, the ultra-small-angle neutron scattering instrument KOOKABURRA, the Laue diffractometer JOEY, the Be Filter spectrometer for chemical analysis and ^{3}He Polarisation Project. In addition to this, the Taiwanese government enabled participation of their neutron science community in OPAL by investing at the Bragg Institute and constructing the cold triple-axis spectrometer SIKA.

The most significant portion of the funding for the new instruments, however, was secured during the time of the global financial crisis when Canberra asked ANSTO to propose possible infrastructure areas where additional funding might be productively allocated. The Bragg Institute proposed a significant *Expansion Plan for Neutron Science in Australia*. The proposal was successful and funding was publicly announced on 12 May 2009 as part of the Super Science – Future Industries initiative. This enabled the construction of the TOF-based small-angle neutron scattering instrument BILBY, the backscattering spectrometer EMU, the radiography/tomography instrument DINGO, two new cold neutron guides CG2A and CG2B as well as new sample environment equipment, including a 12 Tesla split-coil high-field magnet. The description of the $37 million NBI2 project reads as follows: "*This project will significantly enhance the research capabilities of the Australian neutron science facilities at ANSTO. It will provide university, government and industry based users of the new ANSTO research reactor OPAL with new world-class facilities for investigating the structure and dynamics of condensed matter with particular emphasis on the areas nanoscience, soft matter*

[1] To whom any correspondence should be addressed.

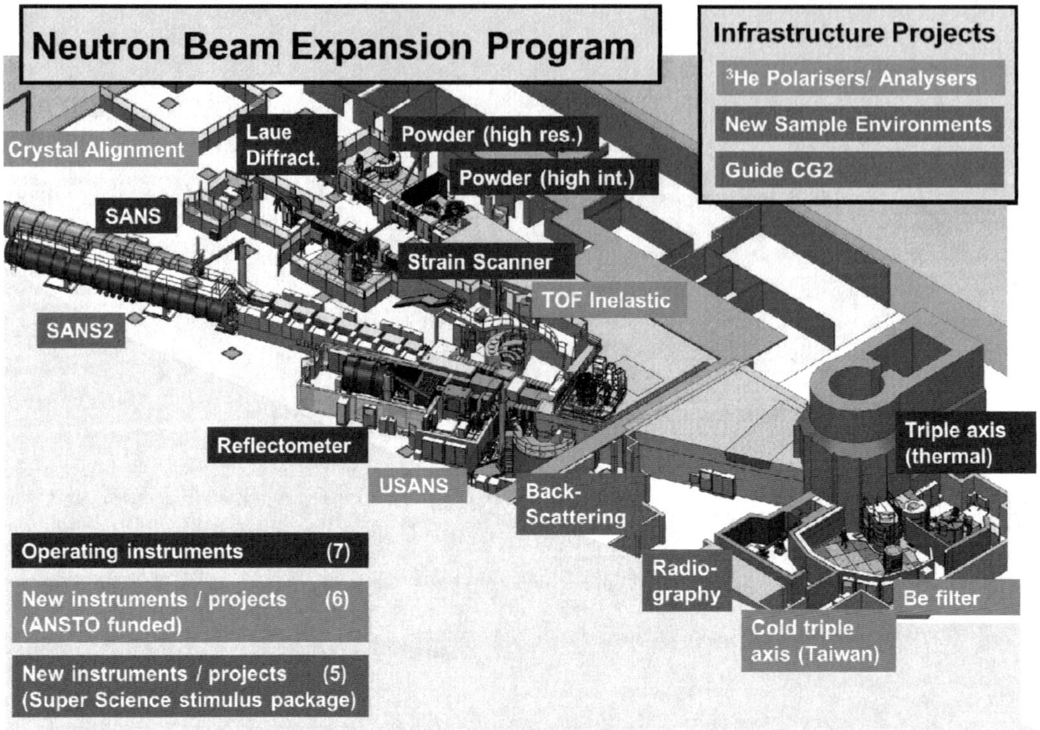

Figure 1. New instruments and projects constructed during the Neutron Beam Expansion Project (2008 - 2013).

Instrument/project name	Instrument/project type	Budget (AUD)	Construction complete
PELICAN	TOF inelastic spectrometer	8.6 M	May 2012
KOOKABURRA	Ultra small-angle scattering	3.2 M	Jun 2013
SIKA	Cold triple axis spectrometer	8.2 M	Oct 2012
JOEY	Laue diffractometer	0.5 M	Sep 2013
DINGO	Radiography/tomography	3.8 M	Jun 2013
BILBY	TOF small-angle scattering	10.1 M	Feb 2014
EMU	Backscattering spectrometer	8.9 M	Jun 2014
	CG2 neutron guides	5.2 M	Apr 2013
	New sample environments	2.8 M	Jun 2013
	^{3}He polarisers/analysers	4.2 M	Feb 2013
	Be filter (option for TAIPAN)	3.2 M	Oct 2014

Table 1. List of new instruments and projects constructed during the Neutron Beam Expansion Project (2008 - 2013)

dynamics and biology which are key areas for future technology and industrial developments in Australia". The total investment for the Neutron Beam Expansion Program at the Bragg Institute exceeds AUD 65M (see Figure 1 and Table 1 for a listing of all new instruments and neutron infrastructure projects).

The choice of the new instruments was based on earlier requests by the scientific users of the Bragg Institute (e.g. the Blue Mountains Workshop in 2005) and initial specifications were developed at the NBI2 Scoping Workshop in 2009. More detailed scientific specifications, schedules and costings were developed during the NBI2 Conceptual Design Workshop which was generously hosted by the Center for Neutron Research of the National Institute of Standards and Technology, USA.

During the peak period from 2011 to 2013, well over 40 staff worked on the Neutron Bean Expansion Program. Our Beam Instruments Advisory Group (BIAG) reviewed the progress of the instrumentation projects on a bi-yearly basis. Major installations of the new instruments started in 2008 and major changes to the OPAL reactor (i.e. the insertion of the new cold guides) were completed in December 2012. In December 2013, PELICAN, SIKA JOEY, DINGO and KOOKABURRA are in the hot-commissioning stage. Early in 2014, these instruments will obtain their operational licenses and will later commence the regular user operation mode. Almost all major components of BILBY and EMU have now been installed and these two instruments are scheduled to commence their hot-commissioning phases in Feb 2014 and June 2014, respectively.

2. Main characteristics of the new instruments and projects [1]

2.1. PELICAN - TOF inelastic spectrometer

PELICAN is a cold-neutron time-of-flight spectrometer. It is designed for studies of polycrystalline, glassy and liquid samples mainly in chemistry and in studies of soft matter. In order to achieve the required pulsed mono-energetic beam of neutrons onto the sample, a focussing HOPG crystal monochromator and a Fermi chopper are used. Changes in neutron energy induced by interaction with the sample, is detected by time-of-flight measurement between sample and detector (energy resolution: 50 μeV to 1 meV; ~2.5% of E_i). PELICAN will be OPALs "first" large time-of-flight spectrometer capable of polarisation analysis (polariser: FeCo/Si supermirrors; analyser: ^{3}He cell), and we expect that the instrument will operate in this mode up to 30% of the time.

Figure 2. PELICAN - TOF inelastic spectrometer

Figure 3. KOOKABURRA - USANS

2.2. *KOOKABURRA - Ultra small-angle scattering*
KOOKABURRA uses the classical Bonse-Hart method for detection of small-angle scattering at very low q (q_{min} = 3 x 10^{-5} A^{-1}), and will be an optimised version of the currently best reactor-based USANS instrument, BT-5 at NIST. The premonochromator consists of a vertically and horizontally focusing PG crystal reflecting at 90°. The instrument has two interchangeable sets of five-bounce channel-cut crystals, Si(111) and Si(311), allowing operation at two distinct wavelengths (λ = 4.43 Å and λ = 2.32 Å, respectively), for optimally accommodating weakly and strongly scattering samples.

2.3. *SIKA - Cold triple axis spectrometer*
SIKA is a cold triple axis spectrometer directly viewing the OPAL cold source. Using the (002) reflection of a doubly focusing HOPG monochromator, the available incident neutron wavelengths range is 1.62 Å $\leq \lambda \leq$ 5.8 Å, corresponding to an incident energy range of 2.43meV $\leq E_i \leq$ 31 meV. For neutron detection, both single tube and PSD ^{3}He counters are available.

2.4. *JOEY - Laue diffractometer*
JOEY is a Laue diffractometer using a CCD camera for neutron detection. This instrument uses the remaining beam transmitted by the instrument KOALA and will be mainly used for pre-orientation of single crystals. The instrument concept is similar to OrientExpress at the Institute Laue - Langevin.

Figure 4. SIKA - Triple axis spectrometer

Figure 5. JOEY - Laue diffractometer

2.5. *DINGO - Radiography/tomography*
The DINGO instrument is located at the reactor beam hall. It uses the direct thermal beamport HB2 of OPAL. DINGO has two experimental configurations, a high-flux mode realised by a 40 mm collimator pinhole (estimated flux at sample: 4.75 * 10^7 n cm^{-2} s^{-1} (for L/D = 500)) and a high-resolution mode realised by a 10 mm pinhole (estimated flux at sample: 1.15 * 10^7 n cm^{-2} s^{-1} (for L/D = 1000)). DINGO uses a CCD camera with 200 mm x 200 mm neutron detection area. The maximum image resolution will be about 25 µm.

2.6. *BILBY - TOF small-angle scattering*
BILBY is a 40 m SANS machine, installed side by side with the operating initial SANS instrument QUOKKA. The essential difference between these instruments is that BILBY uses a chopper-based time-of-flight data collection mode (an optional velocity selector operation mode is planned to be added in the coming year). The instrument concept is similar to D33 at the Institute Laue - Langevin. Using TOF, BILBY can cover a very wide q-range in one single experiment setting (total q-range with normal pinhole collimation: 10^{-3}Å^{-1} - 1.2 Å^{-1}). Using a slit collimator, q_{min} will be 2x10^{-4}Å^{-1}. By combining different pairs of choppers (4 choppers are available in total) the wavelength-resolution can

be adjusted over a wide range of 4% - 30%. This allows to trade q-resolution for intensity. BILBY's detection system consists of a back detector (area: 64 cm x 64 cm) and four 32 cm x 64 cm front detector panels, all comprised of 8 mm diameter ^3He tubes.

Figure 6. DINGO - Radiography/tomography

Figure 7. QUOKKA (grey) and BILBY (red) - SANS

2.7. EMU - Backscattering spectrometer

The EMU back-scattering spectrometer will be OPAL's fourth spectrometer/instrument that can measure inelastic scattering, after our TAIPAN and SIKA 3-axis spectrometers and our PELICAN time-of-flight spectrometer. Its distinctive feature is that it has much tighter energy resolution: between 1 and 30 micro-eV, as opposed to a minimum of ~50 micro-eV for PELICAN or SIKA. The instrument itself is a very specialised 3-axis spectrometer, like TAIPAN or SIKA, in which the incident neutron energy is determined by back-scattering from a single crystal, and the final scattered energy is determined using a huge spherical array of crystals reflecting and focussing these neutrons back to sample position, behind which there is an array of detectors.

Typically, the exact same crystal reflection (e.g. Silicon 111) is used for both monochromator and analyser, and the energy scan is achieved by moving the monochromator backwards and forwards at high speed. This results in a Doppler shift to the incident neutron energy, sufficient to reach ± 30 micro-eV. EMU it will be similar in design to the IN16 backscattering spectrometer at the Institute Laue Langevin, the HFBS at NIST or the SPHERES instrument at FRM-II.

2.8. CG2 - Neutron guides

The new cold neutron guide CG2 is split into two individual guides CG2A (cross-section: 40 mm width; 100 mm height) and CG2B (50 mm x 160 mm) right from the entrance close to the cold moderator. Both are curved supermirror guides (top/bottom: m = 3.0 and left/right sides: m = 2.5/2.0; radii: 1500 m (CG2A) and 1300 m (CG2B)). CG2A will supply BILBY with neutrons and CG2B will provide neutrons to future instruments downstream of the Platypus reflectometer. For accommodating CG2, major changes were made to the OPAL reactor in-pile structures, namely the replacement of the cold guide in-pile plug and primary shutter. With 70 staff involved in this project, this was the most challenging task of the entire Neutron Beam Expansion Program.

2.9. New sample environments

The sample environment upgrade project features a 12 T split coil magnet with variable temperature insert (1.4 K - 350 K) and 15 mK dilution insert. This magnet can be operated in asymmetric field mode which makes it suitable for operation with polarised neutrons. Other new sample environments are a vapour and sorption system which will allow material studies in controlled gas atmospheres and a closed cycle refrigerator cryostat which is compatible with the 15 mK dilution insert.

Figure 8. EMU - Backscattering spectrometer

Figure 9. CG2A and CG2B - Neutron guides

2.10. ^{3}He polarisers/analysers

This project aims to provide polarised Helium-3 based neutron polarisers and polarisation analysers to 6 OPAL instruments to enable full polarisation analysis for magnetism research. The key equipment are the Helium-3 polarising station (contracted from the Institute Laue-Langevin), added equipment on the 6 instruments and the infrastructure to enable the operation of the station and the neutron instruments. Up to December 2012, the station has been partially commissioned at ANSTO and instrument equipment have been incorporated into 4 instruments (TAIPAN, WOMBAT, QUOKKA, PLATYPUS). Neutron tests were successfully carried out on TAIPAN and WOMBAT.

2.11. Be filter option for TAIPAN

The Be Filter provides an alternative measurement mode for using the TAIPAN triple-axis spectrometer which is already in full user operation at the Bragg Institute. The instrument is a high throughput angle-integrating neutron energy filter spectrometer designed specifically to measure vibrational density of states. This additional mode on TAIPAN provides a totally new way for Australian scientists to study atomic and molecular motions. A measurement proceeds by varying the incident neutron energy E_i which is delivered by a HOPG(002) or Cu(200) double-focussing monochromator, and then detecting these energy filtered neutrons scattered from the sample, by an arc of ^{3}He detectors placed behind the filter assembly. The energy transfer is given as $\hbar\omega = E_i - E_f$ where E_f is the mean energy of the scattered neutrons. For the beryllium/graphite filter combination used, E_f is ~1.2 meV, and the measured Gaussian-equivalent energy resolution is ~1.1 meV FWHM. The spectrometer is based on the design from NIST (FANS).

3. References

[1] More details can be found here:
http://www.ansto.gov.au/ResearchHub/Bragg/Facilities/Instruments/index.htm

Acknowledgements

The authors acknowledge contributions by the instrument scientists and project leaders D. Yu, R. Mole, C. Rehm, C. Wu, G. Deng, C. Ulrich, U. Garbe, A. Sokolova, N. De Souza, A. Eltobaji, P. Bentley, P. Imperia, W.T. Lee, A.P.J. Stampfl and the technical/engineering staff of the Bragg Institute.

exTAS: a new concept in three axis spectroscopy for small samples

A Piovano, S Roux and J Kulda

Institut Laue-Langevin, 6 rue Jules Horowitz, 38042 Grenoble Cedex 9, France

piovano@ill.fr

Abstract. The progress in neutron delivery systems and in neutron focusing techniques has made possible neutron studies of excitations in sub-cm^3-sized single crystals, which are still much larger than crystal sizes needed for standard laboratory characterization techniques. In an effort to further reduce this gap we are proposing the *exTAS* project, which intends to stimulate a paradigm shift towards the use of mm^3-sized samples in neutron spectroscopy. The *exTAS* project aims to boost the TAS sensitivity limits by combining sharp mm-sized focal spots, minimizing penumbra effects in sample environment illumination, with a spectrometer layout downscaled to a tabletop size and enclosed in a shielding casemate. The reduced spectrometer dimensions will provide enhanced flexibility in adapting the momentum resolution to the problem being studied and will facilitate the sub-millimetre positioning accuracy of its components, matching the reduced focal spot size.

1. Introduction

For many materials of current interest single crystal samples cannot be grown large enough to reach present standards of 0.1 – 1 cm^3. Even if large single crystals can be grown, this may introduce significant delays before neutron spectroscopy experiments can be performed. Moreover in many materials the relevant parameters are 'fine-tuned' by doping, whose chances for homogeneity increase with decreasing sample volumes. It should be noted that all laboratory measurements, such as specific heat and susceptibility, are performed on well-characterized sub-millimeter sized samples. In this respect neutron results from large volumes or multi-crystal mounts can exhibit sample-dependent features and often remain disputed for a while.

Therefore it is of utmost importance to maximize the amount of neutrons, which can contribute to the measured signal, at the sample location. The intensity of the real-space focal spot can only be augmented by increasing the density of incident neutron flux and by optimizing its momentum distribution [1]. The first way, involving the neutron source brilliance, moderator efficiency, crystal reflectivity and detector efficiency is close to being exhausted by technological progress accumulated since the early days of the technique. The second way has received an increased attention in the last two decades and has lead to quite spectacular gains by a combined implementation of horizontal and vertical focusing optics in three-axis spectrometers (TAS) [2-4], which has made possible efficient studies of excitations in sub-cm^3-sized single crystals [5,6]. After completion of the *ThALES* project [7] (end of 2014) the portfolio of the ILL TAS instruments will represent the state of the art in this respect worldwide. Nevertheless, due to much larger extent of possible parameter combinations, this way still offers a scope for new developments and the purpose of the *exTAS* project is to explore them.

2. The *exTAS* concept

The angular resolution (beam divergence) of the present state-of-the-art TAS is determined by their primary spectrometer focusing distances (usually about 2 m) in combination with lateral dimensions of their monochromators (about 250 x 200 mm^2), resulting in divergence angles of 2 - 3° and in momentum resolution components of about 0.05 to 0.15 Å^{-1} in the thermal energy range. For many practical purposes this resolution still could be relaxed by a factor of 2 – 3 within the framework of a compromise solution permitting to investigate an otherwise inaccessible system, thanks to a gain in count-rate proportional to the square or cube of the linear resolution scale factor.

This gain can be even more impressive in the cases of cold neutron TAS, which use the same construction elements as the thermal ones and where the same beam divergences lead to even finer momentum resolution because of the shorter (2 – 4 times) neutron wave-vectors. With the present design of the cold and thermal neutron TAS the divergence increase into the range of 5 - 10° is impractical due to huge lateral sizes of the crystal devices imposed and due to the desperate background levels resulting from opening up the windows of primary and secondary spectrometers.

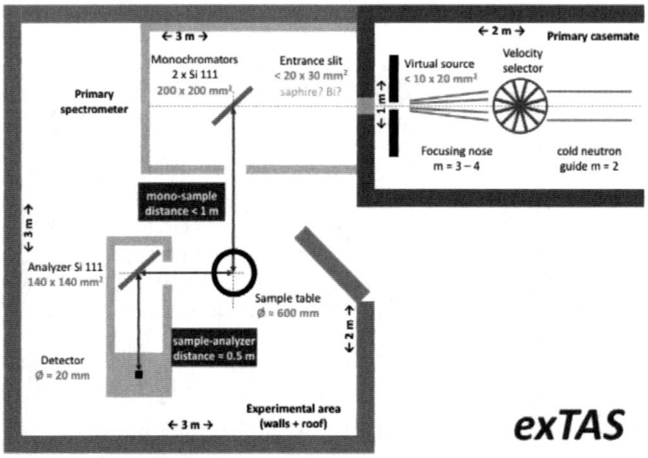

Figure 1: Schematic drawing of the *exTAS* set-up, including the most important dimensions and distances.

We believe, however, that such a step can be explored on a newly built and dedicated instrument, strictly following a number of assumptions and precautions:

- neutron beam from a cold guide, carefully filtered by a velocity selector, and possibly also by transmission filters, in order to minimize the need of shielding on the instrument parts;
- compact instrument design with minimum focusing distances (0.5 - 1 m), permitting to match standard optical devices with high beam divergence;
- sharply focusing optics (perfect bent Si, Ge, diamond crystals) to avoid smearing of focal spots and resolution volumes;
- general layout minimizing exposure of the analyzer/detector chain to parasitically scattered neutrons from the instrument neighbourhood as well as from its own beam.

The combined implementation of these principles should provide an increase in signal count rate of at least one order of magnitude, whenever a relaxed resolution could be sensibly used, and a largely improved signal-to-noise ratio in all cases.

3. Technical outline

The essential idea of *exTAS* is to provide an as intense sample illumination as possible while reducing to the strict minimum the intensity of parasitically scattered neutrons and of all other sources contributing to the signal background. These goals can be achieved by precise real- and reciprocal-space focusing combined with careful filtering of neutron beams and shielding of the whole instrument, implemented in a compact and highly flexible *exTAS* setup schematically represented in Figs. 1 and 2.

The primary casemate of *exTAS* will be best placed at an end position of a high-flux cold neutron guide, whose $m = 2$ super-mirror coating would allow for a significant sub-thermal (warmish) spectral component. A focusing nose will transform the large-area low-divergence beam of the guide into a high-flux beam spot of maximum section 10 x 20 mm^2 (w x h) at the virtual-source slit position with a divergence optimized to illuminate a 200 x 200 mm^2 active monochromator area over a distance of 1 meter. It is essential to filter the neutron beam by a velocity selector before it leaves the primary casemate to minimize radiation load on the monochromator and on its environment and to reduce background in the experimental area. In order to further reduce the neutron- and γ-background, the output window of the primary casemate may be filled with a crystal filter (Bi, sapphire). This will permit to keep the incoming neutron flux in the primary spectrometer close to the level of 1 x 10^{10} ns^{-1} so that its shielding would not need to contain lead or any other heavy element components.

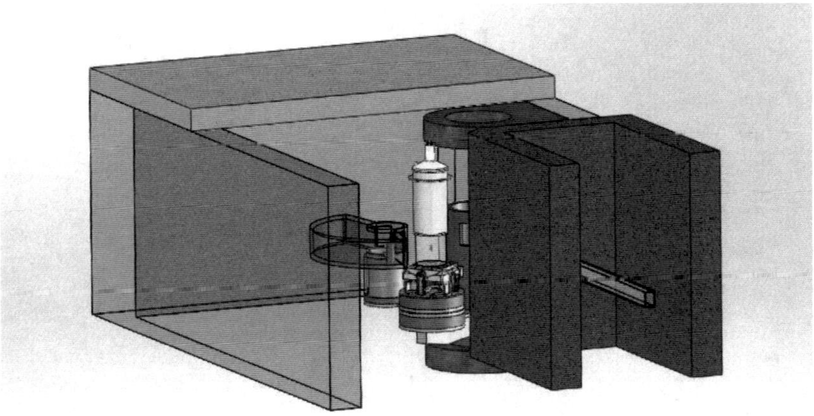

Figure 2: Prototype sketch of exTAS pointing out the compact instrument layout together with its most important shielding parts.

The monochromator will consist of two double-focusing faces (on a rotational changer), both using elastically bent Si 111 crystals but differing in the thickness of the crystal blade stacks, typically 1 – 2 and 10 – 15 mm, to permit quick switching between the low- and high-resolution modes. Its active area will remain about unchanged as compared to the design adopted for ILL TAS instruments, permitting to increase its solid angle acceptance by a factor of four thanks to the two times shorter focusing distance. A mobile slit system will provide a choice of an adequate momentum resolution by limiting its active area according to the needs of a given experiment.

The sample table will be kept as compact as possible but still designed to support equipment mass of up to 120 kg with maximum outer diameter of 600 mm to accommodate standard ILL sample environment equipment as well as the dedicated TAS *VacBox* (vacuum box). The whole secondary spectrometer can be sustained by means of robust arms connected to the primary spectrometer or move on air pads on a Tanzboden platform, elevated with respect to the experimental hall floor.

Thanks to the more compact layout and reduced weight a support plane quality closer to optical benches will be achieved, making possible sub-millimeter positioning accuracy.

The secondary spectrometer in its baseline version is an analogy of a classic single channel analyzer-detector system, employing a double-face silicon analyzer of the same design as the monochromator. Nevertheless sample-analyzer and analyzer-detector distances are reduced to about 50 cm to draw benefits from an increased solid angle. Precise focusing in combination with small sample sizes will permit to minimize the detector size so that its active volume can be kept below 20 x 50 mm^2 (Ø x h), reducing the requirements on the shielding volume of the secondary spectrometer.

The whole experimental area, hosting the primary and secondary spectrometers, will be enclosed in a shielding hutch (walls and roof) built of standard materials used for this purpose at the ILL (light concrete, Pb, B_4C) and minimizing the interaction with neighboring experimental areas in terms of radiation background. The access to experimental zone would be from

- ground level for maintenance and experimental equipment installation,
- top level (roof) for inspection, sample changing, He and N_2 filling.

The top access, through an opening just over the sample table area will further reduce chances of γ-exposure in the presence of lighter monochromator shielding.

4. Ray-tracing simulations

A complete ray-tracing optimization of a new instrument is a vast and complex task, having to take into account correlations of many design parameters. Here we present some preliminary results, obtained with the SimRes code [8], concerning the two optical elements playing a key role in the performance of the proposed instrument: the focusing supermirror nose, providing interface between the neutron guide and the virtual source, and the doubly focusing silicon monochromator, projecting a monochromatic image of the virtual source on the sample.

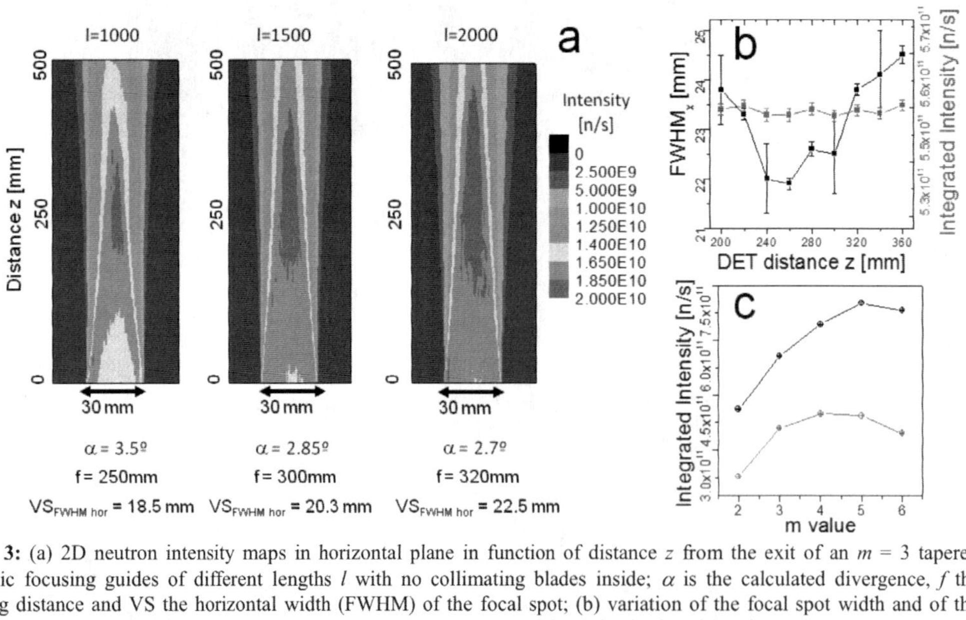

Figure 3: (a) 2D neutron intensity maps in horizontal plane in function of distance z from the exit of an m = 3 tapered parabolic focusing guides of different lengths l with no collimating blades inside; α is the calculated divergence, f the focusing distance and VS the horizontal width (FWHM) of the focal spot; (b) variation of the focal spot width and of the integrated intensity with distance z from the guide exit; (c) integrated intensity in function of m value for a guide with l = 1000 mm and no collimating blades (black) and l = 1000 m and 8 x 8 double side coated collimating blades (red).

The neutron guide system providing neutrons for *exTAS* is catered by the ILL horizontal cold source (cf. the *ThALES* project [4]) and has the following configuration:

- a cold neutron guide 90 m long with a cross-section of 60 x 120 mm^2 (width x height) with $m = 2$ and curvature of 4 km;
- a horizontally and vertically tapered parabolic focusing guide of length l, covered by supermirrors with an index M and consisting of h x k blades (horizontal x vertical); the blades have a thickness of 0.5 mm and may be coated on both sides;
- a heavy input slit acting as a virtual source (VS), placed at the optimum focusing position of the guide nose and having a maximum opening of 10 x 20 mm^2.

Figure 3 collects part of the results of the simulations of the tapered parabolic focusing guide to optimize the m value of the coating, the length l, the focusing distance f and the divergence angle α. From the simulations of focusing noses with different length l, we can infer that a length around 1000 mm is the optimum in terms of sharpness of the beam profile at the focal point (cf. Figure 3a), while keeping the focal distance above the technical minimum required by the *exTAS* layout.

To find out the position of focal point f we made use of a combined intensity and profile width plot as presented in Figure 3b (only horizontal for clarity). The integrated intensity does not vary significantly in the range of some centimeters around the focal point, while the profile width has a minimum at around 250 mm from the end of the guide of length 1000 mm.

Table 1: Comparison between the white beam integrated neutron flux from a straight guide and from a focusing nose with 8x8 blades of $m = 4$, length l = 1000mm and an exit window of 30x60 mm^2, at the guide exit and just after the VS. Integrated intensity calculated over the entire exit window of the guide in line 1 and over the 10 x 20 mm^2 area of the VS in line 2.

Integrated flux	Straight guide	Parabolic l = 1000 mm	ratio
Guide exit	1.00 x 10^{12}	0.40 x 10^{12}	0.40
VS 10x20 (w x h)	0.28 x 10^{11}	0.13 x 10^{12}	4.5

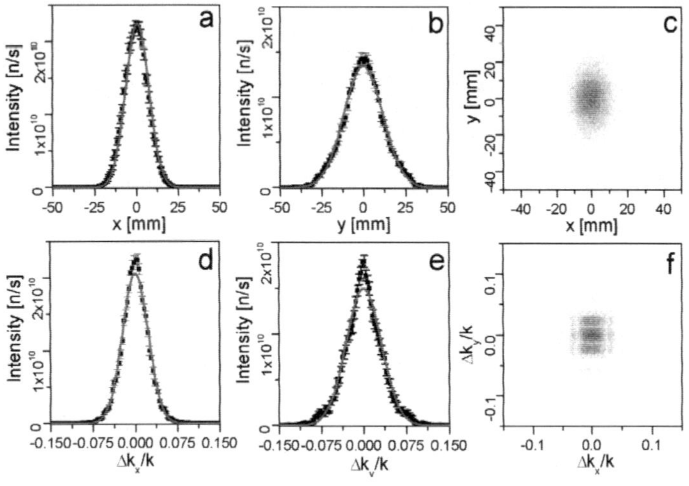

Figure 4: Real space (a-b) and momentum space (d-e) beam profiles together with a neutron density map (c) of the real space focal point and with the momentum density map (f) obtained from the central area 5 x 5 mm^2 of (c). The beam profiles have been fitted with Gaussian curves (red lines) and corresponding calculated FWHM are: (a) 16.5 mm, (b) 27.4 mm, (d) 0.054, (e) 0.073. x axis represents the horizontal direction perpendicular to beam path, y axis represents the vertical direction perpendicular to beam path.

The plot of the variation of integrated intensity versus m value is reported in Figure 3c. Black points refer to simulations without collimating blades while red points refer to simulations with 8x8 blades. In the first case the $m = 5$ value is the most performing solution, while the inclusion of collimating blades makes m = 4 the best compromise due to absorption effects. Even if these effects should be prevented in general, a compromise solution to obtain also a decent real and reciprocal space profiles is necessary in reality.

The profile of the beam at the VS position in real and reciprocal spaces is reported in Figure 4 for a 2D multichannel focusing guide with 8 x 8 blades of $m = 4$ and length $l = 1000$ mm. The profiles have been fitted with a Gaussian distribution to verify the level of homogeneity obtained. The results appear satisfactory and also in the 2D maps of neutron events in Figure 4 c,d display smooth and almost featureless intensity distributions.

Integrated intensity has been calculated and compared with a similar setup where the last 1000 mm of the guide have been kept straight in Table 1. Even if the absorption in the internal blades of the focusing nose halves the output beam flux with respect to the straight guide, the focusing effect allows to concentrate useful neutrons at the VS position and to obtain a gain factor close to 5 in brilliance. This increase is accompanied by an increase in divergence to about 3.5°.

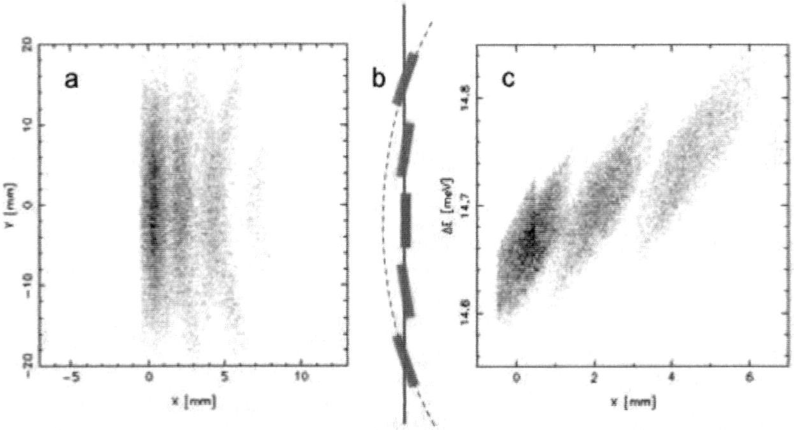

Figure 5: (a) Real space image of an *exTAS* VS slit 1 mm wide at the sample position by a doubly focusing Si 111 monochromator consisting of a vertical array of 9 blades 200 x 20 x 1 mm^3 set for optimum horizontal and vertical curvatures; the fringes are due to departures from perfect sagital focusing (cf. scheme (b)); (c) displays the corresponding shifts in neutron energy on an enlarged scale. x axis represents the horizontal direction perpendicular to beam path, y axis represents the vertical direction perpendicular to beam path.

Beam divergences of several degrees can be efficiently handled and refocused on the sample by large area horizontally and vertically focusing monochromators. In Figure 5 we present SimRes results for an image of a VS slit 1 mm wide at the sample position, projected by a conventional doubly focusing Si 111 monochromator using a vertical array of 9 elastically bent perfect silicon crystals. Thanks to the absence of mosaic structure the image is sharp in the horizontal plane (cf. the millimeter-scale in Figure 5). However, the vertical focusing of the present systems is only approximate, because for practical reasons it uses tilts of the vertical monochromator segments around axes lying in the vertical plane of the central segment. The lateral displacements with respect to the cylindrical surface, which would provide ideal sagital focusing, give rise to lateral aberrations of the focal spots and to slight deviations from ideal monochromatic focusing (cf. right-hand pane in Figure 5). This defect can be neglected when using mosaic crystals (PG, Cu), but cannot be tolerated in high-resolution conditions as those required for *exTAS*.

5. Concluding remarks

The combined implementation of advanced focusing techniques presented in this paper should provide a platform for boosting the TAS performance, when studying mm-sized samples, by at least an order of magnitude. To achieve this goal, the challenges in the development and realization of two key *exTAS* components have to be addressed. The first one is the multi-blade configuration of the two-dimensional tapered guide nose [8], up to now mostly considered only in its 1D form [9] and never implemented in an instrument, even at a prototype level. The other one is the mechanical realization of the true sagital focusing geometry in a 2D variable focusing monochromator/analyzer system required in case of mm-sized focal spots. Our preliminary ray-tracing simulations confirm the key role of these components in terms of optical accuracy and of increased luminosity of the instrument and motivate further development work along these directions. Most of the remaining ideas and elements of the *exTAS* project, outlined in this paper, represent relatively straightforward extrapolations of the present neutron instrumentation and neutron optics technologies, based on experience with design and construction of existing instruments.

Acknowledgement

We would like to mention the creative input of A. Hiess in the early stages of the project, notably concerning the sample environment part of *exTAS*, in the meantime materialized in the *VacBox* sample chamber, which become available to the ILL TAS users recently.

References

[1] Popovici M, 1975 *Acta Cryst.* A **31** 507-513
[2] Komarek A C, Boeni P, Braden M, 2011 *Nuclear Instruments and Methods in Physics Research* A **647** 63-72
[3] Janoschek M, Boeni P, Braden M, 2010 *Nuclear Instruments and Methods in Physics Research* A **613** 119-126
[4] Schneider M, Stahn J, Boeni P, 2009 *Nuclear Instruments and Methods in Physics Research* A **610** 530-533
[5] Kulda J, Šaroun J 1996 *Nuclear Instruments and Methods in Physics Research* A **379** 155-166
[6] Hiess A, Jiménez-Ruiz M, Courtois P, Currat R, Kulda J and Bermejo F J 2006 *Physica* B **385-386** 1077-1079
[7] Boehm M, Hiess A, Kulda J, Roux S and Šaroun J 2008 *Meas. Sci. Technol.* **19** 34024
[8] Šaroun J and Kulda J 2006 *Physica* B **385-386** 1250-1252
[9] Čermák P, Boehm M, Kulda J, Roux S, Hiess A, Steffens P and Saroun J 2013 *J. Phys. Soc. Japan Suppl. A* **82** SA026

International Workshop on Neutron Optics and Detectors (NOP&D 2013) IOP Publishing
Journal of Physics: Conference Series **528** (2014) 012028 doi:10.1088/1742-6596/528/1/012028

Neutron Optics Requirements for Neutron Imaging Techniques

E Lehmann, M Morgano and S Peetermans
Neutron Imaging & Activation Group, Spallation Neutron Source Division, Paul Scherrer Institut, CH-5232 Villigen PSI, Switzerland

E-mail: eberhard.lehmann@psi.ch

Abstract. The utilization of X-rays for material research is common in many respects since their discovery at the end of the 19th century. New sources as electron synchrotrons or free-electron lasers push the methodology and the application ranges further. A similar approach started 50 years later with neutrons when sources with reasonable high intensity became available.
Today, there are many similarities and complementarities visible between X-ray and neutron studies and the involved techniques. Therefore, it is worth to compare and to adapt from the advanced X-ray techniques and to translate it into the neutron world. Despite of the lack of neutron intensities compared to the most brilliant X-ray beams, the specific properties of neutrons (contrast, spin, magnetic moment, penetration power) are utilized and they will further play an important role in non-invasive studies on the micro- and macro scale.
This paper wants to encourage to "look over the fence" into activities of the X-ray community as currently running in the COST action MP-1203.

1. Introduction

Today's neutron research is mainly focused onto two areas: neutron scattering and neutron imaging. Other applications from the past like neutron activation analysis (NAA and PGAA), depth profiling, and boron capture therapy gained much less importance and practical usage. The most prominent neutron sources for research purposes are equipped with advanced and sophisticated diffractometers, spectrometers and other scattering research facilities while imaging beam lines are still a minority there.
However, there are initiatives and strong signals to build and to implement newly developed facilities for neutron imaging too. A list for existing and future facilities is provided in the appendix.
As sketched in a historical review in some details below, there are many similarities and synergies in neutron research in comparison to X-ray studies. This holds in particular with respect to neutron optical components, detector options and scientific techniques and their application.
There are advantages and drawbacks when neutron and X-ray investigations are compared. On atomic level, both kinds of radiation can perform diffraction studies because the wavelengths are in the order of magnitude of the lattice distances. As the best suitable X-ray energies are in the keV range, the transmission into bulk metals is much less than in the case of thermal neutrons.
On the other hand, the intensity of the synchrotron light sources is orders of magnitude higher than neutron sources can provide. This result in much shorter acquisition time and the options for studying very fast dynamic processes with highest spatial resolution, sometimes just during the sample destruction by the applied radiation.
With respect to imaging in transmission mode (radiography or tomography) both kinds of radiation can provide similar image quality but with different contrasts, given by the attenuation behavior of the involved materials for the particular kind of radiation. As shown in the Figure 1, a complementary approach is feasible to gain best possible information about the observed object.

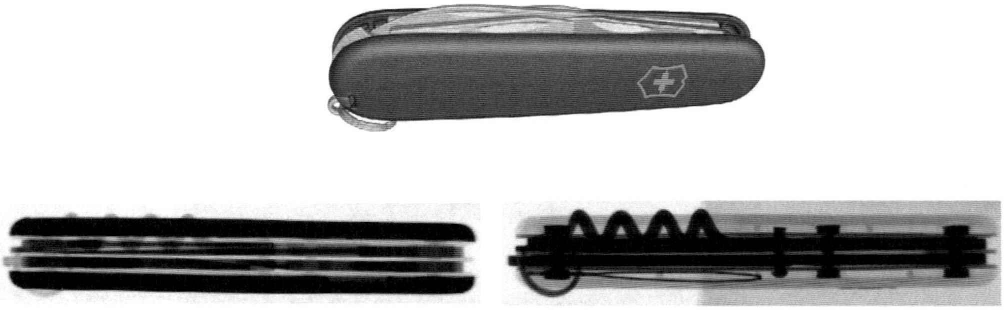

Figure 1: Comparing the neutron and X-ray imaging performance for the case of a Swiss army knife, where the plastic cover is enhanced by neutrons (lower left) and the metallic parts by the X-rays (lower right); the transparency for the two material categories is vice versa. The image on top is the result of a neutron tomography run, not photography.

This paper intends to gain attention for new developments in the field of optics for X-rays with the idea how to translate them for neutrons, in particular for neutron imaging. The considerations are based on existing literature, our own experiences and are linked to the recently started COST action MP-1203, dealing with the study of "Advanced X-ray spatial and temporal metrology" [1].

2. COST MP-1203

Initiated by a group in France (action chair: P. Zeitoun, CNRS), the COST action MP-1203 was started in November 2012 administratively and in April 2013 scientifically [1]. The action is structured into the following Working Groups:

Working Group 1: X-ray spatial metrology of optics
Working Group 2: Spatial and temporal metrology of X-ray sources
Working Group 3: X-ray coherent and incoherent imaging diagnostic
Working Group 4: Damage on X-ray optics
Working Group 5: High brightness and coherent X-ray sources for advanced spatial and temporal metrology
Working Group 6: Dissemination and technology transfer

The topics are mainly based on the work at the leading synchrotron sources, but experiences at free-electron lasers and at common lab sources are also involved. Since the action will last 4 years at least, progress will be reported on-line during the different meetings and mutual exchange opportunities in the COST framework.
Although neutrons are not explicitly mentioned and involved in this action, communications with the X-ray experts will become important and interesting in different respect. As it will be explained and outlined in detail below, there are common fields and synergies between neutron and X-ray physics which should be highlighted and exploited.

3. Historical observation: neutrons vs. X-rays

Looking back into the history of the developments in X-ray and neutron research, a scheme of highlights as shown in Figure 2 can be derived.

While X-ray imaging was the very first application of this newly discovered radiation, neutron images (film radiographs) with the same quality were produced much later. Single event neutron counting was preferentially been done in the first days of neutron research.

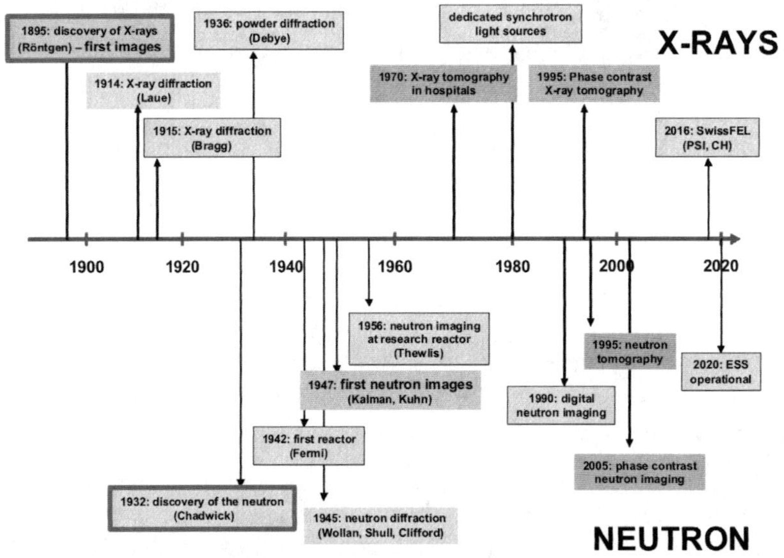

Figure 2. Time scale for important events and developments in the field of X-ray and neutron research, respectively.

From this overview, the following statements can be made:

- Free neutrons were discovered 37 years after the X-rays were found. First detection of the X-rays was done with imaging techniques (using film).
- Neutron imaging started 50 years after first X-ray images were made
- Neutron diffraction comes 30 years later than X-ray diffraction
- Neutron tomography comes 25 years later than X-ray tomography in hospitals
- Phase contrast imaging with neutrons comes 10 years later than with X-rays
- Neutron imaging is now a competitive and complementary method compared to the X-ray techniques

However, there are new approaches and opportunities available at the high intense X-ray sources which are based on coherence, where neutrons have not yet the chance to compete.

The clear challenge for future neutron research will be to use the existing and the few sources under construction with their limited intensity most efficiently in several respects. The improvement of the detector techniques, the use of focusing devices and to narrow energy bands more efficiently are approaches with high importance for future neutron applications.

On the one hand, there are different contrast mechanisms of the neutrons compared to X-rays resulting in alternative and complementary information about the object under investigation. The approach to combine the two beams into one imaging facility is described in [2]. On the other hand, neutrons have unique properties like a spin and magnetic moment that can be used for polarization studies [10] describing magnetic phenomena on the macro-scale.

The COST-action P7 "X-ray and neutron optics" was running from 2002 until 2006 [12, 16] with the aim to understand better the interaction of X-rays and neutrons at flat surfaces and interfaces and to apply this knowledge to create advanced optical elements, mainly to guide neutrons and to enhance the intensities in neutron scattering experiments. Since this time some progress happens, and we focus here more on the neutron imaging aspects.

4. Neutron beams

Neutron sources with the needed high intensity are either reactor based (power 10 to 70 MW) or using the spallation principle at intense particle (mostly protons) accelerators (beam power in the order of 1 MW presently). The initial (fast) flux density cannot be higher than 10^{15} cm^{-2} s^{-1}, mainly for cooling reasons. As the most suitable neutrons for research are in the thermal or cold region, the moderation process is needed before a beam is formed. However, neutrons cannot be manipulated by guiding fields and therefore neutron beams just consist of the neutrons with the right directions. In this respect, a dramatic loss of intensity has to be taken into account and a neutron flux on the order of 10^7 cm^{-2} s^{-1} to 10^{10} cm^{-2} s^{-1} will arrive at the sample.

This order of magnitude has to be compared to photon values (e.g. at the SLS, PSI, CH) where 10^{18} photons/s/mm^2/mrad^2/0.1% band width are common.

A major issue under such circumstances is to use the beam most efficiently by sending the "useful" neutrons to the sample and to collect the neutrons after interaction in the right manner to form a signal which can be analyzed.

Compared to the pioneering time in neutron research, there are many developments done to enable the above mentioned strategy. Neutron guides, other neutron optical elements, analyzing devices, including new detectors have been developed and will be improved further. As an example, the development and implementation of digital imaging detectors replacing the traditional film techniques gave a performance gain by orders of magnitude (30 minutes → some milli-seconds in exposure time).

5. Neutron optics

There are several options for manipulating neutron beams, as summarized in Table 1.

The attempts for understanding of the wave properties of neutrons has started very early after the discovery of the neutron and the availability of suitable beams. Papers like [17] and later overview articles [19, 20] which are recently complemented [18] indicate the high standard of knowledge about neutron optical phenomena from fundamental aspects towards applications in neutron scattering. However, the usage for practical applications in neutron imaging aspects (beam tuning, increase in performance) is still quite limited.

The development of **neutron guides** on the basis of total reflection of (preferentially cold) neutrons and their installation at many prominent sources raised the opportunities for the beam extraction dramatically. Instead of only one or two instruments per beam port, we can now deliver the beams to many facilities at the same time. Compared to the initial Ni based cladding, the "super-mirrors" as multi-layers of Ni-Ti provide higher reflectivity over much wider energy bands now (at the expense of divergence).

Table 1. Neutron interaction options for tuning beams (taken from [3])

Physical interactions and applications

Interaction	Applications
Reflection	Mirrors, capillaries, reflecting lenses
Diffraction	Bragg scattering, monochromators, zone plates
Refraction	Refracting lenses, energy analysers
Absorption	Collimators, filters
Magnetic	Larmor precession

On the other hand, the reflection properties can be used to build **focusing devices** in order to increase the intensity at the sample position. With a parabolic shape in one or two directions a gain by a factor of 10 to 20 was demonstrated with a focal spot in the order of less than a millimeter. Fig. 3 shows an example where adaptive neutron optics devices were used to focus the beam at BOA [4] in one direction.

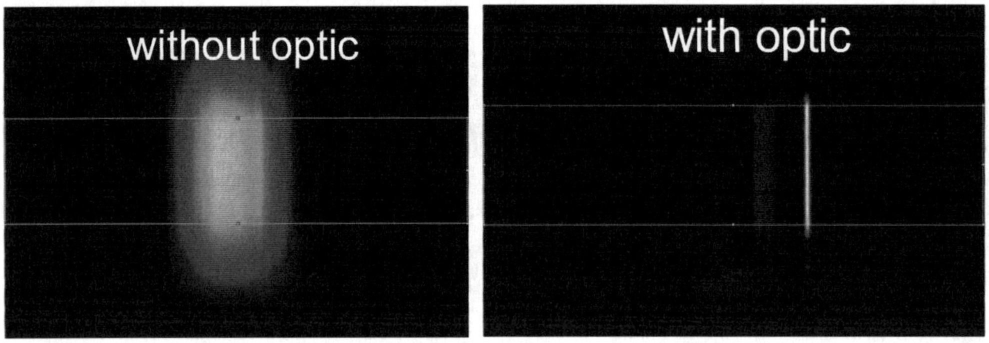

Figure 3. Intensity distributions of a neutron beam at BOA (without focussing device – left) and after focusing with an adaptive optical device in one dimension (right) (see [5]). The gain in the spot was about 12 times the peak intensity of the initial beam.

For *cutting out narrow energy bands* around the wavelength λ from the "white" spectrum in the beam, there are options available with single crystals (C, Si) where the Bragg condition

$$n\lambda = 2\,d\,\sin\theta \qquad (1)$$

is used to separates a specific wavelength. Unfortunately, the beam direction is changed too by θ. Either the experiment is arranged into this direction or a second crystal with same properties is aligned in the manner to bring the scattered beam back into the initial direction (double-crystal mono-chromatizer). The field-of-view and the homogeneity of the beam depend much on the crystal quality. Second order reflections from the crystals can disturb the beam properties. For neutron imaging applications these parameters are essential.

Neutrons can perform refraction, but the corresponding refraction index deviates from 1 only very little (about by 10^{-6}, depending on the refracting material and the wavelength λ of the neutrons):

$$n = 1 - \delta = 1 - N \cdot \sigma_{coh} \cdot \lambda^2 / 2 \cdot \pi \qquad (2)$$

As the coherent scattering cross-section σ_{coh} plays a major role in this relation, preferentially the crystalline solids carry the properties for refraction.

The refraction can be used to build "concave alligator" type *lenses* to focus or to enable magnification of neutrons afterwards. Such lens systems have to be extended along the flight direction due to the small refraction impact per single lens. In this way, the neutron absorption becomes a competing process which will reduce the intensity accordingly. This is the major reason why the concept has not been implemented. For practical neutron imaging, the FOV is too little and the focal spot too large.

Using neutron diffraction, the concept of "zone plates" can be considered in similarity to the X-ray world. However, the dimensions for suitable such plates are very small, given by the wavelength conditions. Furthermore, a high energy resolution and a degree of coherence are required. This limits the applicability of such a concept to neutron applications due to the intensity driven performance.

Grating interferometry is a method to study phase contrast phenomena [13] – described by the parameter δ in (2). It represents also material properties and can be taken to perform a complementary "phase contrast imaging". In the case of neutrons with their magnetic moment, a magnetic phase contribution can be added: $\delta_{mag}=2*\mu_n*B*m*\lambda^2/h^2$.

In the meantime, neutron grating interferometry devices have been found useful to visualize USANS and SANS properties of macroscopic samples by deriving the "dark field" signal from the interference patterns. In this way, large magnetic domains have been visualized in bulk samples [14].

6. Neutron detection

Most of the X-ray detection concepts (starting with films – ending now with digital semi-conductor driven systems) can be transferred from X-rays to neutrons – with some drawback: a conversion nuclear reaction is required before a measurable signal is created. The major reason is the missing charge of the neutrons for a direct excitation.

Common conversion reactions are based on strong neutron absorbers for efficiency reasons; ^3He, ^{10}B, ^6Li, Gd or U-235 are the mostly used detection materials. The secondary processes for the detection are light emission (scintillation), ionization and charge transfer or chemical excitation which can be used for signal readout.

In respect to neutron imaging, this transfer process hinders highest spatial resolution as the distance between the neutron impact and the detection point provides a spread.

Most promising detectors today are based on semiconductor pixel devices [6] with single event counting readout which can be used for imaging and also diffractometer devices in the future. Such detectors have the potential to achieve a 2 to 5μm resolution.

7. Neutron imaging

The usage of neutrons for non-invasive and non-destructive studies of material distributions gained more and more attention in the last years as the performance and the application range has extended in several respects. Depending on the transparency of the object under investigation, the field-of-view can be tuned to the relevant size. Accordingly, the spatial resolution (pixel size of a digital detector) is adapted. As shown in Figure 4, a wide range of objects dimensions can be covered because a beam diameter of 30 cm is quite common at dedicated imaging facilities.

Presently, the highest spatial resolution is on the order of about 10 micro-meters [7]. There are initiatives to improve the spatial resolution further in order to come into the region of a few micro-meters. Than a competition and complement to X-rays is possible while the high contrast for e.g. hydrogen, lithium or boron can be exploited.

Initiatives to create focusing neutron devices on the basis of Wolter optics [15] or mirror capillary (Chumakov) are in the design stage and have not yet proven useful for practical neutron imaging applications.

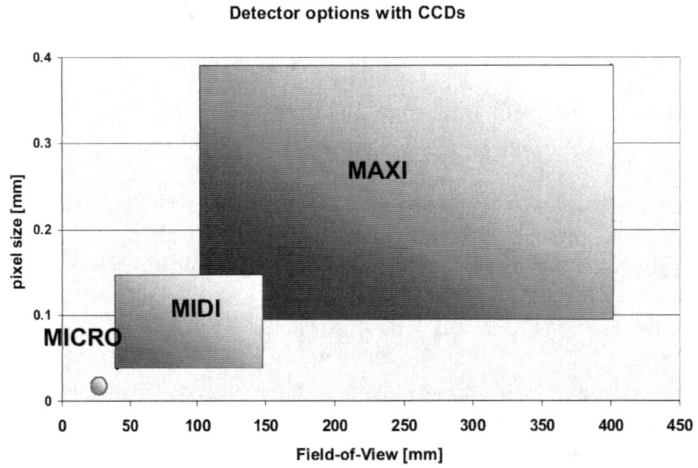

Figure 4. Three detection concepts in use at the PSI's beam lines NEUTRA and ICON [8] with camera based systems covering the sample size of up to 40 cm and the lowest pixel size of 13.5 μm.

Other aspects in the field of neutron imaging (similarly to X-ray imaging where these methods have been set already before) are the implementation of grating interferometers where the phase-contrast is derived simultaneously to the so-called "dark field" image (see above). Both are based on the wave properties of the neutrons and their coherence behavior. One important application is the visualization and study of magnetic domain structures in bulk metal components on the macro-scale.

Because the energy selection plays an important role for some neutron imaging techniques like diffractive imaging [9], polarized neutron imaging [10] and the precise quantification of transmission data [11] the use of pulsed beam in time-of-flight (TOF) mode becomes very important. Therefore, the installations and projects at pulsed sources (IMAT at ISIS (UK), ERNIS at JPARC (Japan), VENUS at SNS (USA)) and ODIN at ESS (Sweden) will provide essential steps forward to establish neutron imaging as a modern research tool.

8. Summary & outlook

As demonstrated at the most advanced X-ray sources, where imaging facilities are well accepted installations next to the spectroscopy and diffraction devices, there is also a trend visible for neutron imaging activities at several places. It is very important, to have a permanent look to the activities in the X-ray community because similarities and complementarities with neutrons can be exploited for our own purpose. Some limitations in respect to the missing brightness, coherence and resolution can partly be compensated by the specific neutron properties. The direct link and comparison between neutron and X-ray data can be used for data fusion on a higher level of knowledge about matter.

9. Appendix

Facilities for neutron imaging around the globe (status 2013):

Country	Site	Institution	Neutron Source	Facility name	Spectrum	Status
Australia	Sydney	ANSTO	OPAL 20 MW	DINGO	thermal	operational
Austria	Vienna	Atominstitut	TRIGA 250 kW	imaging beam line	thermal	operational
Brazil	Sao Paulo	IPEN	IEA-R1M 5MW	imaging beam line	thermal	operational
Germany	Munich	MLZ Garching	FRM-2 20 MW	ANTARES	cold	user lab
Germany	Munich	MLZ Garching	FRM-2 20 MW	NECTAR	fast	user lab
Germany	Berlin	HZB	BER-2 10 MW	CONRAD	cold	user lab
Hungary	Budapest	KFKI	WRS-M 10 MW	imaging beam line	thermal	operational
Japan	Kyoto	Kyoto University	MTR 5 MW	imaging beam line	thermal	user lab
Japan	Tokai	JAEA	JRR3M 20 MW	imaging beam line	thermal	user lab
Japan	Tokai	JAEA	JPARC spallation	ERNIS	pulsed	project
South Korea	Daejon	KAERI	HANARO 30 MW	imaging beam line	thermal	operational
South Africa	Pelindaba	NECSA	SAFARI 10 MW	SANRAD	thermal	user lab
Switzerland	Villigen	PSI	SINQ 1MW spallation	NEUTRA	thermal	user lab
Switzerland	Villigen	PSI	SINQ 1MW spallation	ICON	cold	user lab
United Kingdom	Didcot	Rutherford Lab.	ISIS-TS2	IMAT	pulsed	project
USA	PennState	P S University	TRIGA 2MW	imaging beam line	thermal	operational
USA	Sacramento	Mc Cleallan RC	TRIGA 2MW	imaging beam line	thermal	operational
USA	Gaithersburg	NIST	NBRS 20 MW	imaging beam line	thermal	user lab
USA	Oak Ridge	ORNL	HFIR 70 MW	CG-1D	cold	user lab
USA	Oak Ridge	ORNL	SNS spallation	VENUS	pulsed	project

References:

[1] http://www.cost.eu/domains_actions/mpns/Actions/MP1203
[2] P. Vontobel et al., The XTRA option at the NEUTRA imaging beam line of the spallation neutron source SINQ and its use for neutron-X-ray data fusion, submitted to NIM A
[3] T. Krist, Lecture on Neutron Optics, FU Berlin, 2010
[4] M. Morgano et al., The BOA beam line at Paul Scherrer Institut as a flexible neutron imaging instrument, to be submitted to NIM A
[5] U. Filges, T. Panzner, private communication
[6] E.H. Lehmann, A. Tremsin, C. Grünzweig, I. Johnson, P. Boillat and L. Josic, Neutron imaging — Detector options in progress, 2011 JINST 6 C01050
[7] E.H. Lehmann,G. Frei, G. Kuehne, P. Boillat, The micro-setup for neutron imaging: A major step forward to improve the spatial resolution, NUCLEAR INSTRUMENTS & METHODS IN PHYSICS RESEARCH SECTION A-ACCELERATORS SPECTROMETERS DETECTORS AND ASSOCIATED EQUIPMENT 576 (2-3): 389-396 JUN 21 2007
[8] A.P. Kaestner, S. Hartmann, G. Kühne, G. Frei, C. Grünzweig, L. Josic, F. Schmid, E.H. Lehmann, The ICON beamline – A facility for cold neutron imaging at SINQ, Nuclear Instruments and Methods in Physics Research Section A: Accelerators, Spectrometers, Detectors and Associated Equipment, Volume 659, Issue 1, 11 December 2011, Pages 387-393
[9] S. Peetermans, E.H. Lehmann, Simultaneous transmission and diffraction contrast tomography as a non-destructive 3D method for bulk single crystal quality investigations, Journal of Applied Physics, 2013, 114, 124905

[10] N. Kardjilov, I. Manke, M. Strobl, et al., Three-dimensional imaging of magnetic fields with polarized neutrons, NATURE PHYSICS Volume: 4 Issue: 5 Pages: 399-403 DOI: 10.1038/nphys912
[11] S. Peetermans, E.H. Lehmann, Quantitative energy-selective neutron imaging of structural materials in the absorption range, submitted to NIM A
[12] http://w3.cost.eu/fileadmin/domain_files/PHYS/Action_P7/final_report/final_report-P7.pdf
[13] F. Pfeiffer, C. Gruenzweig, O. Bunk, G. Frei, E. Lehmann, and C. David, Neutron Phase Imaging and Tomography, PHYSICLAL REVIEW LETTERS 96, 215505, June 2, (2006), http://dx.doi.org/10.1103/PhysRevLett.96.215505
[14] C. Gruenzweig et al., Visualizing the propagation of volume magnetization in bulk ferromagnetic materials by neutron grating interferometry, JOURNAL OF APPLIED PHYSICS 107, 09D308 (2010)
[15] B. Khaykovich et al., From X-ray telescope to neutron scattering: using axisymmetric mirrors to focus a neutron beam, Nucl. Instr. & Meth. A 631, 98 (2011); doi:10.1016/j.nima.2010.11.110
[16] A. Erko et al. (ed.), Modern Developments in X-ray and Neutron Optics, Springer 2008, ISBN 978-3-540-74560-0
[17] O. Halpern, Remarks on Some Questions of Neutron Optics, Phys. Rev. 88, 1003-1007 (1952)
[18] M. Utsuro, K. Ignatovich, Handbook of Neutron Optics, John Wiley&Sons, 2010, ISBN 978-3-527-40885-6
[19] V.F. Sears, Neutron Optics, an introduction to the theory of neutron optical phenomena and their applications, Oxford University Press, 1989,
[20] A.G. Klein, S.A. Werner, Neutron Optics, Rep. Prog. Phys., Vol. 46, 1983

On the Goos-Hänchen effect in neutron optics

A I Frank

Frank Laboratory of Neutron Physics, Joint Institute for Nuclear Research, 141980 Moscow region, Dubna, Russia.

frank@nf.jinr.ru

Abstract. The Goos-Hänchen effect, a longitudinal shift of a wave beam at total inner reflection, is a well-known optical phenomenon. It was discovered in 1947 and have been observed many times at reflection of micro- and ultrasonic waves since then. Here we present an elementary theory of the GH. effect to describe reflection of a massive particle and demonstrate how the shift is related to delay time at reflection. It is shown that giant positive or negative longitudinal shifts of the neutron beam may occur at neutron reflection from some specially manufactured planar system. The possibility of delay time measurement at neutron reflection and direct experimental observation of the GH shift are also discussed

1. Introduction

The discovery by Goos and Hänchen [1,2] of the wave beam longitudinal shift at total inner reflection of light called the Goos Hänchen effect has had remarkable influence on the progress in understanding of reflection. It is recognized that the phenomenon is a very general one and its investigation goes beyond the scope of conventional light optics. The GH shift was observed for ultrasonic waves [3-5] and later, for microwaves [6,7] and X-rays [8]. Obviously it was quite natural to suppose that massive particles must also demonstrate the GH. shift at their reflection from the region of the potential.

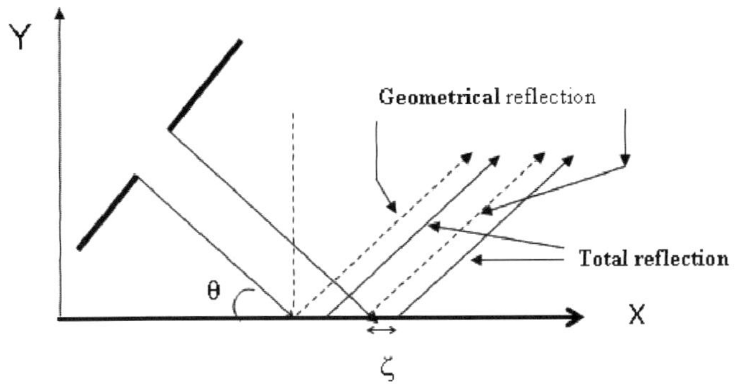

Figure 1. The Goos-Hänchen effect. The longitudinal shift at total reflection

As far as we know the corresponding quantum problem was first analyzed by Hora [9]. Later such approach has become quite common [10,11] especially since the connection of the GH. effect with the quantum problem of reflection time was realized.

Till recently the problem of GH. shift at reflection of a massive particle remained purely theoretical in spite of its more than half-century history. Best conditions for detection of the phenomenon seem to exist in neutron optics. To our knowledge, A.A. Seregin was first to raise the question of the GH. effect at neutron reflection [13]. Later the theory of the effect and possibilities of its observation in neutron experiments were discussed in [14, 15]. The recent work [16] reported that the GH. effect was observed for neutrons for the first time. This gave rise to discussion and made the problem even more actual.

Next section presents main relations concerning the shift of a neutron beam at reflection from matter border. Also, the relation of the shift with the group delay time (GDT) at reflection is discussed. In the third section some examples of possible giant and negative GH. shifts are given. In the final section possibilities of experiments to measure the GDT and the neutron beam shift at reflection are briefly discussed.

2. The Goos-Hänchen shift and neutron reflection time from matter

Below our attention is focused on reflection of neutron waves. The peculiarity of neutron optics is that in most cases interaction of neutrons with matter may be described by introducing the effective potential

$$U = \frac{2\pi\hbar^2}{m}\rho b, \tag{1}$$

where ρ is the volume density of atoms and b is the volume average of the coherent scattering length. To illustrate how the longitudinal shift of the wave beam occurs at reflection of neutron waves from the border of matter we address here to theoretical approach of Artmann [18]. Artmann's method is widely known and with some modification it was multiply used in a number of subsequent papers (see for example [4, 19, 20,14]).

Let a medium with the effective potential U be bordered by the plane y = 0. Considering the reflection of the wave beam from the border and following [21] we accept for simplicity that the beam is formed by just two waves whose wave vectors slightly differ in their direction. For each of the waves the condition of total reflection $E_n < U$ holds. Here

$$E_n = \frac{\hbar^2 k_y^2}{2m}, \tag{2}$$

and k_y is the component of the wave vector normal to the surface.

On the surface of matter x-components of incoming waves are $\exp(ik_x x)$ and $\exp\left[i\left(k_x + \Delta k_x\right)x\right]$. The corresponding reflected waves differ from the incoming ones by phase and their x-components on the surface are $\exp(ik_x x + \varphi)$ and $\exp\left[i\left(k_x + \Delta k_x\right)x + \left(\varphi + \Delta\varphi\right)\right]$, respectively. The phase shift Δφ may be determined in a conventional way from continuity conditions at the border. Such waves interfere and for the x-component of the superposition of the reflected waves we have

$$\Psi_x = \exp(ik_x x + \varphi)\left\{1 + \exp\left[i\left(\Delta k_x x + \Delta\varphi\right)\right]\right\}. \tag{3}$$

The condition of maximum intensity or of constructive interference is

$$\Delta k_x x + \Delta\varphi = 2\pi\nu, \tag{4}$$

where the integer number ν is the order of interference.

In the absence of total reflection, in particular for neutrons with the energy above the barrier $E_n > U$ the phase shift is absent and both φ and Δφ are zero. In this case the condition of maximum intensity is $\Delta k_x x_0 = 2\pi\nu$. Consequently, the phase shift associated with total reflection leads to the beam shift at $\varsigma = x - x_0 = -\Delta\varphi / \Delta k_x$. In the limit of Δk_x going to zero we come to famous Artmann's formula for the longitudinal Goos-Hänchen shift.

$$\varsigma = -\frac{d\varphi}{dk_x}. \tag{5}$$

Above it was believed that the wave number k_x varies only due to variation of the angle of incidence. Taking into account that the total k-number $k_0^2 = k_x^2 + k_z^2$ is constant we write the displacement ζ in equation (5) in terms of the phase derivative with respect to E_n

$$\varsigma = \frac{\hbar^2 k_x}{m}\frac{d\varphi}{dE_n}, \tag{6}$$

Finding the relation for the phase shift of the reflected wave and substituting it into (6) we obtain

$$\varsigma = \frac{2k_x}{k_y\sqrt{k_b^2 - k_y^2}}, \qquad (k_y < k_b) \tag{7}$$

in full agreement with the results [11]. Here $k_b = \sqrt{2mU}/\hbar$ is the border wave number.

Note the following remarkable feature of equation (6). The value

$$\tau = \hbar\frac{d\varphi}{dE_n}, \tag{8}$$

which enters in this equation is the well known group delay time (GDT) introduced by Eisenbud, Bohm and Wigner [22-24] as a measure of interaction time in quantum mechanics. Consequently, equation (6) can be written in the form

$$\varsigma = \tau v_x. \tag{9}$$

With some reservations the GDT may be interpreted as neutron reflection time. The relation between the reflection time and the longitudinal GH shift was first noted by Agudin [12]. Note that in the case under consideration reflection from the potential barrier GDT is

$$\tau = \frac{\hbar}{\sqrt{E_n(U - E_n)}}. \tag{10}$$

Alternative approach to the theory of GH effect based on the fluxes balance in condition of total reflection was developed by Renard [10]. We present here only some based principle of this theory limiting our self only by the case of the total reflection of neutron by matter.

Since at the condition of total reflection a wave penetrates into the region of matter, attenuating there exponentially, a finite particle density and correspondent flux directed parallel to the surface are in this classical forbidden region. This extra flux J_t caused by the partial penetrating of neutrons through interface must be compensated by decreasing of the flux in the region correspondent to the geometrical reflection of the initial beam. Simple calculations lead to the result

$$J_t = \frac{\hbar}{2m}\frac{k_x}{\sqrt{k_b^2 - k_y^2}}\frac{4k_y^2}{k_b^2}, \tag{11}$$

from which it is easy to define the transversal displacement d of the reflected beam and correspondent longitudinal shift ζ along the interface

$$\varsigma = \frac{k_x}{k_b^2}\frac{2k_y}{\sqrt{k_b^2 - k_y^2}}. \tag{12}$$

This result differs from the Artmann - Carter – Hora formula (7) and consequently contradicts to the hypothesis of relation between GH shift ant GDT (9). Nevertheless, the equation (12) was used as a basis for calculation of the results proposed [14] and realized [16] in neutron optical experiments.

For the long time the origin of the discrepancy between the results which follow from these two physically approaches was not properly understood. Only two decades after the publication of Renard's the paradox was resolved by Yasimoto, Oishi and Fedoseev [25, 26]. The point is that it was supposed by Renard, that total reflection differs from ideal-geometrical one only by the presence of the flux of evanescent waves inside the matter. It was not taken into account that phase difference between incoming and reflected waves leads to distortion of the flux in the region of crossing of two beams. Whereas as we could seen from Artmann's approach just this phase shift at total reflection is the origin of the GH. effect.

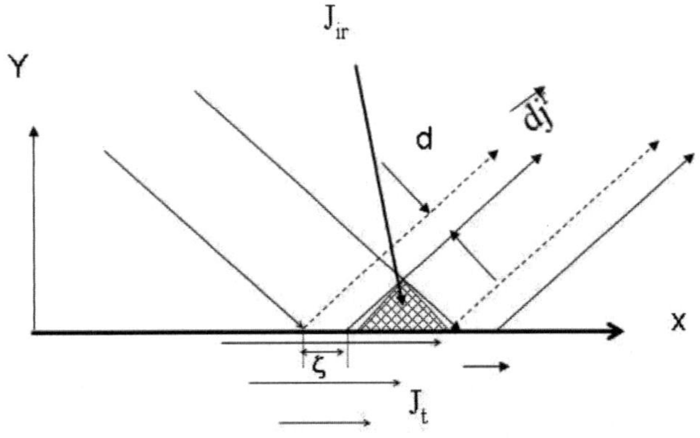

Figure.2. Illustration of the idea of flux balance. The sum of the fluxes J_t and J_{ir} must be compared with the disappearing flux in the region d [25].

Consequently, an extra flux that rises at total reflection is the sum of the flux of evanescent waves (11) inside matter and the additional flux J_{ir} arising due to interference of incoming and outgoing beams (see figure 2). The calculation that takes into account the above circumstance leads to equation (7). Thus, both approaches, namely, the method of Artmann and the one based on flux balance are equivalent. This means that the longitudinal shift ζ of the beam may be represented as a product of longitudinal velocity and GDT.

Taking into account the fact that the effective potential U is typically of the order of 10^{-7} eV and that the condition of total reflection is $E_n < U$, from equation (10) we obtain that with the exception of two narrow regions, i.e., that close to zero energy and the one close to the barrier, the group delay time is $\tau_r \approx 4 \div 6$ ns

3. Giant and negative longitudinal displacements of reflected neutron beam

Direct observation of neutron beam shift at total reflection is quite a difficult task. Smallness of the effect is not the only problem. In the pioneering works of Goos and Hänchen [1,2] the position of the totally reflected wave beam is compared with that of the beam reflected from a metal film because that is very small [27]. In the neutron experiment no reference position which the shift could be counted from was found. In fact, in [13,14] it is suggested that the position of the reflected beam should be compared with the results of calculation.

In addition, [14] suggests measuring the phase of the reflected wave using an interferometer. However, though phase shift at reflection and longitudinal shift of the reflected wave are closely related phenomena, they are different from the physics viewpoint. Therefore, it is not quite correct to interpret rather impressive manifestation of the phase shift effects detected in [16] as direct observation of the GH shift (see discussion in [17]).

The solution to the problem of observation of the GH effect in a neutron experiment will probably be found if one turns to the idea of resonant amplification of the beam shift which may occur at reflection from multilayer structures as proposed by Tamir and Bertony [28]. Later the problem was theoretically investigated [29,30] and a giant shift at light reflection from the waveguide structure was observed [31]. With respect to neutrons a similar assumption was discussed from a slightly different standpoint in [15].

The conditions for the waveguide-type propagation of the neutron flux along the surface of matter may be satisfied for quite various types of planar structures. The simplest one is a homogeneous film with the effective potential U_1 on the substrate with the potential $U_2 > U_1$. It is obvious that neutrons with $E_n < U_2$ will be totally reflected from such a structure. In the case of normal incidence the general expression for the amplitude of the reflected wave is

$$r(k_0) = \frac{(k_1+k_2)(k_0-k_1)+(k_1-k_2)(k_0+k_1)e^{2ik_1\xi}}{(k_1+k_2)(k_0+k_1)+(k_1-k_2)(k_0-k_1)e^{2ik_1\xi}}, \tag{13}$$

where ξ is the film thickness, k_0 is the wave number of the initial wave and $k_1 = \sqrt{k_0^2 - 4\pi\rho b_1}$, and $k_2 = \sqrt{k_0^2 - 4\pi\rho b_2}$ are the wave numbers in the matter of the film and of the substrate, respectively. In the considered case of total reflection from nonabsorbing matter $r = \exp(i\varphi)$, k_2 is imaginary and k_1 may be either real or imaginary depending on neutron energy.

Having obtained from eq. (13) the phase of the reflected wave $\varphi(k_0) = \mathrm{arctg}(\mathrm{Im}(r)/\mathrm{Re}(r))$ it is easy to find the derivative $d\varphi/d(k_0^2)$ and GDT (8) which is directly related to the longitudinal shift (9). In figure 3 the result of such calculation for the film with the thickness 90nm and the effective potential $U_1 = 91$neV (Ag) on the substrate with $U_2 = 245$ neV (Ni) is shown.

As is clearly seen in the figure directly in the resonance the GDT is as large as 260ns. Note that in accordance with (10) in the case of neutron reflection from a pure substrate the GDT is only 5.4 ns. As it follows from equation (9), the longitudinal shift of the reflected beam experiences resonant amplification equal to that of the delay time.

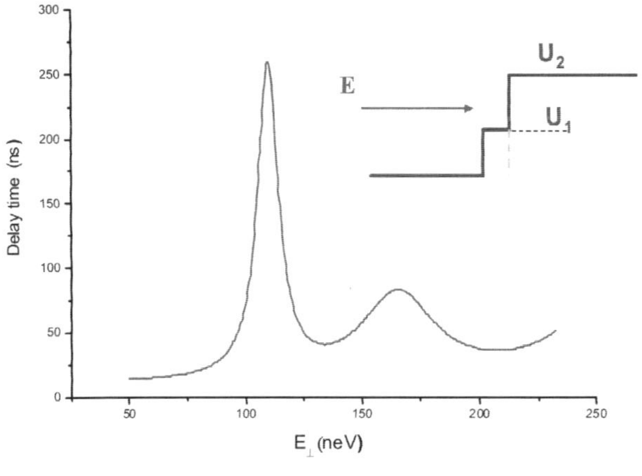

Figure 3. The group delay time at neutron reflection from a thin film on a substrate. In the insert there is shown the potential structure of such a combination. See the text for details.

At neutron reflection from multilayers another remarkable phenomenon may also arise. Under some conditions the group time and the longitudinal shift of the reflected beam may be negative. The possibility of a negative GH shift was first pointed to in paper [28] cited above. Later, the problem of negative beam shift and negative GDT was the subject of many theoretical [32-35] and experimental

[36-39] investigations. There exist a large number of various objects and media reflection from which is accompanied by negative beam shift. It is beyond the scope of the given work to present any detailed review of related papers. We will only dwell on the noted above case of neutron reflection from a simple planar structure and turn back to the simplest one, namely, to a homogeneous film on a substrate.

Differently from the case considered above the film potential, U_1, is now larger than that of the substrate, U_2.The effective potential of such a structure looks like an asymmetric potential barrier. The reflection coefficient of the barrier is then $R\left(k_0\right)=\left|r(k_0\right|^2$ and it can be easily found from equation (13). Above the barrier the coefficient R oscillates and decreases rather rapidly with increasing k_0. The GDT from that asymmetric potential barrier was found in [41] where it is also shown that in the minima of the reflection curve the GDT is negative. In accordance with (9) the reflected beam has to undergo negative shift in such conditions.

On the left figure 4 shows the result of calculation of the GDT and the neutron reflection coefficient from a nickel film 70nm thick deposited on a quartz substrate. It is seen that in the minima of the reflection curve the GDT reaches rather large negative values. In spite of that the reflection coefficient is relatively small in this case one may hope that the negative GDT and the negative longitudinal shift will be measurable.

The variety of multilayers characterized by negative GDT is though quite wide. This permits choosing an optimal relation between an absolute GDT value and the reflected beam intensity.

As an example the results of calculation for a three-layer structure are given. Its effective potential is two barriers not equal in width separated by a well. The data used in the calculation correspond to Ni-Ti-Ni films with the thickness 23, 13, 33 nm, respectively. In this case the reflection coefficient is about 0.4 in the resonance and the negative GDT has a twice smaller absolute value than in the previous case.

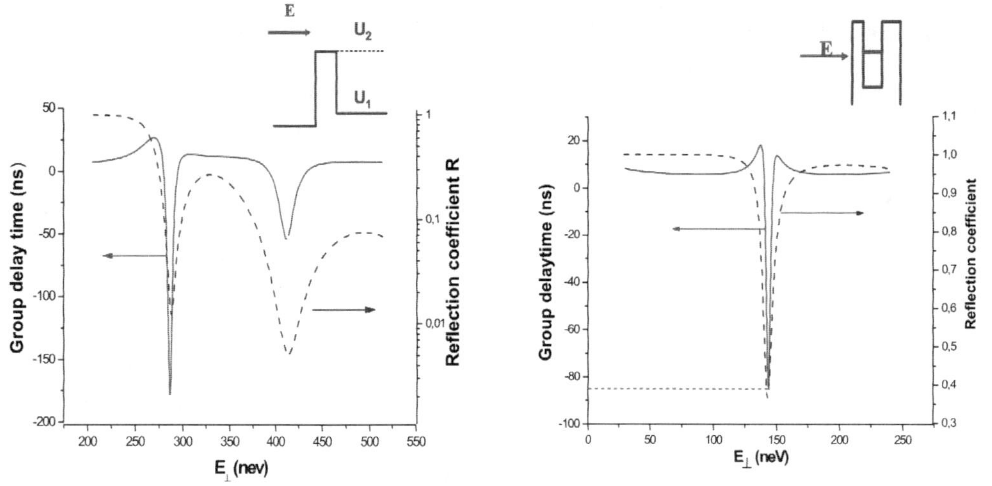

Figure 4. The group delay time (solid red) and the reflection coefficient (dash blue) for two multilayers. On the left hand side – a film on a substrate, on the right hand side – three films. The corresponding potential form is shown in the upper right hand corner of each figure. See the text for details.

4. Possibilities of experimental observation

4.1. Measurement of the group delay time

The possibility of direct measurement of the GDT at neutron reflection was experimentally demonstrated many years ago. The measurement is based on the so-called Larmor clock. The idea is to use the Larmor frequency of neutron spin precession in the magnetic field as a standard of frequency. Baz' [41] was the first to propose using such a clock as an original theoretical method to calculate the time of interaction of a particle with a three dimensional potential. Then Rybachenko [42] used the concept to calculate the particle tunneling time through a potential barrier. The Larmor time appears in many theoretical works dedicated to the problem of interaction time (see, for example [43,44]). It is closely related with the group delay time of Bohm-Wigner. Indeed, the additional angle of Larmor precession arising due to interaction with any object may be identified with the phase difference $\Delta\varphi$ between two spin components of the resulting wave function whose wave numbers are different.

$$k_{\pm} = k_0 \left(1 \mp \mu B/E\right)^{1/2}, \; E = \hbar^2 k_0^2 / 2m \,, \tag{14}$$

where k_0 is the neutron wave number in the absence of the magnetic field, μ is the neutron magnetic moment and B is the magnetic induction. Defining after Baz' the time delay due to interaction as

$$\Delta t_L = \frac{\Delta\varphi}{\omega_L}, \quad \omega_L = \frac{2\mu B}{\hbar} \,, \tag{15}$$

where ω_L is the Larmor frequency and taking into account that $2\mu B = \left(\hbar^2/2m\right)\left(k_+^2 - k_-^2\right) = \Delta E$ we come to the relation $\Delta t_L = \hbar\left(\Delta\varphi/\Delta E\right)$ coincident with (8) in the limit $B \to 0$.

The Larmor clock method was successfully used in neutron experiments for measuring the delay time at refraction, Bragg reflection from a mutilayer mirror and also, at tunneling in the resonance of a quasi-bound state [45-48]. Note that the measured time error was about 0.5ns which is less than GDT at total reflection by an order of magnitude.

4.2. Direct measurement of the Goos – Hänchen shift.

As is shown above the GDT from multilayer structures may reach in the resonance a positive or negative value of an order of 100 - 200 ns. Standard practice of the GH type experiment is that instead of the longitudinal shift $\varsigma = \tau_r v_x$, the transverse beam shift $d = \left(v_y/v_0\right)\zeta$ is measured which is proportional to ζ (see figure 2). By substituting eq. (9) and the relation for the total velocity v_0 we obtain

$$d = \frac{v_y v_x}{\sqrt{v_y^2 + v_x^2}} \tau \approx v_y \tau \;. \qquad \left(v_y \ll v_x\right) \tag{16}$$

As soon as we discuss either total reflection or the case when the energy just slightly exceeds the barrier, the velocity component v_y normal to the matter interface is on the order of some meters per second. Therefore, the inequality in (16) is valid for all neutrons except the slowest neutrons and UCNs. The transverse shift d depends on the normal velocity alone. For the total reflection from homogeneous matter the beam shift is $d \cong v_y \tau_r \approx 10\text{nm}$ and is hardly measurable.

Use of multilayer structures makes possible resonant enhancement of the GDT and of the longitudinal beam shift which is proportional to the former. The transverse shift is then of an order of a micrometer, which provides hope of performing appropriate experiments. The resonant behavior of the effect allows us to solve another serious problem. Because outside the very narrow resonance region the beam shift is small, the position of the beam beyond its limits may be accepted as a reference position with respect to which a relatively large resonant shift can be measured.

It is natural to think that the experiment can be performed with a neutron reflectometer operating in the time-of-flight mode. A narrow neutron beam will then fall on the sample at a fixed incidence angle θ and the normal to surface component of the wave vector $k_y(t) = k_x(t)\theta$ will depend on the time of flight. Thus reflected beam will only shift from its reference position over a short and well-known time interval. The experiment can be performed either with a position-sensitive detector or by using partial screening of the reflected beam with a nontransparent screen with a sharp edge. The beam shift will then result at variation of the count rate.

5. Conclusion

In the paper the well-known Goos – Hänchen effect was analyzed for the case of neutron reflection from matter. It is shown that the longitudinal shift of the reflected beam is always determined by the product of the longitudinal neutron velocity and the group delay time. In the case of total reflection of neutrons from matter the shift is very small and is unlikely to be measurable. However, in the case of neutron reflection from planar multilayer structures resonant enhancement of the effect may take place and in some conditions both the longitudinal shift and the group delay time may be negative. Some possibilities of experimental observation of the group delay time and of the Goos – Hänchen shift in a neutron experiment were discussed.

References

[1] Goos F und Hanchen O 1947 *Ann. der Phys.* **1** 333
[2] Goos F und Lindberg-Hanchen H 1949 *Ann. der Phys.* **5** 251
[3] Schoch A 1950 *Ergeb. Exakt. Naturwiss* **23** 127
[4] Schoch A 1952 *Acustica* **2** 1
1952 Acustica **2** 18
[5] Brekhovskikh L M 1953 *Uspekhi Fiz. Nauk* **50** 539
[6] Akylas V Kaur J and Knasel T M 1974 *Appl.Opt.* **13** 742
[7] Read L A A Wong M and Reesor G E 1978 *J. Opt. Soc. Am.* **68** 319
[8] Tamasaku K and Ishikawa T 2002 *Acta Cryst.* A**58** 408
[9] Hora H 1960 Optik **17**, 409
[10] Renard R H 1964 *J. Opt. Soc. Am.* **54** 1190
[11] Carter J L and Hora H 1971. *Journ. Opt. Soc. Am.* **61** 1640
[12] Agudin J 1968 *J. Phys. Rev.* **171** 1385
[13] Seregin A A 1981 *Sov. J. Nucl. Phys.* **33** 622
[14] Maaza M and Pardo B 1997 *Opt.Comm.* **142** 84
[15] Ignatovich V K 2004 *Phys. Lett.* **A322** 36
[16] de Haan V -O, Plomp J, Rekveldt M Th et al. 2010 *Phys.Rev.Lett.* **104** 010401
[17] Ignatovich V K 2010 *Phys. Rev.Lett.* **105**, 01890
de Haan V -O, Plomp J, Rekveldt M Th et al. 2010 *Phys.Rev.Lett.* **105** 018902
[18] Artmann K 1948 *Ann. der Phys.* **2** 87
[19] Von Fragstein C 1949 *Ann. Phys. (Leipzig)* **4** 271
[20] Brekhovskikh L M 1957 *Waves in layered media* (Moscow Acad. sci. USSR Press)
[21] Puri A and Birman J L 1986 *J. Opt. Soc. Am.* A **3** 543
[22] Eisenbud L 1948 *phd thesisys* (Prinston University),.
[23] Bohm D 1951 *Quantum Theory* (Prentice-Hall, New York)
[24] Wigner EP 1955 *Phys.Rev.* **98** 145
[25] Yasimoto K. and Oishi Y 1983 *J. Appl. Phys.* **54** 2170
[26] Fedoseyev V G 1986 *J. Opt. Soc. Am.* A **3** 826
[27] Merano M., Aiello A., Hooft G.W. 't et al. 2007 *Opt. Express* **15** 15928
[28] Tamir T and Bertoni H L 1971 *J.Opt.Soc.Am.* **61** 1397
[29] Schreier F, Schmitz M and Bryngdahl O 1998 Opt. Lett. **23** 576
[30] Felbacq D, Moreau A and Smaâli R 2003 *Opt.Lett.* **28** 1633
[31] Chen L., Cao Z, Ou F et al. 2007 *Opt. Lett.* **32** 1432

[32] Aniĉin A, Fazlić R and M. Koprić 1978 *J. Phys. A: Math. Gen.* **11** 1657
[33] Wild W and Giles C L 1982 *Phys.Rev.***A 25** 2099
[34] Declercq N F Degrieck J, Briers R and Leroy O 2003 *Appl. Phys. Lett.* **82** 2533
[35] C.-F. Li, Q. Wang 2004 *Phys. Rev.* **E 69** 055601
[36] Breazeale M A and Torbett M A 1976 *Appl. Phys. Lett.* **29** 456
[37] Bonnet Ch, Chauvat D, Emile O et al. 2001 *Opt. Lett.* **26** 666
[38] Yin Xiaobo, Hesselink L Liu Zh et al., 2004 *Appl. Phys. Lett.* **85** 372
[39] Li C-F, Zhu Q-B, Nimtz G et al. 2006 Opt. Comm. **259** 470
[40] Chen Xi and Li Ch.-F. 2005 *Eur. Phys. J.* B **46** 433
[41] Baz' A I 1967 *Soviet J. Nucl. Phys.* **4** 182
[42] Rybachenko V F 1967 *Soviet. J. Nucl. Phys.* **5** 635
[43] Huage E H and Stovneng J A 1989 *Rev.Modern Phys.* **61** 917
[44] Olkhovsky V S, Recami E and Jakiel J 2004 *Physics Reports* **398** 133
[45] Frank A I, Bondarenko I V, Kozlov A V, et al. 2001 *Physica* B **297**, 307
[46] Frank A I, Bondarenko I V, Vasil'ev V V, et al. 2002 *JETP Letters*, **75**, 705
[47] Frank A I, Anderson I, Bondarenko I V, et al. 2002 *Phys. At. Nuc.* **65** 2009.
[48] Frank A I, Bondarenko I V , Kozlov A V et al *Neutron Spin Echo Spectroscopy*, Eds. Mezei F, Pappas C and Gutbertlet T. (Springer) pp164-175

International Workshop on Neutron Optics and Detectors (NOP&D 2013) IOP Publishing
Journal of Physics: Conference Series **528** (2014) 012030 doi:10.1088/1742-6596/528/1/012030

Production of Ultra Cold Neutrons by a Doppler Shifter with Pulsed Neutrons at J-PARC

K Mishima[1], S Imajo[2], M Hino[3], T Ino[4], Y Iwashita[5], R Katayama[6], M Kitaguchi[7,8], T Oda[9], H M Shimizu[7], M Utsuro[10], S Yamashita[1] and T Yoshioka[11]

[1] International Center for Elementary Particle Physics, The University of Tokyo, Tokyo 113-0033, Japan

[2]Department of Physics, Kyoto University, Kyoto 606-8502, Japan

[3]Research Reactor Institute, Kyoto University, Kumatori, Osaka, 590-0494, Japan

[4]KEK, High Energy Accelerator Research Organization, Tsukuba, Japan

[5]Institute for Chemical Research, Kyoto University, Uji, Kyoto 611-0011, Japan

[6]Department of Physics, The University of Tokyo, Tokyo 113-0033, Japan

[7]Department of Physics, Nagoya University, Nagoya 464-8602, Japan

[8] Center of Experimental Studies, Nagoya University, Nagoya 464-8602, Japan

[9] Department of Nuclear Engineering, Kyoto University, Kyoto 615-8530

[10]Research Center for Nuclear Physics, Osaka University, Ibaraki, Osaka 567-0047, Japan

[11] Research Center for Advanced Particle Physics, Kyushu University, Fukuoka 812-8581, Japan

E-mail: mishima@icepp.s.u-tokyo.ac.jp

Abstract. Ultracold neutrons (UCNs) are neutrons whose kinetic energy is around a few hundred nanoelectronvolts. Neutrons with such small kinetic energy can be trapped in a material vessel or magnetic fields. Because of these unique characteristics, UCNs are used for some important experiments of fundamental physics. The Doppler shifter is a device to produce UCN by slowing them down by the reflection on a mirror moving with half of the velocity of incoming neutrons. A Doppler shifter using a quadruple-stack of monochromatic supermirrors that reflects neutrons with a velocity around 68m/s [1, Hino et al.(2010)] was fabricated, and operated with a pulsed neutron source of J-PARC. An important feature of the Doppler shifter is the use of a pulsed neutron beam. Unlike in continuous neutron beams, the neutron velocity can be selected by choosing a time slice in a pulsed neutron bunch. Thus the UCN production improves by ~80 times in the case of J-PARC. We successfully produced the UCNs by the Doppler shifter: the measured UCN production rate is consistent with the simulations.

1. Introduction

Ultracold neutrons (UCNs) are neutrons whose kinetic energy is around a few hundred nanoelectronvolts. Neutrons with such small kinetic energy can be trapped in a material vessel, or potential well of magnetic and gravitational fields. Because of these unique characteristics, UCNs are used for numbers of important experiment of fundamental physics.

A Doppler shifter is a device to produce UCNs by reflecting neutrons by moving materials [2–5]. If the reflector is moving in the same direction as the beam with half of the velocity of the neutron, the velocity in a laboratory system can be slowed down to be zero. Slowing down of neutrons by a Doppler shifter doesn't change the phase space density of neutron beam because the interaction is a conservative force (the Liouville's theorem) unlike by moderators. Thus, an intrinsic UCN density produced by a Doppler shifter is the same as in the initial neutron beam (in the case of perfect conversion). The phase space density, $n(v)$, for a continues beam with the Boltzmann distribution is described as

$$n(v) = \phi \frac{1}{2\pi v_T^4} e^{-\left(\frac{v}{v_T}\right)^2} \quad (1),$$

where ϕ is neutron flux on a surface of a moderator, v_T is a mean velocity corresponding to the neutron temperature, T. In the case of J-PARC, $\phi = 2.1 \times 10^{13}$ cm^{-2}s^{-1} for 1MW proton beam power [6,7], T = 57 K, which means that $v_T = 969$ m/s. Thus, its time averaged neutron density is 0.033 cm^{-3}(m/s)$^{-3}$.

The phase space density of a neutron beam at a pulsed neutron source is concentrated in time. Thus the density in phase space is increased $1/f\Delta\tau$ times compared with a continues neutron source, where f is the frequency of pulses and the $\Delta\tau$ is the effective pulse width [3]. The phase space density of the neutron beam at J-PARC (f=25 Hz, $\Delta\tau$ =500 μsec for neutrons with velocity of 136 m/s) is 2.6 cm^{-3}(m/s)$^{-3}$, which is 80 times greater than at a continuous neutron beam with the same flux. The phase space density can be converted to the UCN density, ρ_{UCN}, as

$$\rho_{UCN} = n\frac{4}{3}\pi v_{UCN}^3 = 3.5\times10^3 \left[\frac{UCN}{cm^3}\right] \quad (2),$$

where v_{UCN} is the critical velocity of UCN and taken as 6.8 m/s in this paper. Thus, a Doppler shifter at an intense, pulsed neutron source has a possibility to produce pulsed UCNs with very high phase space density.

In previous work, a Thermica crystal was used to reflect neutrons with velocity of 400 m/s [3,4]. However, there were some difficulties in the fabrication of the crystal reflector as the mosaic spread of the crystals was too small. In another work [5] intercalated graphite (iHOPG) was used for the reflection crystal which is not for UCN production, but for UCN up-scattering. The reflectivity of the crystal was more than 80% and the mosaic spread of the intercalated graphite was 5.5°. In this paper, we report about UCN production by a new Doppler shifter with a pulsed neutron source in J-PARC. We adopted a stack of monochromatic supermirrors with a large reflection momentum which was newly developed at KURRI [1]. Neutrons reflected from the supermirror have directions perpendicular to the mirror surface unlike the randomized directions produced by mosaic crystals. It is an advantage to convert a monochromatic beam with a small divergence to UCNs.

In this experiment, we used the monochromatic supermirror with a large reflection momentum of 2.2 nm^{-1}, this is the largest reflection momentum available from any supermirror in the world at present. The mirror reflects neutrons with velocity of around 68 m/s at rest, thus incoming neutrons with the velocity of 136 m/s (called very cold neutrons (VCN)), can be used for the slowing down. Because of the relatively slow neutron velocity, the Doppler shifter can be designed as compact and removable.

2. Experimental setup and apparatus

The experiment has been performed at beamline BL05 in J-PARC/MLF [8,9]. The beam line is used for fundamental studies and has three beam branches. The Doppler shifter was set up on the unpolarized beam branch because that produces the most intense neutron beam. The scheme of the experimental setup is shown in **Figure 1**. The branch is inclined upward 2.8°. 4 monochromatic supermirrors (VCN mirrors here after) are set to extract only the VCN component in order to reduce background. The reflecting momentum of the VCN mirrors is 0.65 nm^{-1} with a width of ~10% [10]. The set of VCN mirrors is tilted by angle of 10° from the incident beam, so that the VCNs are reflected downward by18.6°. The set of mirrors is shielded by surrounding B_4C resin to reduce the background of neutrons and gammas. The Doppler shifter is placed 1.0 m downstream of the VCN mirrors, i.e. 17.1 m from the moderator. The VCN beam passes through the Doppler shifter and is recorded by a position sensitive detector [11] at 17.8 m, which is shown as a VCN detector in **Figure 1**. A UCN detector is mounted on the top of the Doppler shifter. The distance from the center of the Doppler reflector to the window of the UCN detector is 241 mm, so that the longitudinal velocity of Doppler shifted neutrons can be measured by the time-of-flight (TOF) method. The cross section of the UCN beam is 82 mm for the VCN beam direction and 57 mm for the horizontal direction. Parameters of the apparatus are summarized in **Table 1**.

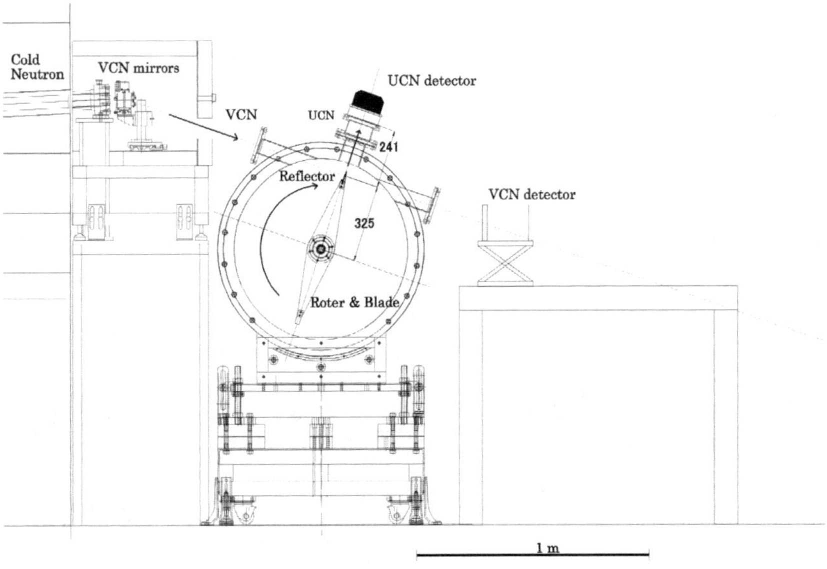

Figure 1. A drawing of the experimental setup of the Doppler shifter and other devices.

2.1. Neutron beam

2.1.1. Cold neutrons and VCNs from J-PARC/MLF BL05

The pulsed neutron source of J-PARC/MLF has three cold moderators; BL05 sees the coupled moderator. The neutron beam from the moderator is divided into three branches by supermirror benders and transported to the exit of the unpolarized beam branch passing through vacuum guide covered by the neutron absorber of 4 m in length. Neutron spectra measured at the exit of this branch are shown in **Figure 2**. A spectrum was measured to longer wavelengths by using intervals between neutron pulses due to operation for Main Ring of J-PARC. The spectra can be fitted by the Boltzmann distribution with attenuation proportional to the wavelength, described as

$$\phi(\lambda) = \frac{4h^3}{m^3\lambda^4}\left(\frac{m}{2k_BT}\right)^{\frac{3}{2}} \exp\left(-\frac{h^2}{2mk_BT^2} - a\lambda\right), \tag{3}$$

where h is the Planck constant, m is the neutron mass, k_B is the Boltzmann constant, T is the temperature, λ is the neutron wavelength, and a is an attenuation coefficient. T = 57.3 K, which was a calculated value by a simulation of the moderator [6,7], is used here. The attenuation coefficient, a, was defined as 0.80 nm^{-1} by fitting, which means that the phase space density is reduced by 45% for 1 nm in wavelength.

2.2. Doppler shifter

2.2.1. The reflection mirrors of Doppler shifter

Monochromatic supermirrors, which can reflect around 2.2nm^{-1} (68 m/s), are used for the reflector of the Doppler shifter. The mirrors were made by an ion-sputter at KURRI on Si substrate of 0.3 mm thickness. The reflectivity for a single mirror is ~20% with width of 10%, which is 2 times larger than the Thermica crystal [3]. Stacking a number of such mirrors can increase the total reflectivity, because the VCNs that cannot be reflected and transmitting by just a single mirror, can be reflected by other mirrors. We use a quadruple stack of the mirrors, whose reflectivity is 40% [1]. The reflectivity spectrum of the reflector is shown in **Figure 3**.

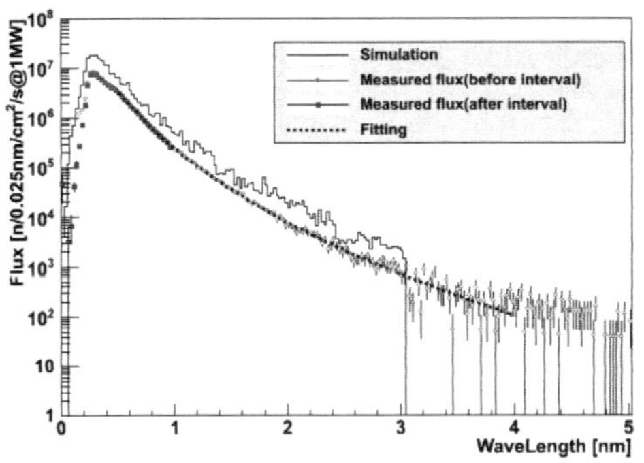

Figure 2. Neutron spectra of the unpolarized beam branch of BL05 at J-PARC/MLF.

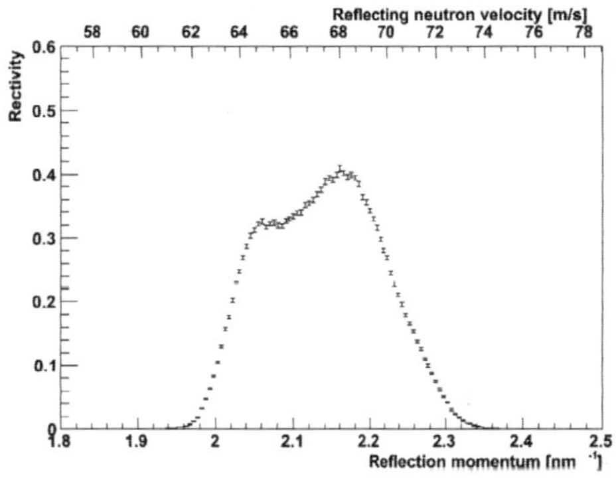

Figure 3. The reflectivity of a quadruple stack of monochromatic supermirrors used for the reflector of the Doppler shifter.

2.2.2. The Rotor of the Doppler Shifter

The reflector is mounted on an edge of an aluminium blade. Its radius to the mirror center is 325 mm. The blade is kept in a vacuum chamber in order to avoid air resistance and UCN losses, and rotated by a servomotor placed outside of the vacuum chamber through a magnetic-fluid seal. The chamber is evacuated to less than 40 Pa during the operation. The rotation speed can be set to between 0 and 3000 rpm. In the present experiment, the rotor was operated with 2000 rpm (30 msec cycle), so that the reflector velocity in the laboratory system was about 68 m/s, which is same as the reflection velocity of the reflector. The pulsed beam frequency is 25 Hz, thus the phases of the beam and the rotor are matching every 120 msec. The phase of the rotor was adjusted to the beam timing by using a resolver of the servomotor, and the deviation of the phase was 0.15° peak to peak, which corresponds to 0.8 mm in the position of the reflector.

2.3. Detectors and DAQ

A proportional gas counter (UCN DUNia-10 FLNP, JINR) was used for the UCN detector. The detector is filled with 1.3 kPa of ^3He, 1.1 kPa of CH_4, and 110 kPa of Ar. A window of the UCN detector is made of Al of 100 μm thickness, whose critical potential is 54 neV [12]. Thus, the detector is insensitive to UCNs with velocity less 3.2 m/s. The detection efficiency is the product of the transmission efficiency of Al window and the reaction efficiency of the detector, described as

$$\varepsilon = \exp\left(-N_{Al}\sigma_{Al}L_{Al}\right)\left(1 - \exp\left[-N_{3He}\sigma_{3He}L_{3He}\right]\right) \quad (4)$$

where *Ns* is the atomic number density, σ_s is the cross section, and *Ls* is the length for each subscript. The reaction efficiency of ^{3}He, which is known to follow the 1/v law, was measured by comparing counting rates for cold neutrons with a "thick" ^{3}He gas detector. By fitting the ratio of counting rates of two detectors from 0.2 nm to 0.8 nm, we obtained the detection coefficient of the UCN detector as a function of wavelength λ:

$$N_{3He}\sigma_{3He}L_{3He} = N\sigma_{3He}(\lambda_0)L\frac{\lambda}{\lambda_0} = (3.11\pm0.04)\cdot10^{-2}\,[\mathrm{nm}] \quad (5)$$

We used transmission efficiency of Al foil measured by A.Steyerl and H.Vonach [13]. The cross section of Al for 10 m/s neutrons consists of absorption cross section of 50.0±1.5 barn and inelastic cross section of 9.2±1.8 barn. The wavelength dependence was calculated using equation (1) in [13]. Using Eq.(4) the detection efficiency was calculated to be 78±1% for 6.8 m/s neutrons (58 nm). Its detection area is I.D. 90 mm, so that all neutrons from the Doppler shifter can be accepted. Note that gamma-rays had been discriminated by the pulse heights.

Table 1. Parameters for the experimental apparatus

Parameters for the neutron beam	
Beamline	J-PARC/MLF BL05, Unpolarized branch
Beam power	120 kW
Repetition	25 Hz
Neutron flux	1.1×10^7 cm^{-2}s^{-1}
Neutron flux in 2.74~3.34nm	2.4×10^3 cm^{-2}s^{-1}
Beam size (vertical)	55 mm
Beam size (horizontal)	45 mm
Beam divergence (vertical)	13.1 mrad
Beam divergence (horizontal)	10.7 mrad
Parameters for the rotor	
Radius to mirror center	325 mm
Rotation speed	2000 rpm
Mirror size	35 mm (horizontal) × 30 mm (vertical)
Mirror speed	64.9 m/s(inner), 68.1 m/s(center), 71.2 m/s(outer)
Phase stability	0.15° (peak to peak)
Distance from the source to the reflector	17.10 m
Parameters for the reflector	
Reflector type	Monochromatic supermirror
Number of mirrors in a stack	4
Reflecting momentum in rest system	2.16 nm^{-1}
Reflecting velocity in rest system	136 m/s
Reflectivity by a stacked mirror	40%
FWHM of reflection peak	10%

A position sensitive detector [11] with ZnS:Ag/^{6}LiF scintillator [14] was used to measure VCNs transmitted through the Doppler shifter. It was used to adjust the tilting angle of VCN mirrors and the rotation phase of the Doppler shifter. The efficiency of VCN detector was estimated 85% for VCN of 3.0 nm.

Data from the detectors were taken by an ADC/TDC module with proton timing pulses which were used for TOF analysis. The time resolution of the DAQ was 1 μsec. Large pulse height events were observed simultaneously with proton pulses - these events were omitted from the analysis.

3. Experimental results

3.1. Transmitted VCN

The TOF spectra at the VCN detector are shown in **Figure 4**. The velocity of VNCs is 136 m/s so that VCNs are expected to be detected 131 msec after the proton pulse, which corresponds to 11 msec in the 4th frame of the neutron pulses of 25 Hz (40 msec). A dip measured at 10-11 msec was caused by the reflection of the Doppler mirror. The phase of the rotor was adjusted to make the dip to be a certain position in the TOF spectrum to reflect VCNs with 136 m/s. The reduction rate of the dip was ~40%, which was consistent with the reflectivity of the Doppler mirror. The phase of the beam pulse and the rotor matched every 120 msec. Three spectra with phase differences of 0°, 120°, and 240° are shown in **Figure 4**. Other dips were observed in every 10 msec due to phase differences. The total neutron count on the VCN detector was 13 kcps at 120 kW.

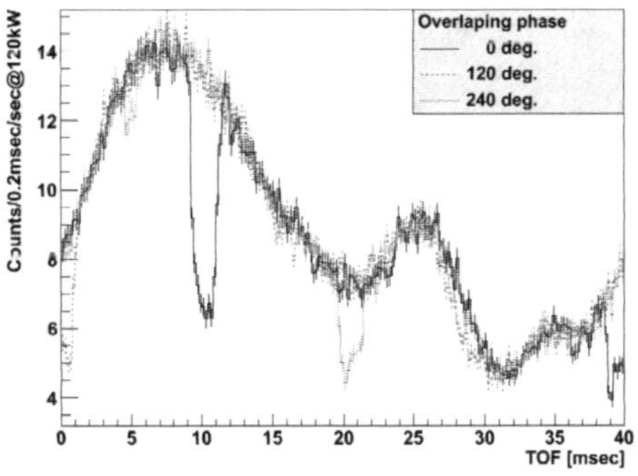

Figure 4. TOF spectra of VCN transmitted the Doppler shifter.

3.2 Doppler shifted neutrons

The TOF spectrum of the Doppler shifted neutrons is shown in **Figure 5**, where the distance from the reflector center to the detector surface was 241 mm. The zero point of the TOF was set to a moment when VCN and the Doppler mirror were crossing. The solid and dashed lines show spectra with the mirror rotating and with the mirror stopping on the VCN path to estimate the background, respectively. A significant peak was observed from 10 to 60 msec in the TOF caused by the rotation. A small peak observed around 45 msec was caused by reflection of the "next" rotation of the mirror.

A spectrum of the Doppler shifted neutrons in a unit of wavelength is shown in **Figure 6** with a Monte Carlo simulation [10], where events with the static mirror in **Figure 5** were subtracted as background. The total count rate on the UCN detector was 1.07±0.05 cps at 120 kW. We adopted an error of 5% as count rate caused by fluctuation of proton beam power. The expected neutron count by the simulation was 1.2 cps so that the observation and the simulation agree to within 11%. According to the simulation, the content of UCNs in the spectrum which have longer wavelength than 58 nm (6.8 m/s) was 28%. It means that the UCN count rate on the detector was 0.30 cps.

4. Discussion and outlook

4.1. UCN density from the phase space

The UCN intensity reaches 0.30 cps with 120 kW of beam power. We can calculate the UCN density in the same manner as reference [4]. By correcting for the detection efficiency for UCNs of 78%, the UCN intensity was 0.046 UCN/pulse. The solid angle of detected UCNs was 8.0×10^{-2} str. This yields that 0.27 UCN exist in the swept volume if the phase space was totally filled by neutrons. The swept volume was estimated as 32 cm^3 by multiplying the reflector size of 3 × 3 cm^2 and effective sweeping length of 3.6 cm, which was deduced by the FWHM of the interacting time of 0.4 ms as a result of the simulation. Finally, the UCN density of the Doppler shifter calculated from the phase space density was 0.22 UCN/cm^3 with 120 kW of the proton beam power. The phase space density of VCN at 120 kW was 9.7 UCN/cm^3 calculated by eq.(2), (3), the transporting efficiency of beam line, and reflectivity of VCN mirrors. Thus, UCN conversion efficiency in phase space by the Doppler shifter was 2.2%.

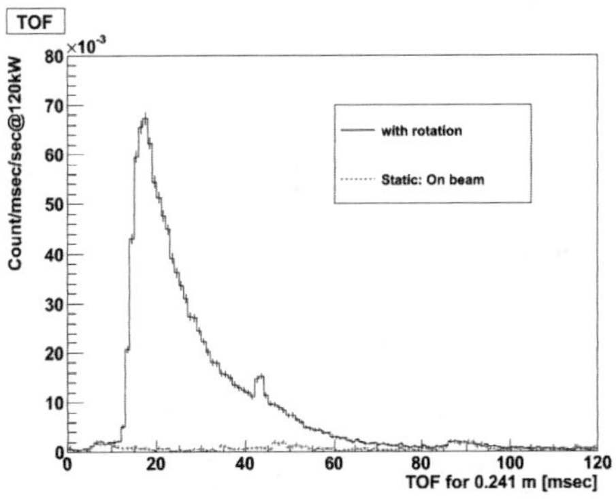

Figure 5. The TOF spectra of the reflected neutrons detected by the UCN detector: The reflector rotating (solid); the reflector at rest on the VCN path (dashed).

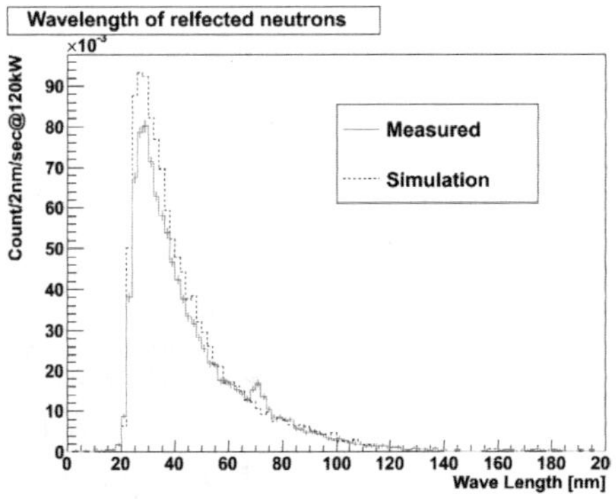

Figure 6. Wave length spectra of longitudinal component of the Doppler shifter neutrons (solid) and Monte Carlo calculation (dashed).

4.2. Future upgrade

The VCN intensity after the Doppler shifter is 13 kcps. The divergent VCN beam is collimated by a vacuum guide with neutron absorber located at 12 to 16 m from the moderator. The vertical and horizontal divergences of the VCN beam are 13.1 and 10.7 mrad, which correspond to transverse velocities of 1.8 and 1.5 m/s, respectively. Because the horizontal velocity is conserved, the phase space of the UCN has a room to be filled up to 6.8 m/s. Thus, UCN intensity can be increased by increasing the VCN divergence. Simulations show that by installing a supermirror guide instead of the vacuum guide in the beam branch one can increase VCN and UCN intensities by factor of about 30.

5. Conclusions

A Doppler shifter with supermirrors having large reflection momentum was fabricated. A quadruple-stack of the monochromatic supermirrors with reflectivity of 40% and width of 10% for reflection momentum of about 2.2 nm^{-1} was used as the reflector. The result of the UCN production experiment at BL05 beamline in J-PARC/MLF was a measured intensity of the Doppler shifted neutrons of 1.07±0.05 cps at 120 kW that within 11% is consistent with the Monte-Carlo simulations.

Acknowledgments
This research was partially supported by the Ministry of Education, Science, Sports and Culture, Grant-in-Aid for Scientific Research (A) 23244047, Scientific Research (B) 20340051, and Creative Scientific Research, 19GS0210.

References

[1] M. Hino, M. Kitaguchi, and Y. Kawabata, Kurri Prog. Rep. 2009, Kyoto Univ. 146 (2010).
[2] A. Steyerl, H. Nagel, and F. Schreiber, Phys. Lett. A **116**, 347 (1986).
[3] T. Dombeck, J. Lynn, and S. Werner, Nucl. INSTRUMENTS METHODS **165**, (1979).
[4] T. O. Brun, J. M. Carpenter, V. E. Krohn, G. R. Ringo, and J. W. Cronin, **75**, 223 (1980).
[5] S. Mayer et al., Nucl. Instruments Methods Phys. Res. Sect. A**608**, 434 (2009).
[6] Y. Ikeda, J. Nucl. Mater. **343**, 7 (2005).
[7] Technical Details, Materials and Life Science Experimental Facility [http://j-parc.jp/researcher/MatLife/en/instrumentation/ns3.html]
[8] K. Mishima et al., Nucl. Instruments Methods Phys. Res. Sect. A**600**, 342 (2009).
[9] Y. Arimoto et al., Prog. Theor. Exp. Phys. **007**, 1 (2012).
[10] S. Imajo, Master's Thesis, Fac. Sci. Kyoto Univ. 1 (2011).
[11] K. Hirota et al., Phys. Chem. Chem. Phys. **7**, 1836 (2005).
[12] V. Sears, Neutron News **3**, 26 (1992).
[13] A. Steyerl and H. Vonach, Zeitschrift Für Phys. **178**, 166 (1972).
[14] M. Katagiri et al., Nucl. Instruments Methods Phys. Res. Sect. A**529**, 274 (2004).

Development of a simulation for measuring neutron electric dipole moment

Ryo KATAYAMA[1], Kenji MISHIMA[2], Satoru YAMASHITA[2], Dai SAKURAI[3], Masaaki KITAGUCHI[4], Tamaki YOSHIOKA[5], Yoshichika SEKI[6]

[1]Department of Physics, The University of Tokyo, Tokyo 113-0033, Japan
[2]International Center for Elementary Particle Physics, The University of Tokyo, Tokyo 113-0033, Japan
[3]Department of Physics, Tokyo University of Science, Noda 278-8510, Japan
[4]Department of Physics, Nagoya University, Nagoya 464-8602, Japan
[5]Research Center for Advanced Particle Physics, Kyushu University, Fukuoka 812-8581, Japan
[6]RIKEN, Wako 351-0198, Japan

E-mail: katayama@icepp.s.u-tokyo.ac.jp

Abstract. The neutron electric dipole moment (nEDM) is sensitive to new physics beyond the standard model and could prove to be a new source of CP violation. Several experiments are being planned worldwide for its high-precision measurement. The nEDM is measured as the ultracold neutron (UCN) spin precession in a storage bottle under homogeneous electric and magnetic fields. In nEDM measurement, the systematic uncertainties are due to the motion of the UCNs, the geometry of the measurement system, and inhomogeneous electric and magnetic fields. Therefore, it is essential to quantitatively understand these effects in order to reduce them. Geant4UCN is an ideal simulation framework because it can compute the UCN trajectory, evaluate the time evolution of the spin precession due to arbitrary electric and magnetic fields, and define the storage geometry flexibly. We checked how accurately Geant4UCN can calculate the spin precession. We found that because of rounding errors, it cannot simulate it accurately enough for nEDM experiments, assuming homogeneous electric and magnetic fields with strengths of 10 kV/cm and 1 μT, respectively, and 100 s of storage. In this paper, we report on its discrepancies and describe a solution.

1. Introduction

The neutron intrinsic electric dipole moment (nEDM) is sensitive to new physics beyond the standard model (BSM). CP violation, which is equivalent to the violation of invariance under charge conjugation and parity transformations, introduces T violation through the CPT invariance theorem. The existence of an EDM violates T symmetry; therefore, measuring the intrinsic nEDM can allow us to effectively probe the CP violation source. The nEDM in the standard model can arise through second-order W exchange interactions, whereas those in several BSM theories result from first-order interactions[1]. The standard model predicts an nEDM of 10^{-31} e·cm, which current experimental techniques cannot detect. On the other hand, several BSM models predict a much larger CP violation, which can be close to the experimental bound. For example, supersymmetry (SUSY) predicts the existence of an nEDM from 10^{-27} to 10^{-28} e·cm, whereas the most precise measurement of the nEDM to date yielded (0.2 ± 1.5 (stat) ±

0.7 (syst))$\times 10^{-26}$ e·cm[2]. Therefore, by improving the precision of the nEDM by one or two orders of magnitude, we could achieve the range predicted by SUSY.

Ultracold neutrons (UCNs), which are cooled to less than about 300 neV, have the property of being reflected at any angle from material surfaces. The nEDM can be calculated accurately by measuring the spin precession of UCNs stored in a material container. The procedure is described briefly as follows:

(i) Polarized UCNs are stored in a container in which steady magnetic and electric fields are applied parallel to each other.

(ii) The spin states, analyzed by the Ramsey method of separated oscillatory fields[1], are measured after a certain time for precession.

(iii) Repeat (i) and (ii) with an electric field antiparallel to the magnetic field.

(iv) If an nEDM exists, the spin precession frequency will vary according to whether the electric field is parallel or antiparallel to the magnetic field (see Figure 1). Thus, by subtraction, we extract the nEDM. Ideally, the magnetic field averaged by the UCN must remain constant.

Then, if any process except for the nEDM changing the magnetic field seen by the UCN exists and is not canceled, it causes a systematic uncertainty.

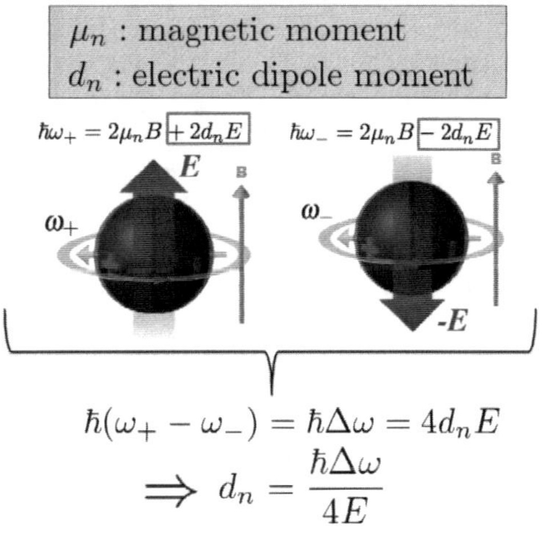

Figure 1. Principle of nEDM measurement

A relativistic magnetic field that occurs when a UCN passes through an electric field causes various systematic uncertainties. The relativistic magnetic field due to a net rotation in conjunction with a radial electric field component results in a spin precession frequency shift[2]. Similarly, the relativistic magnetic field connected with a longitudinal magnetic field gradient $\partial B_z/\partial z$ in a cylindrical trapping material can produce a similar effect[3]. These systematic uncertainties are enhanced when the UCNs are in certain ordered motions; therefore, it is worth studying the mechanism by which ordered UCN motion is diffused. For example, diffuse scattering that occurs when UCNs are reflected at a sidewall strongly inhibits their conservation. Further, a storage cell with a certain geometry is expected to reduce their motion: a cylindrical geometry consisting of a wavy plane should reduce the rotational motion more effectively than

a perfect plane does. Simulation studies are essential to quantitatively understand these effects because they are related to so many factors.
Although there are several simulation toolkits, Geant4UCN[4], based on the Geant4 simulation framework, is particularly good at defining the geometric configuration. For example, an arbitrary geometry can be extracted from CAD data. In addition, it is useful because Geant4 is widely used in particle physics experiments. In those simulations, the trajectory and spin precession are calculated by the Runge–Kutta and adaptive step size methods. We can quantitatively evaluate the suppression of systematic uncertainties by implementing a geometric configuration, a theoretical model for non-specular reflection, and electric and magnetic field distributions in Geant4UCN.
As for other simulations, the two following examples should be shown. The first example is the MATLAB-based simulation used in [5], in which the trajectory is evaluated using the numerical solver for the ordinary differential equation. The reflection points in the geometric configuration are stored from either Geant4UCN or a simple analytic calculation, and the spin precession is computed using the Bloch equation along its trajectory. Another example is the MCUCN used by the PSI collaboration, which is described in reference [6]. In that simulation, the trajectory and spin precession are calculated by a method similar to the calculation in Geant4. By incorporating the geometric configuration into the MCUCN, they calculate the spin tracking.

A Geant4UCN simulation is necessary to calculate a phase of neutron spin with better precision than 6×10^{-7} rad. This is because the nEDM experiments aim for a precision of 10^{-28} e·cm, which corresponds to a 6×10^{-7} rad phase shift from spin precession after 100 s of storage. We checked how accurately the Geant4UCN can calculate the spin precession assuming homogeneous electric and magnetic fields with strengths of 10 kV/cm and 1μT, respectively, and 100 s of storage. In this paper, we report the results. The paper is organized as follows. In Sec.2, we describe how we implemented relativistic spin precession in Geant4 and why the UCN rotational motion enhances the systematic uncertainty of nEDM experiments. In Sec.3, we describe why Geant4 cannot compute the spin precession accurately enough. Then, we describe how to solve these problems. Next, Sec.4 presents a discussion and our future plan. Finally, we summarize the content.

2. Systematic uncertainties associated with UCN rotational motion

As mentioned in Sec.1, several systematic uncertainties are enhanced by certain types of UCN motion. In particular, rotational motion is one of the most interesting situations in simulation studies. Accordingly, in this paper we evaluate the relativistic spin precession phase shift associated with UCN rotation.

In this study, we use Geant4UCN. To compute the relativistic spin precession, we incorporated the equation of motion for spin time evolution, the Bargmann–Michel–Telegdi equation, into Geant4UCN, as follows:

$$\frac{d\vec{s}}{dt} = \vec{s} \times \left[\left(\frac{g_n}{2} - 1 + \frac{1}{\gamma} \right) \vec{B} - \left(\frac{g_n}{2} - 1 \right) \frac{\gamma}{\gamma + 1} (\vec{\beta} \cdot \vec{B}) \vec{\beta} - \left(\frac{g_n}{2} - \frac{\gamma}{\gamma + 1} \right) \vec{\beta} \times \frac{\vec{E}}{c} \right], \quad (1)$$

where $\vec{s}$ is the UCN spin vector; $\vec{E}$ and $\vec{B}$ are the electric and magnetic fields, respectively; g_n is the g factor of the neutrons; β denotes the ratio of the UCN velocity to the speed of light; and γ is $1/\sqrt{1-\beta^2}$. Assuming the particle velocity is less than 7 m/s, which corresponds to the typical maximum velocity of UCNs, the first term can be regarded as a non-relativistic effect, the third term can be considered to be relativistic, and the second term is negligible. Note that Geant4UCN has a new release based on Geant4 ver4.9.6, but in this study we used the older code based on ver4.9.2.

When a UCN rotates with an angular velocity in the plane perpendicular to that in which static electric and magnetic fields are applied, a perturbative rotational relativistic magnetic field occurs in the radial direction. Then, the frequency of the UCN spin precession in the $x-y$ plane, ω_{xy}, becomes the following:

$$\omega_{xy} = \omega_0 + \frac{1}{2}\frac{(\omega_{rel})^2}{(\omega_0 \mp |\omega_r|)}, \tag{2}$$

where ω_0 is the Larmor frequency of the neutrons, and $\mp|\omega_r|$ represents the angular velocity of counterclockwise/clockwise rotation. Further, ω_{rel} is $-\gamma_n B_v$, where γ_n and B_v correspond to the gyromagnetic ratio of neutrons and the third term of (1), respectively. Actually, because of a transverse magnetic field in the experimental facilities, (2) will vary as $(\omega_{rel})^2 = (\gamma_n|\vec{B_v}|)^2 \to (\gamma_n|\vec{B_v} + \vec{B_T}|)^2$, where B_T is the transverse magnetic field in the experimental facilities. Then, the phase shift of (2), like B_v, depends on the electric field direction, and a corresponding systematic uncertainty occurs.

Assuming the nEDM is 10^{-28} e·cm and the electric and magnetic field strengths are 10 kV/cm and 1 μT, respectively, the expected corresponding phase shift amounts to $\frac{4\times10^{-28}\text{e}\cdot\text{cm}\times10\text{kV/cm}\times100\text{sec}\times3\times10^{8}\text{m/s}}{200\text{MeV}\cdot\text{fm}} \simeq 6\times10^{-7}$ rad. Therefore, in an nEDM simulation study, we have to calculate a phase of neutron with better precision than this.

3. Test of Geant4UCN

To check whether Geant4UCN simulates UCNs properly, we should compare the analytic solution predicted by (2) with the Geant4UCN simulation result as the UCN rotates in a plane perpendicular to that along which the static electric and magnetic fields are applied. We have done this as follows:

(i) Neglect the effects of gravity and non-specular reflection so that the UCN can always move in a single plane.

(ii) Apply static electric and magnetic fields of (0, 0, 1 μT) and (0, 0, -10 kV/cm), respectively.

(iii) Choose a geometry consisting of a cylinder with a 6 cm radius as the simulation geometry. In addition, adopt a coordinate system in which the cylindrical axis lies in the z direction, which is perpendicular to the plane in which the UCN moves.

(iv) Throw the UCN from the starting point (5.99 cm, 0, 0), with an initial velocity and unit spin vector of (0, 3 m/s, 0) and (0, 1, 0), respectively. The trajectory that the UCN traces is roughly equivalent to a 54-sided polygon.

(v) To understand the dependence of the simulation on the step length, perform a simulation study using various initial step lengths: 0.02, 0.03, and 0.04 mm.

(vi) Compute the phase of the simulation result after a certain time, $\Phi_{sim}(t)$, by using the following atan2 function, Four-quadrant inverse tangent: $\Phi_{sim}(t) = atan2[spiny(t), spinx(t)]$, where $spinx(t)$ and $spiny(t)$ denote the UCN spin components in the x and y directions, respectively, at a certain time.

(vii) Compute the phase shift of the analytic solution at a certain time, $\Phi_{sol}(t)$, from (2).

(viii) Evaluate the phase difference between the simulation result and the analytic solution after 0.01 s as follows: $d\Phi(t) = \Phi_{sim}(t) - \Phi_{sol}(t)$.

(ix) Repeat procedure (viii) until the simulated time reaches 100 s.

The simulation parameters are listed in Table 1.

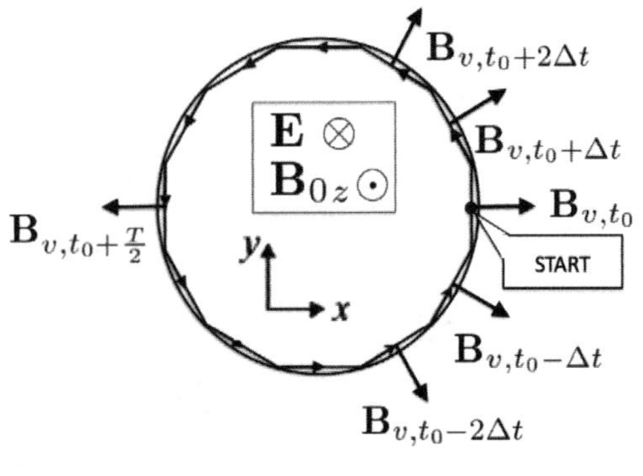

Figure 2. Schematic view of UCN trajectory trace and time evolution of relativistic magnetic field vector

Rotational Motion	Counterclockwise
Gravity Effect	Neglect
Storage Volume	Cylinder
Cylindrical Radius	6 cm
Cylindrical Axis	(0, 0, 1)
Initial Velocity	(3 m/s, 0, 0)
Electric Field	(0, 0, -10 kV/cm)
Magnetic Field	(0, 0, 1 μT)
Starting Position	(0, 5.99 cm, 0)
Initial Unit Spin	(0, 1, 0)
Initial Step Length (mm)	0.02, 0.03, 0.04

Table 1. Profile of simulation study

Figure 2 shows a schematic view of the trajectory the UCN traces in this study. In fact, the UCN traces an N-sided polygon, not a circle. Its angular velocity therefore varies according to the time, and the relativistic magnetic field differs from the ideal one. However, if the radius is 6 cm and $N = 54$, these effects on (2) can be negligible because its phase shift amounts to only 10^{-10}, which corresponds to 10^{-30} e·cm.

The result obtained from the comparison of the phase differences is shown in Figure 3, where the vertical axis corresponds to the difference between the phase shifts obtained by the simulation and the analytic solution at the simulated time, and the horizontal axis represents the simulated time. The solid, dotted, and broken lines denote the results with minimum step lengths of 0.04, 0.03, and 0.02 mm, respectively. According to Figure 3, the phase differences at 100 s for step sizes of 0.04, 0.03, and 0.02 mm reach about 6×10^{-6}, -2×10^{-6}, and -4×10^{-6} rad, respectively, which correspond to false EDMs of about several 10^{-28} to 10^{-27} e·cm. Considering that nEDM experiments aim to detect values of 10^{-28} e·cm, such a large difference is unacceptable.

We think that the rounding errors of the time variable cause these large differences, because the results shown in Figure 3 actually are not improved even though the step length decreases. In the algorithm using the Runge–Kutta method, a differential time δt_n is calculated as $\delta s_n/\delta v_n$, where δs_n and δv_n are the step length and speed, respectively, of the UCN. Thus, if the step length is not divided correctly, a differential step is assigned with an approximate value, namely, the true different time plus a rounding error. These rounding errors accumulate through 100,000,000 steps to the order of 10 ns, which corresponds to several 10^{-28} e·cm. For example, if the time variance due to rounding errors is 10 ns, the resulting false EDM is as follows: $\omega_0 \times 10^{-8}\ \text{s} \times \frac{10^{-28}\ \text{e·cm}}{6\times10^{-7}\ \text{rad}} \sim 3 \times 10^{-28}$ e·cm.

We prepared a new simulation that would avoid the rounding errors by extracting the codes relevant to the essential functions and changing the time variable data type from double to long double. The results are shown in Figure 4,where the vertical axis corresponds to the difference between the phase shift obtained by the simulation and the analytic solution at the simulated time, and the horizontal axis represents the simulated time. According to Figure 4, the difference

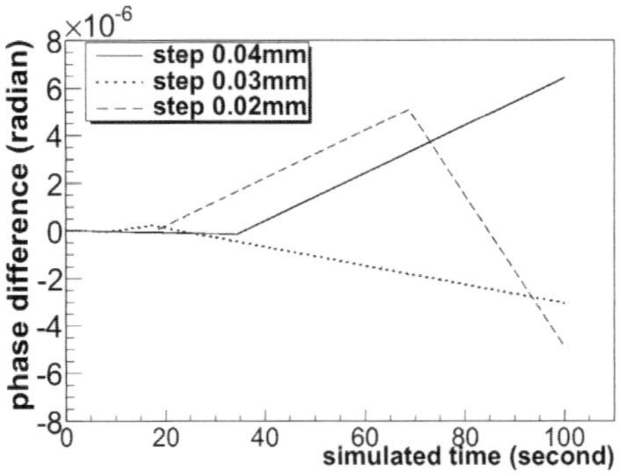

Figure 3. Phase difference between simulation result and analytic solution until 100 s using double time variable for step lengths of 0.02, 0.03, and 0.04 mm

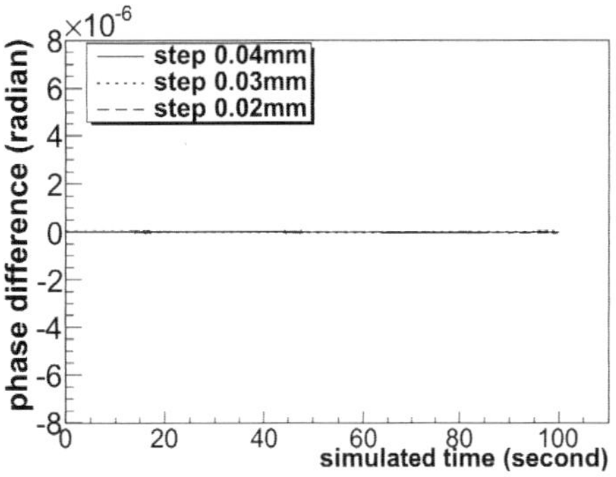

Figure 4. Phase difference between simulation result and analytic solution until 100 s using long double time variable for step lengths of 0.02, 0.03, 0.04 mm

at 100 s is about 2×10^{-8}, which corresponds to a false EDM of about less than 10^{-29} e·cm. Consequently, this proves that the difference is suppressed sufficiently.

4. Discussion and future plan

Rounding errors occur regardless of whether Geant4UCN is based on ver4.9.2 or 4.9.6 because they come directly from the Runge–Kutta algorithm. We need to remove the cause of this problem from Geant4 itself so that Geant4UCN can work properly. Because each UCN does not trace the same path, they each suffered from different time rounding errors. Hence, the rounding errors of the time variable could have serious effects on the simulation results. Further, the problem is not related to whether the CPU is 32-bit or 64-bit because the rounding errors were observed when we used a 64-bit CPU. We consulted a Geant4 specialist about the problem that we reported on in this paper. We are now trying to implement the solution he proposed in Geant4UCN.

Non-specular reflection at the sidewalls can suppress the systematic uncertainties, as described in Sec.1. It generally depends on the surface's state, especially on its roughness. If we can develop an nEDM measurement cell consisting of surfaces that can control the non-specular reflection, it will allow us to manage the systematic uncertainty suppression. However, this means that we need to understand the correct dependence on the non-specular reflection. On the other hand, UCNs can be lost when they are bounced off of material surfaces, which increases the statistical uncertainty. The mechanism of UCN loss, particularly at low temperature, i.e., the UCN anomaly, is still being discussed and remains to be determined, even though several models to explain it have been proposed[7][8][9]. By repeating experiments to obtain a more realistic simulation, we need to try to discover a more precise model describing the reactions upon reflection.

5. Summary

In measurement of the nEDM, systematic uncertainties occur because of a relativistic magnetic field arising in conjunction with inhomogeneous electric and magnetic fields. The systematic uncertainties are enhanced when the UCN ensembles exhibit certain ordered motions. Thus, it is essential to quantitatively understand the effects of diffusing the UCN ordered motion in order to reduce the systematic uncertainties. Geant4UCN is an ideal simulation framework because it can take into account all the essential functions for these effects. Because nEDM experiments aim at detecting values of 10^{-28} e·cm, the simulation study needs to be sufficiently precise.

Rotational motion in inhomogeneous electric and magnetic fields is one of the most interesting situations related to the systematic error enhancement due to certain types of UCN motion. In this situation, a relativistic magnetic field accompanying rotational motion arises, and the UCN spin precession becomes slightly faster. Thus, we can check whether a simulation works properly by comparing the simulation result with the analytic solution for its rotational relativistic magnetic field. Actually, the Geant4 simulation result has rounding errors in the time variable corresponding to a range of several 10^{-28} e·cm to about 10^{-27} e·cm, which is not negligible for nEDM experiments. To avoid this problem, by extracting the set of coding relevant to the essential functions, we prepared an alternative to Geant4UCN that can properly simulate UCNs with errors corresponding to less than 10^{-29} e·cm.

References

[1] Philip Harris, arXiv: 0709.3100.
[2] C. A. Baker et al., Phys. Rev. Lett. 97, 131801 (2006).
[3] J. M. Pendlebury et al., PHYSICAL REVIEW A 70, 032102 (2004).
[4] http://ucn.web.psi.ch/mc/geant4ucn
[5] http://ucn.web.psi.ch/papers/Thesis_Andreas_Knecht_2009.pdf
[6] K.Bodek et al., Phyics Procedia 17(2011) 259-267
[7] E.Korobkina et al., PHYSICAL REVIEW B 70, 035409 (2004).
[8] A.Serebrov et al., PL A 335(2005) 327-336.
[9] G.Danilov, arXiv: 0903.0458.

Application of the MCNPX-McStas interface for shielding calculations and guide design at ESS

E B Klinkby[1,2], E B Knudsen[3], P K Willendrup[3,4], B Lauritzen[1], E Nonbøl[1], P Bentley[2], U Filges[5]

1) DTU Nutech, Technical University of Denmark, DTU Risø Campus, Frederiksborgvej 399, DK-4000 Roskilde, Denmark
2) European Spallation Source ESS AB, Box 176, S-221 00 Lund, Sweden
3) DTU Physics, Technical University of Denmark, DTU Lyngby Campus, Anker Engelunds Vej 1, DK-2800 Kgs. Lyngby, Denmark
4) ESS design update programme - Denmark
5) Paul Scherrer Institute, 5232 Villigen, Switzerland. ESS design update programme - Switzerland

E-mail: `esbe@dtu.dk`

Abstract. Recently, an interface between the Monte Carlo code MCNPX and the neutron ray-tracing code MCNPX was developed [1, 2]. Based on the expected neutronic performance and guide geometries relevant for the ESS, the combined MCNPX-McStas code is used to calculate dose rates along neutron beam guides. The generation and moderation of neutrons is simulated using a full scale MCNPX model of the ESS target monolith. Upon entering the neutron beam extraction region, the individual neutron states are handed to McStas via the MCNPX-McStas interface. McStas transports the neutrons through the beam guide, and by using newly developed event logging capability, the neutron state parameters corresponding to un-reflected neutrons are recorded at each scattering. This information is handed back to MCNPX where it serves as neutron source input for a second MCNPX simulation. This simulation enables calculation of dose rates in the vicinity of the guide. In addition the logging mechanism is employed to record the scatterings along the guides which is exploited to simulate the supermirror quality requirements (i.e. m-values) needed at different positions along the beam guide to transport neutrons in the same guide/source setup.

1. Introduction

The European Spallation Source (ESS), presently starting construction in Lund, Sweden, will be the most intense source of spallation neutrons ever built [3]. Protons from a 5 MW, 2.5 GeV linear accelerator will impact a rotating tungsten target in 14, 2.86 ms long pulses every second. The spallation neutrons hereby created are thermalized in water and some of them are further cooled in liquid para-hydrogen before extracted through individual beam-lines serving 22 neutron scattering instruments[1]. To fully exploit the long pulse characteristics, many instruments benefit from being long - up to 200 m is foreseen. Each instrument will by situated at the end of a guide tailored specifically to its requirements in terms neutron beam parameters. The design of the

[1] 22 is the foreseen number of instruments in the baseline facility design. Additional instruments may by added at a later stage.

guides is challenging and important since guides and in particular their shielding are a major cost driver for the ESS facility.

For these reasons it would be useful to have a tool that:

- Can monitor where in a guide neutrons are lost
- Allows to optimise reflectivity requirements along a guide
- Serves as an input for dose-rate calculations along the guide (n,γ)
- Works within the work-flow accustomed to instrument designers (McStas)

The newly developed tool *McStas Scatterlogger* is the backbone that facilitates this usage. Note that the examples of usage outlined below, to some limited extent are inspired by guide work relevant for the ESS, but the tool is general and can be used to assess neutron scatterings at any neutron scattering instrument.

2. McStas Scatterlogger: Implementation and usage

McStas [4, 5, 6] combines ray-tracing with Monte Carlo methods to accurately and efficiently simulate neutron scattering experiments of all types. As seen from the user perspective (typically instrument designers and users) the program is based upon a so-called *instrument file* in which you place predefined or user-supplied *components* according to the real setup of the instrument in question. The *instrument file* is written in a simple meta language, which at runtime is compiled first to the ansi-C programming language, which in turn is translated to an executable.

The *Scatterlogger* is an additional McStas component which can be placed in an *instrument file* wrapping around any other component - for example a guide component. At each scattering in the guide, the incoming and outgoing neutron state is then temporarily stored, and an iterator is supplied so that the user can analyse the scatterings once the neutron has exited the component. To prevent memory built-up the results of the analysis are stored in histograms, memory is flushed and McStas proceeds to the next neutron - as in any other McStas simulation.

In the intermediate analysis step (event post-processing), the user has access to the full neutron state before the scattering as well as the reflected and transmitted neutron state:

$$\begin{aligned} Incoming\ state : n_{in} &= (\mathbf{x}, \mathbf{v_{in}}, \mathbf{t}, \mathbf{w_{in}}) \\ Transmitted\ state : n_{trans} &= (\mathbf{x}, \mathbf{v_{in}}, \mathbf{t}, \mathbf{w_{trans}}) \\ Reflected\ state : n_{refl} &= (\mathbf{x}, \mathbf{v_{out}}, \mathbf{t}, \mathbf{w_{in}} - \mathbf{w_{trans}}) \end{aligned} \tag{1}$$

Figure 1 contains a visualisation of a guide and the logging utility in operation. At each scattering of the neutron on the guide wall, the un-reflected neutron weight fraction is emitted as a secondary ray, propagated to a monitor component. The reflected neutron proceeds rightwards, traversing the guide with reduced weight.

3. Use-case 1: Lost intensity

This simple example considers a straight guide described by the parameters outlined in table 1 - see figure 1 for a visualization. At each scattering the transmitted weight w_{trans} and the corresponding position along the guide, z, is stored in a histogram and the result is shown in figure 2.

As expected the neutron loss occurs predominately at the first scattering in the beginning of the guide - the strength of the scatterlogging procedure is that it provides a method to quantify this. Obviously the extent at which intensity is lost along a guide depends strongly on the guide geometry and coating as well as the source (ie. energy, divergence and spatial extent). Thus, the

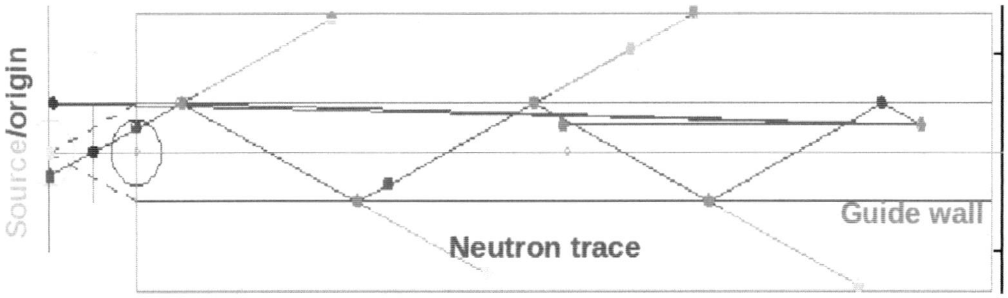

Figure 1. Basic guide geometry - from McStas' *mcdisplay*. Also shown in the figure is an example neutron trajectory. At each encounter with the guide wall, both the reflected and un-reflected neutron ray is stored and both trajectories are visualised. The un-reflected ray is allowed to propagate to a monitor component (green) surrounding the guide where it is detected. The apparent reflection (backscattering) from the right-most scattering back to the origin, is an (unphysical) feature of the visualization procedure.

Parameter	Value
guide length	10 m
guide coating	$m = 2$
cross-section of guide	10 cm × 10 cm
source radius	10 cm
source spectrum	flat in wavelength: [0-10] Å
distance from source to guide entrance	1 m

Table 1. Parameters for the guide and source used in Use-cases 1 and 2. The source is focussed to illuminate precisely the guide entrance and nothing more.

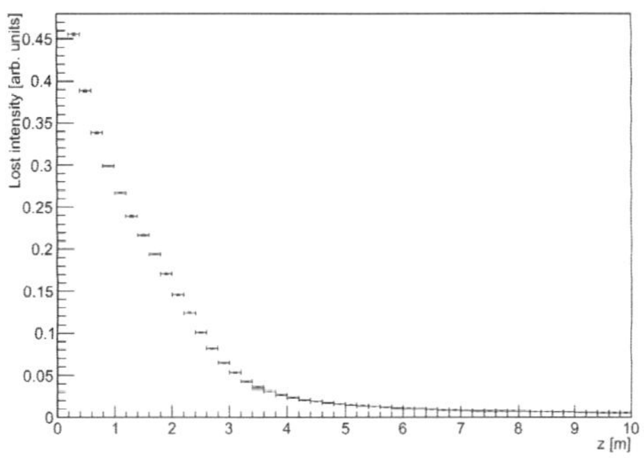

Figure 2. Histogram showing the lost neutron intensity along the straight guide as described in the text.

present example is by no means an attempt to describe guide losses in general, on the contrary the scatterlogging must be utilised individually to aid a particular guide design. Moreover, users should anticipate an iterative process where the results of one iteration (eg. figure 2) gives input to next - eg. increasing the m_{value} in the part of the guide closest to the source.

4. Use-case 2: Reflectivity

Neutrons scattering on a mirror are reflected if the energy/incident angle is sufficiently low. Specifically, the following relations hold (see figure 3):

$$cos(2\theta) = (\vec{v}_{in} \cdot \vec{v}_{out}) \,/\, |\vec{v}_{in}|^2 \tag{2}$$

$$k = |\,\vec{v}_{in}|\; m_h/\hbar \tag{3}$$

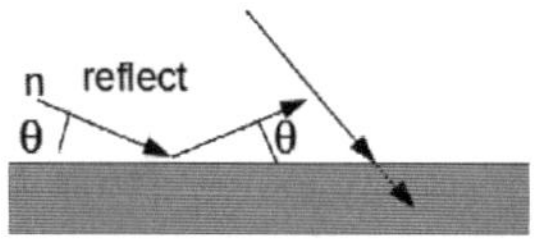

Figure 3. Reflected vs. absorbed neutron.

Given a neutron state and a guide geometry, the minimum mirror reflectivity requirement which would reflect the neutron, m_{min}, can be calculated [6][2]:

$$m_{min} = 2 \cdot k \cdot sin(\theta)/0.0219\ \text{Å}^{-1} \tag{4}$$

Figure 4 shows m_{min} for the first, second and third reflection respectively. The fact that the mirror in this example has an m_{value} of 2, is seen (on the intensity scale on the right-hand side of each figure) to cause significant loss of neutrons, in particular when comparing the first an second reflection.

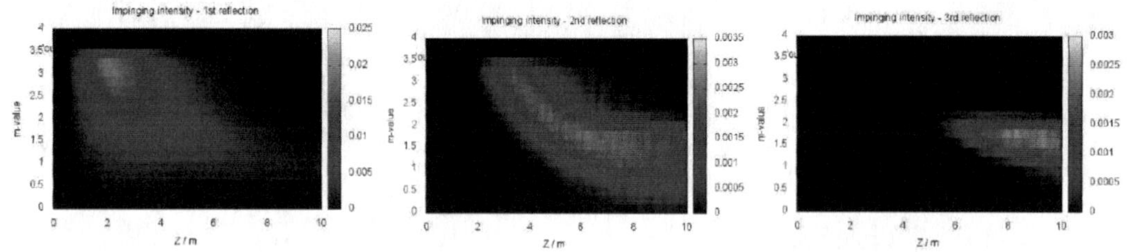

Figure 4. Histograms showing the minimum m-value required for a reflection, as a function of the position z along guide in a simulation of the 10 m straight guide described in the text. The color scale indicates the bin intensity. Note that the simulation was carried out using a $m = 2$ guide, which is seen to cause significant intensity loss in the beginning of the guide.

[2] Note that the value for Nickel reflectivity of 0.0219 is used for consistency with McStas default, despite that the standard value is 0.02175 Å^{-1}.

5. Conclusions and outlook

The McStas Scatterlogger has proved a useful tool in guide design. In particular, it facilitates to monitor where along a guide neutron intensity is lost and allows to optimise where to use high-reflectivity mirror sections, and where less expensive mirrors can be installed with insignificant performance loss. Note that the results depend strongly on the simulated guide geometry and source specifications, wherefore simulations involving the McStas scatterlogger need to be performed based on individual guide\instrument designs.
The software needed to reproduce Use-cases 1 and 2 is available here: [7].

Future foreseen (or possible) extensions include the calculation of dose-rate along guides, backgrounds at the sample and detector position, activation calculations as well as calculation backgrounds and dose-rates at one instrument, caused by neutrons escaping a neighbouring instrument.

References

[1] Klinkby E et al (2013) *Nuclear Instruments & Methods In Physics Research A* **700** 106
[2] Knudsen E B, Klinkby E B and Willendrup P K (2014) *Nuclear Instruments & Methods In Physics Research A* **738** 20
[3] Peggs S (editor) et al (2013) ESS Technical Design Report ESS-doc-274, ISBN 978-91-980173-2-8 URL `http://eval.esss.lu.se/DocDB/0002/000274/015/TDR_online_ver_all.pdf`
[4] P Willendrup, E Farhi, E Knudsen, U Filges, K Lefmann (2014) *J. Neutr. Res.* **17** 35–43
[5] Willendrup P, Knudsen E, Farhi E and Lefmann K (2011) User and Programmers Guide to the Neutron Ray-Tracing Package McStas, Version 1.12c, Risø−R−1416(rev.ed.)(EN)
[6] Willendrup P, Knudsen E, Lefmann K and Farhi E (2011) Component Manual for the Neutron Ray-Tracing Package McStas, Version 1.12, Risø−R−1538(rev.ed.)(EN)
[7] scatter_flog_bundle.zip availble here: `http://mcstas.org/download/share/`

Extension of the VITESS polarized neutron suite towards the use of imported magnetic field distributions

S Manoshin[1], A Rubtsov[1], V Bodnarchuk[1], S Mattauch[2] and A Ioffe [2,*]

[1] Joint Institute of Nuclear Research, Frank Laboratory of Neutron Physics, 141980 Moscow region, Russia,
[2] Jülich Centre for Neutron Science - Outstation Garching, Forschungszentrum Jülich GmbH, Lichtenbergstr. 1, 85747 Garching, Germany

E-mail: a.ioffe@fz-juelich.de

Abstract. Latest developments of the polarized neutron suite in the VITESS simulation package allowed for simulations of time-dependent spin handling devices (e.g. radio-frequency (RF) flippers, adiabatic gradient RF-flippers) and the instrumentation built upon them (NRSE, SESANS, MIEZE, etc.). However, till now the magnetic field distribution in such devices have been considered as "ideal" (sinusoidal, triangular or rectangular), when the main practical interest is in the use of arbitrary magnetic field distributions (either obtained by the field mapping or by FEM calculations) that may significantly influence the performance of real polarized neutron instruments and is the key issue in the practical use of the simulation packages. Here we describe modified VITESS modules opening the possibility to load the magnetic field 3-dimensional space map from an external source (file). Such a map can be either obtained by direct measurements or calculated by dedicated FEM programs (such as ANSYS, MagNet, Maxwell or similar). The successful use of these new modules is demonstrated by a very good agreement of neutron polarimetric experiments with performance of the spin turner with rotating magnetic field and an adiabatic gradient RF-flipper simulated by VITESS using calculated 3-dimensional field maps (using MagNet) and magnetic field mapping, respectively.

1. Introduction

Powerful Monte Carlo (MC) instrument simulation codes play a very important role in the context of the design and optimization of neutron scattering instrumentation. Particularly, comprehensive MC simulations have become a primary tool for developers of polarized neutron instrumentation. Although analytic calculations are sufficient in simple cases of the ideal geometry of magnetic field arrangements, mostly to check new concepts, MC simulations allow considering arbitrary complex geometries of real neutron scattering instruments.

Spin coordinates have been included in three open-source neutron simulation packages, VITESS [1], McStas [2] and NISP [3], during their creation. NISP contains polarization tools allowing for the simulation of the neutron spin precession in constant magnetic field and in magnetic field with variable field direction, however does not allow for simulations with the variable magnitude of a magnetic field. McStas allows for the polarization handling in constant magnetic fields; on-going developments aiming simulations in variable magnetic fields have been reported at this workshop.

---

* To whom any correspondence should be addressed

VITESS has become a very comprehensive polarized neutron simulation package since a new approach allowing for the simulations of neutron spin dynamics in time-dependent, both in magnitude and direction, magnetic fields has been introduced [4]. Now VITESS facilitates simulations of the neutron spin dynamics in practically all polarized neutron devices (the RF neutron flipper, the adiabatic gradient RF flippers and the Drabkin resonator, etc.) [5]. Consequently, it has made simulations of the performance of all kind of neutron spin-echo (NSE) instrumentation using them possible (generic NSE, neutron resonance spin-echo (NRSE), rotating and time-gradient magnetic fields NSE and MIEZE spectrometers, as well as NSE-based diffractometers SESANS and SERGIS) (see [5] and references therein].

However, up to present VITESS modules developed for the simulations with polarized neutrons do not allow for the use of real magnetic field distributions, but only ones given by a few analytical functions (sinusoidal, triangular, rectangular). To achieve flexibility of the VITESS for simulations of *real* instruments it is becoming imperative to have the possibility to load the magnetic field description from an external source (file). Such description can be either the 3-dimensional magnetic field map measured experimentally or the 3-dimensional field distribution that was calculated analytically or simulated by dedicated programs (such as ANSYS, MagNet, Maxwell or similar).

In this article we will present the modification of the VITESS modules allowing for simulations of the performance of a neutron spin turner based on a rotating magnetic field and a real adiabatic RF-flipper, as well as the results of simulations of real devices that are in good agreement with neutron experiments.

2. VITESS modules for simulations with imported magnetic field distributions

A standard VITESS module is defined over the space slab limited by input and output planes. For the VITESS modules handling time-dependent (both in magnitude and direction) magnetic fields this space is divided into rectangular blocks (called "domains") (Figure 1a), in which the magnetic field is considered to be a constant during the neutron propagation across them [4]. Since the neutron spin in VITESS is considered as a classical torque the change in the space orientation of the spin vector is calculated as the result of the Larmor precession around the magnetic field vector $\boldsymbol{B}(k)$ over the propagation time across the k-th domain.

The neutron spin components after the propagation across of the k-th domain are considered as the incoming components of the spin vector for the $(k+1)$-th domain and so on:

$$\left\{s_k^i\right\}_{out} = \left\{s_{k+1}^i\right\}_{in}, \qquad i = x,y,z \qquad k = 1,\ldots,N-1$$

For this approximation to be valid the number of domains N_i in each direction ($i=x$, y or z) should be large enough to satisfy the condition $\boldsymbol{B}(k)=const$ during the time of neutron propagation across these domains, i.e. the size of these domains has to be sufficiently small. Thus, the time-dependent magnetic field is represented as a histogram of $B(k)$ and we neglect changes of the magnetic field vector in domains during the neutron propagation time.

Practically, N_i should be so large that the final result (magnitudes of neutron spin components coming out of the field area) will not depend on the further increase of N_i (Figure 1b) [4]. It should be chosen by an iterative procedure, e.g. by doubling N_i for each subsequent iteration. The procedure is completed when such number of domains N_{max} are achieved that results of two subsequent iterations do not differ within the expected accuracy of the polarization measurements, usually 0.1%.

Up to present, all magnetic field distributions that are used in VITESS modules dealing with polarized neutrons have being given by analytical functions (sinusoidal, triangular, rectangular) calculated using a limited number of parameters. To use the above-mentioned VITESS algorithm for real magnetic fields, they have also to be given as the 3-dimensional array with the sufficiently large number of points. This array of magnetic field values can be either calculated by dedicated programs (FEM software as ANSYS, RADIA, MagNet, Maxwell, etc.) or measured experimentally.

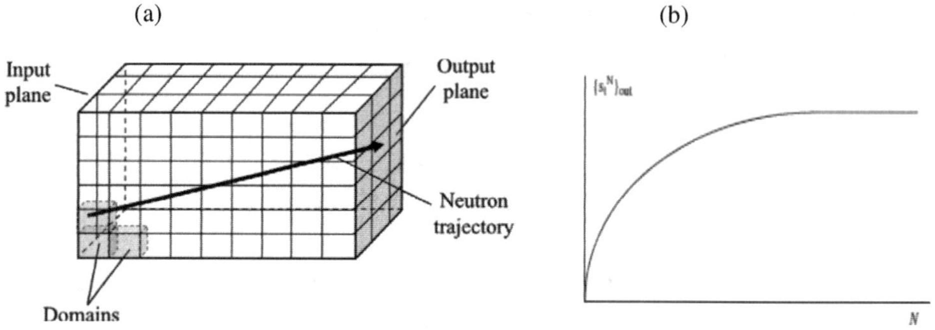

Figure 1. (a) Presentation of the field volume split into 3-dimensional array of domains and (b) the magnitude of simulated outgoing spin components vs. the number of slices *N*.

However, the number N_d of experimental (or calculated by an external software) points may not be sufficient for successful simulations. This is usually a case for spin-handling devices that uses the combination of permanent and time-dependent fields, e.g. RF-flippers. Generally, the steeper the time-dependence $\boldsymbol{B}(t)$ the larger N_d is supposed to be. Therefore, simulations of high-frequency devices require the field space to be split in a large number of domains in the neutron flight direction, usually far exceeding *N*.

We have modified the relevant VITESS modules in a way that allows them to accept external data files written in the format of the MagNet software (Infolityca.com) that we used for calculations of magnetic fields created by different "magnetic elements" (as coils and/or permanent magnets and magnetic screens) (see Figure3b in Sect. 3.1). Note that other FEM programs can also be used as their output files actually have the same structure; however one should pay attention to details of their format, that should be compatible with the MagNet/VITESS input format. In close future we will deliver the modification of the VITESS modules that will automatically choose the format for the import of data files according to the selection of FEM software made by user.

The MagNet software exports the calculated magnetic field as a data file, each line of which contains six columns: *x, y, z* coordinates and B_x, B_y, B_z components of the magnetic field. Each of the coordinates is changed successively from minimum to maximum values (one can imagine such a structure as three nested cycles). These coordinates are considered by VITESS as centres of domains; the number of domains in this case is the same as the number of lines in the file (note that domains in the VITESS structure are identical, so their centres are equidistant). Thus, only three nested cycles could be used to read the magnetic field data.

As it has been mentioned above, the simulations of high-frequency devices require a number of domains N_{max} that is usually exceeding the number of field values N_d measured experimentally or calculated by a FEM program by far. To match these numbers we introduce additional intermediate domains (in the directions of *x, y* and *z)* between two adjacent "main" domains determined by the imported data. The number of intermediate domains should be sufficient for successful simulations; the values of magnetic field in these new intermediate domains are obtained by interpolation (linear or B-spline) between "main" domains.

The physical consideration underlying such a procedure is the expected smoothness of the magnetic field distribution between the centres of "main" domains. In the case of experimental magnetic field maps, points for measurements are selected by an experimentalist in the assumption that the measured field distribution is a representative sampling of the true one, i.e. that no essential features (as jumps or extrema) are omitted. In the case of magnetic field maps calculated by a FEM program, the mesh for calculations is chosen automatically based on two criteria: to have a representative sampling (i.e. to provide the requested accuracy of calculations) and to minimize the number of points where the calculations have to be carried out. As the result the mesh is not homogeneous at all: its density is high at quick changes of the magnetic field vector (magnitude or direction) and low where the decay-like behaviour of the magnetic field magnitude or slow changes of the field direction take place. However, the MagNet software allows user to choose the regular mesh, which is excessive considering the amount of necessary information, but matches the input file format requested by VITESS.

3. Examples of the VITESS simulations with imported magnetic field distributions

3.1. Spin turner for the rotation magnetic fields NSE spectrometer

Neutron spin turners using rotating magnetic fields (RMF) are the key elements of the RMF NSE spectrometer [6]. They are thin devices, where the magnetic field is confined and rotated in the plane *yz* of the spin turner. In the ideal case the magnetic field is fully confined in the spin turner and there is no fringe magnetic field outside it. This is certainly not correct for the real device made of the finite size solenoidal coils (Figure2a), so that the scattered magnetic fields should be confined by the flux closing yokes made of 2mm thick μ-metal plates (Figure2b).

During the experiment carried out at the neutron polarimeter LAP of the Jülich Centre for Neutron Science (JCNS) installed at 4.5Å neutron beam [7], we have been able to measure all three components of the polarization vector passed through the set up consisting of two such spin turners separated by 10cm as a function of frequency ω of the applied sinusoidal electrical current. The result of measurements for the z component is presented in Figure3a together with the simulated (by the VITESS) performance for the case of ideal spin turners – although a sinusoidal dependence is reproduced, the mismatch of the amplitude definitely tells about the presence of fringe magnetic fields. Therefore, the magnetic field distribution in the real spin turner (i.e. two orthogonal coils with the winding density reduced towards the coils' periphery, surrounded by a number of μ-metal plates) was calculated using the MagNet software package and the results have been imported in the VITESS by the procedure described in Sect.2. The result of simulations with such a real magnetic field distribution (Figure3b) shows a good coincidence with the experiment and clearly demonstrates that the magnetic design has to be improved. The great advantage of this experiment is that now the validity of the chain MagNet -VITESS is experimentally confirmed, so that for further magnetic design work no neutron experiments are required till the test of the final version.

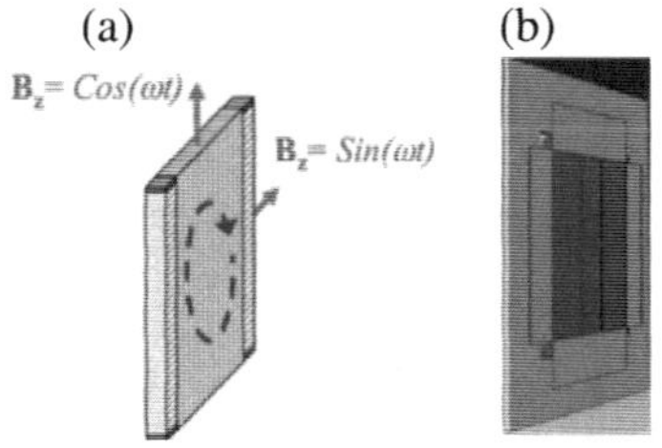

Figure 2. Spin turner made of two orthogonal solenoidal coils (a) and its MagNet model with the surrounding flux closing yokes (b).

3.2. Adiabatic RF flipper

The adiabatic gradient RF-flipper [8] (or adiabatic RF-flipper) consisting of a RF coil immersed in a continiously rising (gradient) permanent magnetic field (Figure 4a) has become one of most often used elements in polarized neutron scattering, as it is effectively flipping the neutron spin for all neutrons with the wavelength above the critical λ^*, that is especially valuable for the use at pulsed sources with basically wide-band spectrum neutron beams.

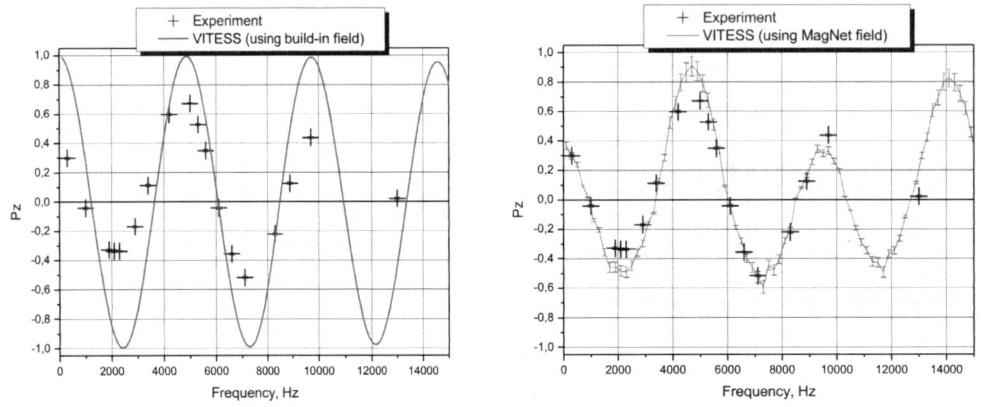

Figure 3. Frequency dependence of the *z*-component of the polarization in comparison with VITESS simulation using built-in (a) sinusoidal and (b) calculated by MagNet magnetic fields.

To maximize the flipper efficiency that may even exceed 99.99%, one should choose most suitable type of the gradient magnetic field. The simplest is the linear dependent field that rises from B_0-ΔB to B_0+ΔB over the flipper's length 2L (Figure 4a), that however results in less spin-flip efficiency than the cosinusoidally dependent field [9].

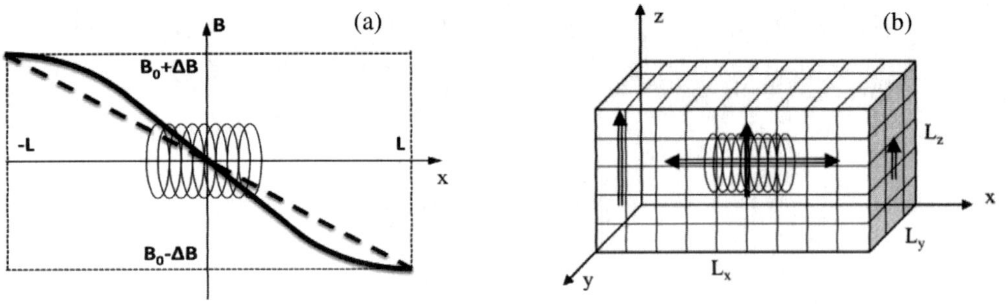

Figure 4. (a) Cosine and linear gradient magnetic fields in the adiabatic RF-flipper. (b) Magnetic field volume split in 9 x 5 x 3 domains used for simulations of the adiabatic RF-flipper.

The VITESS module for simulations of the adiabatic gradient flipper has been containing the built-in generator of both above-mentioned fields (with parameters defined by a user). Now we have also added to it the possibilty for the import of the magnetic field distrbution measured or calculated externally. This option is rather important not only because the exact linear or cosinusoidal magnetic field distributions can be hardly realized, but also because tough experimental conditions quite often do not allow for practical realization of such field distributions. Obviously, the possibility to estimate the expected efficiency of a "real" flipper by simulations is very desirable.

As an example we present here the simulations of a gradient spin flipper for the large cross-section neutron beam, (6 x 19) cm^2, that has been carried out prior to the neutron test of the flipper at the JCNS reflectometer MARIA [10]. The gradient field is created by a set of permanent magnets; the RF field is created by a solenoidal coil supplied by the AC current from a dedicated generator.

The magnetic field volume selected for further simulations – the parallelepiped with size L_x=40cm, L_y= 6cm, L_z=19cm - coincides in the cross-section (6cm x 19cm) with the size of the neutron beam propagating through the flipper along the *x*-axis. This volume is filled with the gradient magnetic field superimposed with the rotating (around *z*-axis) magnetic field created by the RF-coil of radius 2.5 cm and of length 20cm (Figure 4b). The measurements of the gradient magnetic field values have been conducted using a regular grid pattern 9 by 5 by 3 points along *x*-, *y*- and *z*-axes, respectively (Figure 5), and imported to VITESS as an external file. Transverse *y*- and *z*-components of the gradient field were found to be negligibly small with respect to B_z

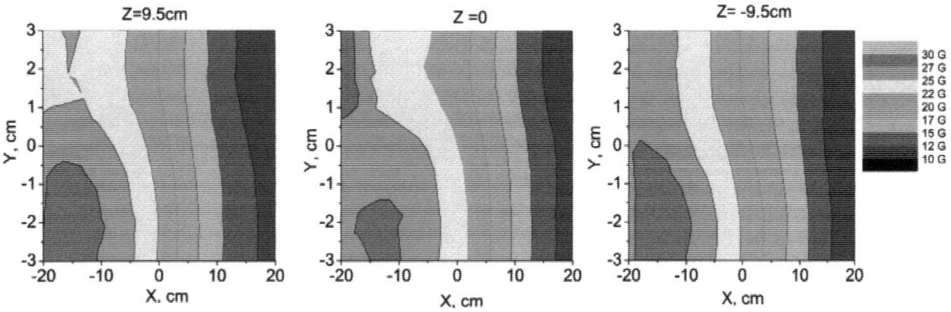

Figure 5. Y-X distributions of the gradient magnetic field in three horizontal cross sections z= 9.5cm, z=0 and z=9.5cm (see Figure 4b).

The amplitude of the rotating field has been calculated using the standard formula describing the magnetic field strength B_z along the solenoid axis *x*:

$$B_z(x) = \frac{\mu_0 nI}{4}\left[\frac{x+l/2}{\sqrt{r^2+(x+l/2)^2}} - \frac{x-l/2}{\sqrt{r^2+(x-l/2)^2}}\right] \tag{1}$$

Here μ_0 is the permeability constant; *n* – the number of turns per unit length; *I*- the current in the solenoid; *r* and *l* – the radius and the length of the solenoid, respectively; *x* -the distance from the centre of the solenoid. Note an extra factor 2 in the denominator of Eq. (1): it reflects that the linear oscillating field in the solenoid is considered as a superposition of two counter rotating fields, whilst in the VITESS we are dealing with only one of these components with the same sense of rotation as the spin Larmor precession (the second component results in the Bloch-Siegert frequency shift [11] and is usually not relevant from the point of view of a neutron scattering instrument). Strictly speaking, one should also take into account transverse components B_x and B_y, that should be calculated using a FEM simulation program. We have neglected them in our case since they are much less than the z-component of the permanent field. The working parameters of the real flipper - the amplitude of the RF (rotating) magnetic field of 25 (12.5) Gauss and the frequency of 60 kHz - have been used for the simulations.

The nine points where the values of the permanent magnetic field were measured correspond to nine domains in x-direction that is no way sufficient for stable simulations. Therefore, the interpolation procedure described in Sect. 2 is applied. The number of domains, *N*, sufficient for simulations is determined from the stability of the simulated result shown in Figure 6, where the values of all three components P_x , P_y and P_z of the polarization vector are depicted in the dependence on *N* (time required for simulations for N≈400-500 and for 5 by 3 domains along *x*- and *y*-axes, respectively (see Figure 5)) at each wavelength is about 10-15 sec at a standard laptop). One can see that their values are not changed anymore beyond $N \approx 450$. Therefore, $N=500$ has been selected for final simulations, those results are shown in Figure 7 together with the result of the neutron test of such flipper at MARIA.

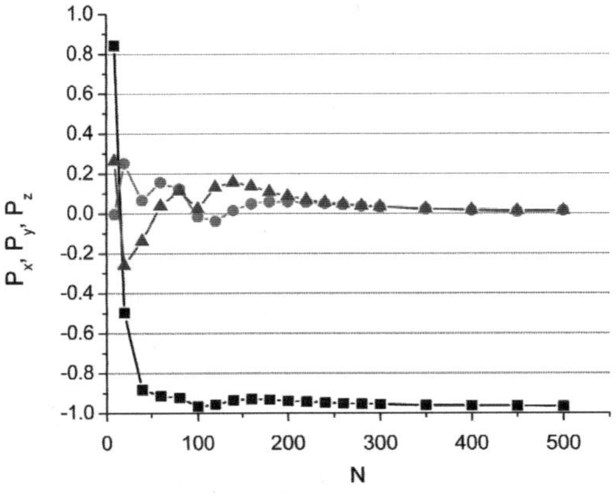

Figure 6. Simulated values of the components of the polarization vector in the dependence on the number of domains in x-direction. P_x –blue triangles, P_y –red circles, P_z – black squares.

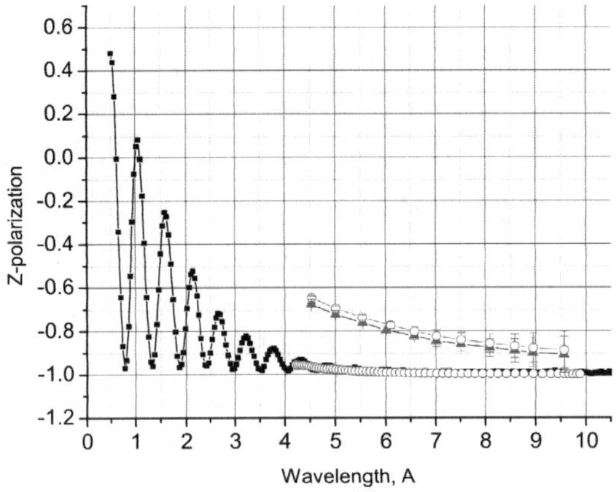

Figure 7. Flipping of the polarised neutron beam by the adiabatic RF flipper in the dependence on neutron wavelength. Full black squares - absolutely monochromatic neutron beam; hollow blue circles – neutron beam monochromatized by the velocity selector ($\Delta\lambda/\lambda$=10%); hollow pink circles - polarization calculated by taking into account the non-perfect polarization of the incident beam and the non-perfect polarization efficiency of the ^{3}He spin filter; full pink triangles - the experimentally measured neutron polarization.

Simulations with absolutely monochromatic ($\Delta\lambda/\lambda$=0) neutron beam (full black squares) show the expected oscillating behaviour of the spin-flipped polarization (Figure 7). In reality, such behaviour is not observed because of the final monochromatization ($\Delta\lambda/\lambda$=10%) of the used neutron beam (hollow blue circles), but one can see that the flipper works very efficiently for neutrons with the wavelength above the critical one $\lambda^* = 4.5$Å. The overall beam polarization (hollow pink circles) can be calculated from the latter curve by taking into account the non-perfect polarization of the incident beam and the non-perfect polarization efficiency of the ^{3}He spin filter used for the polarization analysis of a large neutron beam. This final result of VITESS simulations has to be compared with the experimentally measured polarization (depicted as full pink triangles in Figure 7). One can see rather good coincidence in the wavelength dependences of simulated and experimental curves, when slightly better, about 2%, absolute polarization obtained in the experiment is most probably caused by an error in the knowledge of the absolute polarization efficiency of the ^{3}He spin filter and the incident beam polarization.

4. Conclusions

When "ideal" magnetic field distributions (sinusoidal, triangular or rectangular) are sufficient for the VITESS simulations of the performance of practically all components constituting polarized neutron instrumentation, the main practical interest is in the use of real magnetic field distributions that may significantly influence the performance of real polarized neutron instruments and is the key issue in the practical use of the simulation packages.

This problem is now solved – we have modified VITESS modules in a way that now they allow to import 3-dimensional (space) magnetic field maps, which can be either obtained by direct measurements or calculated by dedicated FEM programs. A very good agreement between neutron polarimetric experiments and simulations carried out by VITESS using magnetic field maps calculated by the MagNet software is achieved.

The next step in this direction will be an extension of the data import option to others popular software packages for the magnetic simulations, such as ANSYS and RADIA. Further on, we will develop an approach allowing for simulations of performance of different instrument components (e.g. mirrors, flippers) deepened into the magnetic field.

Acknowledgements

We acknowledge with gratitude the support by the European Commission under the 6th Framework Programme through the Key Action: Strengthening the European Research Area, Research Infrastructures, Contract Number: RII3-CT-2003-505925, and 7th Framework Programme through the 'Research Infrastructures' action of the 'Capacities' Programme, Contract No: CP-CSA_INFRA-2008-1.1.1 Number 226507-NMI3.

References

[1] Zsigmond G, Manoshin S, Lieutenant K et al,2007 *Physica* **B397** 115.
[2] Seeger P A, Daemen L L 2007 *Nucl. Instrum. and Meth.* **A457** 338.
[3] Willendrup P, Farhi E, Lefmann K 2004 *Physica* **B350** E735; Knudsen E, Udby L, Willendrup P, Lefmann K, Bouwman W G 2011 *Physica* **B406** 2361.
[4] Manoshin S, Ioffe A 2008 *Nucl. Instrum. and Meth.* **A 586** 81.
[5] Manoshin S, Belushkin A, Ioffe 2011 *Physica* **B406** 2337.
[6] Ioffe A 2003 *Physica* **B335** (2003) 169.
[7] Ioffe A, Bussmann K, Dohmen L, Axelrod L, Gordeev G, Brückel Th 2003 *Physica* **B350** E815.
[8] Bazhenov A, Lobashev V, Pirozhkov A, Slusar V 1993 *Nucl. Instr. and Meth.* A**332** 534.
[9] Grigoriev S, Okorokov A, Runov V 1997 *Nucl. Instrum. and Meth.* **A384** 451.
[10] http://www.jcns.info/jcns_maria
[11] Bloch F, Siegert A 1940 *Phys. Rev.* **57** 522.

An Open-Source Software Package for Data Treatment in a MIEZE Experiment

T. Weber[1,2], G. Brandl[1,2], R. Georgii[2,1], P. Böni[1]

[1] Physik-Department E21, Technische Universität München, 85748 Garching, Germany
[2] Heinz Maier-Leibnitz Zentrum, Technische Universität München, 85747 Garching, Germany

E-mail: tweber@frm2.tum.de

Abstract. We present a newly-developed open-source software toolkit for handling and analysing the 4-dimensional datasets generated by modern multi-layer time-of-flight (TOF) position-sensitive detectors (PSD). Two such detectors of the type "Cascade" are in operation at the MIEZE beam line MIRA 2 and at the neutron resonance spin-echo spectrometer RESEDA at FRM II. The software consists of several modules for data analysis and instrument parameter calculations to assist the user during experiments, during simulations and in data evaluation. Specifically for the Cascade detector, it contains modules that correct phase-shifts in the MIEZE signal caused by the detector geometry and layout.

1. Introduction

1.1. *The MIEZE Technique*

MIEZE (Modulation of Intensity Emerging from Zero Effort) [1, 2, 3, 4] is a spin-echo technique related to NSE (Neutron Spin-Echo) [5, 6] and NRSE (Neutron Resonance Spin-Echo) [7, 8]. All three spin-echo techniques are based on encoding the energy transfer ΔE at the sample in a change of the neutron spin phase $\Delta\phi$ that is given by

$$\Delta\phi \;=\; \tau_M \cdot \Delta E/\hbar. \tag{1}$$

The proportionality factor τ_M is called the MIEZE time, or spin-echo time.

The MIEZE setup as shown in fig. 1 consists of two RF π flipper coils placed between two spin polarisers. The coils are operated at two different frequencies ω_A and ω_B.

Such a setup leads to a phase change ϕ of a polarised neutron beam after passing the second coil of [8, 3]

$$\phi \;=\; 2\beta\cdot(\omega_B-\omega_A)\cdot\left(t_b-\frac{L_1+L_b}{v_1}-\frac{L_s}{v_2}\right)+2\beta\cdot\omega_B\cdot\frac{L_1}{v_1}, \tag{2}$$

where v_1 and v_2 denote the neutron velocities before and after scattering at the sample, respectively and t_b is the time the neutrons leave the second coil. The factor β is used to distinguish between non-bootstrap and bootstrap [3] coils with $\beta=1$ for the non-bootstrap and $\beta=2$ for the bootstrap case.

Inserting equ. 2 into equ. 1 and applying a series expansion to first order [3] leads to

$$\frac{\tau_M}{\hbar}\cdot\left(\frac{\partial E}{\partial v_1}\cdot dv_1+\frac{\partial E}{\partial v_2}\cdot dv_2\right) \;=\; \frac{\partial\phi}{\partial v_1}dv_1+\frac{\partial\phi}{\partial v_2}dv_2. \tag{3}$$

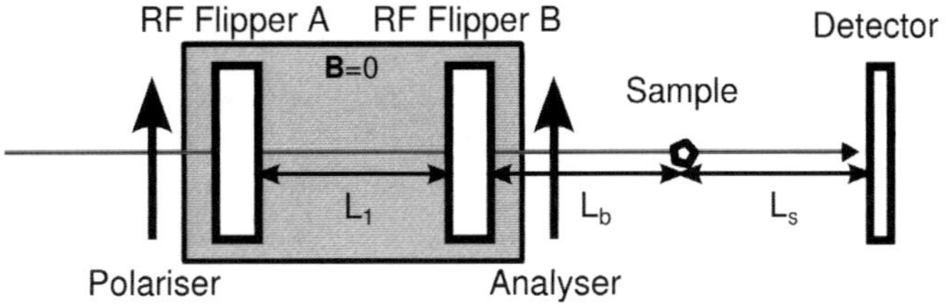

Figure 1. The MIEZE setup consists of a polariser, two flipper coils and an analyser. The distance between the coils is L_1, the distance from the second coil to the sample L_b, the distance from the sample to the detector L_s.

The piecewise equalisation of the corresponding terms leads to the general inelastic MIEZE condition which is used in the software to help the user determine the optimal detector position or – by extension – a correction factor to the coil frequencies prior to the experiment:

$$L_s = \left(\frac{\omega_A \cdot L_1}{\omega_B - \omega_A} - L_b\right) \cdot \frac{\left(v_1^2 + 2\Delta E/m_n\right)^{3/2}}{v_1^3}. \tag{4}$$

The inelastic MIEZE condition, equ. 4, is independent of the bootstrap factor β. For quasi-elastic scattering equ. 4 simplifies to the common MIEZE condition that relates the separation of the coils L_1 and the detector distance $L_2 = L_b + L_s$ to the coil frequencies ω_A and ω_B:

$$L_2 = \frac{\omega_A \cdot L_1}{\omega_B - \omega_A}. \tag{5}$$

After the analyser, the different frequencies of the two coils produce a sinusoidally varying intensity signal with a beating frequency $\omega_M = 2\beta \cdot (\omega_B - \omega_A)$. The so-called MIEZE contrast C is the ratio of the amplitude A to the offset B of the measured signal.

It can be shown [3] that the MIEZE contrast is the Fourier-cosine transform of the scattering function $S(q, \omega)$ of the sample and therefore the MIEZE equivalent of the polarisation P in NSE and NRSE:

$$C = \int d\omega S(q, \omega) \cos(\omega \tau_M). \tag{6}$$

The linewidth Γ of an excitation can thus be determined by measuring the normalised MIEZE contrast C at different spin-echo times τ_M followed by fitting an exponentially decaying function, which is the Fourier-cosine transform of an excitation with a Lorentzian spectral weight function, to the C vs. τ_M data.

1.2. Path Lengths

The MIEZE setup [4] at MIRA [9, 10] uses a Cascade detector system using six GEM foils [11, 12, 13]. Each of the spatially separated foils detects position-sensitive time-of-flight information of the neutron counts.

With its area of $20 \cdot 20\ \text{cm}^2$, the time-of-flight data seen by the foils has to be phase-corrected to prevent a loss of contrast especially at high MIEZE times τ_M. At an offset Δx and Δy away from the center of the detector, the contrast as given by equ. 6 picks up an additional phase $\Delta\phi$ (fig. 2) given by

$$\Delta\phi = \frac{\omega_M}{v_2} \cdot \Delta l = \frac{\omega_M}{v_2}\left(\sqrt{L_s^2 + \Delta x^2 + \Delta y^2} - L_s\right) \tag{7}$$

giving rise to a reduction of the total contrast if ϕ is averaged over the whole detector area.

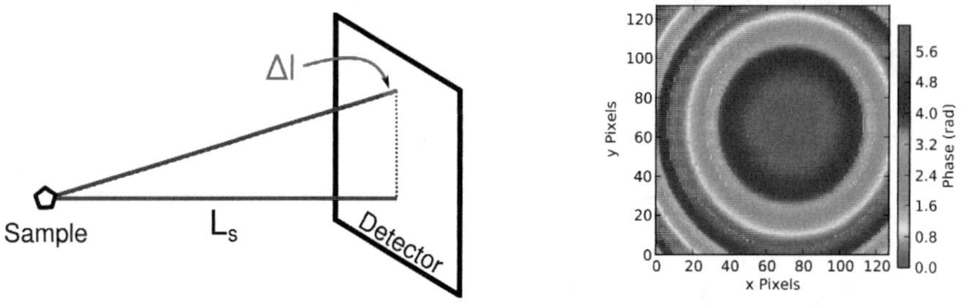

Figure 2. Path length differences on a flat PSD. The additional path length Δl away from the detector centre causes a non-uniform phase-distribution on the PSD area. The right panel shows the theoretically calculated phase distribution which is used for correcting the experimental dataset shown in fig. 4.

Knowing the theoretical phase distribution as given by equ. 7, the phase changes due to the different path lengths can be corrected for by using appropriate software as shown in the next section.

2. Software

The software allows for loading and analysing time-of-flight detector data. Supported data formats include the native data formats used by the Cascade detector, the NICOS[1] data files and the output produced by various McStas [14, 15] components.

The software is written in C++ [16] and Boost[2] for efficiently handling large PSD-TOF datasets. The Qt[3] based graphical front-end employs a multiple document interface (MDI) approach to allow for a simultaneous treatment of multiple detector datasets (fig. 3). The source code for the MIEZE toolkit is available under ref. [17].

2.1. Correction of Path Length Differences

For calculating the total contrast, a correction of the phase-differences encountered on a flat PSD is done using the shift theorem [18] of the Discrete Fourier Transform (DFT):

$$\mathcal{F}\left\{f\left(t+\Delta t\right)\right\} = \underbrace{\mathcal{F}\left\{f\left(t\right)\right\}}_{F(\omega)} \cdot e^{i\omega\Delta t}. \tag{8}$$

Here, $\mathcal{F}$ denotes the DFT. The theorem links a shift Δt in the binning of the dataset in real space to a an additional exponential term in Fourier space. Such a technique is necessary in order to be able to rebin the counts arbitrarily without the constraints imposed by the discrete binning. This operation is done on each pixel of the dataset with the theoretical phase offset as given by equ. 7 used as phase correction factor (see fig. 2, right). For performing the DFT we use the FFTW[4] library.

Additionally, a phase-corrected summation of the signal recorded by each GEM foil of the Cascade detector is performed, yielding a total MIEZE signal. A sine function is fitted to this signal using the Minuit[5] library. In the present step, the phase correction factor for the different

[1] http://trac.frm2.tum.de/projects/NICOS
[2] http://www.boost.org
[3] http://qt.digia.com
[4] http://www.fftw.org
[5] http://seal.web.cern.ch/seal/MathLibs/Minuit2/html/

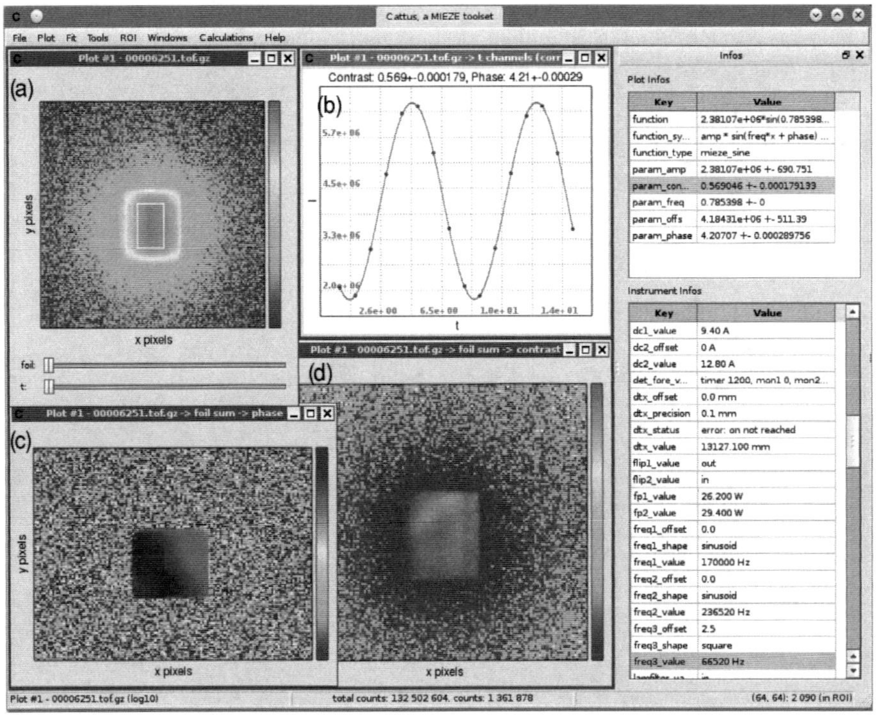

Figure 3. Screenshot of the graphical user interface showing a typical session. Shown in (a) is an example TOF dataset measured at the instrument MIRA, the corresponding phase (c) and contrast (d) images and a total TOF signal inside a selected region of interest (ROI) (b). The software supports loading and analysing of various detector and simulation data formats.

foils can either be determined by an elastic reference measurement or – if statistics are sufficient – by a preliminary fitting step for each foil.

Fig. 4 shows the individual steps of reconstructing the total MIEZE signal, here for an example signal of $\tau_M = 127$ ps ($\omega_M = 200 \cdot 2\pi$ kHz) recorded at MIRA. The left panel shows the raw data for each foil and the corresponding phase distribution on the PSD area. The right panel depicts the area-corrected signal which shows a significantly higher contrast due to the phase-corrected addition of the sinusoidal intensity signals at each pixel of the detector. The total signal using the data of all foils in shown in fig. 5. The higher total counts in each bin of the reconstructed total signal lead to better statistics in the final step of fitting the sinusoidal MIEZE signal.

3. Conclusion

We presented a software system which aids the user in the analysis and correction of measured data in MIEZE experiments. The software is already in use at the MIRA instrument at the FRM II. Measurements show that the implemented contrast correction methods lead to statistically stable results. The advantage of the correction algorithms is their independence of the internal layout of the detector (e.g. the distance of the foils in case of the Cascade detector). Another more elaborate detector-dependent contrast correction technique [19] is currently being adapted and implemented as complement to our general method.

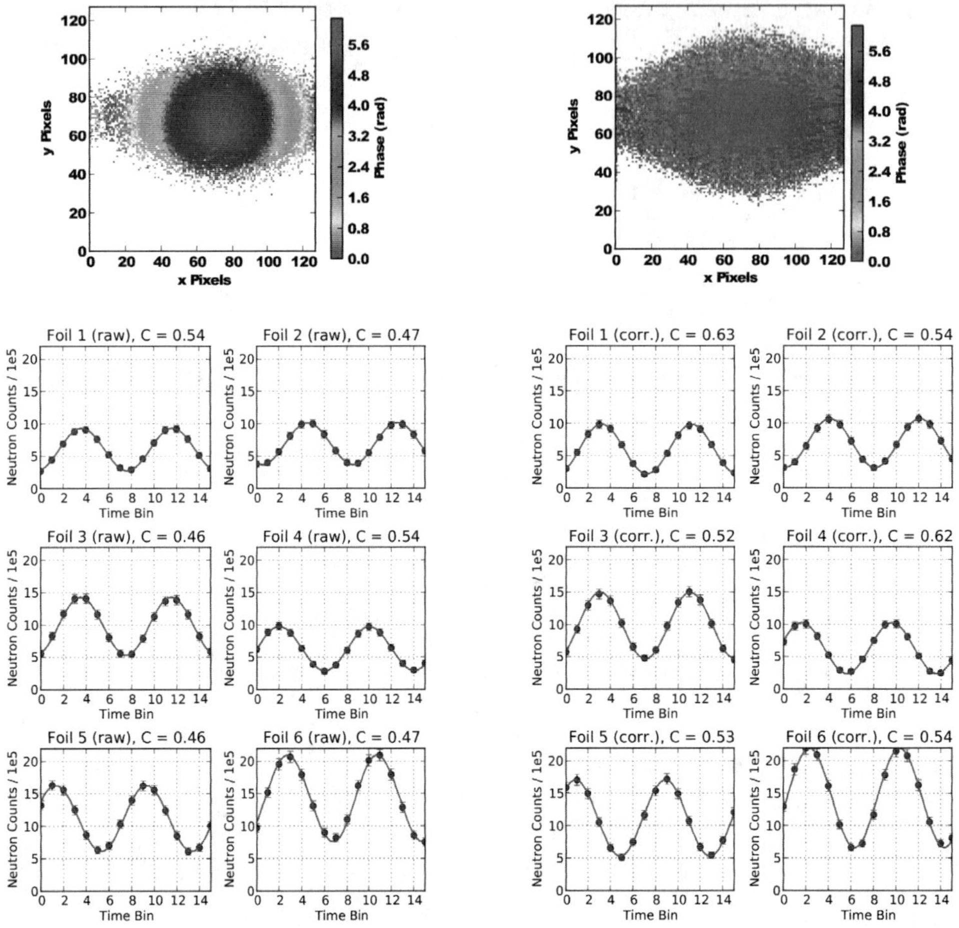

Figure 4. Signal on the individual foils before (left) and after (right) phase correction for an example measurement done with the Cascade detector at MIRA. The phase-corrected signal shows a significantly higher MIEZE contrast C. The top panel shows the phase distribution before (left) and after (right) pixel-wise phase correction.

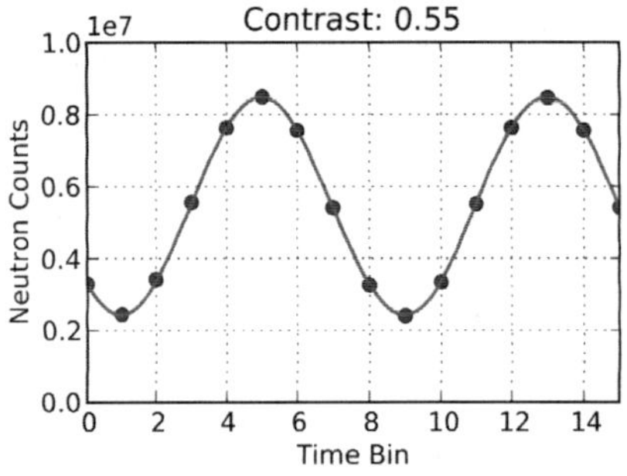

Figure 5. Total MIEZE signal after summation of the data from all six foils (see corrected phase information on the right hand side of fig. 4).

References

[1] Gähler R, Golub R and Keller T 1992 *Physica B: Condensed Matter* **180 - 181, Part 2** 899 – 902
[2] Besenböck W, Gähler R, Hank P, Kahn R, Köppe M, Novion C H D, Petry W and Wuttke J 1998 *Journal for Neutron Research* **7** 65–74
[3] Keller T, Golub R and Gähler R 2002 *Scattering* ed Pike R and Sabatier P (London: Academic Press) pp 1264 – 1286
[4] Georgii R, Brandl G, Arend N, Häußler W, Tischendorf A, Pfleiderer C, Böni P and Lal J 2011 *Applied Physics Letters* **98** 073505 – 073505–3 eid: 073505
[5] Mezei F 1972 *Zeitschrift für Physik A Hadrons and Nuclei* **255** 146–160
[6] Mezei F 1980 *Neutron Spin Echo* (*Lecture Notes in Physics* vol 128) ed Mezei F (Springer Berlin / Heidelberg) pp 1–26
[7] Gähler R and Golub R 1987 *Zeitschrift für Physik B Condensed Matter* **65** 269–273
[8] Golub R and Gähler R 1987 *Physics Letters A* **123** 43 – 48
[9] URL `http://www.mlz-garching.de/mira`
[10] URL `http://www.frm2.tum.de`
[11] Schmidt C J, Groitl F, Klein M, Schmidt U and Häussler W 2010 *Journal of Physics: Conference Series* **251** 012067 – 012067–6
[12] Häussler W, Böni P, Klein M, Schmidt C J, Schmidt U, Groitl F and Kindervater J 2011 *Review of Scientific Instruments* **82** 045101 – 045101–6
[13] Klein M and Schmidt C J 2011 *Nuclear Instruments and Methods in Physics Research Section A: Accelerators, Spectrometers, Detectors and Associated Equipment* **628** 9 – 18
[14] Willendrup P, Farhi E, Knudsen E, Filges U and Lefmann K 2014 *Journal of Neutron Research* **Volume 17**(1) 35–43
[15] URL `http://www.mcstas.org`
[16] Stroustrup B 1997 *The C++ Programming Language, Third Edition* 3rd ed (Boston, MA, USA: Addison-Wesley Longman Publishing Co., Inc.)
[17] URL `http://trac.frm2.tum.de/cgit/cgit.cgi/mira/miezetools.git`
[18] Smith J O 2007 *Mathematics of the Discrete Fourier Transform (DFT)* (W3K Publishing)
[19] Köhli M *Ph.D. thesis in preparation* (Universität Heidelberg)

New developments in the McStas neutron instrument simulation package

PK Willendrup[1,7], EB Knudsen[1], E Klinkby[2,7], T Nielsen[3,6,7], E Farhi[4], U Filges[5,8], K Lefmann[6,7]

[1]NEXMAP, Physics Department, Technical University of Denmark, Denmark
[2]Center for Nuclear Technologies, Technical University of Denmark, Denmark
[3]ESS-DMSC, Copenhagen, Denmark
[4]Institut Laue-Langevin (ILL), France
[5]Paul Scherrer Institute (PSI), Switzerland
[6]Niels Bohr Institute (NBI), University of Copenhagen, Denmark
[7]Danish in-kind contribution toward the ESS
[8]Swiss in-kind contribution toward the ESS

E-mail: `pkwi@fysik.dtu.dk`

Abstract.
The McStas neutron ray-tracing software package is a versatile tool for building accurate simulators of neutron scattering instruments at reactors, short- and long-pulsed spallation sources such as the European Spallation Source. McStas is extensively used for design and optimization of instruments, virtual experiments, data analysis and user training. McStas was founded as a scientific, open-source collaborative code in 1997.

This contribution presents the project at its current state and gives an overview of the main new developments in McStas 2.0 (December 2012) and McStas 2.1 (expected fall 2013), including many new components, component parameter uniformisation, partial loss of backward compatibility, updated source brilliance descriptions, developments toward new tools and user interfaces, web interfaces and a new method for estimating beam losses and background from neutron optics.

1. Introduction

Since wide-spread development of neutron Monte Carlo ray-tracing techniques began in the late 1990s, the technique has supplemented analytic approaches when designing neutron scattering instruments. Before the arrival of todays easy-to-use codes, the instrument scientist would often develop the simulation software from scratch, thereby effectively duplicating existing work. In Monte Carlo ray-tracing, neutrons are transported through space and matter via an efficient and powerful combination of techniques: When complicated physical processes with a stochastic nature occur, Monte Carlo sampling from e.g. source spectra is used, but when deterministic propagation through space is sufficient, ray-tracing on parabolic paths in the gravitational field is applied.

At the RISØ national lab in Denmark, the RITA[1] instrument project was developing during the mid-to-late 1990s. In 1995, the NISP[2] package from Los Alamos National Laboratory was the first and only generally available Monte Carlo tool for neutron instrument design. Initially, NISP was considered for the instrument design of RITA, but was not found optimal for this project, hence an other solution was found. The RISØ department head at that time, Kurt Clausen, decided to seek funding for what would eventually become the McStas[3]-[5] project. From the beginning the idea was to create an open-source collaboration platform where instrument scientists could build on simulation work previously done by others. A collaboration was started between computer scientist Kristian Nielsen and Kim Lefmann, Henrik Rønnow and several other physicists at the RISØ Physics Department during 1997, and the public McStas 1.0 version was released in 1998. In 1997 and 2000 two other simulation codes, RESTRAX[6] and VITESS[7] became available, later joined by the IDEAS[8] code from Oak Ridge National Laboratory. In the years that followed, several collaborative comparison projects proved the value of having a suite of simulation packages available, many bugs were found and corrected, and a situation of friendly competition provided inspiration for new features for all of the packages. Robert McGreevy who was at the time leading the SCANS work package on simulation was very actively supporting the instrument simulation packages in general, but specifically McStas due to its collaborative nature, as described below.

Due to needs for its modernisation projects, the Institut Laue-Langevin (ILL) became a very important McStas partner site, supporting core development of the packages. Later the efforts were also joined by the University of Copenhagen and the Paul Scherrer Institute (PSI).

One significant difference between McStas and the other projects always was its open-source, collaborative nature designed to use component codes written by the individual instrument scientist, later to be shared via the McStas component library. In McStas these small and medium sized component-codes implement the individual parts of a beamline. McStas users need not write components themselves, but always have the option to do so when a component is missing or needs improvement.

For the user operating McStas only knowledge of a simple instrument file containing a logical and easy to read meta-language is needed. The McStas software suite then translates the instrument file into a computation-effective and portable c-code using powerful code-generation and compiler techniques.

One other area where McStas historically had a strong point in regard to the other codes, was the availability of *sample* components, allowing the user to perform a Virtual Experiment[9], where one combines sample and instrument response to give a close-to-physical virtual dataset as output.

Today McStas has an impressive suite of components in all categories, counting 148 pieces of code, out of which around 40% are user contributions. Further, 100 example instruments gives the new user rich inspiration for her or his own projects.

2. Structure and use of the code

McStas was designed with flexibility and modularity in mind, and can be thought of as a framework for neutron instrument simulation.

Structurally, McStas consists of several *layers* and types of code, as illustrated in Figure 1, where the flow of defining and running a simulation should be read from top to bottom. In the figure, solid boxes contain elements of the code that are typically manipulated by the user, and dashed boxes contain elements of the system parts of McStas.

- The user defines an instrument using the mcgui program on the "tool layer", thereby assembling a so-called instrument file containing a logical and easy to read meta-language, positioning components of the beamline in a coordinate frame. The lab- or instrument coordinate system is right-handed, with y defining the vertical axis and z the direction of the neutron beam.
- Each of the components correspond to a McStas component file in the component library, written in a type of structured C-code. A user does not need to write component files, but

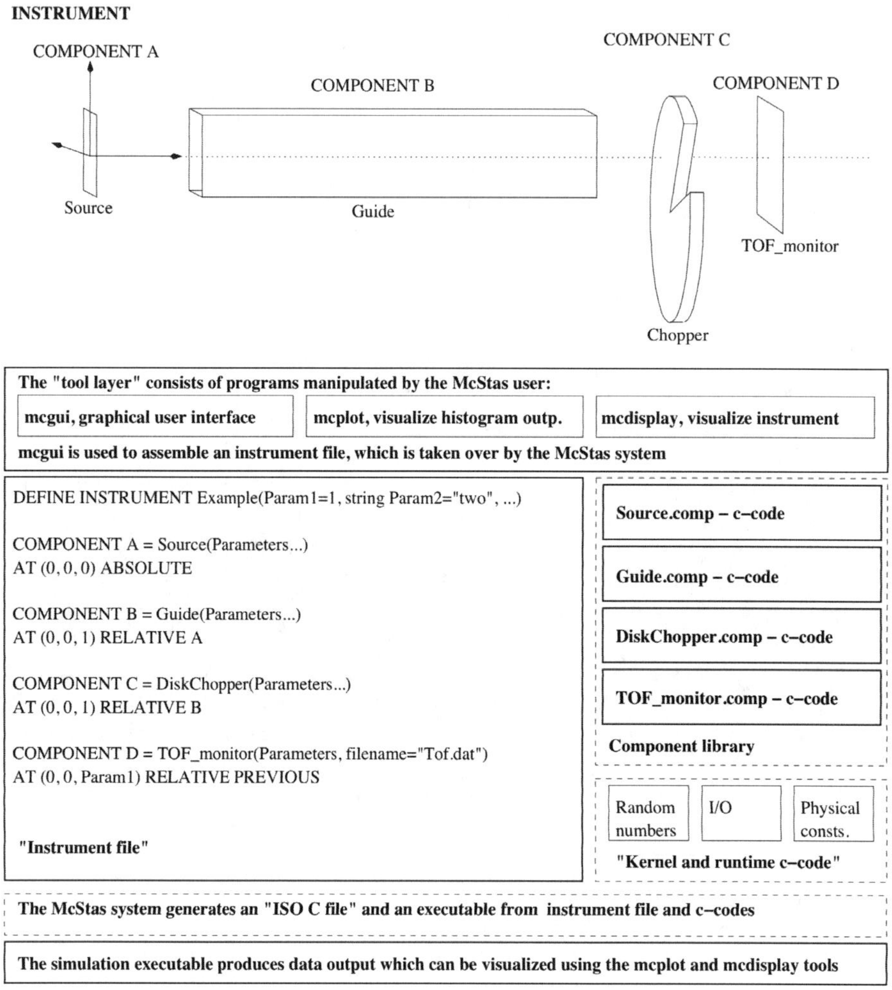

Figure 1. Illustration of a simulation workflow in McStas, to be read from top to bottom. Solid boxes contain elements of the code that are typically manipulated by the user, and dashed boxes contain elements of the system parts of McStas. See the text for details.

has the option to do so. As explained in the introduction, components are often contributed to the development team from the world-wide McStas user community. The components each have their local coordinate frame in which the neutron should be propagated between an *incoming* and *outging* state, e.g. before and after a scattering process. The McStas system takes care of transformation of the neutron state from one component coordinate system to the next.

- From the instrument code, the component codes and special kernel/runtime code containing e.g. a random number generator, propagation routines and physical constants, a computation-effective and portable c-code is generated by the McStas system using a powerful code-generation and technique. Further, a c-compiler is called, generating a binary executable, typically referred to as the *simulation*. These two steps happen transparently and without user interaction.
- Following the compilation process, the simulation can be run to produce output data, in the form of event datafiles, histograms or an interactive visualization of the instrument geometry. The tools layer contains the tools mcplot and mcdisplay, for histograms and instrument geometry data respectively.

3. User contributions - and how we handle them

As mentioned before, many of the components and instruments provided with McStas are contributed from outside the core development team, which we generally feel is an important strength and difference of McStas with respect to the competing codes. To underline the Open Source nature of McStas, we have chosen to distribute the code with contributions under the Gnu Public License [10], version 2.0.

We always welcome contributions and have a relatively informal refereeing procedure: A contributed component requires a contributed test-instrument that demonstrates its use. The in-instrument and in-component documentation must be meaningful, and the code must be relatively readable. If the receiving McStas team member finds that the quality of the code does not live up to the general standard of the McStas code, there is typically an iterative process of mutual communication and improvement of the code.

A contributed component will always start its McStas life within the *contrib* directory of the component library, to signify that it was not written by a McStas core team member.

The McStas team takes care of supporting user requests for both the system- and contributed components, but a contributed component will always contain the name and contact information of the author, who is considered the final responsible of the code.

When a contributor is no longer available to support his code, the McStas team will decide to either retire the component or to promote the component to a fully supported system component, based on estimates of the usability, performance and quality of the component.

4. A quick tour of the example suite

The McStas example instruments are a valuable resource for new users. We currently have 100 example instruments in different categories, out of which a few have been listed

- Large scale structure instruments
  - KWS-II from Jülich/FRM-II, available in several versions with different sample types included.
    SANS-II from PSI - see figure 2
  - Crisp reflectometer from ISIS TS1, including a very nice multilayer-sample (requires GNU Scientific Library)
- Diffractometers

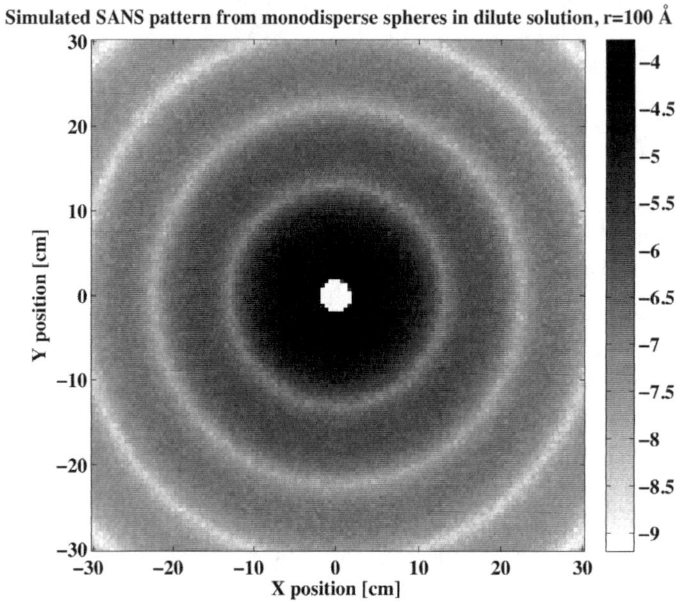

Figure 2. Virtual SANS data, showing the a pattern from monodisperse spheres in dilute solution on a position sensitive detector, logarithmic colorscale. Unpublished simulation data. Instrument file templateSANS.instr by K. Lefmann.

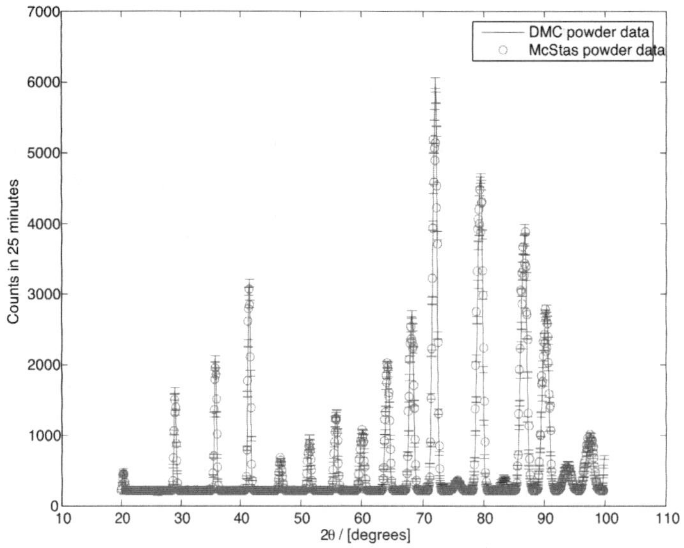

Figure 3. The figure shows a comparison of virtual and measuered diffraction $Na_2Ca_3Al_2F_{14}$ data[11], taken using the DMC powder diffractomer instrument at PSI. The McStas instrument file PSI_DMC.instr by P. Willendrup, U. Filges, L. Keller is included in the McStas distribution.

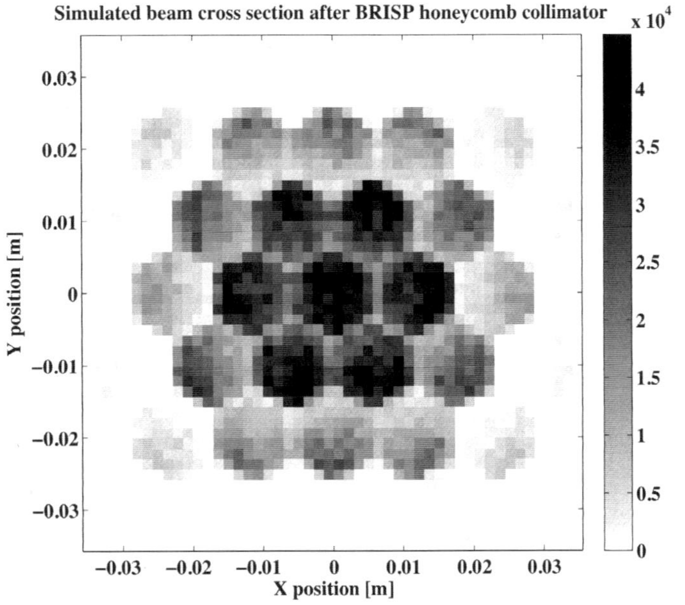

Figure 4. The figure shows a simulation of the beam cross section after the honeycomb-shaped guide at the ILL BRISP spectrometer. Unpublished simulation data. The used instrument file ILL_BRISP.instr by E. Farhi and N. Formissano is included in the McStas distribution.

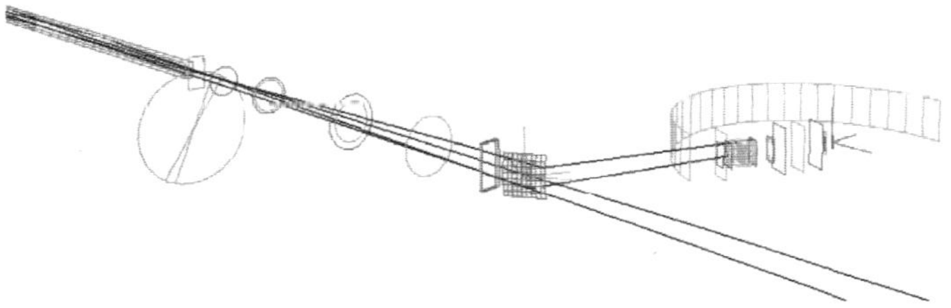

Figure 5. 3D-visualization of the PSI Focus instrument, using the mcdisplay tool. In the left part of the image the final guide-tube a disk-chopper and slits are shown, in the middle the double-focusing monochromator is visible, in the right-most part Fermi-chopper, sample position and detectors are visible. Unpublished simulation data. The used instrument file PSI_FOCUS.instr by U. Filges. See also [14] is included in the McStas distribution.

  - D1A from ILL, a high-resolution powder diffractometer
  - D20 from ILL, a high-intensity powder diffractometer
  - DMC from PSI - see Figure 3
  - Generic Laue diffractometer
- Triple axis instruments
  - IN12 from ILL, a cold triple axis
  - IN22 from ILL, a thermal triple axis on a guide (CRG)
  - IN20 from ILL, a thermal triple axis on a beamtube

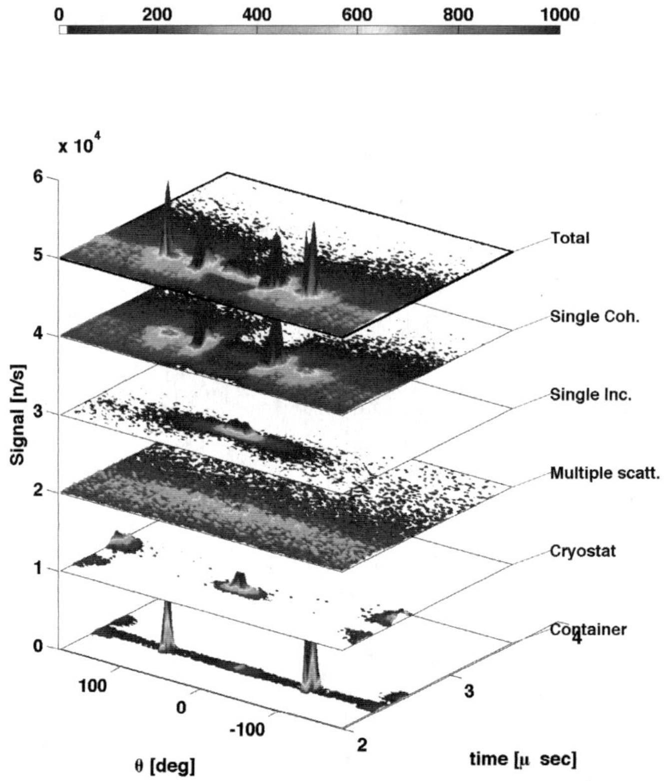

Figure 6. Virtual experiment on liquid Ge, see [15]. Using labeling of individual neutron scattering processes, the full detector signal is decomposed into parts: Coherent, incoherent and multiple scattering contributions from the sample, contaminations from sample container and surrounding sample environment (cryostat). The used instrument file template_TOF.instr by E. Farhi is included in the McStas distribution.

  - h8 from Brookhaven, Brockhouse's original TAS
  - TAS1 from Risø, decomissioned
- TOF inelastic
  - IN5 from ILL, a cold TOF spectrometer
  - IN6 from ILL, a cold TOF spectrometer
  - BRISP from ILL - see Figure 4, small-angle spectrometer (CRG)
  - HET from ISIS TS1, decomissioned
  - Focus from PSI - see Figure 5
  - IN5-type from ESS
  - Generic direct geometry TOF - see Figure 6
- Hybrid inelastic
  - Prisma2 from ISIS TS1, decomissioned
- Spin echo, a simple spin-echo instrument based on transmission pl
  - Generic Spin-Echo instrument - see Figure 7
- Imaging

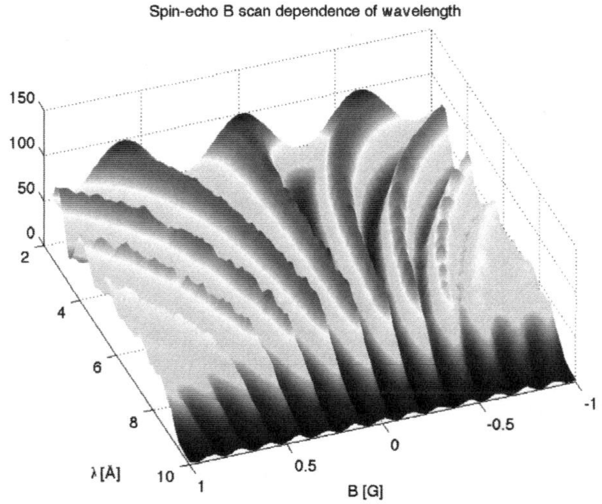

Figure 7. A series of concatenated, virtual spin-echo scans. As a function of varied wavelength, spin-echo groups of varying width are revealed when scanning a simulated correction coil. The instrument file used is SE_example.instr by E. Knudsen and P. Willendrup, is included in the McStas distrubtion. See also [16] and [17].

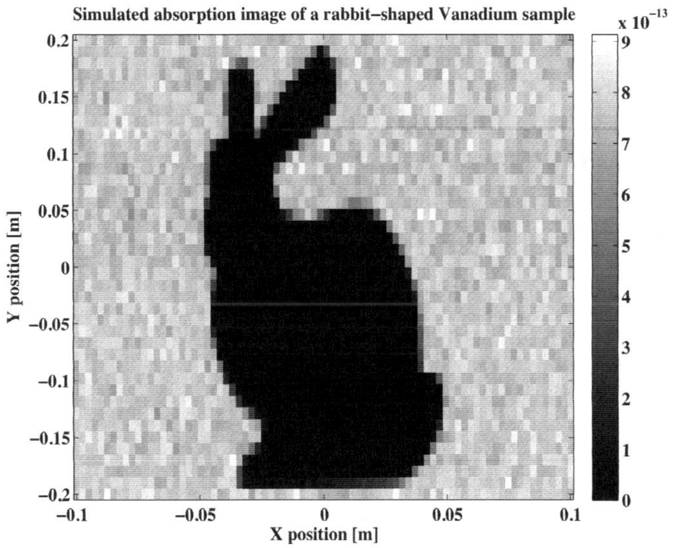

Figure 8. Absorption image / single projection in the tomography example. A rabbit-shaped Vanadium sample is revealed. Unpublished simulation data. The used instrument file Tomography.instr by P. Willendrup and R. Arnerin is included in the McStas distribution.

– Generic tomography station - see Figure 8

Supplementing these, a number of test instruments showing usage of individual components are included, as well as the very special "conversion tool" histogrammer.instr. Built on the general ability to read and use event-datafiles from McStas, VITESS, MCNP/MCNPX[12]-[13] and Tripoli, the histogrammer serves as an illustration of the flexibility of the McStas framework: One, simple, general instrument file provides a generic tool to generate *any* histogram-plot from any type of supported event data. And compiled once and for all, defines a stand-alone, alternative plotting tool e.g. for use with Vitess.

5. McStas releases 2.0 and 2.1

The latest version of McStas was release 2.0, dating from december 21st 2012. Our reason to increment the major version number related to certain loss of backward compatibility. We did however sacrifice the compatibility for at good cause: A nomenclature for parameter naming was defined, allowing much easier exchange of components between e.g. guides of different type, essentially only the component type need to be changed.

The most important general changes in the 2.0 release were a newly defined, uniformized parameter naming across components (which does break some backward compatibility for user-instruments) and a more robust and complete implementation of polarised neutron transport, including transport in user-defined tabularised magnetic fields e.g. from Radia[18] or freefem[19]).

From the user perspective, the addition of a new tool layer based on Python (which will eventually replace Perl) plus interfaces to Mantid[20], MCNP[21] and iFit[22] are important additions. With release 2.0, we have further started an effort together with the Vitess team of documenting how to define source brilliances for seleceted neutron sources with the two packages respectively. Currently, benchmarked parameters have been defined for the ILL, HZB, FRM-II, ISIS and ESS sources.

An important new feature of the sample components and some optics components is the support for abstract geometrical outer shapes, facilitated via the Geomview OFF file format[23].

A short-list of the most important new components added are

- Sample_nxs.comp from Mirko Boin, Helmholtz Zentrum Berlin. Diffraction/imaging oriented sample where one defines the structure by means of the NXS crystallography library[24] which is now also included with McStas. The component comes with the test-instruments Test_Sample_nxs_diffraction.instr and Test_Sample_nxs_imaging.instr.
- Mirror_Curved_Bispectral.comp and Mirror_Elliptic_Bispectral.comp from Henrik Jacobsen, NBI. Mirrors for the bispectral ESS instrument ESS_2001_bispectral.instr. See [25].
- Virtual_mcnp_ss_input.comp, Virtual_mcnp_ss_output.comp and Virtual_mcnp_ss_Guide.comp from Esben Klinkby, DTU Nutech. Allows to read and write MCNP/MCNPX 'source surfaces', for use in simulations where neutrons need transport in both codes, see [26].
- A large suite of SANS-samples from Martin Cramer Pedersen, NBI Various approaches to describing SANS diffraction e.g. using the Protein Data Bank (PDB)[27]. The TestSANS.instr instrument is provided for use cases.

For a full set of release notes for the McStas 2.0 release, please consult the McStas website [5].

The next release McStas 2.1 is expected early 2014, depending on development progress. The main remaining parts are a documentation overhaul and improvements to the new Python-based tool layer, however for many types of usage, the development version available via our Subversion source code repository http://www.mcstas.org/svn is in practice functional. We hereby recommend any experienced user to upgrade to the development version at will.

Users experiencing problems with either the development version or the actual McStas 2.1 release are encouraged to inform the McStas developers via the user mailinglist mcstas-users@mcstas.org.

6. Summary

McStas has during its 16 years long history of development proven to be a healthy and active project. What seperates McStas form its competitors is mainly the applied code-generation technique, the quality of documentation and the strongly collaborative nature of the package, allowing users to contribute.

The package includes an extensive example library consisting of instruments of all types, readily available as inspiration for the user.

The latest release is McStas 2.0 released in december 2012 and McStas 2.1 is planned for early 2014.

Acknowledgements

During its lifetime, McStas has received funding primarily from the sources listed below.

1996-2000	XENNI RTD (EU FP4)
1998-2001	Cool neutrons RTD (EU FP4)
2000-2004	SCANS RTD (EU FP5)
2004-2008	MCNSI JRA, NMI3 (EU FP6)
2006-2009	ISIS TS2 project (EU FP7)
2008-2010	MCNSI7 NMI3/FP7 (EU FP7)
2012-	VNT work package NMI3/FP7-II (EU FP7)
2009-2013	ESS update phase, Danish in-kind contribution

Key individuals

We would like to express a special gratitude to Kurt Clausen for his original idea to start the project and to Robert McGreevy for his extensive support during the various EU programmes as well as the TS2 project. We would also like to thank Helmut Schober and Mark Johnson for their support during NMI3/FP7 and NMI3/FP7-II.

Past team members

We would like to thank Kristian Nielsen for his original design and computer science work for McStas, which has proved to be extremely stable and flexible. We would also like to thank our other former team members Per-Olof Åstrand, Klaus Lieutenant and Peter Christiansen.

User contributions

As written elsewhere, the success of the McStas code is to a large extent due to contributions from users. The following list of people has been extracted directly from our component and instrument code:

Abrahamsen A., Ansell S., Arleth L., Arnerin R., Bernhardt P., Boin M., Campioni G., Carbogno C., Carlsen H., Castan Guerrero C., Champion D., Christensen N., Christiansen P., Dalgliesh R., Daoud-aladine A., Desert S., Euzen G., Farhi E., Filges U., Formissano N., Fortescue B., Frielinghaus H., Giller L., Gallmeier F.X., Granroth G., Hagen M., Hansen T., Hennane C., Hugouvieux V., Jacobsen H., Keller L., Klausen S., Klinkby E., Klenø K., Knudsen E., Lauridsen E., Lefmann K., Lieutentant K., Link P., Mezei F., Monzat C., Nielsen K., Okkels Birk J., Ollivier J., Ostermann A., Panzner T., Pedersen M.C.,, Percival A., Poehlmann M.,

Roberts T., Roennow H., Roth S., Rycroft S., Sales M., Schedler R., Schober H., Stewart R., Udby L., Venturi G., Willendrup P., Zsigmond G., Vickery A., van Vuure T.

Since the list is likely not complete, we would also like to thank our large user base for all of their input, ideas and problems to fix.

References

[1] TE Mason, KN Clausen, G Aeppli et al. Canadian Journal of Physics **73**/11-12 pp 697-702 (1995).
[2] PA Seeger et al. Argonne National Laboratory Report ANL **98**/33, pp 202-218, (1998).
[3] K. Lefmann and K. Nielsen , Neutron News **10**, 20, (1999).
[4] P Willendrup, E Farhi, EB Knudsen, U Filges and K Lefmann, Journal of Neutron Research **17**, 35-43 (2014)
[5] McStas project website http://www.mcstas.org/
[6] J Saroun and J Kulda, Physica B **234**, pp. 1102, (1997).
[7] D Wechsler, G Zsigmond, F Streffer and F Mezei, Neutron News **11**, p 25, (2000).
[8] WT Lee and XL Wang, Neutron News **13** pp. 30-34, (2002)
[9] K Lefmann et al., Journal of Neutron Research **16**, pp. 97-111 (2008)
[10] GNU website http://www.gnu.org/licenses/gpl.html
[11] P Willendrup, U Filges, L Keller, E Farhi, K Lefmann , Physica B-Cond Matt **385** p 1032 (2006)
[12] X-5 Monte Carlo Team, Los Alamos report 5LA-UR-03-1987
[13] LS Waters, GW McKinney, JW Durkee, ML Fensin, J. S. Hendricks, et al., AIP Conf.Proc. **896** (2007)
[14] See http://lns00.psi.ch/mcworkshop/papers/focus_intercomparison.pdf
[15] E Farhi, V Hugouvieux, M Johnson and W Kob, Journal of Computational Physics **228**, pp. 5251-5261 (2009)
[16] EB Knudsen, L Udby, P Willendrup, K Lefmann and WG Bouwman, Physica B-Cond Matt **406** p 2361 (2011)
[17] EB Knudsen, A Tranum-Rømer, P Christiansen, P Willendrup, Lefmann K, Journal of Neutron Research **17**, 27-34 (2014)
[18] Radia website http://www.esrf.eu/Accelerators/Groups/InsertionDevices/Software/Radia
[19] Freefem website http://www.freefem.org/
[20] Mantid website http://www.mantidproject.org
[21] MCNPX website at LANL http://mcnpx.lanl.gov/
[22] iFit website http://ifit.mccode.org
[23] Geomview website http://www.geomview.org/docs/html/OFF.html
[24] M Boin Journal of Applied Crystallography **45** 603-607 (2012)
[25] H Jacobsen, K Lieutenant, C Zendler, K Lefmann NIMA **717** 69-76 (2013)
[26] E Klinkby et al. NIMA **700** 106-110 (2013)
[27] PDB website http://www.rcsb.org/pdb/home/home.do

International Workshop on Neutron Optics and Detectors (NOP&D 2013) IOP Publishing
Journal of Physics: Conference Series **528** (2014) 012036 doi:10.1088/1742-6596/528/1/012036

VITESS 3 - Virtual Instrumentation Tool for the European Spallation Source

C Zendler[1,2], K Lieutenant[1,2], D Nekrassov[1,2] and M Fromme[1]

[1] Helmholtz-Zentrum Berlin für Materialien und Energie GmbH, Hahn-Meitner-Platz 1, 14109 Berlin, Germany

[2] ESS design update programme, Germany

E-mail: carolin.zendler@helmholtz-berlin.de

Abstract. VITESS is a software widely used for simulation of neutron scattering experiments. Although originally motivated by instrument design for the European Spallation Source, all major neutron sources are available. Existing as well as future instruments on reactor or spallation sources can be designed and optimized, or simulated in a virtual experiment to prepare a measurement, including basic data evaluation. This note gives an overview of the VITESS software concept and usage. New developments are presented, including a 3D visualization of instruments and neutron trajectories, a numerical optimization routine and a parallelization tool allowing to split VITESS simulations on a computer cluster.

1. Introduction

Simulation of neutron scattering instruments is important for the development of new beamlines or update of existing instruments by enabling tests of new ideas and optimization of instrument components prior to construction. It can further be useful to conduct a virtual experiment to prepare a planned measurement, analyzing data obtained by a simulation of the instrument in the foreseen configuration including an adequate description of the sample and detector output. Therefore several software programs have been developed to perform such simulations, many of which are dedicated to a certain type of instruments. The Virtual Instrumentation Tool for the European Spallation Source (VITESS) [1, 2] is one of the most widely used of these programs due to its general applicability and easy handling. Despite its name, all major reactor and spallation neutron sources are provided. A detailed implementation of processes involving polarized neutrons [3] is a particular strength of VITESS compared to most other programs, the most common of which are McStas [4] (which recently added a comparable treatment of polarized neutrons [5]), RESTRAX [6], NISP [7] and IDEAS [8]. VITESS is continuously validated by comparison with simulations from aforementioned other programs as well as against data from real experiments [9–12].

VITESS is an open source software available on the VITESS website [13] for Windows, Linux und Macintosh. It is mainly developed at the Helmholtz-Zentrum Berlin (formerly Hahn-Meitner-Institut) in Germany but also includes contributions made by advanced users from other facilities.

In the following, the concepts upon which the VITESS simulation package is build are described in section 2, before the latest developments regarding physical models and software features are introduced in sections 3 and 4.

2. Software concept and program layout

VITESS uses a Monte Carlo ray-tracing technique in which neutrons are modeled as trajectories travelling from a source through an instrument onto a sample in which they are scattered and thereafter measured in a detector. They are assigned with a weight which corresponds to the intensity of the particular phase space element and makes sure that the flux normalisation throughout the instrument is always correct independent of the number of generated trajectories. This weight is modified at each interaction with an instrument componenent. Each component is represented by a module coming as a stand-alone executable in the software package. These modules run independently of each other, with each single module reading in neutrons from the previous one and passing them on to the next one. A whole instrument is simulated by running several modules sequentially in a pipe. This is done in packages of 10000 neutrons such that in a simulation with larger statistics as usual, all modules can run simultaneously. The properties of generated neutron trajectories can be examined by monitors at any point in the instrument, or written to a file to be read in again later by a new simulation continuing the instrument.

VITESS contains a graphical user interface (GUI) from which an instrument can easily be build by assembling the needed modules and which allows to perform complex simulations without knowledge of any scripting language. A VITESS simulation can however also be run either directly from the command line or via a shell, python, perl or tcl script to which an instrument description can be exported from the GUI. This can also be used to run VITESS on a computer cluster, and since release 3.1, a framework is provided to split the simulation on a cluster into several instances to save simulation time.

Different kind of neutrons can be studied separately by ray-tracing, which is done by selecting the neutrons of interest in a first run of the simulation and save them into a file, followed by a second simulation in which only those neutrons are processed. Monitor outputs like phase space diagrams or wavelength spectra are written as ASCII files which can be viewed directly from the GUI or analyzed with a preferred external software. A visualization of the neutron paths through the instrument is possible since release 3.0 and is described in more detail in section 4.1.

The effect of gravity is included in all components apart from the bender module, but can be switched off if the user sees this fit for a particular purpose.

VITESS provides tools to assist the users in performing a simulation by e.g. creating input files or calculating chopper phases. Documentation of all tools and modules as well as of the most important features can be accessed by a help button directly from the GUI, while short descriptions of each parameter of a module can be viewed by clicking on the parameter name. Running a simulation creates a log file in which the pipe command, the module parameters as well as the flux at the end of each module are documented for error diagnostics or later reference.

3. Improved modelling of physics processes

In VITESS 3, the description of physics processes or resulting behavior like emitted source spectra or reflectivity curves has been improved in several modules. This section summarizes the software development connected to a more realistic modelling, while improvements in the program's usage are described in section 4.

3.1. Neutron sources

The description of several sources has been updated, including e.g. new source characteristics of the FRMII reactor in Munich, Germany, and a new source module describing the CSNS spallation source in Dongguan, China. For the spallation neutron sources SNS and ESS, a choice between different moderator descriptions is now available.

3.2. Reflectometer sample

The reflectometer sample module was updated to include both the incoherent and offspecular scattering. The incoherent scattering probability p is calculated taking into account the inverse of the mean free path in the sample material μ provided by the user and the randomly chosen penetration depth of the trajectory in the corresponding sample x: $p = 1 - e^{-\mu x}$. The offspecular reflectivity mode can be used by providing a lookup table, which contains the reflectivity $R(k_{z,i}, k_{z,f})$ as a function of the incoming wave vector $k_{z,i} = 2\pi/\lambda \cdot \sin\theta_i$ and the outgoing wave vector $k_{z,f} = 2\pi/\lambda \cdot \sin\theta_f$.

3.3. Supermirror reflectivity and attenuation models

In VITESS 3, improved models describing the reflectivity of neutrons at supermirror coated plates as well as in case of no reflection their attenuation in the substrate material are implemented.

Before VITESS 3, the reflectivity of supermirrors was modelled as a linear decrease from R_0 at the critical angle θ_C corresponding to a mirror coating $m=1$ to a reflectivity value of R_m at $\theta_m = m\theta_C$, followed by an instantaneous drop to zero for $\theta > \theta_m$ in the `sm_ensemble` module, which describes an arbitrary configuration of flat supermirror plates. Parameters R_m, θ_C and θ_m had to be given by the user. Reflectivity files used in the modules `guide` and `bender` allowed for a slower decrease broadened by a factor $W \neq 0$:

$$R = \frac{1}{2} R_0 \left(1 - \tanh\left(\frac{Q - mQ_C}{W}\right)\right)\left(1 - a\left(Q - Q_C\right)\right) \,, \tag{1}$$

with $a = \frac{R_m - R_0}{mQ_C - Q_C}$ and $Q_C = 4\pi/\lambda \cdot \sin\theta_C$. The linear reflectivity decrease is however a simplification of the behavior seen in measurements of reflectivity curves of supermirror plates, therefore this model has now been extended by an additional term:

$$R = \frac{1}{2} R_0 \left(1 - \tanh\left(\frac{Q - m_2 Q_C}{W}\right)\right)\left(1 - a\left(Q - Q_C\right) + b\left(Q - Q_C\right)^2\right)\right) \,, \tag{2}$$

where W, a, b and m_2 are linear functions of the mirror coating m[1], which is now the only input parameter required. Parameters of the new models have been found by fits to experimental data performed by H. Jacobsen for McStas, and implemented in VITESS within comparison of simulation results in [12].

In the tool `Generate Mirror Files`, this new model is used in VITESS version 3.0 or newer if *quadratic Swiss Neutronics description* is chosen. In the `sm_ensemble` module it is default from VITESS version 3.1 onwards, in which R_0 is also set to $R_0 = 0.99$ instead of formerly $R_0 = 1$. The reflectivity curves obtained with equation 2 for 1 Å neutrons are shown in figure 1(a) for m-values between 2 and 6. Table 1 summarizes which model is used in different VITESS versions.

R formula	VITESS version		
	<3.0	3.0	3.1
`Generate Mirror Files`	1	1, 2	1, 2
`sm_ensemble`	1	1	2

Table 1. Overview of reflectivity models implemented in different VITESS versions.

[1] $W = -0.0002 \cdot m + 0.0022$; $a = 0.1204 \cdot m + 5.0944$ if $m > 3$, else $a = m$; $b = -7.6251 \cdot m + 68.1137$ if $m > 3$, else $b = 0$; $m_2 = 0.9853 \cdot m + 0.1978$

In the `sm_ensemble` module, neutrons that are not reflected can be transmitted with probability

$$p = e^{-(\mu\lambda + \mu_{ic})d'}\ , \tag{3}$$

where μ (μ_{ic}) is the wavelength dependent (independent) part of the macroscopic attenuation and d' the pathlength of the neutron trajectory in the mirror, taking refraction into account. An improved description of the attenuation in sapphire is obtained by using

$$p_{Al_2O_3} = e^{-(\mu\lambda + \mu_{ic}/\lambda - c)d'} \tag{4}$$

instead of (3), where the constant c was again found by fitting this model to experimental data in contex of the McStas study [12]. Since the aforementioned study uses a mirror on sapphire substrate, an additional fit of the attenuation for silicon to data extracted from [14] has been performed for VITESS. The obtained attenuation is shown in figure 1(b) for three different wavelengths.

The old description of equation 3 can still be used by choosing neither silicon nor sapphire as mirror substrate (option *"other"*).

3.4. Neutron detection

3.4.1. Detector geometry and output file The upgraded `detector` module implemented in VITESS 3.1 allows for more flexible geometries: instead of a 2D flat detector surface, a 3D volume detector can be build including several detection layers and a possibly non-perfect resolution in all three dimension. The overall shape of a detector module can either be "flat" (now meaning cubic) or cylindrical. A flat detector can now also be simulated with a substructure of detector tubes. The tube length is equal to either the detector height or width, and a tube geometry can be chosen with either circular or rectangular tube cross-sections. The different amount of detection material depending on the neutron position with respect to the center of a circular tube is taken into account. Space between circular tubes as well as tube walls are treated as vacuum, i.e. no detection is possible in these regions but also no unwanted scattering is simulated[2]. Layers can be shifted with respect to each other by half the tube diameter in order to minimize dead detector regions, as illustrated in the central picture of figure 2(a). The tube option is only available for an overall flat geometry.

In an overall cylindrical geometry, the cylinder axis can now be oriented along all three coordinate axes. Furthermore, the new `const. phi` option allows the pixel size in the dimension along the cylinder axis to be variable such that a constant angular width is covered by each pixel.

Several detector modules can be combined in an `array` to build any complex shape. Figure 2(b) shows an example consisting of two cylindrical and two flat detector modules.

An event mode detector output is written if a file name is given as `Output filename`. It contains a list of detected neutrons with their respective detected position and time as well as neutron weight and color.

3.4.2. Detection efficiency and resolution The efficiency of neutron detection in the `detector` module of VITESS version 3.1 can be calculated from the total cross-section σ of neutron scattering with the chosen active detector material, a linear approximation of which has been extracted from cross-section tables from the National Nuclear Data Center [15] for ^{3}He, ^{10}B and ^{6}Li. For BF_3 or ^{3}He gas, the gas pressure and temperature have to be given as input parameters

[2]Scattering from e.g. detector housing is not implemented in the `detector` module. However, background from a material layer between sample and detection material can be modelled by using the `sample_environment` module.

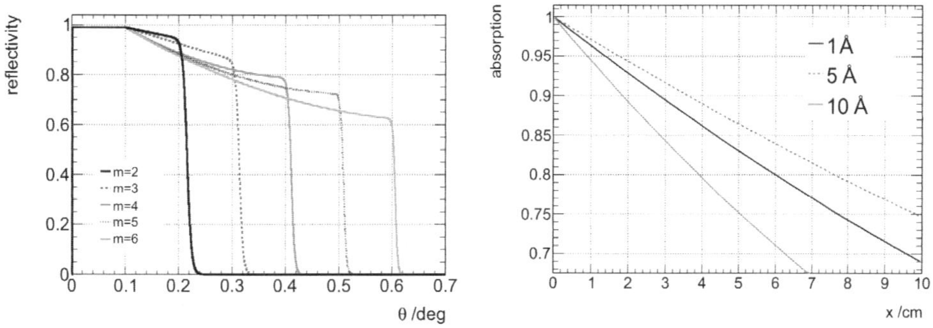

(a) (b)

Figure 1. Reflectivity for $\lambda = 1\,\text{Å}$ neutrons as a function of the scattering angle θ for different super-mirror coatings m as calculated with the new reflectivity model (a) and attenuation in silicon as a function of penetration depth for $\lambda = 1\,\text{Å}$, $5\,\text{Å}$ and $10\,\text{Å}$ (b).

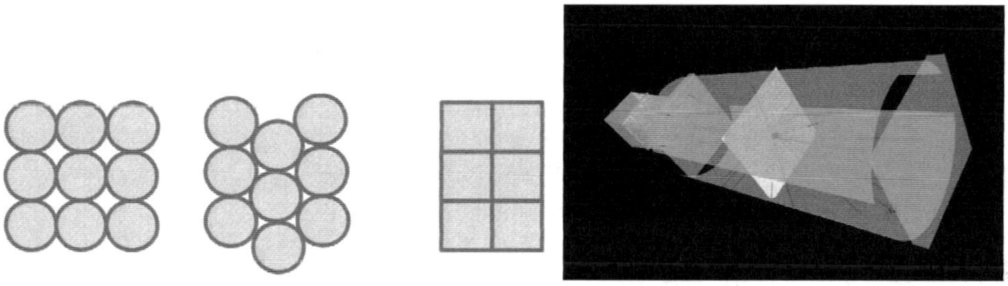

(a) (b)

Figure 2. Side view of tube detector with circular, circular shifted or rectangular tubes (a) and detector array consisting of four detector modules (b).

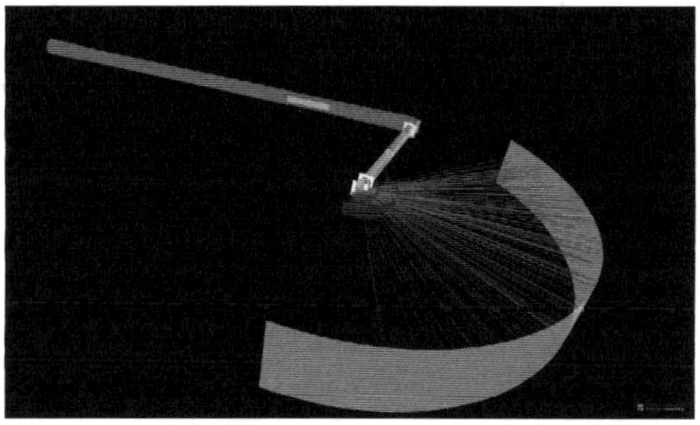

Figure 3. Visualization of the ODIN beamline at IFE, Norway.

from which the particle density N is calculated. For layers of solid material containing ^{10}B or ^{6}Li, the particle density has to be given directly. The probability of a neutron interaction with the detector material is then calculated as

$$p = N\sigma \cdot e^{-N\sigma l} \cdot L/N_{rep} \cdot \epsilon_{mod} \ , \qquad (5)$$

where L is the detector thickness in the direction along the neutron trajectory and l a random number between zero and L following a constant probability density distribution, determining the path length of the neutron in the detector before scattering. Compared to sampling the depth l directly from an exponential decay, this approach follows the VITESS philosophy of generating possible outcomes with the same probability and adjusting the weight accordingly to ensure that the tails of the probability distributions are sufficiently populated. Every neutron trajectory is split into N_{rep} neutrons detected at different possible path lengths l in the detector material, with the number of repetitions N_{rep} given as input parameter by the user. The normalization factor L/N_{rep} can be interpreted as a length segment in which the interaction probability is constant, hence the calculated efficiency is more accurate for large N_{rep} >1. Finally, the detection probability can be modified by the wavelength independent input parameter ϵ_{mod} which can represent e.g. losses due to secondary particle detection after neutron scattering in a solid ^{10}B or ^{6}Li converter layer.

Alternatively, by not choosing a defined detection material ("*other*"), the `efficiency modifyer` ϵ_{mod} can be used as a mean detection efficiency of neutrons traversing the detector perpendicular to the detector surface. In this case the detection probability is wavelength independent.

As a third alternative, the detection efficiency can be set by the user via an input file containing the efficiencies for different neutron wavelengths. The probability of neutron detection is then independent of the path through the detector and the exact detector geometry.

The new module further uses 3D pixelization allowing the cloud of charge carriers to be detected in the wrong pixel by the new input parameters `hor. resolution`, `vert. resolution` and `resolution in x`. If one of these parameters in set to a value $\neq$0, the respective coordinate of the neutron position is changed to a Gaussian random number centered around the true position with width `resolution`. Shifting the output position to the closest pixel center is performed after this Gaussian smear. In case of a tube detector, a non-zero resolution is only allowed in the dimension along the tube axes.

4. New or improved program features

4.1. Visualization of instruments and neutron trajectories

A visualization of the instrument including neutron trajectories can be called from the GUI by the new `Trajectories` button. A prerequisite is an X3D viewer like the recommended `instantreality`. The visualization of the instrument ODIN at IFE, Norway as obtained with the `instantreality viewer` is shown in figure 3 as example.

The information about the location and type of interaction of neutron trajectories used for their visualisation can also be utilized to monitor the intensity loss and thus radiation along the instrument, e.g. in a neutron guide. This constitutes the first step towards a background estimation via an interface to a program that models the interaction of radiation and high energy particles with matter.

4.2. Numerical Optimization

The combination of VITESS simulations with an optimization routine is done such that the latter is the main program and calls the simulation in a subroutine. All files necessary to run a VITESS optimization can be found in the directory `vitess/OPTIMIZATION`, a 7-step recipe is available in `vitess/Concepts/Optimization.pdf`. Three optimization algorithms to choose from are currently implemented: two gradient methods and a metropolis algorithm. The gradient method `opt_grad` determines gradients in the figure of merit by varying a parameter in variation steps defined by the user. This step size should be small enough to get a good estimate of the gradient, but large enough to avoid a false result dominated by statistical fluctuations. The adapted gradient method `opt_grad_mc` is modified in order to be more robust against statistical fluctuations in Monte Carlo simulations: a parameter is changed only if variation in both directions indicate the same gradient, and the step size is varied during the optimization. The `metropolis` algorithm searches the parameter space in a step size set by the user, accepting a new parameter value always if the figure of merit is improved, and if it is worse with a probability according to a kind of Boltzmann factor with a user-defined standard deviation of the measurement as temperature equivalent. Compared to gradient methods, the `metropolis` algorithm is less prone to deliver a local optimum, but more simulation steps are needed.

The default mapping between optimization parameters p_{opt} and simulation parameters p_{sim} is a direct assignment: $p_{opt} = p_{sim}$. Other relations can be set by writing a dedicated function; this procedure is planned to be made more user-friendly in future VITESS versions.

The figure of merit is designed flexible: It is calculated from a signal file containing neutron intensity values $I_{sign,i}$, like e.g. a monitor file recording neutrons with a certain divergence at the sample position, as

$$FoM = \frac{\sum_i \frac{w_i I_{sign,i} \lambda_i^l}{I^m_{ref,i}}}{\sum_i I^n_{noise,i}}, \text{ or } FoM = \frac{\langle \frac{w_i I_{sign,i} \lambda_i^l}{I^m_{ref,i}} \rangle}{\langle I^n_{noise,i} \rangle}. \qquad (6)$$

All parameters apart from $I_{sign,i}$ are optional. The powers l, m and n can be float values. The parameters $I_{ref,i}$, $I_{noise,i}$ and w_i can be given via names of a reference, noise or weight file, where the former two are like the signal file created as output of the simulation and the latter is provided by the user. In order to optimize for example the integrated flux at the sample position, the sum in the figure of merit is chosen, an according monitor file has to be given as signal file, $l = m = n = 0$ and the reference, noise and weight file are set to `no_file`. If equation 6 is not appropriate, the figure of merit can also be calculated by a user-defined function.

4.3. Parallel processing

Until VITESS version 3.1, parallel processing was only possible via so-called helper threads using several cores of the same computer [16]. With the most recent VITESS release, simulations can be split into several instances on a computer cluster by calling `vitess/TOOLS/gridrun`. The simulation is then divided into several pipes with less statistics, each of which is sent to the cluster by either SGE or ssh. A work directory is created in which all input files are copied, as well as subdirectories for each pipe in which the simulation results are stored. After completion of all simulations, the output files can be merged and the subdirectories are deleted.

4.4. Further improvements

New generic monitors `monitor1D` and `monitor2D` can be used to view any neutron parameter, and combine any two parameters in the latter. Furthermore, two filter parameters can freely be chosen and combined with a logical `AND` or `OR`. The `monitor1D` can create up to three monitor output files per module (to which the same filter settings are applied). These new generic monitor modules are planned to replace the sum of old monitor modules, but all are available for a transition period in VITESS 3. In the old one-dimensional monitor modules, the option `all files [yes/no]` has even been added which allows to create a separate monitor output file for each color present in the neutron beam. The use of color has been changed such that the value 0 is the default color value and -1 stands for all colors.

A new brilliance monitor `mon_brilliance` is available that writes the brilliance instead of the intensity. It can also monitor the brilliance transfer if a `reference file` is given, and calculates the average and peak brilliance which can be written into a `flux file` (but can also be found in the logfile).

A new guide module called `guide_ideal` was written to describe guides with a perfect elliptic shape without the use of segmentation. The location of reflection points of neutron trajectories off the guide walls is calculated analytically. The module runs significantly faster compared to the regular guide module with a fine segmentation.

Creating a `sm_ensemble` input file is made easier by the new tool `Generate Extraction System` in which position and orientation of supermirror plates can be given in a user-friendly, intuitive way. Although written with a bi-spectral extraction system in mind that consists of several stacked and bended mirror plates, this tool can be used for any configuration of supermirrors.

In the monochromator module `monochr_analyzer`, a transition mode has been added so the transmitted neutron beam can be analyzed alternatively to the reflected neutrons.

The module `writeout` has been expanded such that trajectories can now be written in McStas format, and supplemented by a new module `read_in` that can read trajectories from both VITESS and McStas output files.

Handling of the graphical user interface has further been improved: It is now possible to import an instrument description from a file containing the pipe command, to give modules a short descriptive name that appears in the module list to ease distinguishing between similar instrument components, and to select between different programs (gnuplot, shell) with which monitor files are viewed directly from the GUI.

Note that only the most important program improvements could be mentionen here; a complete list can be found in the release notes separately available for every VITESS release. Detailed descriptions of new and old features are further available in the documentation within the VITESS software package.

5. Summary

The software package VITESS is an important tool for the simulation of neutron scattering instruments. It is easy to use due to its graphical user interface, the flexible modular design as well as user-friendly components and tools. The program is under constant development and has made large progress in releases 3.0 und 3.1. New sources have been added and existing moderator descriptions updated, the reflectometer sample can simulate off-specular scattering and incoherent background, and more accurate reflectivity and attenuation models are used in neutron guides and supermirror plates. The detector module has been updated such that it allows more flexible geometries including an array of several detector banks, the detection efficiency can be calculated from the scattering cross-section of a chosen material and resolution effects can be taken into account. Visualization of the simulated instrument including neutron trajectories is now possible. A numerical optimization routine is provided to optimize instruments with respect to a flexibly designed figure of merit. Furthermore, parallel processing on a computer cluster is enabled to save simulation time. Several smaller improvements and new features further ensure VITESS staying a state-of-the-art program contributing to the development of cutting-edge neutron instrument design.

Acknowledgments

We thank A. Houben for contributions to the `detector` module update. We are further indebted to H. Jacobsen for sharing his reflectivity and attenuation model and comparing McStas simulation results to VITESS.

This work was funded by the German BMBF under "Mitwirkung der Zentren der Helmholtz Gemeinschaft und der Technischen Universität München an der Design-Update Phase der ESS, Förderkennzeichen 05E10CB1".

References

[1] Lieutenant K, Zsigmond G, Manoshin S, Fromme M, Bordallo H N, Champion D, Peters J and Mezei F 2004 **Proc. SPIE 5536** 134–145

[2] Zsigmond G, Lieutenant K and Mezei F 2002 *Neutron News* **13** 11–14

[3] Manoshin S and Ioffe A 2008 *Nucl. Instrum. Meth. A* **586** 81 – 85 proceedings of the European Workshop on Neutron Optics - NOP '07

[4] Willendrup P, Farhi E, Knudsen E, Filges U and Lefmann K 2014 *Journal of Neutron Research* **17** 35–43

[5] Knudsen E, Tranum-Rømer A, Christiansen P, Willendrup P and Lefmann K 2014 *Journal of Neutron Research* **17** 27–34

[6] Saroun J and Kulda J 1997 *Physica B* **234-236** 1102–1104 proceedings of the First European Conference on Neutron Scattering

[7] Seeger P A and Daemen L L 2004 *Proc. SPIE* **5536** 109–123

[8] Lee W T, Wang X L, Robertson J, Klose F and Rehm C 2002 *Applied Physics A: Materials Science & Processing* **74**(0) s1502–s1504

[9] Seeger P, Daemon L, Farhi E, Lee W, Wang X, Passell L, Saroun J and Zsigmond G 2002 *Neutron News* **13** 24–29

[10] Filges U, Saroun J, Juranyi F and Zsigmond G 2006 Presentation at the International Workshop on Applications of Advanced MC simulation URL `http://lns00.psi.ch/mcworkshop/papers/focus_intercomparison.pdf`

[11] Klenø K H, Lieutenant K, Andersen K H and Lefmann K 2012 *Nucl. Instrum. Meth. A* **696** 75 – 84 ISSN 0168-9002

[12] Jacobsen H, Lieutenant K, Zendler C and Lefmann K 2013 *Nucl. Instrum. Meth. A* **717** 69 – 76 ISSN 0168-9002

[13] http://www.helmholtz-berlin.de/vitess vitess website

[14] Freund A K 1983 *Nucl. Instrum. Meth. A* **213** 495 – 501 ISSN 0167-5087

[15] http://www.nndc.bnl.gov/sigma/index.jsp?as=10&lib=endfb7.1&nsub=10 national Nuclear Data Center

[16] Lieutenant K, Zendler C, Manoshin S, Fromme M, Houben A and Nekrassov D 2013 *Journal of Neutron Research* Doi:10.3233/JNR-130005

Design and development of position sensitive detectors for neutron scattering instruments at National Facility for Neutron Beam Research in India

Shraddha S Desai
Solid State Physics Division, Bhabha Atomic Research Centre, Mumbai, 400085, India

E-mail: ssdesai@barc.gov.in

Abstract. Various neutron scattering instruments at Dhruva reactor, BARC, are equipped with indigenously developed neutron detectors. Range of detectors includes proportional counters, beam monitors and linear position sensitive detectors (PSD). One of the instruments is recently upgraded with multi-PSD system of high efficiency and high resolution PSDs arranged in stacking geometry. These efforts have resulted in improving the throughput of the instrument and reducing experiment time. Global scarcity of ^{3}He has made essential to explore other options like BF_3 gas and ^{10}B coatings.

PSDs with coaxial geometry using BF_3 gas and ^{10}B coating (90% enriched) are fabricated and characterized successfully. These PSDs are used as the alternative to ^{3}He PSD in equivalent geometry. Though efficiency of PSDs in similar dimensions is lower than that with ^{3}He, these large numbers of PSDs can be arranged in multi-PSD system. The PSD design is optimized for reasonable efficiency. An array of 60 BF_3 filled PSDs (1 m long) is under development for the Time of Flight Instrument at Dhruva. Further improvement in efficiency can be obtained with novel designs with complex anode-cathode geometry. Various challenges arise for long term operation of PSDs with BF_3 gas, in addition to complexity of data acquisition electronics. Study of gas aging with detector fabrication materials has been carried out. PSDs with ^{10}B coating show advantage of non toxic nature but have low efficiency. Multiple ^{10}B layers intercepting neutron beam are used to increase the efficiency. PSD designed with small anode-cathode spacing and array of multiwire grids placed between double sided ^{10}B coated plates are being fabricated. Assembly is arranged in curvilinear geometry with zero parallax. Overview of these developments is presented.

1. Introduction

National Facility for Neutron Beam Research at Dhruva reactor consists of 12 beam lines, 8 in the reactor Hall and 4 in Guide Lab. The user community is from BARC, National laboratories and various Universities. The instruments are supported with indigenously developed detectors. Range of detectors consists of neutron beam monitor, neutron proportional counters and position sensitive detectors (PSD) (Figure 1). High efficiency high resolution He^{3} filled PSDs in stacking geometry are developed successfully. But recent scarcity of ^{3}He globally has made it essential to find an alternative. It is essential to explore the use of other alternatives as BF_3 and ^{10}B coating, particularly for neutron scattering applications. No other currently available detector technology offers the stability, sensitivity and convenience of ^{3}He filled detectors. Among various feasible alternatives, BF_3 gas has most potential. BF_3 filled PSDs were seldom used, but were not popular due to toxic and corrosive nature of BF_3. Handling the generation and distillation system and minimise impurity level is a challenge. Performance of BF_3 filled position sensitive detectors (PSD) is evaluated and novel designs are being tried for gain in efficiency. Performance of ^{10}B coating based PSD is also evaluated. Though efficiency of ^{10}B coated detector in equivalent dimensions is lower than that with BF_3, advantage of ^{10}B coated PSD is its non-toxic nature. The overview of these developments in detector instrumentation supporting the neutron scattering facility is presented.

2. Neutron detectors mounted at various neutron scattering instruments, Dhruva

All spectrometers are equipped with in-house made BF_3 filled beam monitor counters with 0.03 cps sensitivity. These monitors are used to normalise the scattering spectra for any fluctuations in neutron beam intensity with time. A small PSD (200 mm long and 50 mm dia) is mounted at Neutron Reflectometer. Charge division is carried out through resistive carbon coated quartz filament as anode.

Small Angle Neutron Scattering instrument is equipped with a single linear PSD of 900 mm sensitive length and 5 mm position resolution. Powder Diffractometer is mounted with a multiPSD system with 5 PSDs [^{3}He (4 bar) + Kr (2 bar)] covering the 2θ range of 3° to 140°. Triple Axis Spectrometer, Single Crystal Diffractometer and Polarised Neutron Spectrometer are mounted with high efficiency counting detectors in end-on position with scanning mode.

2.1 Up gradation of Hi-Q Diffractometer using high efficiency PSDs

High Q Diffractometer was initially mounted with 5 PSDs [^{3}He (4 bar) + Kr (2 bar)] for data collection covering the 2θ range of 2° to 140° with some angular overlap. In order to improve the performance of instrument, high efficiency PSDs with various cathode diameters, fill pressures and additive gases were developed [1] and characterized. Design parameters and performance are optimized for high efficiency PSD [cathode diameter 37 mm and fill gas He^3 (10 bar) + Kr (2 bar)]. The performance of these PSDs is tested at Hi Q Diffractometer. Figure 2 shows the comparison of diffraction data of an old PSD He^3 (4 bar) and a new PSD He^3 (10 bar) at High-Q spectrometer. Spectra indicate gain of 2.2 in counting statistics. Later this instrument was upgraded with 15 such PSDs, arranged in a pattern of 3 stacks with 5 PSDs in each stack. This setup with PSD in stacking geometry has improved the throughput of the instrument considerably.

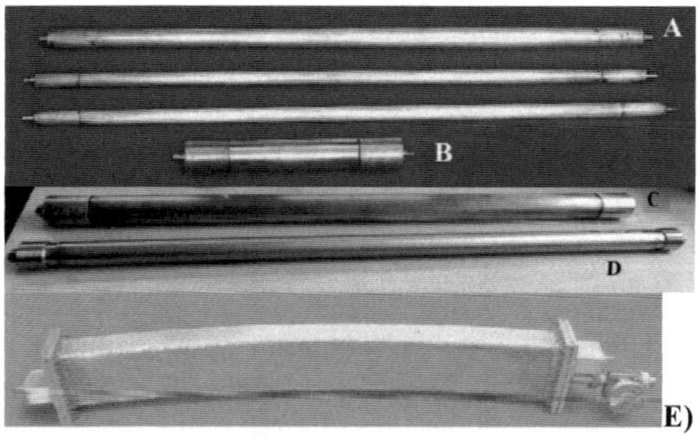

Figure 1 Picture of various linear 1-D PSDs developed for use at neutron spectrometers, at Dhruva A) High efficiency PSD for Hi Q Diffractometer B) He^3 PSD for Neutron Reflectometer C) B^{10} coated PSD D) BF_3 PSD for TOF instrument E) Curvilinear Multigrid PSD

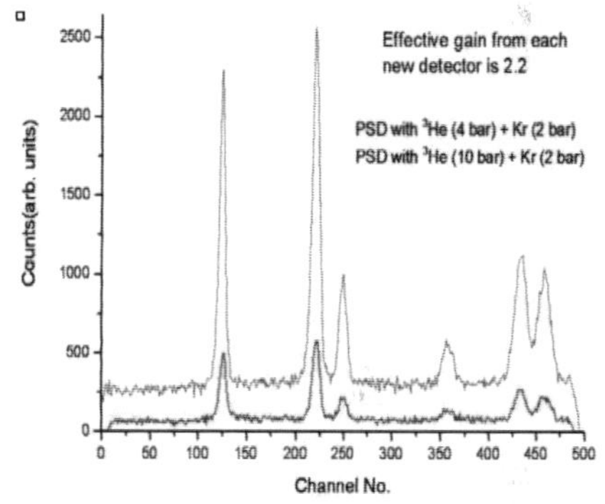

Figureure 2 Comparison of Diffraction spectra recorded using various PSDs 1) He^3 (4 bar) + Kr (2 bar) and 2) He^3 (10 bar) + Kr (2 bar) with same geometry and angle covered.

3. PSDs with BF_3 gas and B^{10} coatings as ^{3}He alternatives

Further up gradations of instruments were hindered due to non-availability of ^{3}He gas. Efforts are on to find suitable alternatives using BF_3 gas and solid B^{10} layers. In-house facility for generation and distillation of BF_3 gas and fabrication of detectors has facilitated various characterization of the gas behaviour in detector.

3.1. BF_3 based PSD

Generation of BF_3 gas from $CaF_2(BF_3)$ complex involves various processes, such as moisture extraction, gas generation by thermal decomposition of complex and distillation of gas at triple point temperature. Gas purification is carried out with repeated distillations. A PSD with 1 m long, 37 mm diameter cathode and 10 μm anode wire is assembled and filled with BF_3 gas at 0.8 bar pressure. Diffraction spectra of Fe is recorded with the PSD at Powder diffractometer using λ = 0.783 Å [2]. Comparison of spectra from BF_3 PSD (0.8 bar) with ^{3}He + Kr (10 + 2 bar) PSD is carried out with reference to peak intensity and peak width. Position resolution of the BF_3 PSD is found to be ~7 mm, and is acceptable for the present resolution of the Spectrometer. However, the efficiency of BF_3 PSD is less than that of ^{3}He PSD by a factor of 20 (Figure 3). Efforts are towards improving efficiency by high pressure gas filling and use of stacking geometry. Test on PSD filled with BF_3 at 1.6 bar were also carried out and resulted in efficiency lower than ^{3}He filled PSD by factor of 16, whereas, it resulted in broadening of peaks. Thus operating PSDs with high pressure in equivalent geometry do not help to gain efficiency. The nature of BF_3 gas makes the avalanche formation very sensitive to small traces of impurities [3] like SiF_4, HF, and SF_6. Pulse height distribution deteriorates with the pressure. Efforts are taken towards reducing impurity level by repeated distillations of BF_3 gas. It has a limited effect.

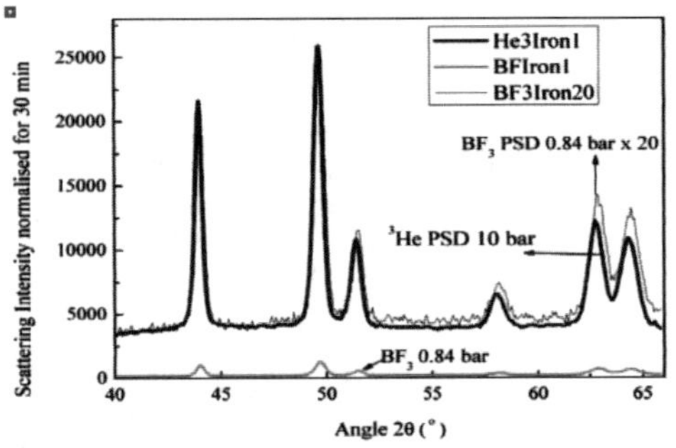

Figure 3. Comparison of Diffraction spectra using ^{3}He and BF_3 filled PSDs mounted at Powder Diffractometer

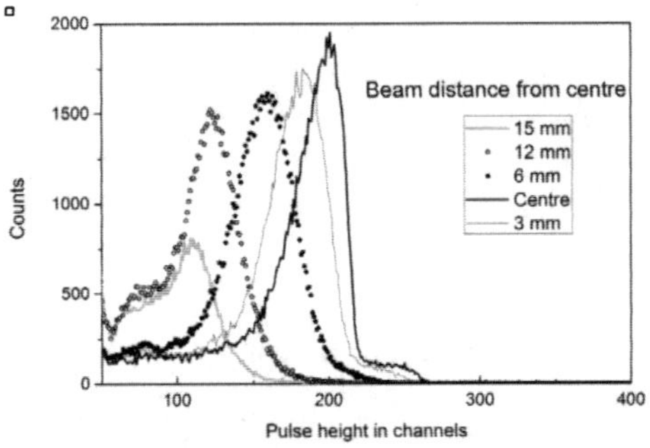

Figure 4 Effect of position of neutron interaction in drift region on pulse height distribution

Effect of electronegative impurities in drift region is evaluated using a fine neutron beam parallel to anode. Beam is shifted along radial direction towards cathode. Effect of beam position on pulse height distribution using 2 mm point beam is shown in Figure 4. As the beam is shifted away from anode, drift length for all the primary charge clouds formed along the track of neutron is increased. Peak position of the pulse height spectrum is reduced and it is due to loss of primary electrons in the drift region. This loss of electrons is further amplified in the avalanche region resulting in lower charge collection. Recombination and electron attachment is dominant at large cathode radius and higher gas pressure. This effect can be reduced using higher drift field. Such behavior is not observed in He^3 filled detectors.

3.2. PSDs for the proposed Time of Flight (TOF) Spectrometer at Dhruva

Various tests on effect of fill gas pressure, cathode drift region and anode dimensions on detector performance are carried out. These results are implemented in the PSD design. PSDs with higher gas pressure show deterioration in pulse shape, due to the increased impurity concentrations. They are well suited with lower drift region and high drift voltage. For similar gains, higher drift field is attained using thicker anode, whereas for PSDs based on charge division encoding, it is essential to use a resistive anode. Thus resistive anode of 25 μm diameter is used.

The PSDs for detector bank of TOF spectrometer are optimised, considering the performance of BF_3 detector. Dimensions of PSD is 38 mm diameter cathode, chosen to have higher beam height interception and gas depth in direction of neutron as compared to 25 mm dia cathode tube. Resistive anode used is 25 μm NiCr and total anode resistance is ~3 kΩ. Drift field in the drift region at 1 cm from the anode is calculated for both the PSDs with anode bias needed for similar gas gains. Drift field increased from 5.8×10^3 V/cm for 10 μm anode to 1.4×10^4 V/cm for 25 μm anode. PSD in cylindrical geometry is mounted with two weldable, alumina ceramic feed thrus. BF_3 gas does not show any aging with alumina ceramic. Inner wall of cathode is coated with activated charcoal. One of the PSDs filled at 1 bar BF_3 gas, shows reasonable good pulse height distribution. It is tested for gas aging over few weeks. Figure 5 shows the pulse height spectrum of the BF_3 PSD recorded over a span of 28 days from gas filling. No aging or deviation in the performance over the holding time is noticed. Out gassing of PSDs for 15 days at 110°C, has been useful in maintaining the purity of gas. The multi-PSD system is designed to cover the detection area of 2.5 m^2 with 2.5 m arc length (70°), 1 m height (28°), consisting of an array of vertically arranged 60 PSDs (1 m long). PSD designs considering all these parameters have proved to be successful in replacing ^{3}He PSDs. In spite of non availability of He^3 gas, it has made possible the development of new TOF Spectrometer at neutron scattering facility.

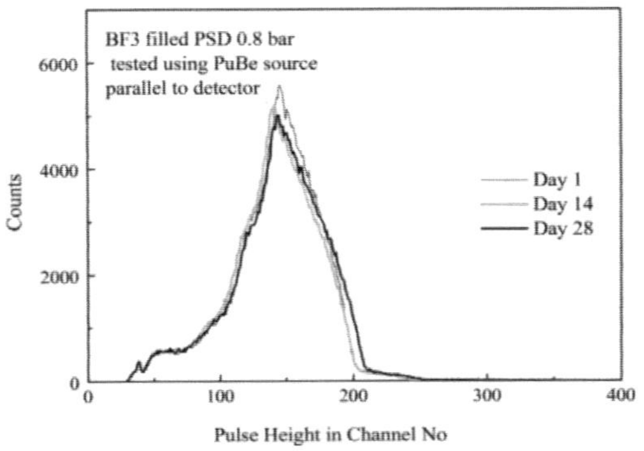

Figure 5 Pulse height spectra of BF_3 filled PSD over the time of 28 days after gas filling.

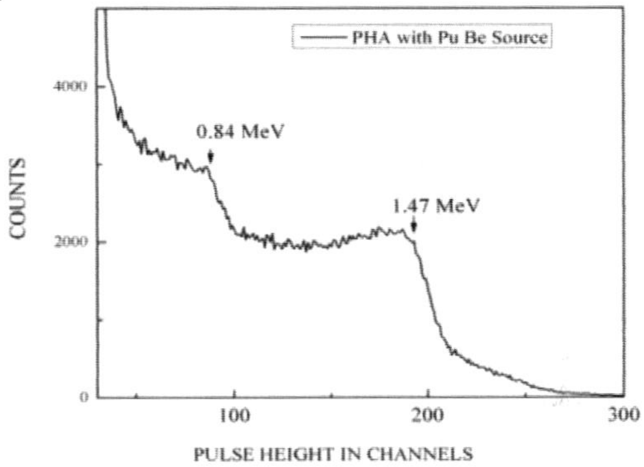

Figure 6 Pulse height distribution of a ^{10}B coated PSD using Pu-Be neutron source

3.3. BF_3 filled PSD with complex geometry

Considering these limitations on cathode diameter and fill gas pressure, gain in efficiency is attained using longer gas length using different geometry. It is possible with combination of multiple anode-cathode geometry. In such PSD, drift field and anode resistance can be designed independently. Complex geometry needs multiple wire supports and resistors for charge division. A multigrid PSD fabricated by ILL, Grenoble, was filled with BF_3 gas at our facility and tested using neutron beam at Dhruva. Structure of the PSD, anode-cathode geometry is similar to the B^{10} coated Multigrid PSD [4] with 10 cells in direction of neutron beam. Gain in efficiency was recorded but gas aging within the PSD was noticed. Extensive out gassing of the PSD helped up to an extent.

In addition to complexity of data acquisition electronics, design and fabrication of a chamber and maintaining the gas purity over long time is a challenge. Restrictions are on use of insulators for wire assembly and surface mount resistors within the sensitive volume of PSD. Considering corrosive nature of BF_3 gas, materials need to be chosen carefully to maintain good life of a PSD. Aging studies were carried out on various materials as Macor, PCB and Teflon. Teflon is found suitable with BF_3. Further design of multitube is included with Teflon supports for mounting anode wires. Though BF_3 gas has limitations with impurity concentration, electro-negativity and aging with various materials, these efforts of were helpful for design of new PSD.

4. ^{10}B thin film based cylindrical PSD

B^{10} coating is also a promising alternative to ^{3}He with non toxic nature. Enriched B^{10} coating can be introduced on PSD cathode wall. Self absorption of charged particles, Li and α within the coated layer, limits the neutron detection efficiency. Ionization resulting from emitted charged particles is recorded. Two prototype PSDs (PSD-1 and PSD-2) with ^{10}B coating and coaxial cylindrical geometry are fabricated and characterized. Cathode dimensions are 5 cm diameter and 90 cm long and anode is 25 μm diameter NiCr wire. Boron coating, 90% ^{10}B enriched with coating thickness of 1 mg/cm^2 on inner lining of cathode, is obtained commercially with the painting and baking process. PSD-1 is filled with gas Ar + CO_2 at 0.5 bar and evaluated for the pulse height, plateau characteristics and position resolution. Figure 6 shows the pulse height distribution of B^{10} coated PSD. The position resolution was 10 mm. Thus PSD-2 is filled at 2 bar and position resolution is improved to 7 mm [4]. Figure 7 shows the position scan of 2 mm beam at 10 cm spacing, over the sensitive length of the PSD. Position resolution is acceptable for present instrument resolution. Uniformity of coating thickness was observed with the flooded neutrons and variation in intensity of pattern was within 2%. Neutron detection at each position in the spectrum is an integral contribution from all interaction with coating along circumference of cathode and projected on the anode. Sensitivity comparison of ^{3}He PSD (2 bar), BF_3 PSD (1 bar) and ^{10}B coated PSD was carried out using uniformly flooded neutrons. Sensitivity of the PSDs varies as 1:0.48:0.16 respectively at thermal neutron energy.

Characterization of these PSDs was useful for further designs of large area PSD with multiple anode-cathode geometry. Advantage of non toxic nature of ^{10}B allows the choice of suitable proportional gas for fast drift velocity. Splitting of peaks is observed with inclined beam by 10°, as parallax is dominant at this cathode diameter. Parallax in He^3 PSD is observed as broadening of peaks at higher θ, whereas, distinct position peaks are observed with ^{10}B coated PSD. This is due to that neutron interaction takes place at two points of cathode walls across neutron flight path and projection of these charge clouds on anode is recorded as position peaks. Further gain in efficiency can be obtained by introducing multiple ^{10}B coated layers along neutron path.

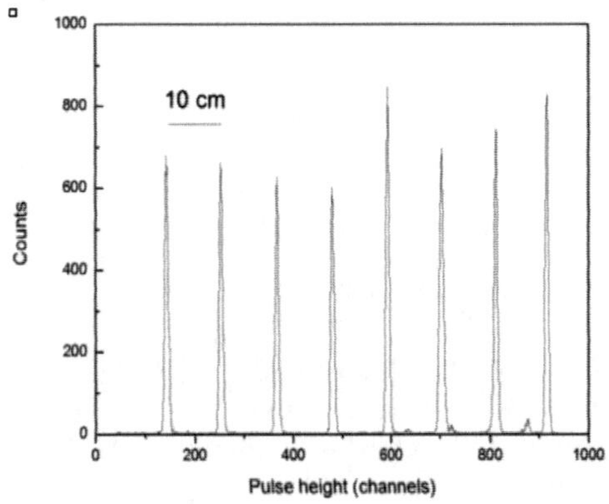

Figure 7 Position scan of B^{10} coated PSD-2 using a 2 mm fine collimated beam and 10 cm spacing.

Figure 8 Schematic side view of a Teflon holder assembly of four ^{10}B coated plates and three multiwire grids.

4.1. ^{10}B coated Multigrid Curvilinear PSD

Multiple ^{10}B layers intercepting neutron beam and associated anodes for charge collection are essential to increase the neutron detection efficiency. A novel design with multiple ^{10}B coated surfaces and curved design to avoid parallax is presented. An assembled curvilinear PSD is shown in Figure1E and curved Teflon holder is used to support anode-cathode structure. Four ^{10}B coated plates, with two single and two double sided coating, are arranged in curved casing. Six ^{10}B coated surfaces are intercepted by the incident neutron beam. Height of coated plate is 6 cm and it adds to the height of neutron beam intercepted by detector. This can be further increased, considering the broadening at smaller angle. Each coated surface is facing a multiwire grid along the curvature of cathode plates. Mounting structure of anode grids and cathode plates is shown in Figure 8. Spacing between anode grid and cathode plate is 5.5 mm. Pitch of wire on anode grid is 5 mm. Position readout is carried out using charge division encoding method.

An arc length of 0.7 m (detector sensitive length) covers the angular range of 23°. The active gas thickness is 35 mm. Complete PCB and Teflon assembly is guided through a curved cathode casing. Three multiwire grids with resistive wires are connected in series. Central anode grid is mounted at radius 2000 mm and other two grids at ± 12 mm radii from central position. At constant sensitive length of 700 mm for all three grids, deviation in 2θ for other two grids is ± 0.12° ~ ± 4 mm when central anode wires on multiwire grids are aligned. This deviation in position can be normalized. Presently PSD with single grid is tested and position resolution obtained is 5 mm.

5. Conclusions

The neutron scattering instruments at Dhruva are supported with indigenous development of neutron detectors and PSDs. Up gradation of Hi-Q Diffractometer using high efficiency and high resolution PSDs in stacking geometry was carried out with a considerable gain in throughput of Instrument. Efforts are initiated to find alternatives to ^{3}He, using BF_3 gas and B^{10} coatings. Prototype PSDs in coaxial geometry using these materials show encouraging results. In-house facility for generation and distillation of BF_3 gas has facilitated characterization of gas behaviour in the detectors. PSDs for Time of Flight instrument are successfully designed and fabricated with acceptable efficiency and durability.

Aging of BF_3 gas due to various construction materials of complex geometry detector was noted. Results were useful for further designs of PSD with multiple anode-cathode structure. PSD with ^{10}B coatings though show low efficiency, gain in efficiency is desired by introducing multiple ^{10}B coated layers along neutron path. Design aspects of Curvilinear Multigrid PSD with zero parallax are mentioned. This novel design is expected to show 3 times gain in efficiency as compared to cylindrical PSD.

References

[1] S S Desai, A M Shaikh, Rev. Sci. Instr. **78**, 023304, 1-6 (2007).
[2] Shraddha S Desai, Shylaja Devan, and P S R Krishna, AIP Conf. Proc. **1349**, 489 (2011).
[3] J. Davilla Aponte and S A Korff, *Rev. Sci. Inst.* Vol **31**, No **5** (1960) 532.
[4] Shraddha S. Desai and Shylaja Devan, AIP Conf. Proc. **1512**, 524 (2013).
[5]http://www.ill.eu/fileadmin/users_files/Annual_Report/AR-12/page/pg_contents.htm?rub=4_31

Prompt γ radiation measured with a NaI scintillation detector: a beam monitor for neutron scattering instruments which needs no space in the beam

O. Holderer[1], M. Zamponi[1], M. Monkenbusch[2], R. Engels[3]

[1]Jülich Centre for Neutron Science (JCNS) at Heinz Maier-Leibnitz Zentrum (MLZ), Forschungszentrum Jülich, Lichtenbergstr. 1, 85747 Garching, Germany
[2]Jülich Centre for Neutron Science (JCNS) and Institute for Complex Systems, Forschungszentrum Jülich, D-52425 Jülich, Germany,
[3]Zentralinstitut für Engineering, Elektronik und Analytik, Systeme der Elektronik (ZEA-2), Forschungszentrum Jülich, D-52425 Jülich, Germany

E-mail: o.holderer@fz-juelich.de

Abstract. We investigate the possibility of using the prompt γ rays emitted by aluminum windows in order to monitor the neutron flux of the beam. A NaI scintillation detector is used to detect the prompt γ rays. No additional material apart from the unavoidable Al windows along the flight path is placed in the beam. The performance of the monitor is compared to that of a standard BF3-monitor placed in the beam. Influences of a magnetic field on the photomultiplier of the NaI monitor is discussed, as well as the influence of activation gammas. At an instrument using a beam chopper the time behaviour is discussed.

1. Introduction

Usually, beam monitors are placed directly in the beam, with a low sensitivity of the order of $10^{-4} - 10^{-5}$, for example He3- or BF3 monitors. Two Al-windows are unavoidably in the beam, producing some absorption, small angle scattering and prompt γ radiation. Prompt γ rays from absorption of neutrons in aluminum, which is the main material used for neutron windows, have a significant contribution of high energetic gammas, e.g. for aluminum at 7.7 MeV [1] [2], compared with the activation γ's of Al which have the main contribution at 1.7 MeV [3]. With a scintillation detector sensitive also to high energy gammas, where the lower energies are filtered and discriminated away, a beam monitor without additional material besides the unavoidable Al windows can be realized. Fig. 1 shows the absorption coefficient of lead as a function of energy. At the J-NSE neutron spin echo spectrometer [4, 5, 7] at the FRM II research reactor of the TU München this type of monitor has been tested and showed that it is a relieble and easy to handle possibility as a beam monitor. Time behaviour has been investigated at the SPHERES backscattering spectrometer [8], where the phase space transform chopper imposes a time structure on the incoming neutron beam.

2. Experiment

Neutron capture releases the total binding energy of a neutron in the order of 8-10 MeV mostly in the form of one to several γ-photons. A fraction of them has energies close to the total binding

Journal of Physics: Conference Series **528** (2014) 012038 doi:10.1088/1742-6596/528/1/012038

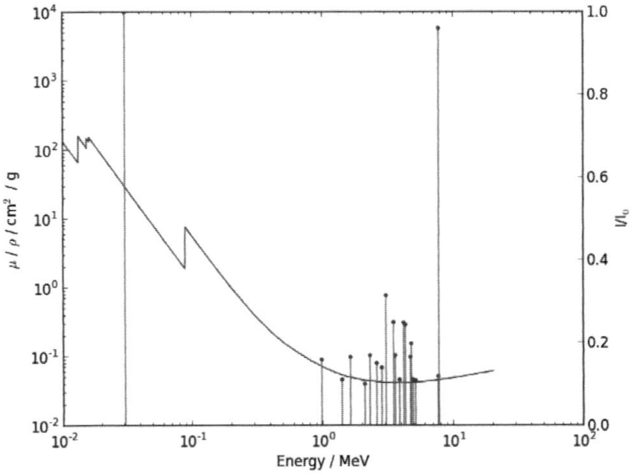

Figure 1. Absorption coefficient of lead as a function of γ energy and the relative intensities of the prompt γ lines of Al (with more than 10% rel. intensity). Lead acts as a filter for the low energy γ's from the produced radioactive Al-28 isotope and others. A very strong prompt γ line is above 7 MeV. Data from [9] and [2].

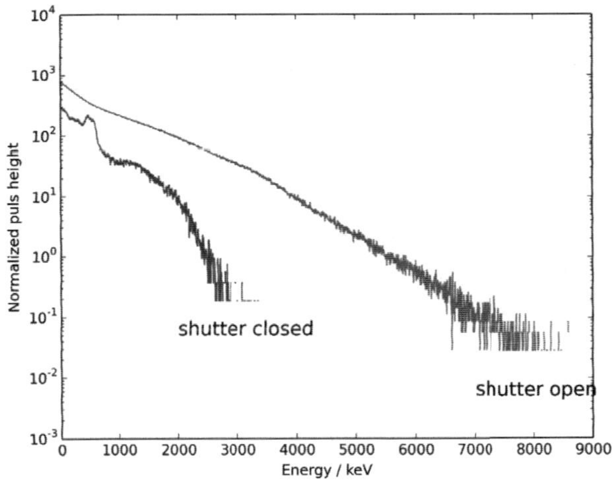

Figure 2. Pulse height spectra of the NaI monitor. The peak at closed beam is attributed to the 511 keV line of positron-electron annihilation. The low energy part contains contributions from activation decay, the high energy part is used for monitoring the beam intensity.

energy. To detect them with a scintillation detector, a rather large scintillation crystal is needed. In the present experiment, we used a SCIONIX scintillation detector with a NaI crystal with 76 mm diameter and 102 mm height to detect the prompt γ's emitted from the aluminum windows in front of the first main coil of the J-NSE spectrometer.

A first data acquisition of the analog output has been taken with the monitor placed

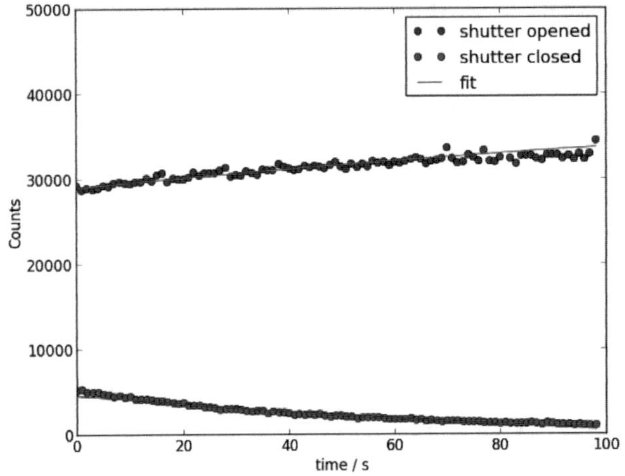

Figure 3. Monitor at a position about 1.4m away from the beam, without lead shielding. Count rate at the analog output with open beam (blue) and closed beam (green) as a function of time. Counting time was 1 sec. The activation gammas are counted with these settings. Fits are made with the Al decay time of $T_{1/2}$=2.246 min.

underneath the lead shielding of the polarizing unit. The built up and decay of the activation γ's from Al is clearly visible in the analog pulses of the monitor (see Fig. 3). Without the lead acting as a high energy filter, the NaI scintillator alone is therefore not suitable as a beam monitor. The situation can be improved by pulse height discrimination, by this means the obvious Al-28 decay signal with a half life time of 2.2414 minutes (BNL data) (β-decay with associated 1.78 MeV γ [6]) can be suppressed. However, the high count rate of low energy γ's leads to saturation and dead-time effects.

2.1. Low γ discrimination

The use of the NaI-counter as a beam monitor requires that the low energy γ's are cut away. As a filter, the lead hutch of the second polarizer of the J-NSE spectrometer has been used with its 50 mm thickness to absorb preferentially the low energy γ rays. Fig. 2 shows pulse height spectra of the analog signal of the NaI-monitor with the shutter of the neutron beam open ("beam open") and closed ("beam closed") respectively. With the beam closed, there is still a significant number of pulses at lower energies. They can be filtered to a large amount by some (typically 5 cm) Pb. With an adequate threshold for the digital signal it is possible to be only sensitive to the high energy part of the spectrum, where most of the prompt γ-lines from Al are found.

Measurements with a Pb filter realized by placing 50 mm lead between the NaI detector and the beam (i.e. placing the detector outside the lead hutch of the long wavelength polarizer) lead to a significantly reduced count rate of the analog output by one order of magnitude. The ratio of the count rate with beam open and close is about 15 and can be largely improved for the digital output in combination with the appropriate trigger level. A good sensitivity to the prompt γ rays could be achieved in this way. In this monitor position, the count rate with the reactor not running is at about 2-3 cps. The remaining dark count rate is due to the unshielded position towards other instruments in the neutron guide hall.

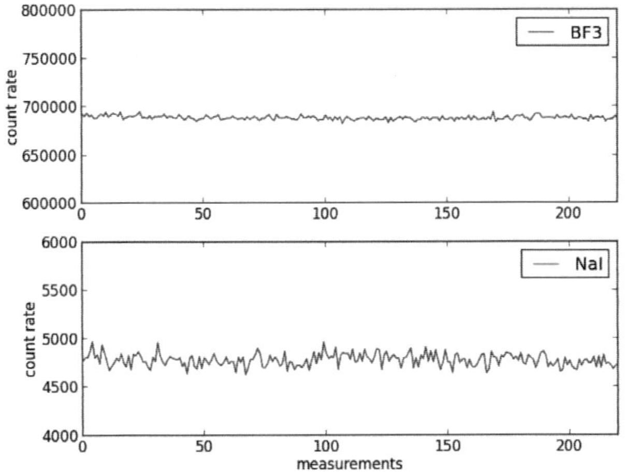

Figure 4. Detector counts versus number of measurement. Top: BF3-monitor, Bottom: NaI-monitor. Spectra are time normalized (counts per second), resulting in the different levels of noise.

2.2. Reliability

The time behaviour of the NaI monitor has been compared to the J-NSE standard BF3 monitor which covers the whole beam cross section. A series of 220 measurements is shown in Fig. 4. The standard deviation has been compared to the square root of the count rate as the expected deviation. The standard deviation of the BF3-Monitor is 2369, the square root of the count rate 829, while for the NaI monitor (with a much lower overall absolute count rate) the two values are 73 and 69 respectively. It has to be pointed out that above a magnetic field of 2 Gauss at the position of the monitor the photomultiplier tube of the NaI counter has been disturbed despite of it's μ-metal shielding. The count rate drops for a field of 4 Gauss by 1/3 compared to the undisturbed count rate. A suitable magnetically stable position is therefore needed.

2.3. Time behaviour

At the backscattering spectrometer SPHERES the time response of the monitor has been tested. At the end of the neutron guide, inside the housing of the spectrometer, is the phase space transform chopper with added Pb shielding on the front side. The monitor was placed on top of the neutron guide shielding towards the chopper. Shown are counts versus chopper phase for monitor (shifted by 20000) and a large angle detector tube (separated into elastic and inelastic counts). The count rate of the NaI monitor followed exactly the opening and closing of the chopper. The Al windows in front of the chopper produce a constant high count rate which is then modulated with the frequency of the chopper. The response is as fast as the response of the detectors just behind the chopper (Fig. 5). The steps in the first ~24 degree of the chopper open and reflecting phase corresponds to a flight distance of ~1.6 m. We attribute it to prompt γ radiation from neutrons passing the Al window in front of the Doppler monochromator. The wiggling structure during the reflection phase of the chopper is in our opinion due to slightly varying absorption of the chopper wheel and reflection of the different crytals. The average count rate is nontheless a measure of the incoming neutron flux. A really time resolved operation needs a good shielding towards all unwanted sources of γ radiation.

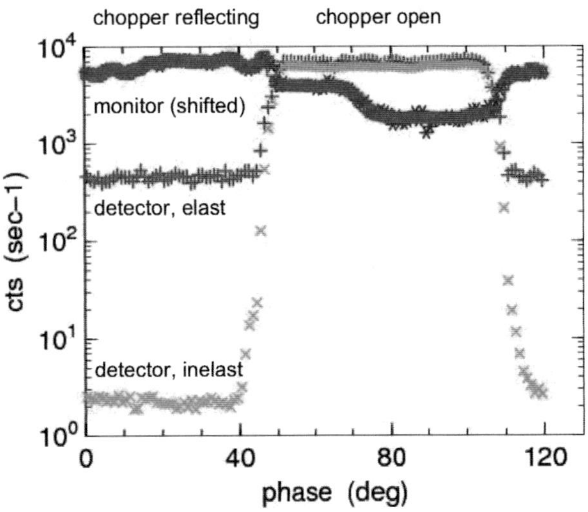

Figure 5. NaI monitor at a chopper instrument, the backscattering spectrometer SPHERES. The time structure of the monitor follows the opening and closing of the chopper (see text).

3. Practical setup as a beam monitor

Fig. 6 shows a sketch of how the prompt γ radiation can be used. A thin lead shielding (depending of the γ intensity) of about 5 cm filters all low energy γ rays, a thicker lead shielding (about 10 cm) blocks the main part of the remaining γ rays from other sources. The general hall background can be substracted with a second background monitor, which has to be placed in a equivalent throughout thick shielding (without the thin lead filter towards the beam).

4. Conclusion

The NaI monitor is an easy to use plug-and-play device which performed similarly to the currently used traditional BF3-beam monitor for the 60x60 mm^2 beam. The stability and reproducibility at the same settings of the instrument seems to fulfil the requirements of the NSE spectrometer. It has the advantage that no additional material is needed in the beam since Al-windows producing the required prompt γ's are available anyway and thus no losses of intensity due to the monitor are encountered. The monitor has a slight dependence on the magnetic field of the main coils, it is visible at the present position as a drop of count rate as the B-field is at about 4 Gauss. The signal-to-background ratio at the current position is 20. The monitor was not shielded against γ's coming from other experiments or neutron guides. A better placement and a lead shielding around the monitor will certainly improve the background situation. The magnetic field sensitivity might be reduced with a μ-metal shielding. One only needs to ensure that the influence of the shielding on the magnetic field homogeneity of the instrument is negligible. Between neutron guide and second polarizer would be such a position at the J-NSE spectrometer, which is further away from the main coils with two Al-windows present. The temperature dependence has not been studied, here the monitor has been operated in an air conditioned stable environment. Variations in temperature will have a certain influence on the efficiency of the monitor. The sensitivity to external background could be further reduced by a differential measurement with a second NaI-monitor, placed in an equivalent position concerning the γ sources from other experiments, but shielded against the prompt γ's of the own instrument.

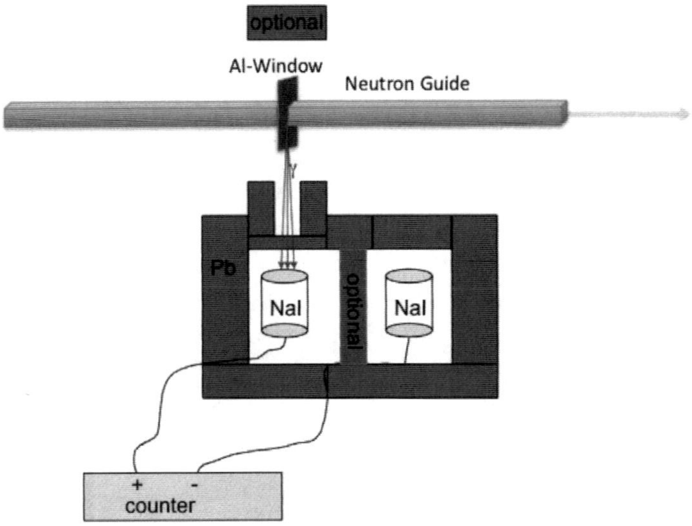

Figure 6. Measurement principle for the NaI beam monitor. A thin lead shielding of about 5 cm filters all low energy γ rays, a thicker lead shielding (about 10 cm) blocks the main part of the γ rays from other sources. The general hall background can be substracted with a second background monitor.

5. Acknowledgement
We thank Karl Zeitelhack (TU München) for fruitful discussion.

[1] http://www-nds.iaea.org/pgaa/tecdoc.pdf.
[2] http://www.nndc.bnl.gov/capgam/byn/.
[3] G. Pfennig, H. Klewe-Nebenius, W. Seelmann-Eggebert, Karlsruher Nuklidkarte, 6th edition (1998).
[4] F. Mezei, ed., *Neutron Spin Echo*, no. 128 in Lecture Notes in Physics Vol. 128 (Springer, Berlin, Heidelberg, New York, 1980).
[5] F. Mezei, P. C., and G. T., eds., *Neutron Spin Echo Spectroscopy*, no. 601 in Lecture Notes in Physics (Springer, Berlin, Heidelberg, New York, 2003).
[6] H.T. Motz and D.E. Alburger, Phys. Rev. 86 (1952) 165.
[7] M. Monkenbusch, R. Schätzler, and D. Richter, Nuclear Instruments & Methods In Physics Research Section A-accelerators Spectrometers Detectors And Associated Equipment **399** (1997) 301.
[8] J. Wuttke, A. Budwig, M. Drochner, et. al, Rev. Sci. Instrum **83** (2012) 075109.
[9] http://physics.nist.gov/PhysRefData/XrayMassCoef/ElemTab/z82.html.
[10] D. Richter, M. Monkenbusch, A. Arbe, and J. Colmenero, Adv. Polym. Sci. **174** (2005) 1-221.
[11] M. Kerscher, P. Busch, S. Mattauch, H. Frielinghaus, D. Richter, M. Belushkin, and G. Gompper, Phys. Rev. E **83** (2011) 030401(R).
[12] H. Frielinghaus, M. Kerscher, O. Holderer, M. Monkenbusch, and D. Richter, Phys. Rev. E **85** (2012) 041408.
[13] A.G. Zilman, R. Granek, Phys. Rev. Lett. **77** (1996) 4788-4791.
[14] H. Frielinghaus, V. Pipich, A. Radulescu, M. Heiderich, R. Hanslik, K. Dahlhoff, H. Iwase, S. Koizumi, and D. Schwahn, J. App. Cryst. **42** (2009) 681.

Precision neutron flux measurement with a neutron beam monitor

T Ino[1], H Otono[2], K Mishima[3], and T Yamada[3]

[1]KEK, High Energy Accelerator Research Organization, Tsukuba, Japan

[2]Kyushu University, Fukuoka, Japan

[3]University of Tokyo, Tokyo, Japan

E-mail: takashi.ino@kek.jp

Abstract. Neutron beam monitors are regularly used in various neutron beam experiments to compare two or more sets of data taken in different experimental conditions. A neutron lifetime experiment at BL05, the NOP beamline, in J-PARC requires to monitor the initial neutron intensity with an precision of 0.1% to measure the neutron lifetime with the same accuracy. The performance of a thin ^{3}He gas neutron beam monitor used for the experiment was studied to estimate the systematic uncertainties in the neutron lifetime measurement.

1. Introduction

A precision neutron lifetime measurement is underway at BL05, the NOP beamline, in J-PARC [1]. It is aimed to measure the neutron lifetime with an accuracy of 0.1% by detecting the neutron decays in flight. In such a precise measurement, it is crucial to monitor the neutron intensity with high precision.

At a spallation neutron source, the neutron intensity fluctuates with time depending on the characteristics of the proton beam to the neutron target, the current, the position, and the profile. The moderator temperature should also affect the neutron intensity as well as the energy spectrum. Figure 1 shows such time variation of the neutron intensity measured at the NOP beamline in J-PARC. Short-term and long-term fluctuations of the neutron flux that exceed far beyond 0.1% are clearly observed. A rough correlation between the proton current and the neutron intensity can be seen in the first half period, but later the discrepancy increased more than 1%. This fact tells that it is definitely necessary to continuously keep track of the incident neutron flux for the neutron lifetime experiment.

2. Systematic uncertainties in neutron beam monitoring

Beam monitoring with a gas proportional counter is affected by the change of the gas gain that depends on the high voltage, the temperature, the atmospheric pressure, and the beam flux (space charge effect). However, in a sealed gas detector such as a ^{3}He neutron detector, the gas pressure or the gas density remains the same with atmospheric pressure changes, and therefore no gas gain change. The gas gain should also vary with the incident beam position because of the electric field distribution inside the detector. The data acquisition electronics such as amplifiers and ADC may be responsible for the instability. All these characteristics regarding the systematic uncertainties in the neutron flux measurement are to be examined.

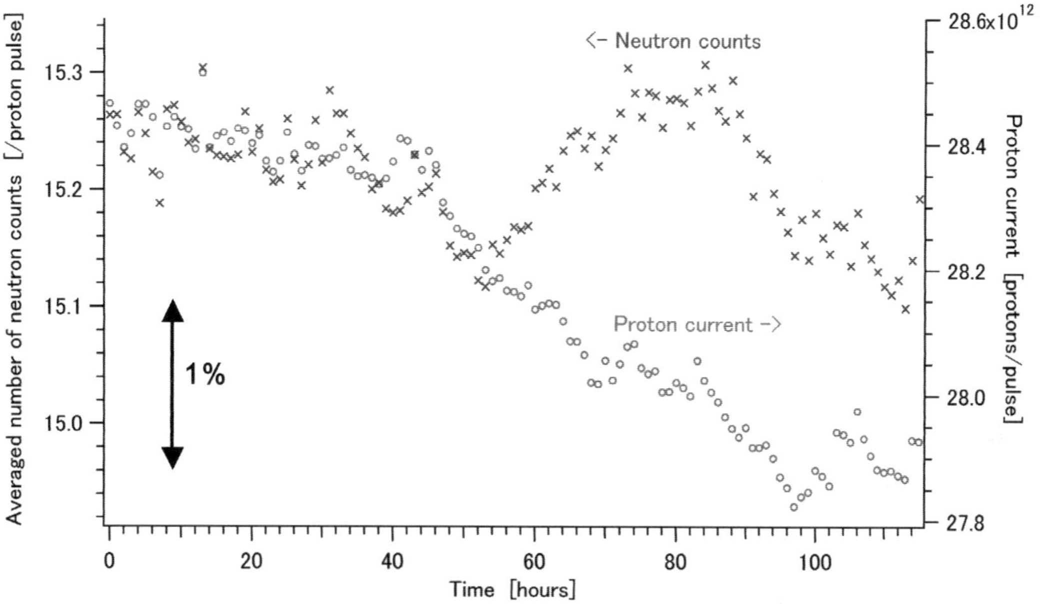

Figure 1. Time variation of the incident neutron flux was measured at the NOP beamline in J-PARC (crosses). Each measurement point corresponds to the average over one hour. The proton beam current to the neutron target is also plotted in circles (right axis). The left and right axes are properly scaled.

3. Neutron beam monitor and data acquisition system

The neutron beam monitor MNH10/4.2F used in the neutron lifetime experiment was purchased from CANBERRA (Figure 2). It is a thin ^{3}He gas detector with a neutron detection efficiency of $\sim 4\times10^{-5}$ at the thermal energy. The specifications of the beam monitor are listed in Table 1.

The data acquisition (DAQ) electronics are schematically shown in Figure 3. The detector output is connected to a charge amplifier CANBERRA 2006E followed by a shaping amplifier ORTEC 570 with a decay time constant of ~ 5 μs. The pulse heights are digitally recorded in the list mode with their trigger times by a peak detect ADC module IWATSU A3100. The nominal high voltage (HV) of 1000 V for the beam monitor is provided by a HV power supply ORTEC 556. The specifications of each module are summarized in Table 2.

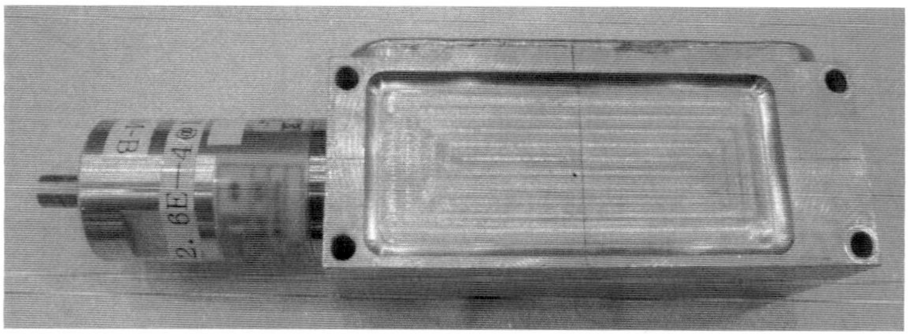

Figure 2. Neutron beam monitor used for the neutron lifetime measurement at the NOP beamline in J-PARC.

Table 1. Specifications of the neutron beam monitor.

Mechanical structure	material	A5 aluminum
	active length	100 mm
	active width	42 mm
	active thickness	40 mm
	window thickness	2 mm (each side)
Gas	1.3 bar of Ar + 10% of CH_4 + little ^{3}He	
Mean high-voltage	1000 V	
Neutron detection efficiency	$\sim 4\times 10^{-5}$ (thermal neutrons)	

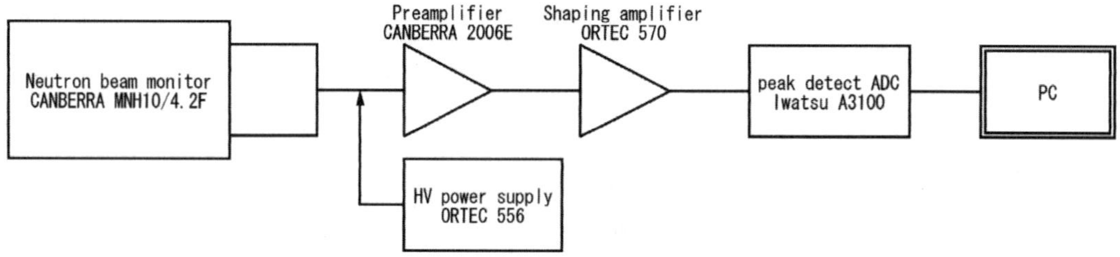

Figure 3. The data acquisition electronics for the neutron beam monitor.

Table 2. Selected specifications regarding the stability for the DAQ electronic modules.

High voltage power supply ORTEC 556	Temperature coefficient	< ±50 ppm/°C
	Long-term drift	< 0.01% / hour < 0.03% / 24hour
Preamplifier CANBERRA 2006E	Integral nonlinearity	< ±0.02%
	Temperature coefficient	< ±0.01%/°C
Shaping amplifier ORTEC 570	Integral nonlinearity	< ±0.05%
	Temperature coefficient	≤ ±0.0075%/°C
	DC Level	≤ ±50 μV/°C
Peak detect ADC IWATSU A3100	Integral nonlinearity	≤ ±0.025%
	ADC resolution	13 bits for 10 V full scale

4. Neutron beam measurements

Neutron beam tests were performed at the NOP beamline in J-PARC. The experimental setup was essentially the same as the neutron lifetime measurement [1] except for the HV and the position dependence tests where the neutron beam was tightly collimated. The cross sectional profile of the neutron beam was a square of $20\,\text{mm}\times 20\,\text{mm}$, and incident at the center of the beam monitor. The neutron flight path from the neutron source to the beam monitor was 18.4 m.

The pulse height distribution of the beam monitor signal is shown in Figure 4. The spectrum looks quite strange for a ^{3}He neutron detector. It turned out that the detector somehow contains boron or boron compound that makes several peaks and a shoulder in addition to the one by the ^{3}He(n,p) reactions:

$$^3\text{He} + \text{n} \rightarrow \text{p (573 keV)} + {}^3\text{H (191 keV)}\ .$$

The highest peak denoted as **I** in Figure 2 corresponds to the total energy of p (573 keV) + ^{3}H (191 keV).

Peaks **II**, **III**, and **IV** are from the neutron absorption reactions of ^{10}B:

$$^{10}B + n \rightarrow {}^{7}Li\ (1015\ keV) + \alpha\ (1777\ keV) \quad 6\% \quad \text{and}$$
$$^{7}Li\ (840\ keV) + \alpha\ (1470\ keV) + \gamma\ (482\ keV) \quad 94\%,$$

and they correspond to ^{7}Li (1015 keV), α (1470 keV), and α (1777 keV), respectively. The 482 keV γ rays are not detected. The ^{7}Li (840 keV) energy depositions are contained in peak **I** and do not make a separate peak due to the limited energy resolution. There is a shoulder denoted as **V** in the spectrum, and it is most likely made up by ^{7}Li (840 keV) + α (1470 keV) where the total deposition energy becomes 2310 keV. It is obvious that boron or boron compound is not in the gas state but adheres to the inner walls of the beam monitor because each reaction product, ^{7}Li or α, forms a peak individually, and the accompanying produced nuclei are absorbed in the detector walls. If boron compound were in the gas state, there would appear two peak at 2310 keV and 2792 keV by the ^{7}Li and α pairs. The peaks/shoulder and the corresponding particle(s) energy are summarized in Table 3.

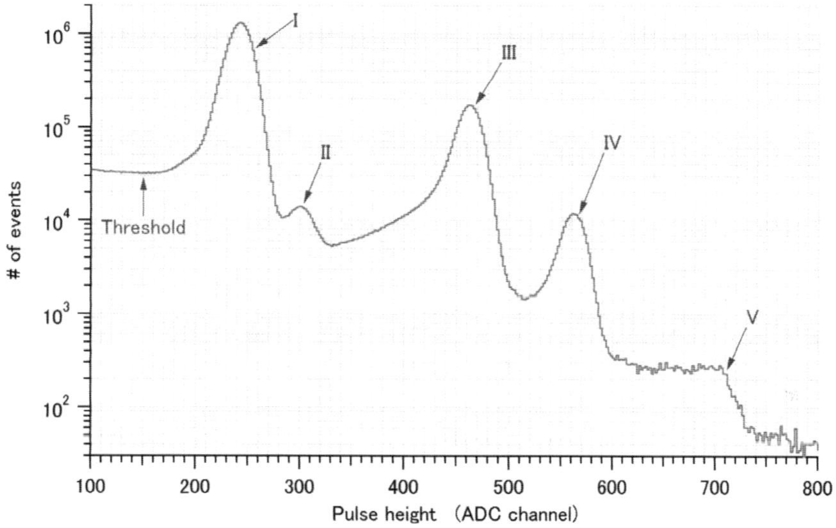

Figure 4. Pulse height distribution by the neutron beam monitor. Four peaks **I, II, III,** and **IV** are observed together with a shoulder (**V**). The neutron events are identified as the pulse heights over the nominal threshold of 150.

Table 3. The peaks and the shoulder in Figure 4 and the corresponding particle(s) energy.

Peak / shoulder	Reaction and particle(s) energy
I	p (573 keV) + ^{3}H (191 keV) from $^{3}He(n,p)$ ^{7}Li (840 keV) from $^{10}B(n,\alpha)$
II	^{7}Li (1015 keV) from $^{10}B(n,\alpha)$
III	α (1470 keV) from $^{10}B(n,\alpha)$
IV	α (1777 keV) from $^{10}B(n,\alpha)$
V	^{7}Li (840 keV) + α (1470 keV) from $^{10}B(n,\alpha)$

The instability of the neutron detection efficiency should not be affected by the presence of ^{10}B since the energy deposition by the $^{10}B(n,\alpha)$ reactions are much higher than the nominal threshold of 150 for the neutron identification. However, it has position dependence as detailed in the next chapter.

The time-of-flight (TOF) spectrum of the incident neutron beam in the neutron lifetime experiment apparatus is shown in Figure5. The peak counting rate was observed to be ~ 22000 cps around a tof of 21 ms, which corresponds to a neutron wavelength of ~ 0.45 nm, when the proton accelerator was operated at ~ 300 kW with a repetition rate of 25 Hz. It is noted that Figure 5 shows the neutron counting rate in which the neutron detection efficiency of the beam monitor is convoluted.

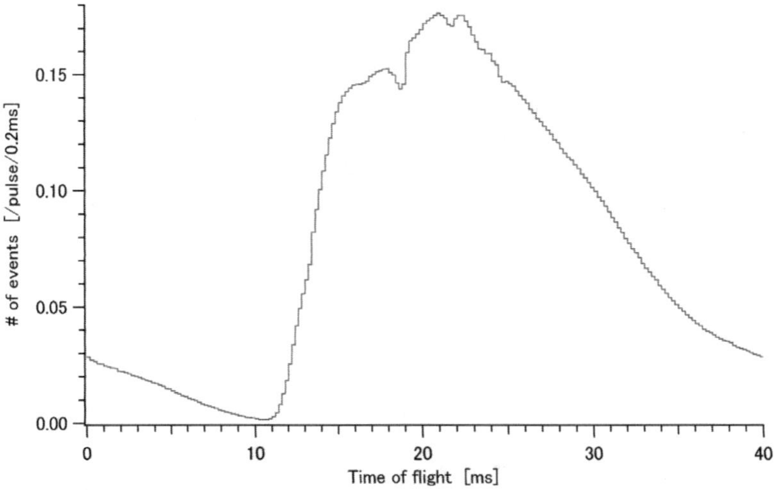

Figure 5. TOF spectrum of the incident neutrons in the neutron lifetime experiment. Only cold neutrons are delivered to the apparatus by the use of neutron supermirror benders (see Ref. 2 for details). The highest intensity appears around TOF of 21 ms. Frame-overlap neutrons are seen in TOF less then ~ 10 ms.

5. Results and discussions

5.1. Gain fluctuation and neutron detection efficiency

The change of the neutron detection efficiency due to the fluctuations of the gas gain and the amplifier gains was estimated with the pulse height spectrum in Figure 4 by changing the energy threshold around the nominal value of 150. Figure 6 shows such dependency to be $\Delta\varepsilon / \Delta G = -0.104$, where $\Delta\varepsilon$ and ΔG are the normalized values of the efficiency and the gain changes, respectively. This leads the tolerance of the gain variation to be 1% to keep the efficiency change within 0.1%. The temperature stabilities of the electronic modules are much less than 1% in normal room temperature as shown in Table 2; the amplifiers and the ADC have good enough linearities as well.

5.2. High voltage

The high voltage (HV) dependence of the detector gas gain was measured by actually changing the HV to the beam monitor and turned out to be $\Delta G / \Delta V = 3.7 \times 10^{-3}$ / Varound the nominal HV of 1000 V (.Figure 7). The corresponding tolerance is ± 2.7 V or $\pm 0.27\%$ to keep $\Delta G < 1\%$. HV power supplies commonly used in physics measurements, including ORTEC 556, have better stabilities and smaller temperature coefficients compare to this tolerance (Table 2).

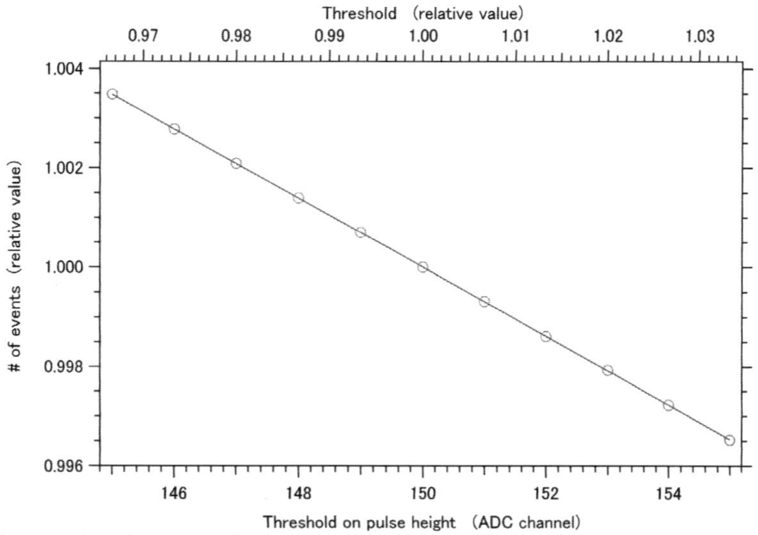

Figure 6. The relation between the neutron detection efficiency and the gas/amplifier gain was estimated by changing the ADC threshold level.

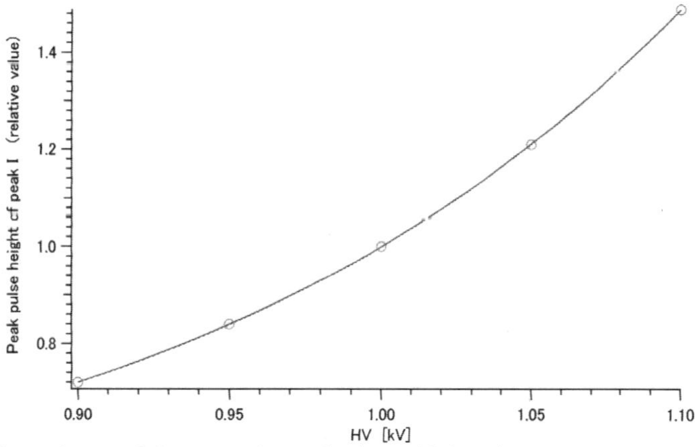

Figure 7. The dependence of the gas gain on the input high voltage.

5.3. Neutron flux effect

The detector gas gain may drop with increasing counting rate because of the space charge effect where a huge amount of gas ions is created near the anode wire by the electron avalanche and disturbs the electrostatic field. Such nonlinearity of the gas gain was studied by changing the neutron intensity with attenuators. Aluminum plates were placed ~ 2 m upstream the beam monitor as neutron attenuators. Figure 8 shows the attenuation in the neutron intensity by ~10% with 12 mm thick aluminum, corresponding to an attenuation factor of 0.84% for an aluminum thickness of 1 mm. The gas gains for the different neutron intensities were measured as the peak pulse height of **I** in the pulse height spectrum. The gas gains measured with different aluminum thicknesses are plotted in Figure 9. No clear dependence on the neutron intensity is observed. The gas gain variation is less than 0.1% for a change of the neutron intensity of 10% in the neutron lifetime experiment apparatus.

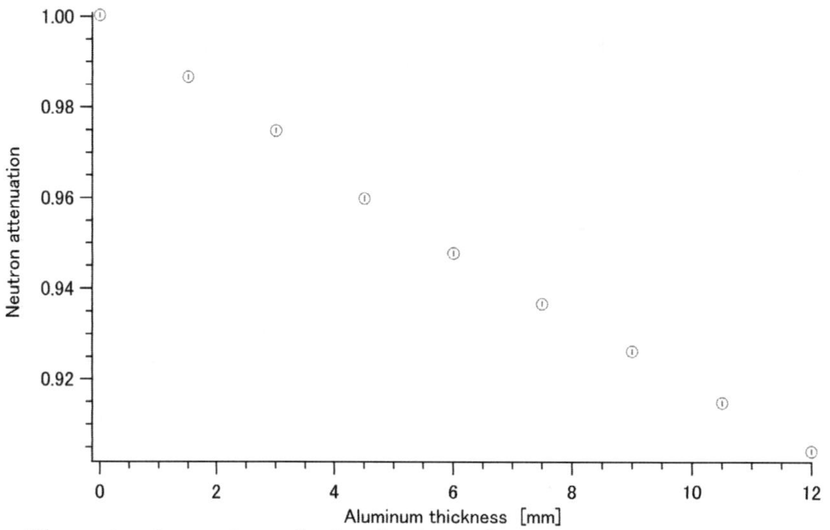

Figure 8. The neutron beam attenuation by aluminum plates.

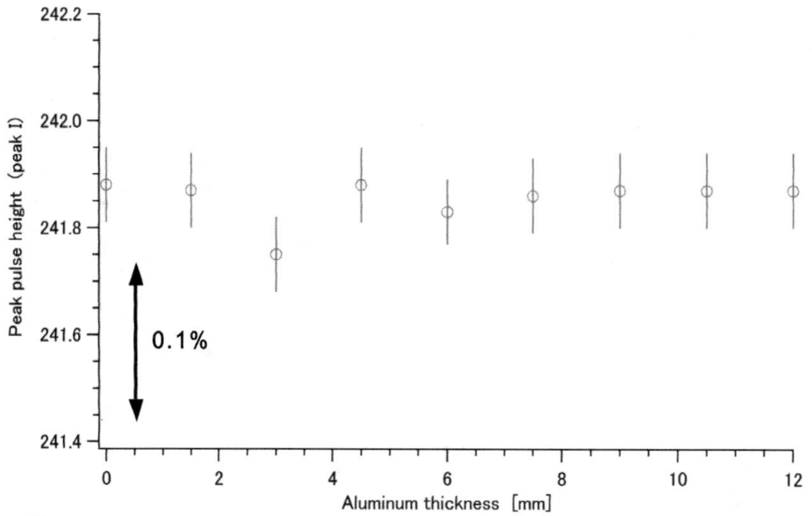

Figure 9. The gas gains were measured as the peak pulse heights **I** in the pulse height spectra with different neutron attenuations.

5.4. The temperature dependence of the gas gain

The temperature dependence of the gas gain in sealed proportional counters was studied by Vanha-Honko [3], and it was found to be $\sim 100\,\mathrm{ppm/^{\circ}C}$, which is small enough compared to the requirement of 1%.

5.5. Position dependence

The position dependence of the neutron detection efficiency was measured by scanning the beam monitor position to the fixed neutron beam of ϕ 10 mm. The position was moved with 8 mm steps in the longitudinal (12 positions) and transverse (5 positions) directions. The signal pulse heights were measured at each detector position to see the position dependent gas gain. Figure 10(a) shows the peak pulse heights of **I** in the pulse height spectra as the gas gains, normalized with that at the detector center. A systematic change of the detector gas gain is clearly seen due to the electrostatic field distribution inside the beam monitor. Sharp drop and increase near the edges are visible in the longitudinal direction. For the transverse direction, the gas gain change is rather limited, and the minimum appears at the center. The position dependence of the gas gain around the detector center is ~0.3%/cm and ~1.5%/cm for the longitudinal and transverse directions, respectively.

The neutron detection efficiency map, normalized at the center, is shown in Figure 10(b). Apparently, there is no systematic change in the efficiency, and the variation is more than 10% and randomly distributed. The ^{10}B adhesion should be responsible for it, but the reason of the randomness in the ^{10}B thickness is not known. This position dependence is not acceptable for the neutron lifetime experiment, and therefore the beam monitor has to be fixed at a definite position.

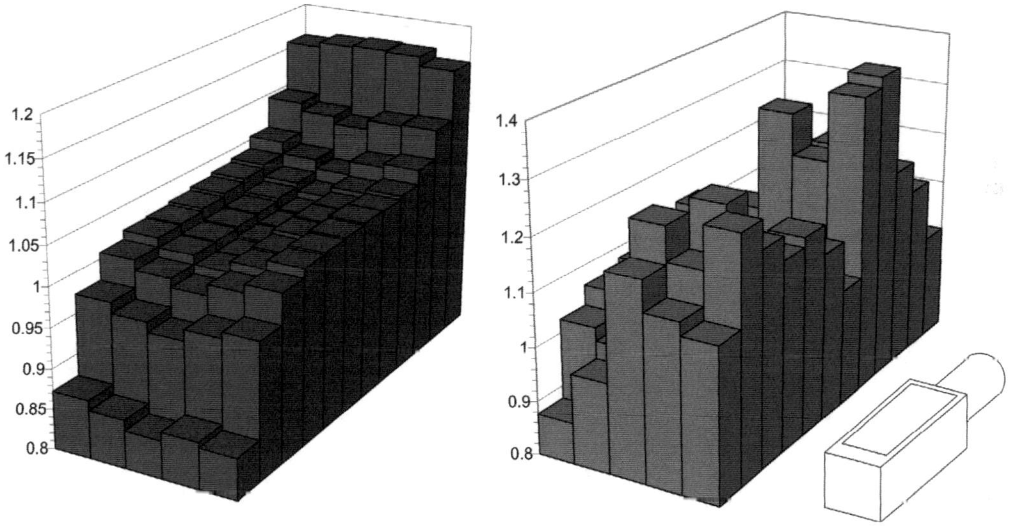

Figure 10.
Left : (a) the position dependence of the peak **I** pulse height (~ ch240) in Figure 4 is shown as the gas gain variation, normalized at the detector center.
Right : (b) the neutron counting efficiency is plotted for each incident neutron position. The efficiency is normalized at the detector center. The beam monitor sketched at the right bottom corner shows the positional relation between the plots and the beam monitor.

Finally, we note that background events including gamma rays, neutrons coming from other than the beamline, the electronics noise, and other sources were measured by placing a $^{10}B_4C$ plate or a ^{6}LiF plate in front of the beam monitor for every beam test, and each number of the background events was confirmed to be negligible small.

6. Conclusions
We have studied the performance of a neutron beam monitor MNH10/4.2F manufactured by CANBERRA. The neutron counting efficiency of the detector with the data acquisition electronics was measured its stability to the high voltage drift, the temperature variation, and the neutron intensity fluctuation, and it was found to satisfy the requirement in the neutron lifetime experiment. The position dependence of the detector gas gain was also studied and turned to have a systematic trend with a limited change. It was discovered, however, that the neutron detection efficiency varies randomly with the incident neutron position probably due to the adhesion of boron or boron compound

on the inner walls of the beam monitor. The variation is more than 10%, which far exceeds the tolerable level, and the beam monitor has to be used with a fixed position to the incident neutron beam.

References

[1] Y. Arimoto *et al.*, "Present status of neutron fundamental physics at J-PARC," Prog. Theor. Exp. Phys. (2012) 02B007.
Hidetoshi Otono, "New detector system for the precise neutron lifetime measurement using pulsed cold neutron beams," Ph.D thesis, University of Tokyo, 2011.
[2] K. Mishima *et al.*, "Design of neutron beamline for fundamental physics at J-PARC," Nucl. Instrum. Methods A600, 342 (2009).
[3] Vesa Vanha-Honko, "The temperature dependence of the gas gain in sealed proportional counters," Nucl. Instrum. Methods 176, 213 (1980).

International Workshop on Neutron Optics and Detectors (NOP&D 2013) IOP Publishing
Journal of Physics: Conference Series **528** (2014) 012040 doi:10.1088/1742-6596/528/1/012040

In-beam test of the Boron-10 Multi-Grid neutron detector at the IN6 time-of-flight spectrometer at the ILL

J. Birch[1], J.-C. Buffet [2], J.-F. Clergeau[2], J. Correa[2], P. van Esch[2], M. Ferraton[2], B. Guerard[2], J. Halbwachs[2], R. Hall-Wilton[3], L. Hultman[1], C. Höglund[1,3], A. Khaplanov[2,3], M. Koza[2], F. Piscitelli[2], M. Zbiri[2]

[1] Linköping University, Thin Film Physics Division, IFM, SE-581 83 Linköping, Sweden
[2] Institute Laue Langevin, Rue Jules Horowitz, FR-38000 Grenoble, France
[3] European Spallation Source ESS AB, P.O Box 176, SE-221 00 Lund, Sweden

E-mail: anton.khaplanov@esss.se

Abstract.
A neutron detector concept based on solid layers of boron carbide enriched in ^{10}B has been in development for the last few years as an alternative for ^{3}He by collaboration between the ILL, ESS and Linköping University. This Multi-Grid detector uses layers of aluminum substrates coated with $^{10}B_4C$ on both sides that are traversed by the incoming neutrons. Detection is achieved using a gas counter readout principle. By segmenting the substrate and using multiple anode wires, the detector is made inherently position sensitive. This development is aimed primarily at neutron scattering instruments with large detector areas, such as time-of-flight chopper spectrometers. The most recent prototype has been built to be interchangeable with the ^{3}He detectors of IN6 at ILL. The ^{10}B detector has an active area of $32 \times 48\,cm^2$. It was installed at the IN6 instrument and operated for several weeks, collecting data in parallel with the regularly scheduled experiments, thus providing the first side-by-side comparison with the conventional ^{3}He detectors. Results include an efficiency comparison, assessment of the in-detector scattering contribution, sensitivity to gamma-rays and the signal-to-noise ratio in time-of-flight spectra. The good expected performance has been confirmed with the exception of an unexpected background count rate. This has been identified as natural alpha activity in aluminum. New convertor substrates are under study to eliminate this source of background.

1. Introduction

The Multi-Grid is a novel detector for thermal neutrons. It is based on thin solid converter layers of $^{10}B_4C$. The detector concept has been developed at the ILL in collaboration with the ESS in order to meet the requirements of upcoming neutron scattering instruments where detector areas of many square meters will be required. Cold and/or thermal Time-of-Flight (ToF) spectrometers in particular require a sensitive area of the order of $30\,m^2$, and it is at these instruments that the Multi-Grid design is particularly targeted. As can be seen in the ESS Technical Design Report [1], 3 instruments of this type may be expected to be built at the ESS. While previously such detector arrays [2] have been built with ^{3}He as the converting

Figure 1. In this image of the detector, the B_4C layers are horizontal and neutrons arrive from the top.The 6 modules of 16 grids each form an assembly of 360 gas-filled tubes, 24 tubes wide and 15 deep, which are visible on the front side. Each tube is readout by an anode wire.

medium, this option is likely unfeasible in the future due to a very limited world-wide supply of ^{3}He [3, 4, 5, 6], therefore new technology is required.

1.1. Multi-Grid detector

Descriptions and characterization of the previous Multi-Grid prototypes have been published [7, 8, 9, 10]. Here, we report on the first test of this technology in the time-of-flight spectrometer IN6 at ILL.

The present detector is composed of 6 modules of 16 grids as shown in figure 1. Each grid is composed of 14 so-called *blades* which are coated on both sides with $1\,\mu m$ of $^{10}B_4C$ [12] and mounted orthogonally to the direction of the incoming neutrons. The other surfaces are not coated. The unit cell of a grid has a $20 \times 20\,mm^2$ sensitive area and $10\,mm$ depth. When a stack of 16 grids is assembled, 60 rectangular tubes are formed. These are readout by anode wires. Signals are readout from both the anode wire and the individual frames. Coincident signals are recorded and define 3-D positions of detected neutrons. The depth of the interaction is necessary since in ToF applications, the exact flight distance of each neutron must be known in order to reconstruct its velocity.

Grids were connected 3-by-3 together with their neighbours from the adjacent modules and readout by a single amplifier. The anode wires were connected using resistances in charge division chains. Each such chain contained 30 wires which run from the back of the detector to the front, connect to a neighbouring set of tubes and then run from the front of the detector to the back. In this arrangement, the discrete channel information corresponding to the detector depth was converted to analogue information and some ambiguity may arise as to the exact wire that has detected an event. Note, however, that no such ambiguity exists in the other 2 dimensions. In total, only 32 channels were required to readout all frames and 24 channels to readout all wires (2 ends of 12 charge division chains). With this setup, the new detector was readout with a similar number of channels as the present IN6 detectors.

In order to limit the effect of neutron scattering in the detector, the 6 modules were shielded

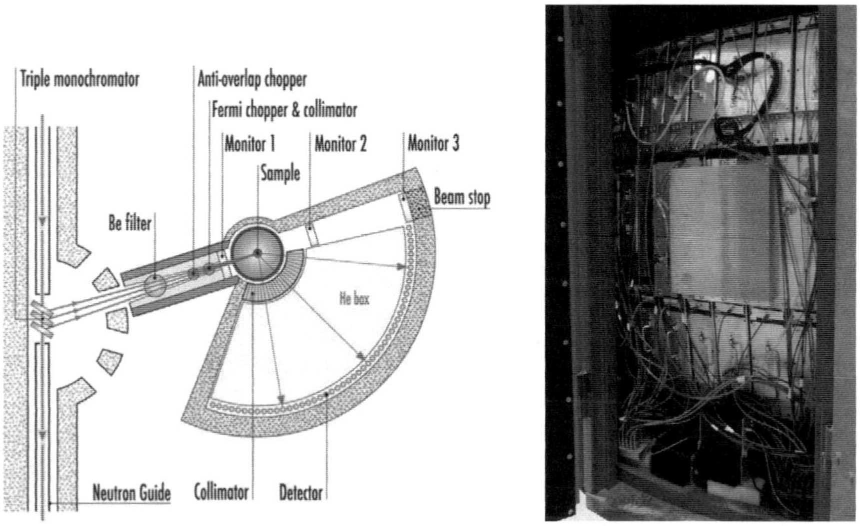

Figure 2. **Left:** The geometry of IN6. **Right:** The Multi-Grid detector installed at the instrument. ^{3}He modules can be seen above and below the ^{10}B detector.

from each other by sheets of cadmium. The sides of the detector were additionally shielded. The detector was filled with $1\,bar$ of $ArCO_2$ 90/10 mixture. The gas was continually flushed during the operation.

1.2. IN6 instrument

IN6 is a cold neutron time-focusing time-of-flight chopper spectrometer at the ILL. Its layout is shown in figure 2. The incident neutron wavelength can be selected by adjusting the monochromator orientation to the Bragg condition for 4.1, 4.6, 5.1 and 5.9 $\AA$. A chopper system defines pulses incident on sample. The detector array has a sensitive area of $4\,m^2$ and detects neutrons in the scattered angular range of $10° - 115°$. The neutron flight path from the sample to the detectors is $248\,cm$. By measuring both the angular and temporal spread of the detected neutrons, both momentum and energy transfer between scattered neutrons and the sample can be determined. More information about the instrument is available at the ILL website [11].

The IN6 detector comprises 337 ^{3}He tubes arranged in 3 tiers and angled so that each tube is tangential to the cone of neutrons scattered at the same angle. The tubes are compressed so that their width is $32\,mm$ and the depth is $16\,mm$. The pressure of ^3He is $3\,bar$. The ^{10}B detector replaced a total of 25 ^{3}He tubes in the middle tier at scattering angles of approximately $95 - 105°$ as shown in figure 2. This configuration was operated for several weeks, including 2 weeks of beam time, during which regular experiments were performed using the wavelength setting of $5.1\,\AA$. Additionally, dedicated detector tests were carried out at wavelength 4.1 and $4.6\,\AA$.

2. Results of measurements at IN6

2.1. Efficiency

Efficiency of a neutron detector based on multiple layers of solid converters depends on the thickness of the layers. In order to obtain an optimal efficiency, two parameters – the probability of conversion and the probability to detect secondary particles – must be balanced. The

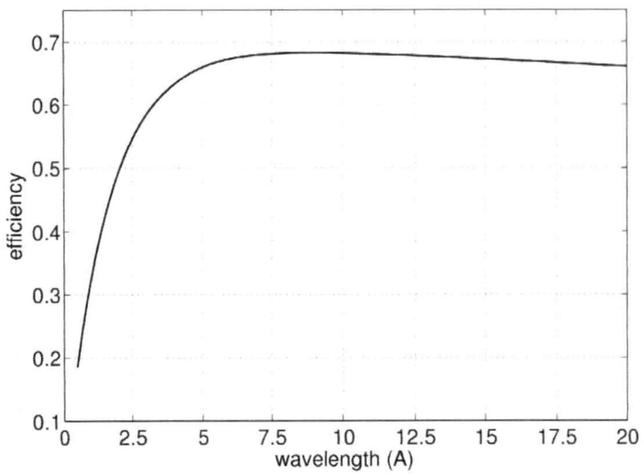

Figure 3. Calculated efficiency for the Multi-Grid detector with 28 $^{10}B_4C$ layers of $1\mu m$ as a function of neutron wavelength.

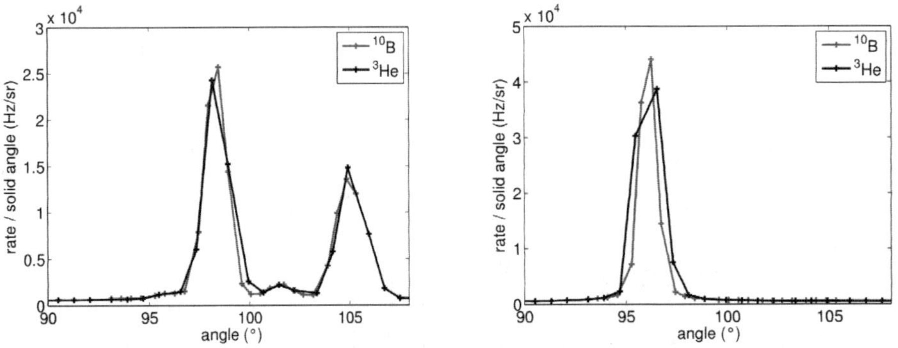

Figure 4. Angular distributions obtained with ^{3}He tubes and Multi-Grid detector. **Left:** 4.1 $\mathring{A}$ and **right:** 4.6 $\mathring{A}$ incident wavelength.

optimal value depends on the neutron wavelength. With $^{10}B_4C$ layers of $1\mu m$, the detector is optimized for 2.5 $\mathring{A}$. We therefore expect that the Multi-Grid should have an advantage at these lower wavelengths. Measurements with monochromatic neutrons of 2.5 $\mathring{A}$ at the CT1 and CT2 beam lines at the ILL have shown that the efficiency of modules used in the IN6 prototype is $\approx 50\%$ [7, 8, 9]. The efficiency can be calculated for a range of neutron wavelengths [13], this is shown in figure 3.

During the IN6 test, it was also possible to make a relative comparison to the efficiency of the ^{3}He detectors. Measurements were carried out at 4.1 $\mathring{A}$ and 4.6 $\mathring{A}$ incident wavelengths, using samples of yttrium garnet and silicon respectively. These combinations of wavelengths and samples resulted in Bragg scattering of neutrons into the Multi-Grid detector. In this way the wavelength of the majority of detected neutrons is known and the rate is significantly higher than for neutrons scattered inelastically.

Figure 4 shows the neutron rate as a function of scattering angle for the Multi-Grid prototype and the ^{3}He tubes in the same angular range. Here we consider the rate per solid angle. Each point in figure 4 shows the counts registered in a tube divided by the solid angle of that tube as

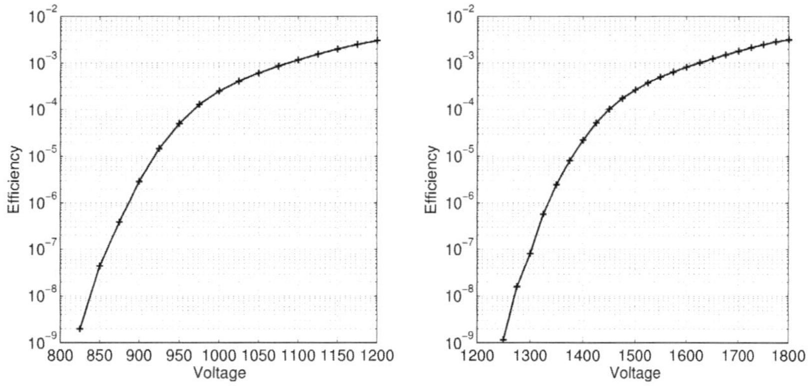

Figure 5. Measurement of the γ-ray efficiency with the Multi-Grid (left) and a ^{3}He detector (right) as a function of bias voltage. The nominal operating voltages for these detectors are $850\,V$ and $1350\,V$ respectively (as determined prior to this measurement).

a function of the angular position of the tube with respect to the sample.

Next we compare efficiencies including the effects of the dead spaces specific to each detector for a given wavelength. The counts are integrated over the range of tubes in which a Bragg peak is emitted. The ratio $rate(^{10}B)/rate(^{3}He)$ is 1.09 at 4.1 $\AA$, and 0.97 at 4.6 $\AA$. This means that the Multi-Grid detector is globally only 3% less efficient than the IN6 detection system at 4.6 $\AA$, and 9% more efficient at 4.1 $\AA$. This is in agreement with the optimization of the Multi-Grid for lower wavelengths.

2.2. Gamma-ray sensitivity

The sensitivity of neutron detectors to gamma-rays is an important characteristic since neutron scattering instruments are subject to a strong gamma flux originating from the instrument itself and from other sources such as the surrounding neutron guides and other instruments. γ-rays may interact in the detector, inducing unwanted signals and decreasing the signal-to-background ratio.

Gamma-ray discrimination in both the ^{10}B and ^{3}He detectors is based on the fact that the gamma photons are not likely to deposit more than a few tens of keV in the detector, whereas neutron interactions lead to larger energy deposits (up to $1.8\,MeV$ in the case of ^{10}B converter). This is because γ-rays are detected through secondary electrons while neutrons are detected through secondary ions interacting with the gas, therefore leading to a large difference in the energies deposited over a finite distance.

Gamma sensitivity measurements have been performed on a number of Multi-Grid prototypes [14]. Figure 5 shows the efficiency to detect a $662\,keV$ γ-ray measured with a $164\,MBq$ ^{137}Cs source, as the function of the high voltage, and, therefore, gas amplification gain, with a fixed value of the threshold. We see that at the typical bias voltage of $850\,V$ (for $1\,bar$ of $ArCO_2$), γ-ray efficiency is below 10^{-7} and can be improved by another order of magnitude if the voltage is lowered by $25\,V$. For comparison, figure 5 also shows the γ-ray efficiency for a ^{3}He detector measured with the same source (this is not an IN6 detector). While the γ-ray energy does influence the detection efficiency, the dependance is rather weak [14].

Taking advantage of the ToF information provided by the IN6 instrument allows validating the low γ-ray sensitivity in an alternative way. A large portion of the γ background inside the

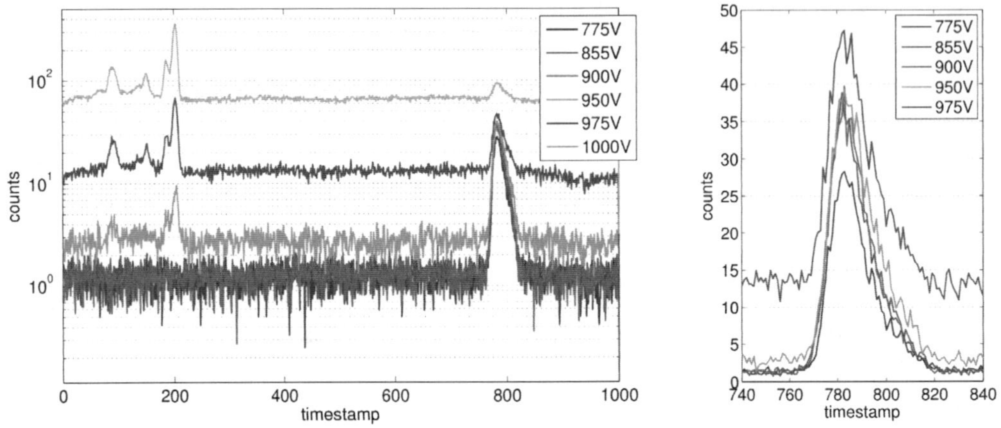

Figure 6. Time spectra measured for different values of the high voltage. **Left:** the full time scale corresponding to 6 ms. Peaks on the left were verified to correspond to γ-rays using a NaI detector. The peak on the right is due to the elastic neutrons. **Right:** the neutron peak in linear scale. Note the loss of neutron counts for the lowest voltage.

IN6 chamber comes from the components of the instrument itself, due to nuclear interactions of neutrons with the different elements of the instrument (choppers, sample, beam stop). This gamma background is therefore associated with a well defined time structure, as was observed in the measurement and can be seen on figure 6. The ToF spectrum in this figure was measured for several values of the high voltage. It is not corrected for the detector depth, since here we are interested in photons which travel at the speed of light. At each voltage, an elastic neutron peak is seen at about timestamp 800. The peaks at timestamp 200 and below can be seen only when the high voltage reaches $950\,V$. These are due to γ-rays. The time structure of these peaks implies that they are generated by the pulsed neutron beam of IN6 as it passes the absorbing components of the instrument. Of the constant γ rate, $\approx 2/3$ is due to the instrument, possibly emitted in the chopper closed state when the entire beam is absorbed.

The time spectrum in figure 6 was verified using a NaI scintillator detector placed adjacent to the ^{10}B prototype. The neutron peak cannot be seen in the photon detector, but the γ peaks are identical. This detector also allowed to estimate that on the order of 10^6 γ-rays impinge on the Multi-Grid prototype every second. No peaks corresponding to these γ-rays were found in any neutron time of flight spectra with high voltage of $900\,V$ and below.

2.3. Background

Photons are not the only source of background in a neutron detector. Indeed, in during the IN6 test, it became clear that a different background was present – one corresponding to significantly higher pulse heights than γ-rays. Figure 7 compares ToF spectra taken simultaneously with the ^{10}B and ^{3}He detectors. While the physical signal is evident in the Multi-Grid data and matches that measured by ^{3}He, the background level is higher. This background is not modulated when the beam is pulsed and is present irrespective of the instrument's shutter state or even the reactor operation. Subtracting a constant background, corresponding to $4.4\,Hz$ distributed over the whole detector, from the ^{10}B data, reproduces the ^{3}He data, although of course, with a higher statistical noise level.

The anode wire pulse height spectrum of the background is compared to the spectrum

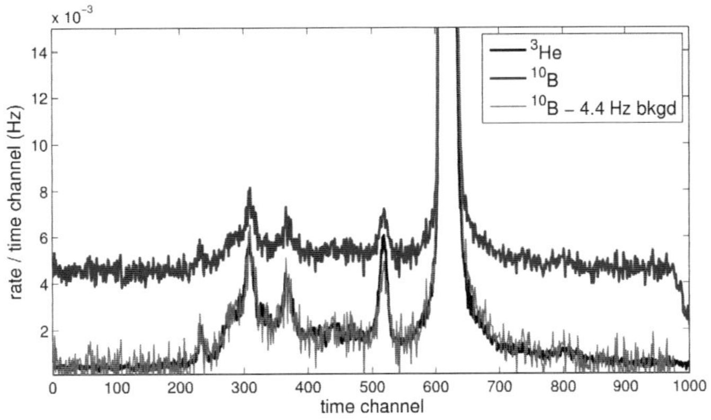

Figure 7. Time spectra obtained with ^{3}He detectors (black curve) and Multi-Grid detector (blue curve). The red curve is obtained by subtracting a flat 4.4 Hz background from Multi-Grid data in order to match the ^{3}He data.

corresponding to neutrons in figure 8. We see that energy of the majority of background events overlaps with the neutron events and some have higher energies. The background spectrum has a shape characteristic of a strong wall-effect (*i.e.* of particles stopped by the walls surrounding the gas volume before depositing the full energy in the gas). This is inconsistent with secondary atmospheric particles created by cosmic rays, since at Earth's surface, most of these are minimum ionizing and the expected energy deposit is similar to that of γ-rays – at most tens of keV for this size of the gas chamber. Furthermore, cosmic ray particles are more likely to be traveling vertically, and therefore would deposit more energy when a gas tube is also oriented vertically. The rate of the background, however, was constant for both vertical and horizontal orientations. We therefore exclude cosmic rays as the source of this background.

The background pulse height spectra do, however, closely resemble spectra of α particles. Figure 8 also shows simulated spectra of α particles emitted from the frames, where α are emitted from a random depth in the aluminium and in a random direction. The energies of the α's simulated correspond to those emitted by decay chains starting at 4 isotopes. The two most common primordial α-emitters, ^{232}Th and ^{238}U are shown as well as the relatively long-lived radium isotopes produced as a result of their decay. We see that the simulated spectra are qualitatively consistent with measurement, in particular the shape of the high-energy tail of the spectrum and its end-point at about $3\,MeV$. The likely source of such emissions are the naturally-occurring concentrations of uranium and thorium in aluminium. These undergo a chain of decays before arriving at a stable isotope, emitting 7-8 α in the process. The alpha energies vary between 4 and $9MeV$.

Contamination of aluminum by uranium and thorium isotopes and their decay daughters has been studied in the past [15, 16, 17]. Corresponding rates for standard aluminium alloys were found to be between 0.05 and $0.25\,cm^{-2}h^{-1}$, which for this configuration of Multi-Grid detector, leads to total rates over the whole detector between $0.96\,Hz$ and $4.8\,Hz$, consistent with what was measured experimentally.

This background level is presently the main concern in the framework of developing the Multi-Grid technology for large area position sensitive detectors for inelastic instruments. Two solutions are currently being investigated to eliminate this background. One is to deposit a thin nickel layer at the surface of aluminium in order stop the alpha particles before they reach the gas. Another solution is to use U/Th-free aluminium for which the concentration of uranium

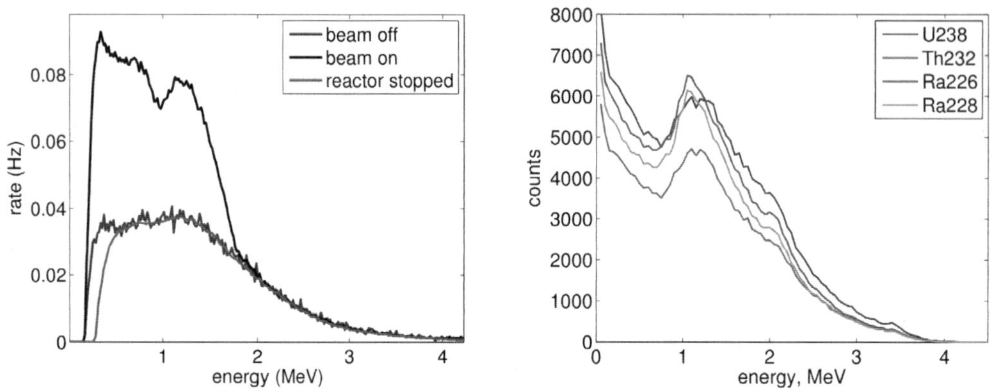

Figure 8. Left: energy spectra obtained in the Multi-Grid detector in IN6. With beam on, a characteristic spectrum of secondary particles emitted from the conversion layer is evident, however, it is superimposed with a higher-energy spectrum. With beam stopped, the spectrum no longer resembles that of neutrons. Note that a different threshold was used in the *beam on/off* and *reactor stopped* measurements. **Right:** Monte-Carlo simulation of energy spectra produced by α particles in the gas, emitted in the walls of the detector for different primary contaminants.

and thorium is 2-3 orders of magnitude lower than in standard aluminium alloys.

2.4. Neutron scattering in aluminium

The advantage of using aluminium for construction of the Multi-Grid detectors is its transparency for thermal neutrons. However, the scattering cross section in aluminium cannot be totally neglected, and may induce undesirable geometrical and temporal effects. Two measurements were carried out at 4.1 $\mathring{A}$ and 4.6 $\mathring{A}$ to observe these effects. For these wavelengths, Bragg reflection is allowed in aluminium. These measurement are compared to those at 5.1 $\mathring{A}$ in figure 9, where count rate is shown as a function of time and depth of the detector. The maximum of the elastic peak is moving toward larger time-of-flight values with increasing depth, due to an increased flight path inside the detector. The slope of the curve giving the position of the maximum of the elastic peak as a function of depth is directly correlated with the velocity of elastically scattered neutrons.

A secondary distribution of a much lower intensity moves toward higher ToF values with decreasing depth, which is characteristic of neutrons traveling with a negative velocity (from the back to the front of the detector). These neutrons are therefore identified as elastically scattered neutrons in the back of the frames. There is a large block of aluminium at the rear of the frame, which leads to this visible scattering effect. Figure 10 shows the time spectrum obtained for the whole detector after time correction. The broadening of the base of the elastic peak toward larger ToF values is a consequence of this effect. Note, that as expected, it is not present for neutron wavelength of 5.1 $\mathring{A}$. In the future versions of the Multi-Grid, cadmium shielding can be used directly behind the last boron layer in order to minimize this effect. Additionally, other aluminium parts of the frame will be thinner. Furthermore, absorbing material can be considered on the elements of the frames parallel to the neutron flux.

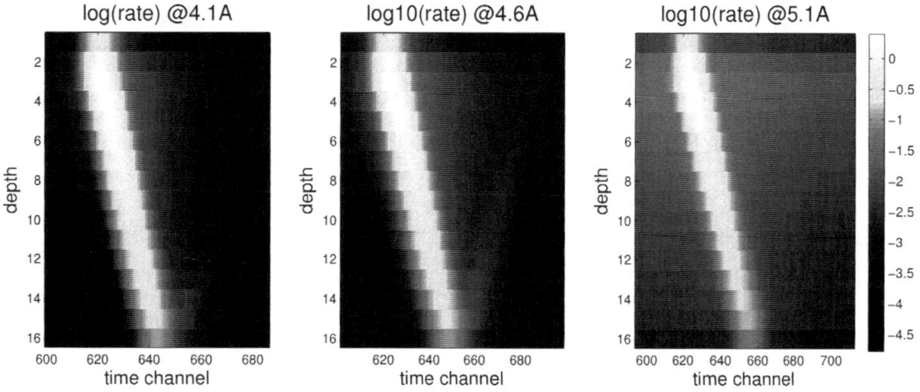

Figure 9. Time spectra measured as a function of depth (cell number) for 4.1 Å, 4.6 and 5.1 Å incident neutron wavelengths.

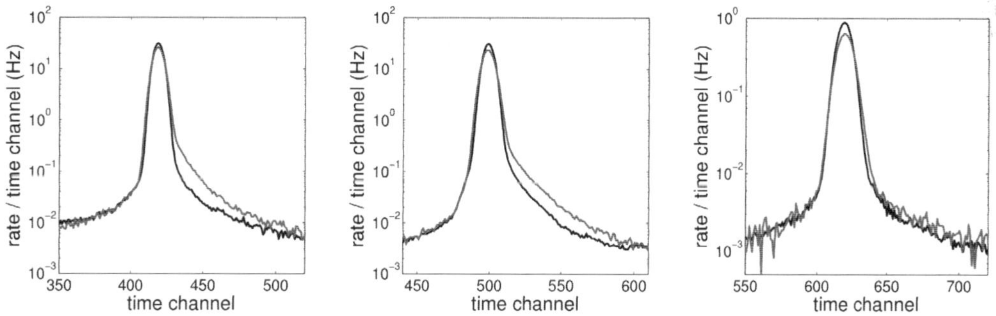

Figure 10. Time spectrum measured in the Multi-Grid detector (red) compared to the IN6 setup (black), from left to right – for 4.1, 4.6 and 5.1 Å. The enlargement of the base of the elastic peak is a consequence of scattering in aluminum.

3. Conclusion and Outlook

In combination with previous characterization studies of Multi-Grid prototypes, the in-beam test at IN6 at the ILL provides encouraging results. We have demonstrated that the new detector is able to reproduce the measurements currently done with ^3He detectors. In particular, efficiency, time resolution and γ-ray sensitivity are all already competitive. The contribution of scattered neutrons has been identified as expected. There are clear possibilities for improvement here, although some amount of scattering must be accepted as it must in any detector.

Due to the multi-layer nature of the ^{10}B detector, there are two aspects in which the Multi-Grid differs from conventional ^{3}He detectors. Firstly, the ratio of the internal surfaces to the active area of the detector is about an order of magnitude greater than for a single tube. Therefore, also the impact of radioactive contamination of the materials is likewise enhanced. Aluminium, the material chosen due to favourably low cross sections and cost, naturally contains more α emitters than other common metals. In order to reduce the background to that seen in ^{3}He detectors, two methods are currently investigated – nickel plating of the aluminium to stop the emitted α's, and the use of specially purified aluminium. Respective costs will likely be the deciding factor between these two approaches.

A second significant difference with ^{3}He detectors is that the detector can be optimized for a specific wavelength or distribution of wavelengths [13]. The current prototype has 30 $^{10}B_4C$ layers of $1\,\mu m$, which is the optimal for $2.5\,\mathring{A}$ neutrons. This may or may not be the best configuration for a given instrument. For example, efficiency can be significantly increased using thinner layers, if only cold neutrons are of interest, however, in this case the efficiency for thermal neutrons will decrease. In case of ^{3}He by contrast, the only parameter is the total amount of the gas per detector area, and the trade-off in the example above does not occur.

As the following step in this project, a larger demonstrator is currently being built. This detector will have the dimensions and specifications of a module of IN5 [2]. With 15 times the sensitive area ($80 \times 300\,cm^2$), this detector will demonstrate the feasibility and cost of large-scale production of both the mechanical parts as well as the $^{10}B_4C$ layers required.

Acknowledgments

The work has been supported by the CRISP project (European Commission 7th Framework Programme Grant Agreement 283745).

References

[1] S. Peggs, *ESS Technical Design Report* (2013) http://europeanspallationsource.se/scientific-technological-documentation
[2] H. Mutka, J. Ollivier and L. Didier., *The new cold neutron time-of-flight spectrometer IN5*, *Neutron News* 21(2) (2010).
[3] R.T. Kouzes, *The 3He supply problem. Technical Report 18388, PNNL* (2009) page.
[4] GAO, *Neutron detectors: an alternative to using Helium-3. Technical Report 11-753, US Government Accountability Office* (2011).
[5] D. A. Shea and D. Morgan, *The Helium-3 shortage: Supply, demand, and options for congress. Technical Report R41419, Congressional Research Service* (2010).
[6] K. Zeitelhack, *Search for alternative techniques to helium-3 based detectors for neutron scattering applications*, Neutron News, 23:4 (2012), 10-13.
[7] A. Khaplanov *et al.*, *Multi-Grid Boron-10 detector for large area applications in neutron scattering science*, *arXiv:1209.0566* (2012).
[8] J. Correa *et al.*, *10B4C Multi-Grid as an Alternative to 3He for Large Area Neutron Detectors*, *Trans. Nucl. Sc.* (2013) DOI: 10.1109/TNS.2012.2227798.
[9] B. Guerard *et al.*, *10B multi-grid proportional gas counters for large area thermal neutron detectors*, *Nucl. Instr. Meth. A* (2012) http://dx.doi.org/10.1016/j.nima.2012.12.021.
[10] Patent no. 20110215251
[11] Y. Blanc, *Le Spectrometre A Temps De Vol IN6, characteristiques, techniques et performances*, report 83B121G (1983) http://www.ill.eu/nc/instruments-support/instruments-groups/instruments/in6/more/infofilestodownload/
[12] C. Höglund, *et al.*, *B_4C thin films for neutron detection*, *J. Appl. Phys.* 111(104908) (2012).
[13] F. Piscitelli and P. van Esch, *Analytical modeling of thin film neutron converters and its application to thermal neutron gas detectors*, JInst, 8, P04020, (2013).
[14] A. Khaplanov *et al.*, *Investigation of gamma-ray sensitivity of neutron detectors based on thin converter films* arXiv:1306.6247 (2013) *submitted to J. Inst.*
[15] D.P. Bouldin, *The measurement of alpha particle emissions from semiconductor memory materials* Journal of Electronic Materials, Vol. 10, No. 4 (1981).
[16] M. Leroy, *Alpha Rays Emitting Impurities in Ultra Pure Aluminum Evolution Through the Successive Refining Steps* Journal de Physique 111, Volume 5,(1995)
[17] J. Hofmann *et al.*, *"Nuclear Safety and the environment: Natural Radionuclide Concentrations in Materials Processed in the Chemical Industry and the Related Radiological Impact"*, DG-Environment Report EUR 19264 (2000)

Upgrade of the POLDI diffractometer with a ZnS(Ag)/^{6}LiF scintillation detector read out with WLS fibers coupled to SiPMs

J.-B. Mosset[1], **A. Stoykov**[2], **V. Davydov**[1], **M. Hildebrandt**[2], **H. Van Swygenhoven**[1,3], **W. Wagner**[1]

[1]Paul Scherrer Institut, Spallation Neutron Source Division, Villigen, Switzerland
[2]Paul Scherrer Institut, Laboratory for Particle Physics, Villigen, Switzerland
[3]Ecole Polytechnique Fédérale de Lausanne (EPFL), NXMM, Lausanne, Switzerland

E-mail: jean-baptiste.mosset@psi.ch

Abstract. A thermal neutron detector based on ZnS(Ag)/^{6}LiF scintillator, wavelength-shifting fibers (WLS) and silicon photomultipliers (SiPM) is under development at the Paul Scherrer Institute (PSI) for upgrading the POLDI instrument, a pulse-overlap diffractometer. The design of the detector is outlined, and the measurements performed on a single channel prototype are presented. An innovative signal processing system based on a photon counting approach is under development. Its principle of operation is described and its performances are evaluated on the basis of a Monte Carlo simulation.

1. Introduction

POLDI (Pulse-Overlap Diffractometer) [1] is a neutron time-of-flight diffractometer commissioned in 2002 at the Swiss neutron spallation source (SINQ) at PSI. The instrument is designed to function as a strain scanner for the investigation of residual stress in engineering materials and components, and for in-situ deformation measurements. To allow a simultaneous measurement of the axial and transverse strain components during in-situ measurements, POLDI will be upgraded with two oppositely placed detector banks based on ZnS(Ag)/^{6}LiF scintillator replacing the actual single ^{3}He detector. As shown in Fig. 1, it is foreseen to install five modules per detector bank. However, due to the limited space in the POLDI area, the number of modules could be reduced, depending on the packing fraction of the modules. The minimum requirement is two modules per detector bank.

The actual high price of ^{3}He makes the scintillator technology much more attractive in terms of cost. Moreover, from a technical point of view, the scintillator technology is well established and is widely used for diffractometer instruments [2, 3]. Unlike the actual detector, the new detector will have a thin detection volume and it will not need to be in time-focusing geometry. This is a clear advantage of the scintillator over the ^{3}He technology.

As illustrated in Fig. 2, the detector module has a curved geometry with a 2 m radius of curvature. It is one dimensional position sensitive and its sensitive strips of 200 mm × 2.5 mm are parallel to its longitudinal axis. A detector bank consists of 5 detector modules stacked together in a 2 m radius sphere centered on the sample. Table 1 shows the specifications of the current and the new detectors.

Each single channel of a module may consist of several sandwiches made of an array of wavelength shifting (WLS) fibers squeezed between two ZnS(Ag)/^{6}LiF screens. Fig. 3 shows a schematic view of a module made of 3 sandwiches with 12 WLS fibers per channel. The fiber pitch and the thickness of the scintillating layers are geometrical parameters which need to be optimized in terms of light yield and neutron absorption to maximize the neutron detection efficiency. In order to reach the required detection efficiency (65% at 1.2 Å), the detector module will be designed so that it will provide a neutron absorption probability of at least 80% (The module will very likely consist of 3 or 4 sandwiches). The WLS fibers belonging to the same channel are grouped toghether and coupled to a silicon photomultiplier (SiPM). The use of SiPMs instead of photomultiplier tubes (PMT) is motivated by their high packing fraction and their insensitivity to magnetic fields. The space in the POLDI experimental area is very limited and tests of samples in high magnetic fields are foreseen.

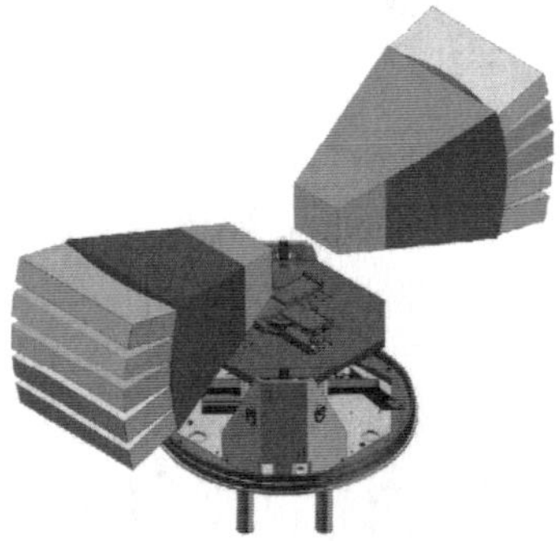

Figure 1. View of the aimed new POLDI detector.

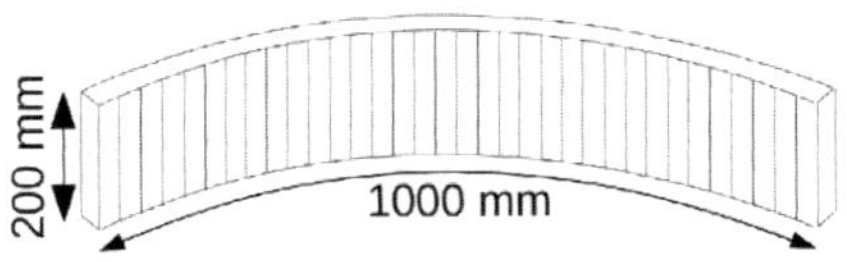

Figure 2. Schematic view of the detector module.

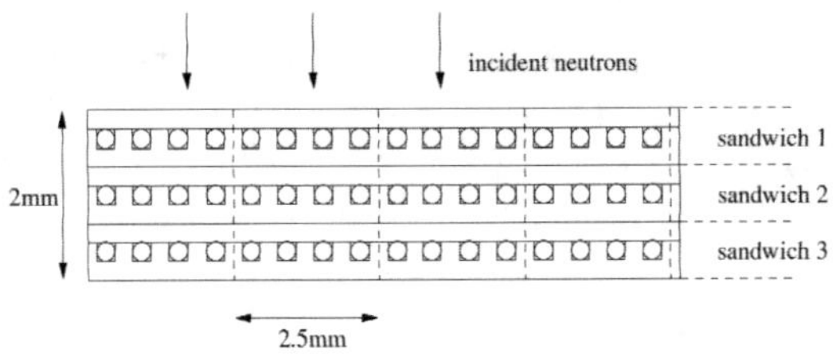

Figure 3. Cross-sectional view of a possible detector module. The channels are delimited by the dashed lines. One channel consists of 12 WLS fibers.

2. Test of a single channel prototype

2.1. Experimental-set-up

Fig. 4 shows the single channel prototype which has been tested. It consists of one sandwich made of 2 scintillating layers (ND(2:1) from Applied Scintillation Technologies [4]) and four WLS fibers (Y11(200)MC from Kuraray [5]) with a diameter of 250 μm. The fiber pitch is 0.6 mm. In the 440 μm thick layer (labelled bottom layer), four grooves of 300 μm width and 300 μm depth are machined. The fibers are glued into the grooves with an optical epoxy (EJ500 from

Table 1. Specifications of the current and new detectors.

	actual detector [1]	new detector
Detector technology	^{3}He	ZnS(Ag)/^{6}LiF
Size of one module	1×0.2 m^2	1×0.2 m^2
Number of modules	1	2×5
Vertical coverage	$\pm 3.2^\circ$	$\pm 16^\circ$
Horizontal coverage	$2\theta = 75^\circ$ to 105°	$2\theta = 75^\circ$ to 105°
Secondary flight path	2 m	2 m
Angular channels	400	400
Spatial resolution	2.5 mm	2.5 mm
Wavelength	1.2 Å - 6 Å	1.2 Å - 6 Å
Detection efficiency (at 1.2 Å)	65%	65%
Time resolution	< 2 μs	< 2 μs
Sustainable count rate	4 kHz/ch	4 kHz/ch
Gamma sensitivity	No issue	$< 10^{-6}$
Intrinsic noise	< 0.003 Hz/ch	< 0.003 Hz/ch
ΔQ/Q resolution	1×10^{-3} - 2×10^{-3}	1×10^{-3} - 2×10^{-3}

Eljen technology [6]) and the top scintillating layer 220 μm thick (labelled top layer) is glued with the same optical epoxy.

On one side, the fibers are cut along the edge of the sandwich and polished. Then, an aluminum foil acting as a mirror is glued to improve the light yield on the other side of the fibers where the photodetector is connected. On this side, the fibers are glued together into the hole of a plexiglas holder and polished. Fig. 5 and 6 show a picture of the sandwich before and after polishing of the fiber-end.

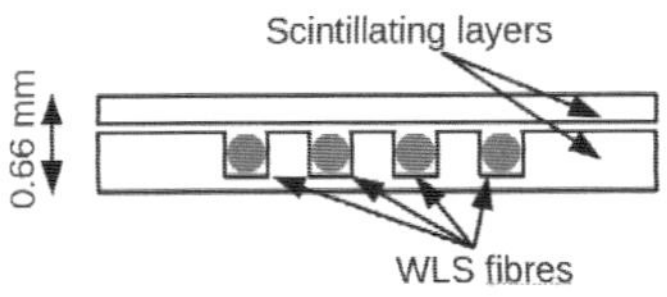

Figure 4. Schematic of the sandwich cross-section.

Figure 5. The single channel sandwich before polishing of the fiber-end.

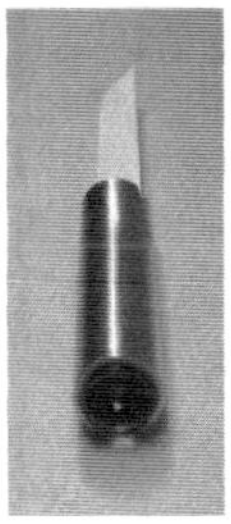

Figure 6. The single channel sandwich ready to be connected to a photodetector.

In order to study the intrinsic characteristics of the scintillator sandwich, the light yield distribution and the decay curve, the fibers were coupled to a low noise photodetector (H10682-210 from Hamamatsu [7]), this allowing to set the trigger threshold at a very low level. Fig. 7 shows a schematic of the experimental setup. The sandwich is irradiated with an ^{241}Am source. The energy of α particles emitted by the source (5.48 MeV) is close to the energy left in the scintillator by the triton (2.73 MeV) and α particles (2.05 MeV) resulting from the absorption of

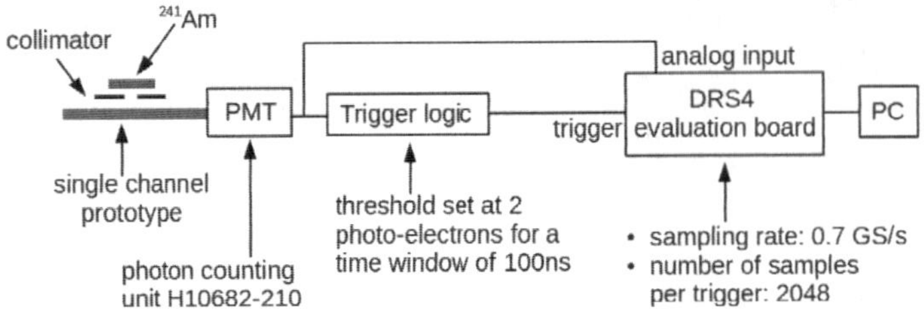

Figure 7. Schematic of the experimental setup used to characterize the intrinsic characteristics of a scintillator/fiber sandwich.

a neutron by a ^{6}Li nucleus. The use of an α-source is less constraining than the use of a neutron source (higher rate, no need for radiation shielding) and it has the advantage that α particles are stopped close to the scintillator surface, within the first 22 μm, this allowing to measure the light yield distribution as a function of the distance from the fiber by varying the thickness of the scintillating layer exposed to the source. Furthermore, the higher is the distance between the interaction point and the fibers, the lower is the light yield. So, the light yield measured with α particles is expected to be smaller than the one measured with neutrons, the neutrons being absorbed at any depth in the sandwich. In this sense, the use of an α-source provides conservative results.

The partially dead area located on the edge of the sandwich, where there are no fibers, was covered with a 50 μm thick aluminum foil, to prevent the interaction of α particles in this region. The optical fibers are coupled without glue or grease to the photon counting unit H10682-210 from Hamamatsu. This unit consists of a PMT and a discriminator which triggers at the level of one photoelectron providing a TTL signal on its output. This counting unit has a dark count rate of 50 Hz at 25°C, a quantum efficiency of about 20% at 476 nm (the peak emission wavelength of the fibre) and a pulse-pair resolution of 20 ns.

The output of the photon counting unit is sampled with the DRS4 evaluation board [8] at a frequency of 0.7 GS/s. For each trigger, 2048 samples are saved, this allowing to study the train of photons over a time window of about 2 μs. The threshold of the trigger logic is set at 2 photo-electrons for a time window of 100 ns.

2.2. Results

Fig. 8 shows the decay curve of the scintillator. The limited sampling duration restricts the time range to 2 μs. To study the slow components of the scintillation, additional measurements are performed with different delays of the trigger: 4, 8, 40, 80 and 400 μs. Fig. 9 shows the sum of this time spectra (1000 triggers per spectra). In the 0 to 400 μs time range, the decay curve is analytically well described by the following function:

$$A(t) = N \cdot (C + \sum_{i=1}^{4} A_i \cdot e^{-t/\tau_i}) \tag{1}$$

where N is the number of photons measured, C is the background, A_i and τ_i are the amplitude and the decay time of the component i of the decay. Table 2 shows the result of the curve fit.

Fig. 10 and Fig. 11 show the photo-electron spectrum when the top layer and when the bottom layer are irradiated. The fibers are located 220 μm away from the surface of the top layer whereas they are only 140 μm away from the surface of the bottom layer. As expected, the average number of photo-electrons is higher when the bottom layer is irradiated. The peak at 2 photo-electrons appearing on both spectra is due to the afterglow which is still present 20 min after a day light exposure of the scintillator.

Table 2. Result of the decay curve fitting.

Component	1	2	3	4
Decay time τ_i (μs)	0.008	0.118	1.224	15.2
Amplitude A_i	0.078	0.012	0.010	0.0010
Fraction of photons from the component i ($\frac{A_i \cdot \tau_i}{\sum_{i=1}^{4} A_i \cdot \tau_i}$)	2%	5%	41%	52%

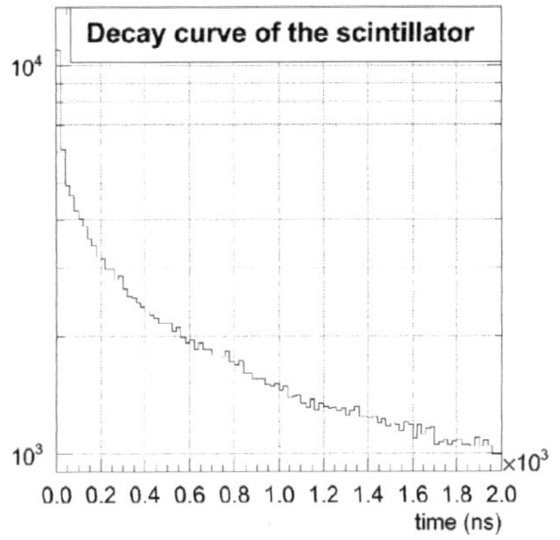

Figure 8. Decay curve in a time range of 2 μs.

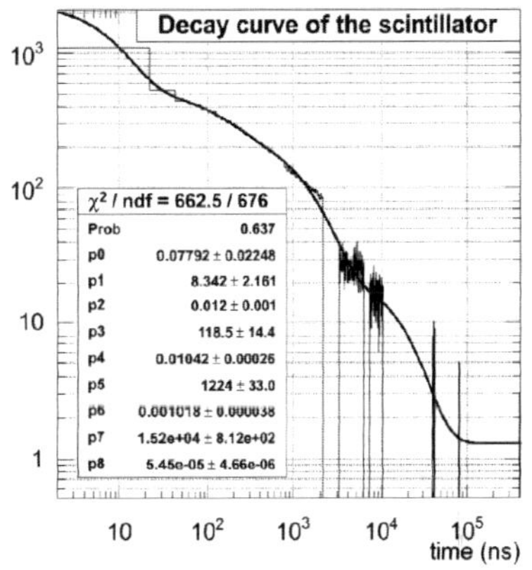

Figure 9. Decay curve in a time range of 400 μs. The solid line is the fitted curve following Eq. 1.

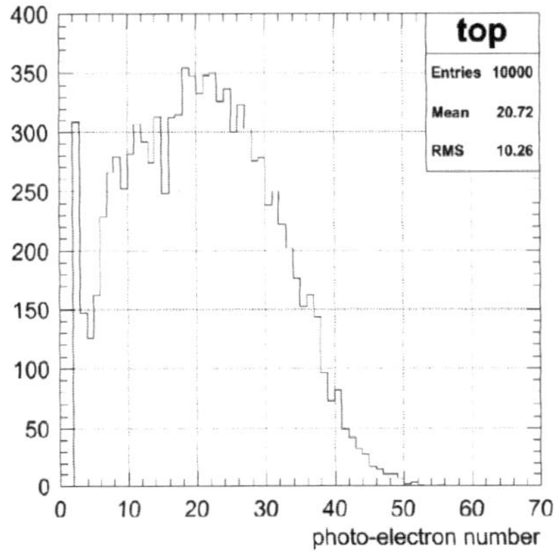

Figure 10. Photo-electron spectrum measured when the top layer is irradiated.

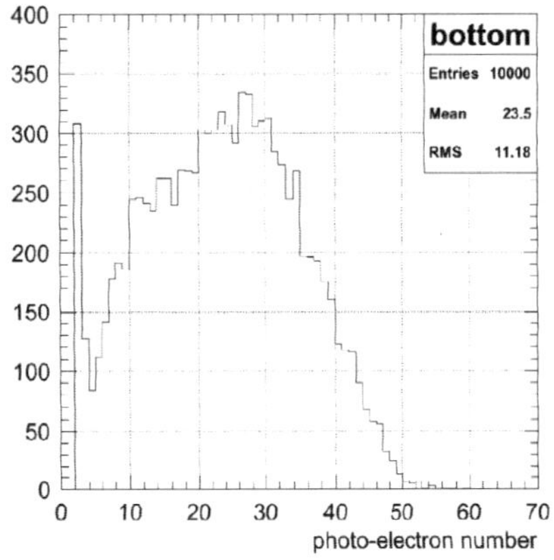

Figure 11. Photo-electron spectrum measured when the bottom layer is irradiated.

3. Development of an innovative signal processing system

The signal processing system (SPS) which is under development for the new POLDI detector is based on a photon counting approach. Two reasons makes this approach feasible. Firstly, the ZnS(Ag)/^{6}LiF scintillator is relatively slow and consequently, the photons to detect are sufficiently spaced out in time to allow the detection of almost each individual photon emited in a given time window. Secondly, SiPM have an excellent photon counting capability. Fig. 12 and Fig. 13 show the block diagrams of the SPS.

The SPS measures (almost) continuously the number of photo-electrons over a time window of 2 μs. To do that, counting experiments over 200 ns are performed at a frequency of 5 MHz. Every 200 ns, the sum (SUM) of the last 10 values measured by the counter is calculated. If the sum is higher than a threshold value, the readout of the last 10 values is triggered, as well as the next 9 (to be sure to get a sampling of the neutron signal over 2 μs). Among the 10 samples producing the trigger of the readout, the sample with the maximum count number provides the time stamping of the event. It is called S_0. The following 9 samples are called S_1, ..., S_9. The variable $S_0 + S_1 + ... + S_9$ is used for neutron/noise discrimination and the variable $(S_0 + S_1)/(S_2 + ... + S_9)$ is used for neutron/γ discrimination. Time stamping, n/noise and n/γ discriminations are performed online by a FPGA.

The pulse processing algorithm consists of the following sequence of operations:

(i) The FPGA is waiting for $SUM > threshold$ (trigger of the measurement).
(ii) The FPGA reads 19 samples.
(iii) The FPGA performs the online analysis: time stamping, n/γ and n/noise discrimination.
(iv) If the event is neutron tagged, the data (time + channel information) are sent to the PC.
(v) The FPGA is waiting for $SUM < end_{threshold}$ (end signal). As soon as it measures an end signal, the FPGA return in the state (i).

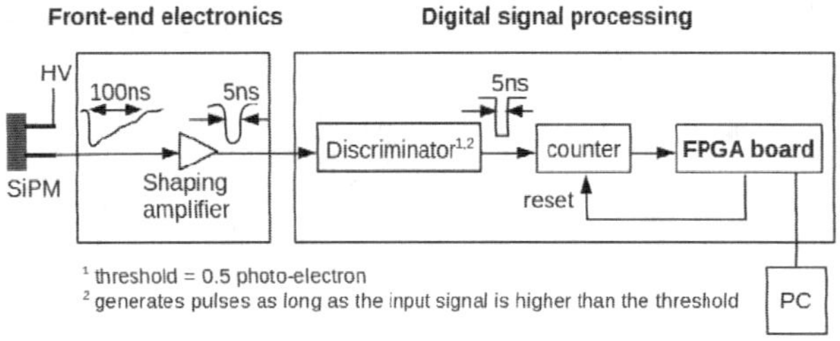

Figure 12. Block diagram of the SPS.

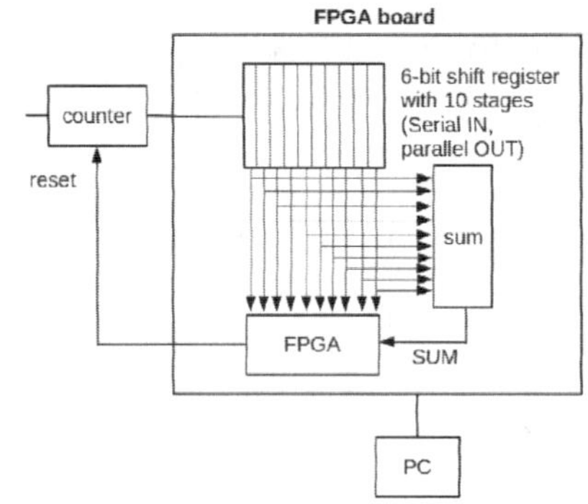

Figure 13. Block diagram of the FPGA board.

This signal processing approach offers many advantages compared to a simple signal processing based on a CR-RC pulse shaping network:

- When several pixels of a SiPM are fired at the same time due to optical crosstalk, only one count is recorded. This allows to get rid of the SiPM crosstalk and this is a crutial point in the reduction of the noise.

- The pulse shape discrimination becomes straightforward.
- Numerous parameters of the readout electronics are set in the FPGA firmware, as for example, the clock frequency, the depth of the shift register, the pulse shape criteria. This provides a large flexibility and makes the optimization of the readout electronics easier.

4. Monte Carlo simulation of the signal processing system

A Monte Carlo simulation has been performed to evaluate the background rejection power and the neutron detection efficiency of the SPS. The input parameters of the simulation are the following:

- SiPM dark noise: 300 kHz (afterpulses are neglected)
- neutron rate: 100 Hz
- decay curve of the scintillator (Fig. 9)
- photo-electron spectrum (Fig. 10)

The Monte Carlo output is a data file which provides, for each 200 ns counting experiment, the number of SiPM dark counts, the number of neutrons absorbed, the number of photo-electrons and the value of SUM. From this data file, the detection efficiency and the background rate can be determined as a function of the threshold values for the two variables SUM and $S_0 + S_1 + ... + S_9$.

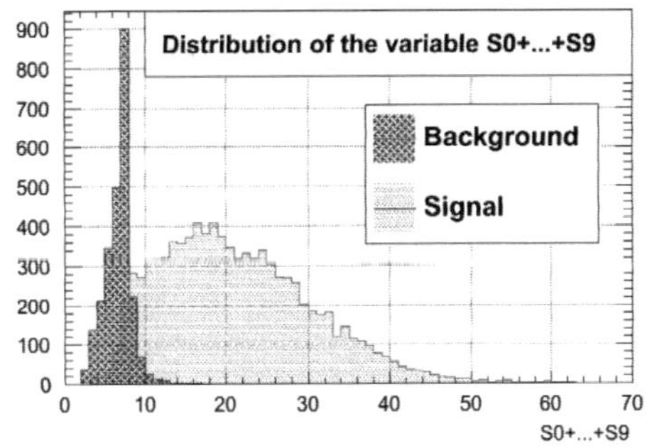

Figure 14. Distribution of the variable $S_0 + S_1 + ... + S_9$ for background and for neutron events.

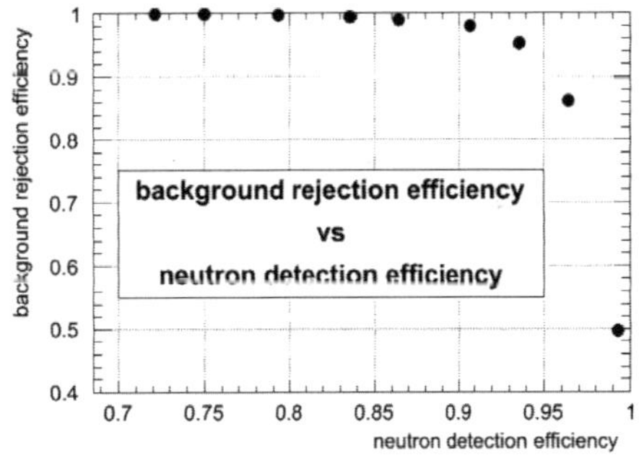

Figure 15. Background rejection as a function of the neutron detection efficiency, for events which trigger the readout.

For the condition $SUM > 6$, the trigger efficiency on neutrons amounts to 90% and the rate of trigger on the noise amounts to 25 Hz. Fig. 14 shows the distribution of the variable $S_0 + S_1 + ... + S_9$. The background rejection and the neutron detection efficiency have been calculated as a function of the threshold value for the variable $S_0 + S_1 + ... + S_9$. Fig. 15 shows the background rejection as a function of the neutron detection efficiency. For a threshold value of 10 on the variable $S_0 + S_1 + ... + S_9$, the background rejection amounts to 98 % and the neutron detection efficiency amounts to 90%. It must be stressed that the neutron/noise discrimination is performed only on the events which trigger the readout. Hence, the overall efficiency of the SPS amounts to 81 % (90%×90%) and the background rate amounts to 0.5 Hz (25 Hz×2%). Preliminary measurements tend to show that the light yield of our test detector is 1.5 times higher when it is exposed to neutrons than when it is exposed to α particles. Hence, the presented results are extremely conservative.

5. Outlook

Important next steps are the development of the signal processing system and the test of prototypes read out with SiPMs in a neutron beam. Radiation hardness tests of different SiPMs are underway. The optical isolation between channels or alternatively the processing in the SPS of the optical crosstalk between channels will be investigated.

References

[1] U. Stur, "Time-of-flight diffraction with multiple frame overlap Part II: The strain scanner POLDI at PSI", *Nucl. Instr. and Meth. A*, vol. 545, pp. 330-338 (2005).
[2] N.J. Rhodes, "Status and future development of neutron scintillation detectors", *Neutron News*, vol. 17, pp. 16-18 (2006).
[3] N.J. Rhodes, "Scintillation detectors", *Neutron News*, vol. 23, pp. 26-30 (2012).
[4] http://www.appscintech.com
[5] http://kuraraypsf.jp/index.html
[6] http://www.eljentechnology.com/
[7] http://www.hamamatsu.com/jp/en/index.html
[8] S. Ritt, "Design and performance of the 6 GHz waveform digitizing chip DRS4", *Conf. Rec. IEEE Nucl. Sci. Symp.*, pp. 1512-1515 (2008).

Development of a wavelength-shifting-fibre-based scintillator neutron detector as an alternative to ^{3}He at J-PARC/MLF

T Nakamura[1], K Toh[1], K Honda[1], A Birumachi[2], M Ebine[2], K Sakasai[1], K Soyama[1] and M Katagiri[3]

[1] J-PARC center, Japan Atomic Energy Agency, Tokai, Ibaraki 319-1195, Japan
[2] Nuclear Science Directorate, Japan Atomic Energy Agency, Tokai, Ibaraki 319-1195, Japan
[3] Frontier Research Center for Applied Atomic Sciences, Ibaraki University, Tokai, Ibaraki 319-1106, Japan

E-mail:nakamura.tatsuya@jaea.go.jp

Abstract. A wavelength-shifting-fibre-based scintillator detector has been developed as an alternative detector to ^{3}He gas. The detector is intended for use in an inelastic neutron-scattering instrument at J-PARC. The detector being developed, which is based on the one made for a SENJU diffractometer, is designed to cover a large scattering angle with a moderate pixel size as well as exhibiting a high detection efficiency, low gamma-ray sensitivity and low background count rate. A prototype detector with a pixel size of 20 × 20 mm and a neutron-sensitive area of 320 × 320 mm^{2} has been constructed. This paper describes the design of the detector and its performance relative to the original SENJU detector.

1. Introduction

The recent worldwide shortage of ^{3}He gas has accelerated the development of neutron detector technologies that could be alternative to conventional ^{3}He-gas-based detectors [1, 2]. The neutron detectors used in a neutron-scattering experiment should exhibit position sensitivity, high detection efficiency, low gamma-ray sensitivity and the capability of detecting high count rates. A particular requirement of detectors for use in an inelastic neutron-scattering (INS) instrument is that they should cover a large solid angle around a sample, with a typical requirement of several tens of square metres.

Several gas-based and scintillator-based neutron detectors are commercially available, such as those using a BF_3 gas, a thin film of ^{10}B or ^{6}Li, or a ZnS/^{6}LiF scintillator, but none of them satisfies the detector specifications required for an INS instrument. Current international research and development collaborations are investigating the use of such instruments in neutron-scattering experiments [3]. These projects can be categorized into three lines of development: a gas-based detector with a ^{10}B converter film [4], a BF_3 gas-based detector [5] and a scintillator-based detector [6].

This paper presents the development status of our detector incorporating a scintillator based on wavelength shifting (WLS) fibre, as well as the design and detection performance of a prototype detector.

2. Detector design for an INS instrument

Our detector employs detector technologies developed for the SENJU instrument [7]. SENJU is a single-crystal Laue instrument constructed in the Materials and Life Science Experimental Facility at the Japan Proton Accelerator Research Complex (J-PARC/MLF) [8]. Table 1 summarizes the typical detector specifications for a position-sensitive ^{3}He-gas tube used in an INS instrument and the original SENJU scintillator detector installed at J-PARC/MLF. To make a detector that would be a suitable alternative to a ^{3}He-gas detector, the SENJU detector should be redesigned to fulfil the requirements for detection efficiency, background count, pixel size and detector spatial coverage, together with a low manufacturing cost. Moreover, the anticipated detector for an INS instrument should be modular in design and cover an area of several tens of square metres.

Figure 1 is a diagram of a large-area detector unit for an INS instrument. The detector unit covers a total neutron-sensitive area of 1.64 m^2 with a packing density of 79%, with each detector unit comprising 16 detector modules. One detector module has a neutron-detecting area of 320 × 320 mm^2 with a pixel size of 20 × 20 mm^2.

Table 1. Specifications for a ^{3}He-gas tube and the SENJU detector. The latter is a position-sensitive scintillator detector used in the SENJU diffractometer at J-PARC/MLF.

Detector characteristics	^{3}He detector*	SENJU *detector*
Neutron efficiency	70% at 1 Å	20% at 1 Å
Gamma sensitivity	10^{-6}	10^{-6}
Background	~10^{-5} 1/cm^2/s	2×10^{-4} 1/cm^2/s
Spatial resolution (*X*)	25 mm	4 mm
Spatial resolution (*Y*)	15 ~ 25 mm	4 mm
Local rate capability	50 kHz on a pixel	<50 kHz
Global rate capability	50 kHz on a tube	<50 kHz
Time resolution	1 μs	2 μs
Area/det	0.025 m^2	0.065 m^2
Environment	Cryogenic vacuum	In air

* ^{3}He 10 bar, 25-mm diameter, 1 m long

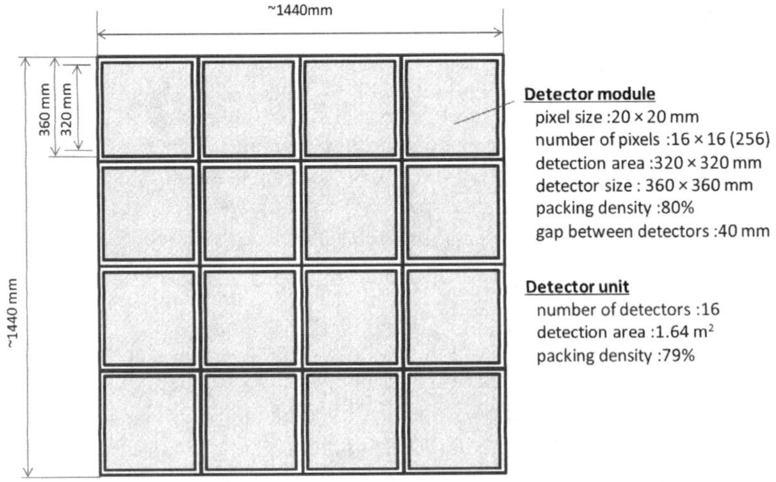

Figure 1. The design of a detector unit for an INS instrument.

3. Development of a large-area scintillator detector module

3.1. Increasing the pixel size and detection area

Figure 2 shows a schematic view of the neutron-detecting head of the scintillator/WLS-fibre-based detector module. A pitch of the WLS fibres is 5 mm, in contrast to 4 mm in the original SENJU detector, which increases the neutron-sensitive area of the detector by a factor of 1.56. The pixel size can be made larger (i.e. 20 × 20 mm^2) by reading out four WLS fibres as one channel while maintaining a reasonable light-collection efficiency. Technically these four WLS fibres are coupled to one pixel by of a multianode photomultiplier tube (MA-PMT).

The light-collection efficiency of the WLS-fibre-based detector was evaluated by measuring the photoelectron (PE) number distribution (PND) detected with mock-up detectors. Detectors constructed with WLS fibres at pitches of 4 mm and 5 mm were prepared for this test. A scintillator and MA-PMTs with the same parameters were used to facilitate the comparison of PNDs.

Figure 3 shows the PNDs measured with the detectors. A $ZnS/^{10}B_2O_3$ scintillator screen was implemented in the detectors [9, 10]. To reject false triggering events that occurred due to afterglow, the PNDs were measured only when more than three PEs were detected in both the x and y WLS fibre arrays within a 2-μs gated window. This operating condition was also used in the original detector [7]. The measured PNDs had modes at 13 PEs and 12 PEs for the detectors with pitches of 4 mm and 5 mm, respectively. These modes varied with the threshold setting of each WLS fibre array, but the PNDs actually represented distributions of the detected number of PEs that were accepted by the signal processing and encoder electronics.

The mean number of detected PEs was calculated as

$$(\text{mean number of PEs}) = (\Sigma C_i\, i)/(\Sigma C_i)$$

where i is the number of detected PEs and C_i is the count at an i. The corresponding mean numbers of detected PEs were 23.3 and 21.5, respectively. The mean light-collection efficiency decreased by only ~8% when the pitch of WLS fibres in the detector increased from 4 mm to 5 mm.

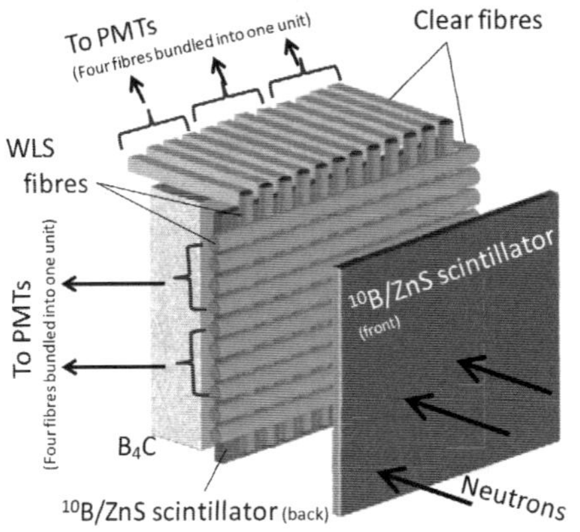

Figure 2. Schematic view of the neutron-detecting head of the scintillator/WLS-fibre-based detector.

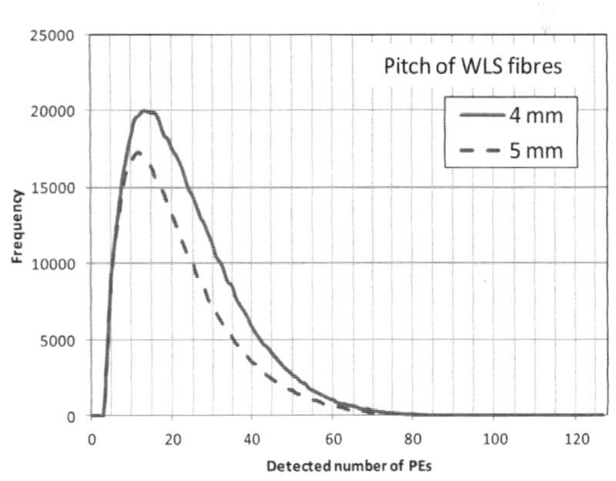

Figure 3. Distribution of the detected number of PEs measured with the WLS-fibre-based detector that was implemented with a $ZnS/^{10}B_2O_3$ scintillator screen.

3.2. PMTs for the detector

The prototype detector should be equipped with a PMT that has sufficient gain and uniformity, and a low background count. An MA-PMT can fulfil these requirements and is particularly effective in reducing the manufacturing cost of a detector. Four candidate MA-PMTs for a prototype detector were selected, and their adaptabilities were evaluated. Table 2 summarizes the specifications of the selected MA-PMTs. These MA-PMTs were chosen because their photocathode area is capable of accommodating four 1-mm-diameter fibres (corresponding to an area exceeding 4 mm^2). The R12445 and R11265-100 PMTs, which have been released on the market, have a larger pixel size than the conventional H8711-300 and H8804 MA-PMTs, which would ease the requirement to combine four fibres into one pixel. Their photocathode size of 6 × 6 mm would also be effective in reducing signal crosstalk between neighbouring pixels.

Table 2. Specifications of the evaluated multianode PMTs.

PMT type	No. of pixels	Pixel size	Max rating voltage
H8711-300	16	4 x 4 mm	-1000 V
H8804	64	2 x 2 mm	-1000 V
R12445	16	6 x 6 mm	-1100 V
R11265-100	64	3 x 3 mm	-1100 V

Figure 4a and b show the uniformity of the scanned PMT gains over the photocathode and the mean PMT gain as a function of the PMT voltage, respectively. The gain uniformity was measured by illuminating the photocathode area via a 0.2-mm-diameter optical fibre by a blue LED placed close to (~0.5 mm) the end of the fibre. The PMT gains at the centre of the pixel varied over the range 0.8–1.2 for all of the tested PMTs. This range of variations was acceptable for our detector. The mean PMT gain was much lower for the R12445 and R11265-100 PMTs than for the other devices (Figure 4b), which made them unsuitable for a prototype detector since a gain of at least 5×10^6 is required at the first stage of our photon counting electronics. Among the tested MA-PMTs, the H8711-300 device exhibited the best gain uniformity and overall gain. Its gain reached 5.5×10^6 when using a reasonable applied voltage of −900 V (corresponding to 90% of its maximum rating). This PMT has an ultra-bialkali photocathode, which is suitable for detecting shifted green light that has propagated through a WLS fibre, and hence the H8711-300 was used to construct a prototype detector.

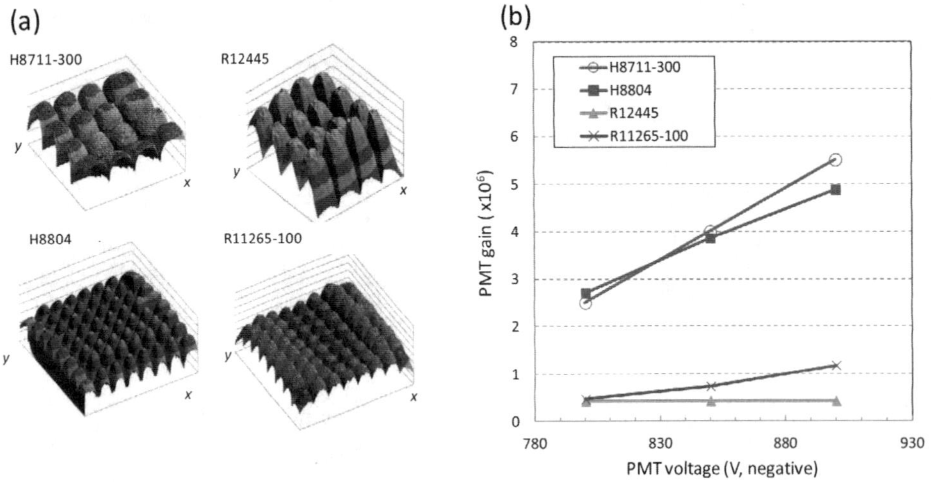

Figure 4. (a) Pixel-by-pixel PMT gain uniformity and (b) mean PMT gain versus the applied PMT voltage.

Journal of Physics: Conference Series **528** (2014) 012042 doi:10.1088/1742-6596/528/1/012042

3.3. Detector electronics cards for a prototype detector

New detector electronics cards were produced for a prototype detector: (a) comparator cards, (b) a signal processing and encoder card and (c) a high-voltage card for driving the PMTs. All of these cards were designed to conform to the VME-3U size standard. Each of these cards was implemented in the prototype detector.

The comparator board that handles 32 analogue inputs and is equipped with a fast comparator chip (ADCMP605, Analog Devices) ensures a delay time of only ~1.6 ns. The card dissipates 7.4 W with a power supply of ±5V (corresponding to ~230 mW/channel).

The signal processing and encoder card accepts 2 of 32 digital signals. Field-programmable gate arrays (Cyclone II, ALTERA) operates at a clock frequency of 100 MHz. The functions of the board are similar to those in the original detector [7]: photon counting on each input channel, *x*-*y* coincidence, position calculation and data transfer to/from data-acquisition electronics. An optical-fibre link is implemented for data transmission, which makes the system robust in an electrically noisy environment.

The high-voltage card provides two outputs to bias the PMTs. Each output can supply a voltage up to −1250 V (at a maximum of 0.5 mA). The use of a dedicated high-voltage card decreases the cost of a medium-size detector system that requires less than 100 high-voltage channels compared to the commercial system. The card can be operated in either local or remote mode. The local mode enables a detector to operate in a stand-alone manner, where the setting of the PMT voltage is recorded in the on-board memory.

4. Detection performance of the prototype detector

The detection performance of the prototype detector was evaluated in terms of its detection efficiency, gamma-ray sensitivity and background count rate. The prototype detector was irradiated with neutrons from a ^{232}Cf source with a radioactivity of 1.83×10^{7} Bq that was set in the centre of a graphite moderator block of size 800 × 800 × 800 mm^3. The detector was placed 900 mm from the source. The gamma-ray sensitivity was measured using a ^{60}Co source with a radioactivity of 8.3×10^{5} Bq. The ^{60}Co source was positioned 200 mm from the scintillator, where the incident fluence was estimated to be ~1.7×10^{5}/s/detector. The gamma-ray sensitivity is defined as the probability of counting a gamma ray for a single incident photon. The background count rate was measured with no specific shielding implemented for the environmental radiation.

Figure 5a and b show photographs of the front and rear views of the prototype detector, respectively.

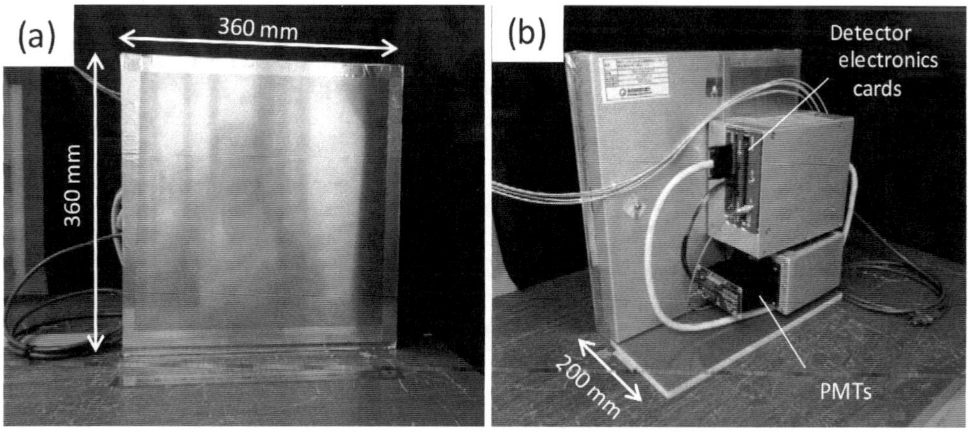

Figure 5. Photographs showing (a) front and (b) rear views of the prototype detector.

The detector was designed to be compact and light, with a physical size of 360 × 360 × 200 mm^3 and a weight of 11.6 kg. All of the detector electronics—including PMTs, the amplifier/discriminator card, the high-voltage card and the signal processing/encoder card—were installed behind the case housing the neutron-detecting head. The detector can be powered using either ±5 V DC or 100 V AC. The data were transferred using four optical fibres, with a maximum data transfer rate of 160 kcps at present. This transfer rate could be improved to more than 500 kcps by replacing the LED driver with a device with a faster time response.

Figure 6 shows the detector metrics, with the neutron count (detection efficiency) plotted versus the ^{60}Co gamma-ray sensitivity. The data measured with the original detector are included for comparison. At the ^{60}Co gamma-ray sensitivity of 10^{-6} (the normal operating condition), the prototype detector exhibited larger neutron counts than the original detector, with a calibrated detection efficiency of about 40%. The PND was evaluated with the prototype detector using a method similar to that described in Section 3.1. The mode and mean numbers of detected PEs were 16 and 27.6, respectively. These values are about 30% larger than those measured with the original detector, and indicate the better detection efficiency of the prototype detector, which could be due to its higher operating PMT gain.

Figure 7 shows the background count rate plotted versus the ^{60}Co gamma-ray sensitivity. In both detectors the background count rate decreased as the gamma-ray sensitivity decreased. For a ^{60}Co gamma-ray sensitivity of 10^{-6}, the background count rates were 2×10^{-4} cps/cm^2 and 4×10^{-4} cps/cm^2 for the original and prototype detectors, respectively. Although these values are typical of our WLS-fibre-based scintillator detectors, they were about an order larger than that of the conventional position-sensitive ^{3}He detector (~1×10^{-5} cps/cm^2). Possible sources of the background count are cosmic rays and contamination of the radioactive elements in the detector components. This should be clarified in future experiments.

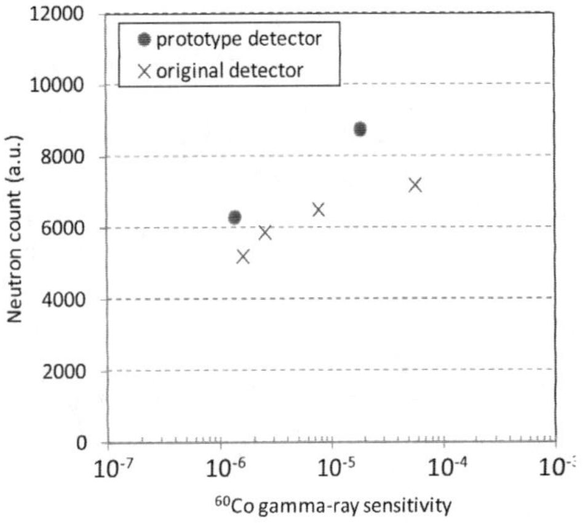

Figure 6. Neutron count versus ^{60}Co gamma-ray sensitivity.

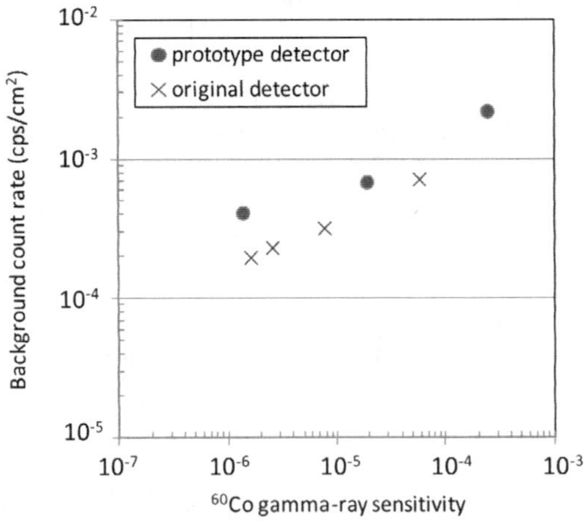

Figure 7. Background count rate versus ^{60}Co gamma-ray sensitivity.

Figure 8a and b show a flood image measured with a ^{232}Cf source and count frequencies in the *x* and *y* projections, respectively. An entire detector area was irradiated with thermal neutrons; in this experiment scintillator screens with dimensions of 256 × 256 mm^2 were implemented in the detector. The count uniformity was evaluated as 15.1±1.5% (mean±SD) by Gaussian fitting, which would be acceptable for most neutron-scattering experiments.

Finally, the specifications and detection performance of the prototype detector are summarized in Table 3.

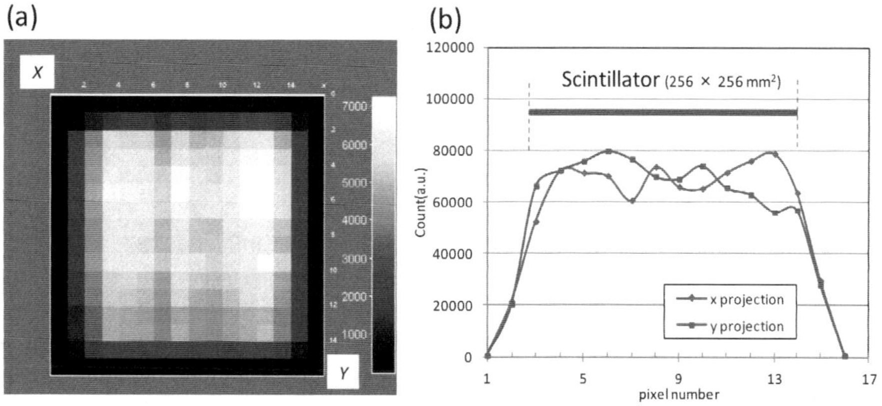

Figure 8. (a) Flood image and (b) count frequency in x and y projections measured with the prototype detector.

Table 3. Specifications and detection performance of the prototype detector.

Neutron-sensitive area	: 320 × 320 mm^2 (256 × 256 mm^2 at present)
Pixel size	: 20 × 20 mm^2
Detector efficiency	: 40% for 1.8 A
Gamma sensitivity	: 1 × 10^6
Background count	: 4 × 10^{-4} cps/cm^2
Count uniformity	: 15 ± 2%
Physical size	: 360 × 360 × 200 mm^3
Weight	: 11.6 kg
Power dissipation	: 13 W

5. Summary

A prototype scintillator/wavelength-shifting-fibre-based detector has been developed for an INS instrument as an alternative to a ^{3}He-gas detector. The detector parameters are designed specifically for an INS instrument, with a pixel size of 20 × 20 mm^2 and a neutron-sensitive area of 320 × 320 mm^2. The detector exhibited a detection efficiency of 40% for thermal neutrons and a gamma-ray sensitivity of 10^{-6}. Further work will include reducing the intrinsic background signal of the detector to the level of conventional position-sensitive ^{3}He detectors.

References

[1] United States Government Accountability Office 2011 *Managing critical isotope* GAO-11-472
[2] Kouzes R T 2009 *The ^{3}He supply problem* Report PNNL-18741 Pacific Northwest National Laboratory
[3] Zeitelhack K 2012 *Neutron News* vol. **23** No 4 10
[4] Bigault Y, et al 2012 *Neutron News* vol. **23** No 4 20
[5] Wilpert T 2012 *Neutron News*, vol. **23** No 4 14
[6] Rhodes N J 2012 *Neutron News*, vol. **23** No 4 26
[7] Nakamura T, et al. 2012 *Nucl. Instr. & Meth.* A **686** 64
[8] Tamura I., et al. 2012 *Journal of Physics: Conference Series* **340** 012040
[9] Kojima T, Katagiri M, Tsutsui N, Imai K, Matsubayashi M and Sakasai K, 2004 *Nucl. Instr. & Meth.* A **529** 325
[10] Nakamura T, et al. *Development of a ZnS/$^{10}B_2O3$ scintillator with low-afterglow phosphor Journal of Physics: Conference Series* in these proceedings

Development of a $ZnS/^{10}B_2O_3$ scintillator with low-afterglow phosphor

T Nakamura[1], M Katagiri[2], N Tsutsui[3], K Toh[1], N J Rhodes[4], E M Schooneveld[4], H Ooguri[5], Y Noguchi[5], K Sakasai[1] and K Soyama[1]

[1] J-PARC center, Japan Atomic Energy Agency, Tokai, Ibaraki 319-1195, Japan
[2] Frontier Research Center for Applied Atomic Sciences, Ibaraki University, Tokai, Ibaraki 319-1106, Japan
[3] Chichibu Fuji Co. Ltd, Chichibu, Saitama 368-0193, Japan
[4] ISIS, Rutherford Appleton Laboratory, Chilton, Oxfordshire OX11 0QX, UK
[5] NICHIA, Anan, Tokushima 774-8601, Japan

E-mail:nakamura.tatsuya@jaea.go.jp

Abstract. A low-afterglow, ^{10}B-doped neutron-sensitive $ZnS/^{10}B_2O_3$ scintillator was developed. The developed ZnS phosphor has a primary decay time constant of about 60 ns with a low afterglow. The developed scintillator exhibited a mean afterglow height of 4% relative to the peak at 1 μs after the peak, which is half that of a commercial $ZnS/^{6}LiF$ scintillator manufactured by Applied Scintillation Technologies. The count-rate capability of a wavelength-shifting-fibre-based detector was increased to 30,000 cps by implementing the developed scintillator from 5,000 cps with the commercial scintillator.

1. Introduction

^{6}Li-doped ZnS scintillators have been used extensively in thermal-neutron detection in many neutron-scattering instruments at various pulsed-neutron facilities, including at the ISIS in the UK and the SNS in the USA [1, 2]. The properties of these scintillators of a high light yield with an excellent capability to discriminate neutrons and gamma rays make them potentially suitable for constructing large-area position-sensitive detectors for use in neutron-scattering experiments. Also, $ZnS/^{6}LiF$ scintillators can easily be constructed in sheet form, making them economically efficient. This is particularly important for neutron-scattering instruments that require many detector modules to cover a large scattering angle around a sample.

One of the challenges in using a ZnS scintillator is improving its count-rate capability. The commercial ZnS:Ag phosphor is known to have a high light yield of ~160,000 photons in the nuclear reaction $^{6}Li(n,\alpha)T$, but it also exhibits a high afterglow that lasts for more than 100 μs [3, 4]. The pulse pileup associated with this afterglow hinders neutron detection at a high count rate, and becomes particularly problematic for a detector used at a MW-class pulsed-neutron source, such as in the Materials and Life Science Experimental Facility at the Japan Proton Accelerator Research Complex (J-PARC/MLF).

There are several neutron-sensitive scintillator materials that exhibit a faster time response than the $ZnS:Ag/^{6}LiF$ scintillator. A Ce-activated ^{6}Li glass scintillator is known to have a typical decay time of 50–70 ns [5]. However, the gamma-ray sensitivity of a ^{6}Li glass scintillator is one to two orders of magnitude higher than that of a ZnS/^{6}LiF scintillator. Other types of Ce-doped scintillator have been developed with moderate gamma-ray sensitivities, while the capability for gamma-ray rejection can be increased using pulse-shape discrimination, such as with $Li_6Gd(BO_3)_3$ [6], Cs_2LiYCl_6 [7] or $LiCaAlF_6$ [8]. However, the light yields of such materials are more than an order of magnitude lower than for ZnS, which makes them unsuitable for constructing a large-area position-sensitive detector that is coupled to optical fibres in neutron-scattering instruments.

This paper presents a new neutron-sensitive ZnS:Ag-based scintillator that achieves a high light yield, low afterglow contamination and low gamma-ray sensitivity with a reasonable manufacturing cost. The count-rate capability of a wavelength shifting fibre (WLS)-based detector implemented with a ZnS/$^{10}B_2O_3$ scintillator was also evaluated.

2. Neutron-sensitive ZnS:Ag scintillator with low afterglow

2.1. Sample preparation

Table 1 summarizes the details of the samples used in experimental evaluations. Two types of neutron-sensitive ZnS scintillator screens were prepared for the evaluations using the low-afterglow ZnS phosphor: ZnS:Ag/$^{10}B_2O_3$ and ZnS:Ag /^{6}LiF. The fabrication methods are summarized below; further details are available elsewhere [9].

A ZnS:Ag powder with a low afterglow and a particle size of about 5 μm was prepared in-house [10] and mixed with $H_3{}^{10}BO_3$ at a mass ratio of 1.0~1.5. The composite powder was poured onto a 0.3-mm-thick aluminium base plate that was placed in an electronic furnace and sintered for 3 hours at 540 °C. $H_3{}^{10}BO_3$ becomes glassy when it melts, water escapes and it becomes $^{10}B_2O_3$. The $^{10}B_2O_3$ melts at 480°C and it acts as a binder material. The resultant material is ZnS/$^{10}B_2O_3$ ceramic scintillator. The nominal ^{10}B enrichment rate exceeded 90%.

A polycrystalline ^{6}LiF powder with a mean particle diameter of about 10 μm was used as a converter material when constructing the ZnS/^{6}LiF scintillator. An inorganic binder was added to the ZnS powder, and the mixed composite was poured onto an aluminium base plate to form a scintillator sheet. The nominal ^{6}Li enrichment factor was 90%.

Table 1. Properties of the scintillator samples.

Scintillator type	Sample number	ZnS type (manufacturer)	ZnS: $H_3{}^{10}BO_3$ or ^{6}LiF (by weight)	Coating amount (mg/cm²) $H_3{}^{10}BO_3$ or ^{6}LiF	Total	N*t value [b] (x10^{20}, 1/cm²)
ZnS/$^{10}B_2O_3$	No 1	tst13-no3 (JAEA)	1.00 : 1	30.0	60.0	2.5
	No 2	tst13-no1 (JAEA)	1.14 : 1	26.3	56.0	2.1
	No 3	tst14-no1 (JAEA)	1.33 : 1	22.5	52.0	1.8
	No 4	tst13-no2 (JAEA)	1.50 : 1	20.0	50.0	1.6
	No 5	tst14-no3 (JAEA)	1.15 : 1	32.5	70.0	2.7
	No 6	1109-041 (NICHIA)	1.25 : 1	30.0	67.0	2.5
	No 7	amt6-no3 (JAEA)	1.36 : 1	27.5	65.0	2.3
	No 8	amt6-no3 (JAEA)	1.50 : 1	25.0	62.5	2.1
ZnS/^{6}LiF	No 1	amt6-no3 (JAEA)	1.00 : 1	30.0	60.0	5.9
	No 2	amt6-no3 (JAEA)	1.20 : 1	30.0	66.0	5.9
	No 3	amt6-no3 (JAEA)	1.50 : 1	30.0	75.0	5.9
References [a]	AST(2:1)	--	2.00 : 1	~24	~72	~4.7
	AST(4:1)	--	4.00 : 1	~15	~80	~2.8

a) ZnS/^{6}LiF from AST. Coating amounts were determined from the experimental results.
b) N*t values is the product of the density of the neutron-conveter material and the thickness of the samples.

The commercial product, a $ZnS/^6LiF$-based scintillator manufactured by Applied Scintillation Technologies (AST), was also prepared as a reference. This $ZnS/^6LiF$ neutron screen is constructed from polycrystalline ZnS mixed with a ^6LiF converter material, which provides thermoplastic characteristics.

2.2. Experimental setup

A bialkali photomultiplier tube (PMT; Model 9102B, ET Enterprises) was used to record neutron-induced signals. An iris mask constructed from a 15-μm-thick aluminium foil was attached to the PMT to give a neutron-sensitive area of 10×10 mm^2. A test sample was attached directly to the PMT through the iris mask using optical grease. The scintillator sample and PMT detector were irradiated with neutrons from an Am-Be source with a radioactivity of 3.7×10^{11} Bq. The neutrons from the Am-Be source were thermally moderated with a polyethylene moderator that had a volume of $1200 \times 1200 \times 1200$ mm^3. Neutron-induced signals were recorded with a 12-bit digitizer (Model U1065A, Acqiris), and the signal waveforms were analyzed. The count rate on the detector was typically below 100 cps, and hence the count loss of the measurement system was negligible.

3. Results and discussion

3.1. Signal waveform

Figure 1 shows averaged neutron-induced signal waveforms of the samples (signal waveforms of the typical three samples were shown in the figure). These samples exhibited similar primary emission decays. An exponential fit of the primary decay indicated a 1/ε time constant of ~60 ns. The afterglow was much lower for both the developed $ZnS/^{10}B_2O_3$ and $ZnS/^6LiF$ scintillators than for the reference AST scintillator, which demonstrates that the prepared low-afterglow ZnS:Ag phosphor was effective at reducing the afterglow.

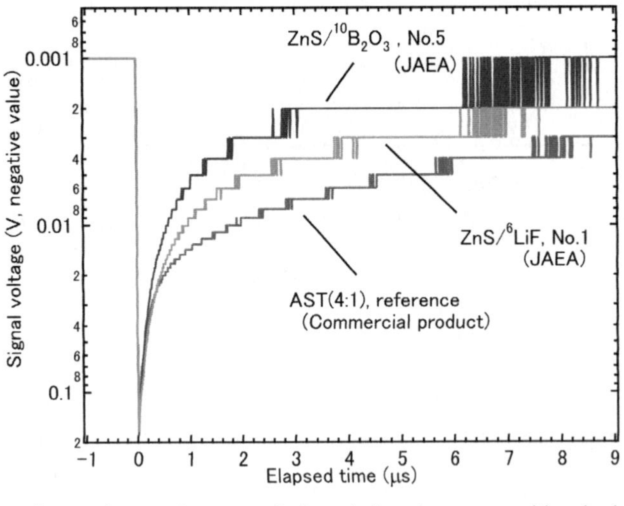

Figure 1. Averaged waveforms of neutron-induced signals measured by the bialkali PMT (Model 9102B, ET Enterprises). The signals shown are means of 9400 waveforms.

Figure 2a, b and c show the (a) signal pulse height at the peak, (b) signal pulse height at 1μs after the peak time and (c) the ratio of the signal heights at 1 μs after the peak time and at a peak time (called the "afterglow rate" hereafter). The peak height was similar among all of the samples, including the reference AST(2:1) scintillator (Figure 2a), at 0.14–0.20 V. On the other hand, the pulse height at 1 μs after the peak time was significantly lower for the developed samples (Figure 2b): the afterglows of the $ZnS/^{10}B_2O_3$ and $ZnS/^6LiF$ scintillators were about 40% and 60% of the reference value, respectively. This decrease in the afterglow was clearly evident in the pulse shapes normalized by their own peak heights, with afterglow rates of 2.7–4.0%, 5.5–5.7% and 9.0–10.1% for the $ZnS/^{10}B_2O_3$, $ZnS/^6LiF$ and reference AST scintillators, respectively (Figure 2c). The afterglow rates varied slightly with the converter materials and mixing ratios of the composite materials; these effects should be clarified in future studies.

The neutron absorption probability, ε, is described as follows,

$$\varepsilon = 1 - e^{-N \cdot t \cdot \sigma}$$

where, N is a density of neutron-converter material, t is a thickness of the scintillator, and σ is a neutron-absorption cross-section. The neutron absorption probabilities of each sample were calculated by using the product of N and t shown in the Table 1.

Figure 3 shows that the measured neutron count rate as a function of the calculated neutron absorption probability. The measured count rate was higher for samples with the larger N and t

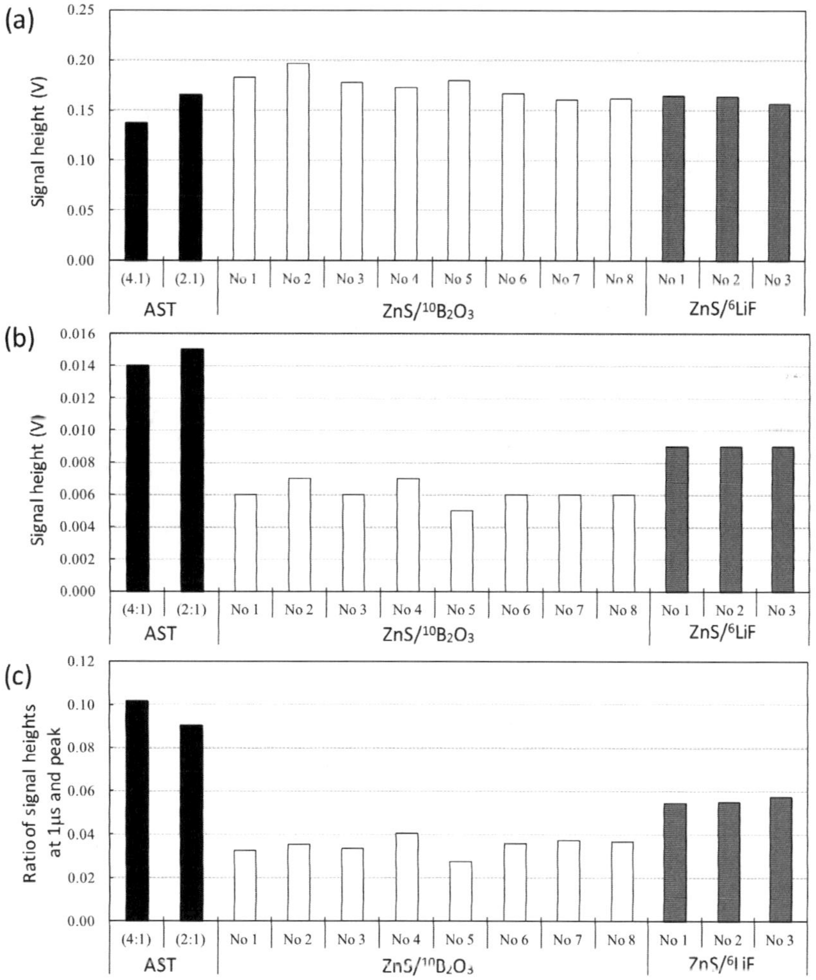

Figure 2. Signal heights (a) at the peak and (b) at 1 μs after a peak. (c) Ratio of signal heights in (a) and (b).

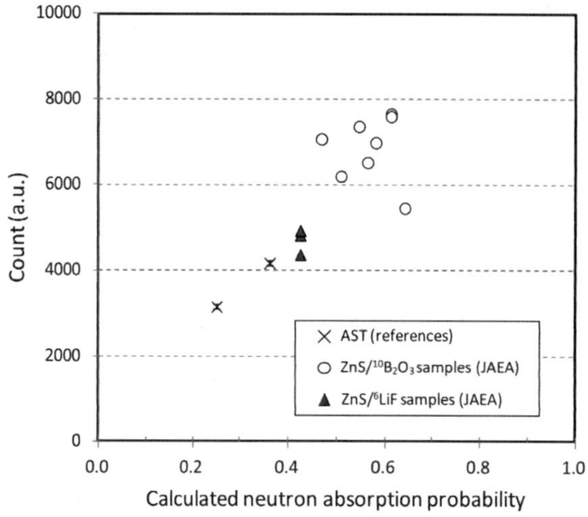

Figure 3. Scatter plot of the measured neutron count versus the calculated neutron absorption probability for the samples.

product. Moreover, the neutron count was higher for the ^{10}B-doped ZnS scintillator than for scintillators doped with ^{6}LiF. This result indicated that using the ZnS/$^{10}B_2O_3$ allowed the construction of a detector with a high detection efficiency in our particular samples.

In general the detection efficiency of a ZnS scintillator is strongly affected by the neutron absorption rate and the light transmittance of a sample since ZnS is opaque to its own emission light. Predicting the light transmittance is also of particular difficult since it is contributed by many factors such as scintillator thickness, grain size, light attenuation length, particle track length. In practice precise measurements of light transmission were also difficult due to the use of an aluminium base plate in our samples. The experimental finding of a highly linear correlation between the neutron absorption rate and measured counts indicated that the prepared samples had similar light

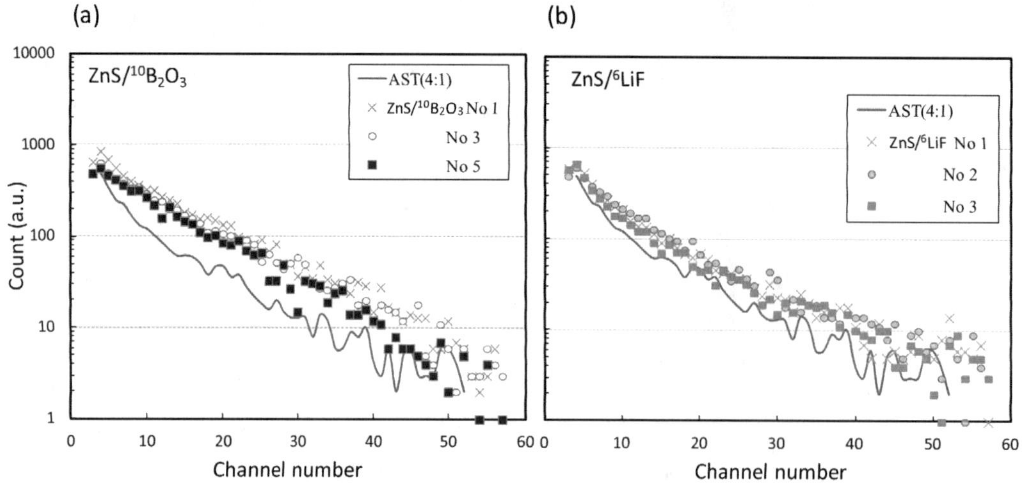

Figure 4. Pulse height distribution of (a) ZnS/$^{10}B_2O_3$ and (b) ZnS/^{6}LiF constructed using the ZnS scintillator with a low afterglow. Pulse height distribution of the AST(4:1) scintillator is included as reference.

transmittances. Figure 4 shows measured pulse height distributions of typical samples measured while irradiating with neutrons. All of the other samples also had similar neutron-induced signal heights, which supports the above interpretation.

Figure 5 is a scatter plot of the measured neutron count versus the afterglow rate, which represent the detection efficiency and count-rate capability of the implemented detector, respectively. The data for our samples were clearly separated each other, with the ZnS/$^{10}B_2O_3$ scintillator exhibiting the highest neutron count rate and the lowest afterglow rate, thereby ensuring a detector with a high detection efficiency and good count-rate capability.

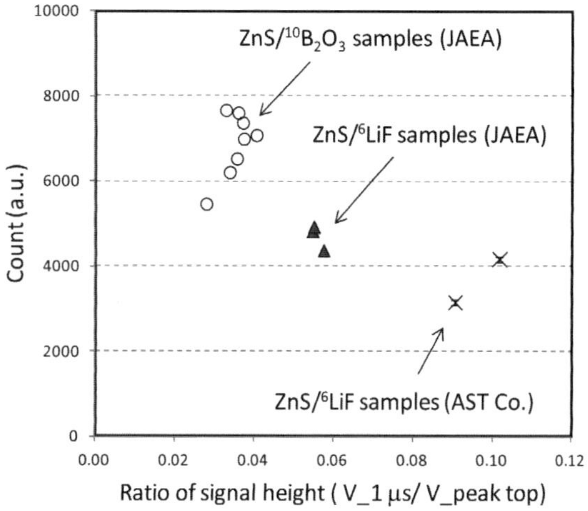

Figure 5. Scatter plot of the measured neutron count versus the afterglow rate, defined as the ratio of the signal height at 1 μs after the peak time to that at the peak time.

3.2. Detector performance with the ZnS/$^{10}B_2O_3$ ceramic scintillator

The count-rate capability of the ZnS/$^{10}B_2O_3$ scintillator was evaluated using a WLS-fibre-based

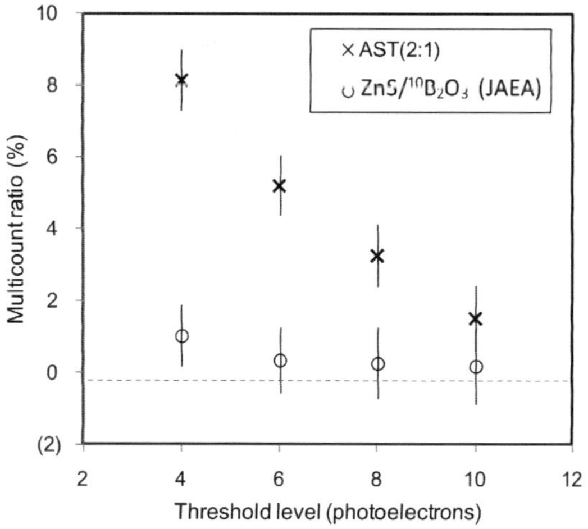

Figure 6. Multicount ratio of the ZnS/$^{10}B_2O_3$ (JAEA, No1) and the AST(2:1) scintillators measured by the scintillator/WLS-fibre detector that was developed for the SENJU instrument at J-PARC/MLF.

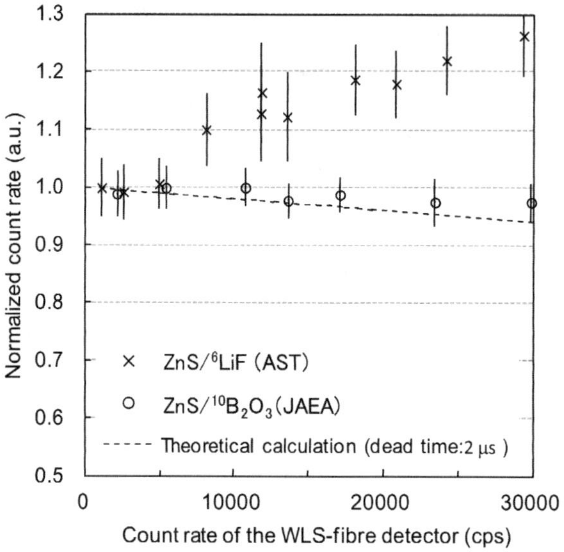

Figure 7. Normalized count rate of the WLS-fibre detectors implemented with the reference scintillator and the $ZnS/^{10}B_2O_3$ scintillator screen, respectively.

detector. The detector comprised 64 × 2 WLS fibre arrays, with the WLS fibres placed at a regular pitch of 4 mm in both the *x* and *y* directions [11]. The $ZnS/^{10}B_2O_3$ scintillator screen constructed from the composite indicated as No. 1 in Table 1 was used for this test. Two scintillator screens sandwiched the WLS fibre arrays in the detector.

Prior to the test at a reactor, the multicount rate of the detector was evaluated using a ^{252}Cf neutron source. The multicount rate is defined as the probability of counting single neutrons more than once, and such an event can result from the afterglow of the scintillation light, making it a good indicator of the count-rate capability of a detector. To evaluate the effect of the afterglow to the count rate no specific signal processing method was implemented to system such as inserting a dead time.

Figure 6 shows the multicount rates for the detectors implemented with the $ZnS/^{10}B_2O_3$ and AST(2:1) scintillators as functions of the threshold level. The counts produced by trigger pulses with widths of 100 ns and 20 μs were compared; the latter counts were considered the true counts without multicount contamination. The multicount rate was in the range of 2–8% for the reference AST scintillator when measured with a threshold level of 4–10 photoelectrons, and was about an order of magnitude lower for the developed $ZnS/^{10}B_2O_3$ scintillator, at 0.1–0.5%. Multicount events occur mainly due to a high afterglow during ZnS scintillation, and hence the developed low-afterglow $ZnS/^{10}B_2O_3$ scintillator was quite effective.

The count-rate capability of the detector was evaluated using a cold-neutron beam at the CHOP port in Japanese Research Reactor 3 at Japan Atomic Energy Agency. The cold-neutron beam had a peak wavelength of 4Å and was collimated into a size of 5 × 5 mm^2, with the beam illuminating the centre region of the neutron-sensitive area. Figure 7 shows the normalized count rates of the detectors implemented with the reference scintillator and the $ZnS/^{10}B_2O_3$ scintillator screen. The count rate of the detector was divided by the count measured by a detector monitoring the incident neutrons (a ^3He detector with a detection efficiency of $\sim 10^{-4}$ for thermal neutrons), with the resultant normalized ratio facilitating comparisons with the theoretical calculations. Since this monitored count is proportional to the true incident rate (i.e. with negligible count loss), the count ratio should be

$$n_{wlsf} / n_{monitor} \propto 1 - n_{wlsf} \cdot \tau_{wlsf}$$

where n_{wlsf} and $n_{monitor}$ are the count rates measured by the WLS-fibre detector and the monitoring detector, respectively, and τ_{wlsf} is the dead time of the WLS detector (which was ~2 μs in the test detector system). The count ratio for the reference AST scintillator deviated from the calculated value when the incident rate exceeded 5,000 cps, which was due to the presence of significant afterglow. This deviation increased approximately proportionally with the incident count rate, reaching about 30% for a count rate of 30,000 cps. In contrast, the count rate of the developed $ZnS/^{10}B_2O_3$ scintillator followed the calculated value up to 30,000 cps within experimental errors. This clear decrease in multicount events can be attributed to the lower afterglow of the developed scintillator.

4. Summary

We have developed a $ZnS/^{10}B_2O_3$ ceramic scintillator constructed using a ZnS phosphor that has a low afterglow for neutron-scattering instruments at J-PARC/MLF. The $ZnS/^{10}B_2O_3$ ceramic scintillator exhibited the best performance of the prepared samples, exhibiting both a high detection efficiency and a low multicount rate. The count-rate capability of the $ZnS/^{10}B_2O_3$ ceramic scintillator was evaluated in combination with a WLS-based detector, which demonstrated that the developed scintillator was effective at minimizing the multicount event rate and also improving the count-rate capability of the detector.

References

[1] ISIS facility: http://www.isis.stfc.ac.uk/
Regarding the detectors installed in the ISIS, for example, Rhodes N J, Schooneveld E M and Eccleston R S 2004 *Nucl. Instr.& Meth.* A **529** 243
[2] SNS facility: http://neutrons.ornl.gov/
[3] Rhodes N J 1996 *Proc. of SCINT95 (Delft)* (Delft: Delft University Press) p 73
[4] van Eijk C W E 2001 *Nucl. Instr. & Meth.* A **460** 1
[5] Fairley E J and Spowart A R 1978 *Nucl. Instr. & Meth.* **150** 159
[6] Czir J B, MacGillivray G M, MacGillivray R R and Seddon P J 1999 *Nucl. Instr. & Meth.* A **424** 15
[7] Bessiere A, Dorenbos P, van Eijk C W E, Krämer and Güdel H U 2005 *Nucl. Instr. & Meth.* A **537** 242
[8] Getkin A, Shiran N, Neicheva S, Gavrilyuk, Bensalah A, Fukuda T and Shimamura K 2002*Nucl. Instr. & Meth.* A **486** 274
[9] Kojima T, Katagiri M, Tsutsui N, Imai K, Matsubayashi M and Sakasai K 2004 *Nucl. Instr. & Meth.* A **529** 325
[10] Katagiri M. in private communication
[11] Nakamura T, et al 2012 *Nucl. Instr. & Meth.* A **686** 64.

International Workshop on Neutron Optics and Detectors (NOP&D 2013) IOP Publishing
Journal of Physics: Conference Series **528** (2014) 012044 doi:10.1088/1742-6596/528/1/012044

Digitized signal acquisition of gas-filled, multiwire, neutron detector with delay-line readout

J Orban[1], L Rosta[1], A Nagy[2]

[1] Wigner Research Centre for Physics, Budapest, Hungary
[2] Mirrotron Ltd., Budapest, Hungary

E-mail: orban@wigner.mta.hu

Abstract. Multiwire counters are likely to remain a reliable and cost-efficient option for a large class of instruments. Delay-line position encoding is widely used in multiwire position sensitive particle detectors. Improvement of several detector parameters is a continuous demand such as count rate or position resolution. Digitization of the analog detector signals retains all information about the detected events at the highest count rate. Neutronic measurement results on our standard detector with 200mm x 200mm active area are presented with different acquisition algorithms. The analog signals were digitized by a high speed, multichannel Acqiris digitizer with 1 ns sampling resolution.

1. Theory of delay-line encoded multiwire proportional chamber

Several spectrometers at BNC are equipped with our standard, multiwire, 2D position sensitive gas-filled neutron detector with delay-line readout. These gas detectors contain a mixture of ^3He and CF_4 [1]. In these detectors, the neutron is sensed by the well known reaction:

$$^3He+n=p+T+764keV$$

The CF_4 limits the ionization particles by its high stopping power. This effect is the primary limitation of the position resolution of the detector.

Our multiwire detectors contain three parallel wire frames. The central anode and the two surrounding cathodes are mounted in parallel. The wires of one cathode frame are orthogonal to the other cathode frame. An electric field is created between the anode and cathode layers by applying same high voltage on the all anode wires. The small diameter of the anode wires creates large electric field around the anode wires. The high electric field multiplies the ionized electrons. This improves the signal to noise ratio. During this gas gaining effect, signals are induced on the close cathode wires providing the position of the captured neutron [2].

All cathode wires can be lead out from the chamber (individual readout). More cost-efficient solution is the connection of the cathode wires by delay-line elements. This serial connection of the delay-line elements creates a delay-line. In this way the position of the neutron is coded to the time difference between the anode and delayed cathode signals. The connected anode wires and two ends of the two delay-lines give five output signals [3]. Fast current amplifiers are used for these outputs.

The starting-point of these signals correlates with the neutron position. Simple threshold triggering causes a dependence of the trigger time on pulse height. The constant fraction discrimination method can reduce this dependence. Careful fine-tuning is necessary to minimize this dependence as any kind of signal dispersion in time degrades the position resolution of the detector.
Finally, the delays between the CFD output signals are measured by a time to digital converter (TDC). Many readout system follow this solution [4,5]. The usual time resolution of a TDC is around few hundred of picoseconds. The total length of a delay line is between 100-200 ns.

2. Digitized signal acquisition results

The conventional delay-line readout electronic converts the delay-line analog outputs to digital signals causing loss of information like pulse height, rise time or fall time. Sufficiently fine digitizations of the analog signals preserve all the possibly useful information about the detected neutron event. We investigated different data acquisition procedures that can improve several detector parameters but further offline improvements on the stored data are also possible. Fast, multichannel digitizer was used for storing the data. The data were processed in Matlab.

2.1. CFD software replacement

The operation of the conventional analog CFD is well known and published in many articles [6]. Our first improvement is the numerical realization of the constant fraction discrimination. This means a Matlab-simulated CFD. This algorithm performs the same operation as a real analog CFD. Two further numerical discriminator algorithms have been also tested. Either of them calculates the starting point from the falling edge of the signal because this seems the simplest way to calculate the starting

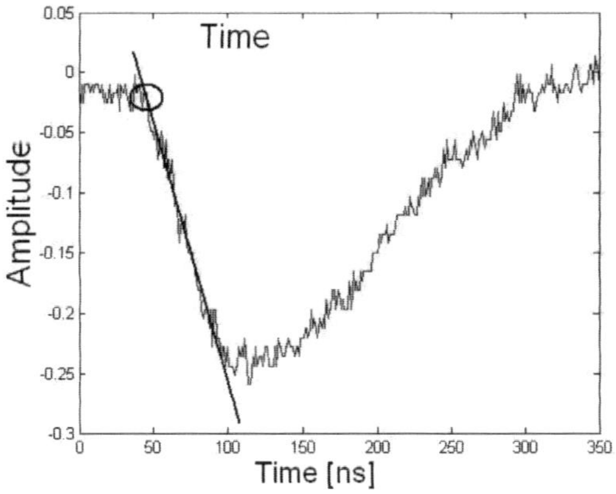

Figure 1. Linear fitting on the analog output signal of a delay-line detector

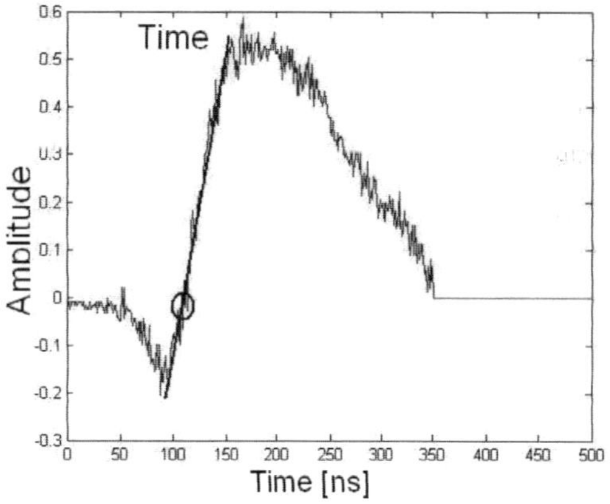

Figure 2. Linear fitting on the CFD signal of a delay-line detector

point. This method is illustrated in Figure 1. The other algorithm follows the original CFD method simulation but the zero cross finding on the rising edge is improved by linear fitting on 10 digitized points. This method decreases the effect of the noise on the rising part of the signal. This is illustrated in Figure 2.

The three numerical algorithms have been compared with the real, analog PS715 constant fraction discriminator. The comparators are compared by the time dependence mentioned before. The sum of the travelling times of the cathode signals on the delay line should be equal to the length of the delay line. This sum is dispersed in time by the CFD. The discriminators are characterized by the FWHM of the curves represent this dispersion. The results are displayed in Figure 3. The values are summarized in Table 1. The result of the simulated CFD algorithm is very close to the real, analog CFD. The simple linear fitting on the falling edge of the signal provides bad result. This indicates the necessity of constant fraction discrimination. The improved method provides the best results.

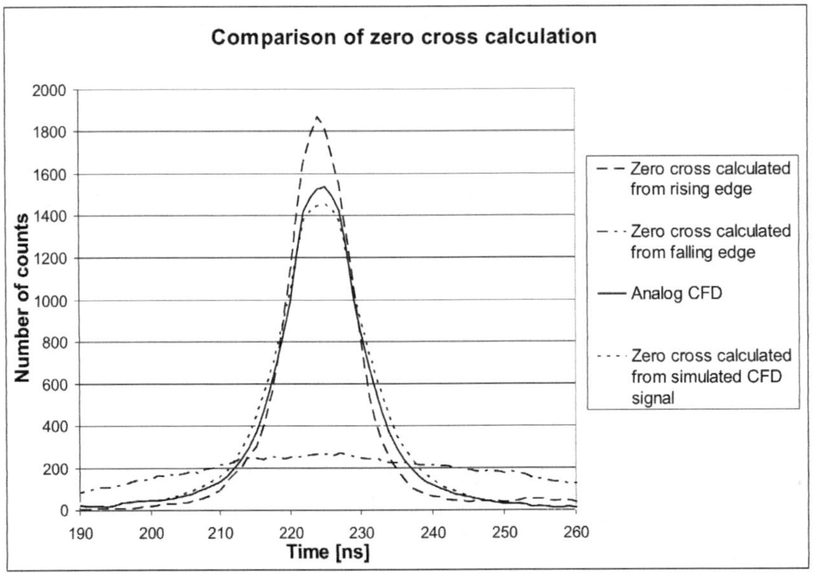

Figure 3. Comparing the different methods

Table 1: Quantitative comparison of the CFD methods

Type	Time dispersion FWHM [ns]
Analog CFD	12,2
Simulated analog CFD	13
Improved simulated CFD	10,5
Linear fitting on falling edge	63

2.2. Variable threshold

Further improvement is the position dependent discriminator threshold that can improve the homogeneity and also the gamma sensitivity. The solution is similar to the individual readout where every wire has an individual threshold but in this solution an optimal threshold is assignable for every pixel that can separate the neutron signals from the noise and gamma signals. The method is demonstrated on Figure 4 where effect of higher gain due to a low-tensioned wire is compensated between the 120th and 127th channels. The tuning of the individual virtual thresholds set takes a long time but the numerical way can be automated by a complex algorithm. We should mention that the conventional threshold tuning by potentiometers in individual readout detectors takes also a long time.

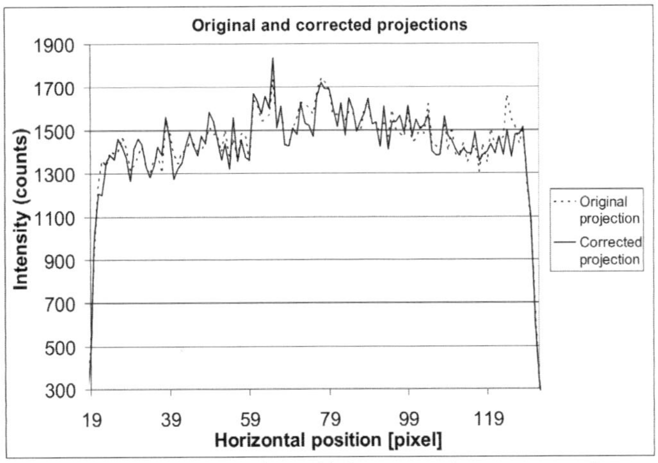

Figure 4. Demonstration of inhomogeneity compensation

2.3. Signal integration

Another practical application is the real-time numerical integration of the digitized signals that improves the energy resolution and gamma sensitivity of the detector keeping the high count rate capability of the detector. Delay-line detectors generally use fast current amplifiers because the fast falling time gives lower jitter and better time and position resolution. The analog integration of the detector current signals gives better energy resolution because the linearity between the pulse height and total charge is better but the count rate capability of the detector is degraded due to the slower integrated signals. Numerical integration keeps the high count rate capability of the detector meanwhile the charge is integrated.

The better separation of neutron and gamma events is illustrated on Figure 5. The first peaks belong to the noise and gamma events, the seconds to the neutron events. The curves show the pulse height spectra of the original and the numerically integrated anode signals. After calibration the neutron peaks are at the same position for the better comparability of the curves. The methods are characterized by FWHM of the peaks belongs to the neutron events. The FWHM of the curve belongs to the original current signals is 243mV, the FWHM of the curve belongs to the integrated signals is 198mV.

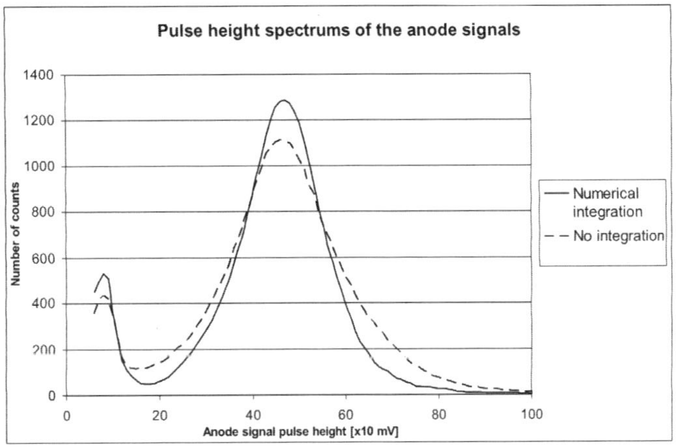

Figure 5. linear fit on the analog output signal of a delay-line detector

3. Conclusion

Effect of an acquisition method on digitized detector signals of a delay-line detector is presented. The data were recorded on-line by an Acqiris digitizer and processed off-line by Matlab for demonstration. Precise acquisition of the delay-line analog signals improves several detector parameters. New acquisition algorithm on numerically realized CFD signals improves the time dispersion comparing with a real, analog CFD. This leads to better position resolution. The homogeneity is also improvable by set an individual discrimination threshold for every pixel. The numerical signal integration improves the energy resolution by the better signal to noise ratio without degradation of the count rate capability. This signal digitization method provides a flexible system because the recorded signals retain all information and can be processed off-line with different algorithms.

References

[1] L Cser, A Gabriel, M Kocsis, G Messing, L Rosta, Gy Török, Neutron measurements on the MK 200N-1 type two-dimensional delay-line position- sensitive detector with the new OTDCV1.11 time-to-digital converter, *Physica B* **276** (2000) 216-217

[2] G Charpak, R Bouclier, T Bressani, J Favier, Č Zupančič, *Nucl. Instr. and Meth.* **62** (1968) 262-268

[3] C J Boulin, R Kempf, A Gabriel, M H J Koch, *Nucl. Instr. and Meth. A* **269** (1988) 312

[4] F V Levchanovski, B Gebauer, E I Litvinenko, A S Nikiforov, V I Prikhodko, Ch Schuly, Th Wilpert, A PCI DAQ board for MWPC detectors with high-rate 2D delay line position readout

[5] W Shang, B Robrahn, F Golding, M H J Koch, A versatile data acquisition system for time resolved X-ray scattering using gas proportional detectors with delay line readout

[6] Glenn F Knoll, *Radiation detection and measurement*, ISBN 0-471-49545-X (1979) 683

Performance evaluation of high-pressure MWPC with individual line readout under Cf-252 neutron irradiation

K Toh[1*], T Nakamura[1], K Sakasai[1], K Soyama[1] and H Yamagishi[1,2]

[1]J-PARC Center, Japan Atomic Energy Agency, 2-4 Shirakatashirane, Tokai 319-1195, Japan

[2]Nippon Advanced Technology 3129-45 Hibara Muramatsu, Tokai 319-1112, Japan

E-mail: toh.kentaro@jaea.go.jp

Abstract. A multiwire proportional chamber (MWPC) neutron detector system was developed for the Materials and Life Science Experimental Facility at the Japan Proton Accelerator Research Complex. Its basic performance was evaluated by an irradiation experiment using a Cf-252 neutron source. A short response time and high spatial resolution can be obtained using an individual line readout method. The detector system exhibited a one-dimensional uniformity of response of 4.8% and 3.8% in the x- and y-directions, respectively. The uniformity of all pixels in the two-dimensional image was 7.9%. The average intrinsic spatial resolution was 1.55 mm full width at half maximum in the sensitive region calculated by taking into account the track lengths of secondary particles. The signal intensity of the system remained constant during the operation for 500 min under Cf-252 neutron irradiation.

1. Introduction

The high-intensity pulsed neutron facility at the Materials and Life Science Experimental Facility (MLF) at the Japan Proton Accelerator Research Complex (J-PARC) has 23 beam lines (BLs), and time-of-flight (TOF) measurements are performed using 25-Hz pulsed neutrons [1]. Some neutron scattering experiments performed at the J-PARC/MLF require advanced two-dimensional neutron detectors with short response time, good spatial resolution, and high detection efficiency. For this purpose, we are currently developing a two-dimensional position-sensitive neutron detection system consisting of a two-dimensional detector element and capable of individual signal line readout [2–4]. In a two-dimensional gas-based neutron detector, the thickness of the conversion gap should be made as small as possible to prevent the occurrence of the parallax effect of incident neutrons. Therefore, it is necessary to increase the gas pressure to achieve detection efficiency greater than 70%. Superior spatial resolution can be obtained by using a buffer gas with a high stopping power such as CF_4 used in this study, and by increasing the gas pressure.

Multiwire type two-dimensional neutron detectors with helium-3 (^3He) gas are widely used in neutron scattering experimental facilities. These detectors are known to have high detection efficiencies, low gamma-ray sensitivity, and excellent long-term stability [5–8]. These gas-based neutron detectors measure the ionized charges generated by secondary particles in nuclear reaction $^3He(n,p)t$. At the J-PARC/MLF, we develop an advanced multiwire type two-dimensional neutron detector system without sacrificing the excellent performances of the conventional detector. A short response-time and

* To whom any correspondence should be addressed.

better spatial resolution can be achieved by our system using an individual line readout method. Optical devices have also been developed and incorporated in the system for long-distance signal transmission and for isolating the detector head in the sample irradiation area from the signal processing circuits in the data acquisition room. In this paper, we present our multiwire-type detector system developed for the J-PARC/MLF, and discuss its performance based on neutron irradiation experiments using a Cf-252 neutron source.

2. Experimental

The multiwire element used in our detector system comprises 128 anode wires, 128 cathode wires along y-axis, and 128 cathode strips along x-axis. The pitch between anode/cathode wires is 1.0 mm, and the distance from the anode wires to the cathode wires and strips are 1.0 and 1.5 mm, respectively. The multiwire element was placed inside a pressure vessel. To prevent outgassing from the detector element, the pressure vessel was vacuumed to below 10^{-5} Pa for at least 60 h prior to the gaseous mixture filling. Hereafter, we refer this detector head as a multiwire proportional chamber (MWPC).

The system consists of a 256-channel detector element (x: 128 lines, y: 128 lines), a pressure vessel, amplifier-shaper-discriminator (ASD) boards, optical signal transmission devices, position encoders with field-programmable gate arrays (FPGAs), and a data acquisition device. A schematic view of the system and a photograph of the detector head are shown in Figure 1. The multiwire element was installed in the pressure vessel with a conversion gap of 20 mm. The pressure vessel could withstand pressures of up to 0.8 MPa and with a helium leakage of less than 10^{-8} Pa•m^3/s. Experiments were performed using a gaseous mixture of ^{3}He and CF_4 which functioned as a neutron converter and a stopping gas, respectively. The performance of the detector was strongly dependent on the gas conditions in the pressure vessel. In this experiment, the pressure vessel was filled with He/(20%)CF_4 at 0.6 MPa. Under this gas condition, the thermal neutron detection efficiency was 74% determined from the nuclear reaction cross section of ^{3}He gas and the size of the conversion gap. Taking into account the gas pressure, the track lengths of secondary particles, protons (L_p) and tritons (L_t), calculated by the TRIM code [9] were 2.68 and 0.94 mm, respectively. Therefore, it can be estimated that the MWPC has a few output signal lines since the pitch of both axes is smaller than the track length.

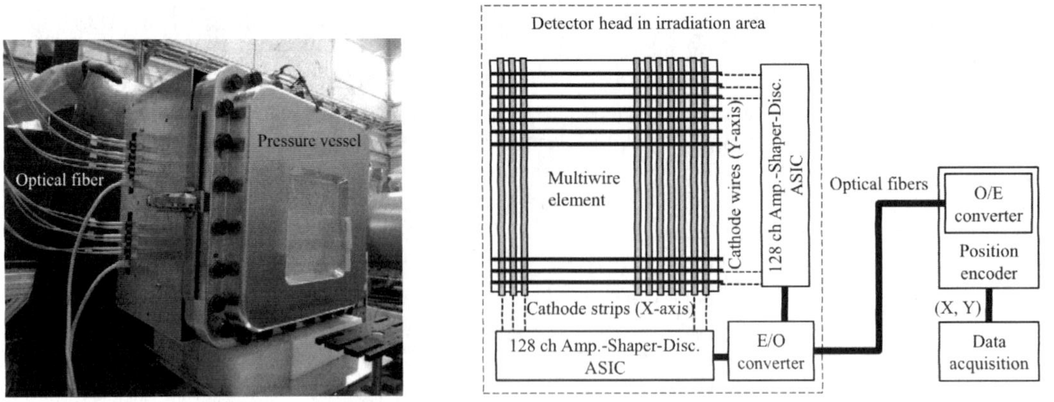

Figure 1. A photograph of the detector head (left) and a schematic view of the detector system (right). The detector head which includes the detector element and amplifier-shaper-discriminator boards are isolated from the position encoder and data acquisition device, and connected together using optical fibers.

In the MWPC, a charge signal is generated as a result of the gaseous amplification around each anode wire induced by the strong electric field between the electrodes. Cathode wires and strips are arranged orthogonally. The signal from each cathode is individually amplified, shaped, and discriminated by amplifier-shaper-discriminator application-specific integrated circuits (ASD-ASICs), which nominal settings are an amplification factor of 3.1 V/pC and a decay time of 90 ns. Digital signals from the ASD-ASIC are transmitted to the position encoders through optical fibers as optical signals converted by a specially fabricated electrical/optical (E/O)-O/E converter. The optical fiber is insensitive to electromagnetic fields and electrical noises. Further, optical fibers can significantly extend signal transmission distance due to the low transmission loss compared to that of conventional electrical cables.

Neutron irradiation on the MWPC was performed by embedding a Cf-252 neutron with an intensity of 100 MBq in a graphite cube with sides measuring 80 cm. The MWPC was placed at a distance of 30 cm from the surface of the graphite cube. Upon measurement with an absolutely calibrated BF_3 counter, the neutron fluxes at the surface of the cube and detector position were found to be 1.2×10^7 and 5.6×10^6 n/m^{2}s, respectively.

3. Results and discussion

Figure 2 shows the flat-field image measured by the detector to confirm the spatial homogeneity of the detector element. Figure 3 shows the one-dimensional projections to the x and y axes, and the histogram of the pixel contents of the flat-field image. The counts of one-dimensional projections

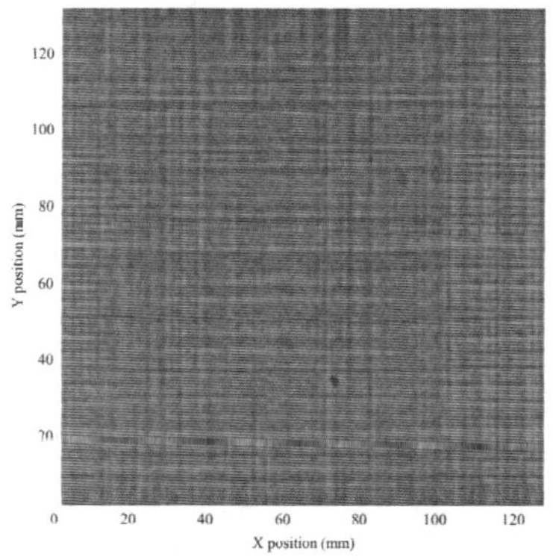

Figure 2. Two-dimensional flat-field image.

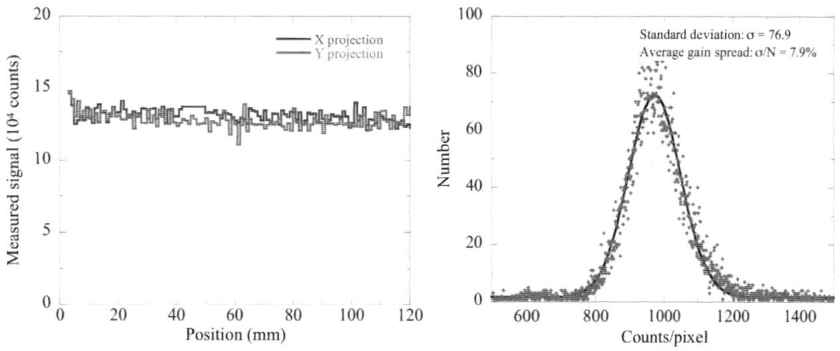

Figure 3. One-dimensional response projected in both axes (left), and the histogram of the count distribution of the flat-field image (right).

exhibit excellent uniformity, as shown in Figure 3(a). The relative standard deviations were 3.4% and 4.3% for x- and y-projections, respectively. The histogram of all pixel contents was evaluated by a

Gaussian fit and showed good homogeneity. The average pixel count was 972 with a standard deviation of σ = 76.9, which corresponded to an average gain spread of 7.9%.

In our MWPC, incident position of a neutron, N(x,y), is defined as

$$N(x,y) = (\frac{x_h + x_l}{2}, \frac{y_h + y_l}{2}) \tag{1}$$

where x and y are the channel numbers and the subscript l and h represent the start and end channels of signals. The errors are arisen from the differences of the track length between protons (L_p) and tritons (L_t) because the center of the track is defined as the position of the incident neutron in our position detection. Figure 4 shows a schematic view of our position detection method and the intrinsic error arisen from the difference of track lengths.

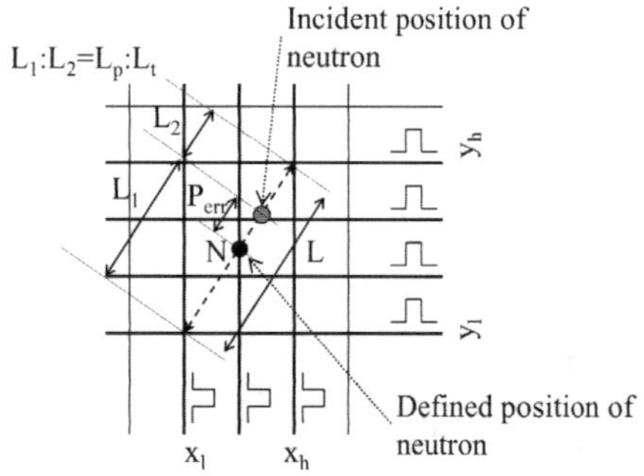

Figure 4. Schematic view showing the method and its intrinsic error for position detection of incident neutron.

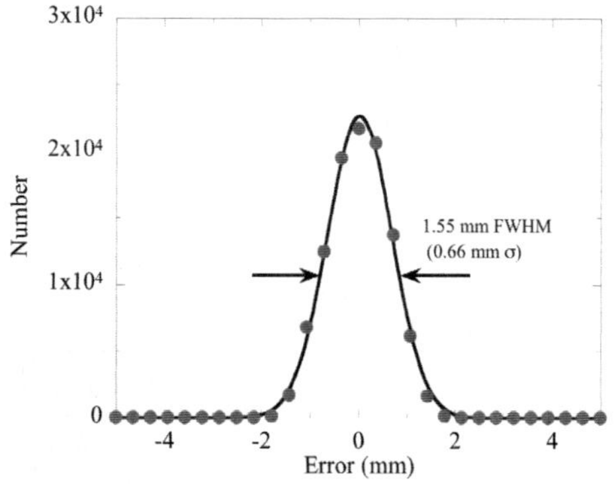

Figure 5. Errors in position detection of our detector.

In our detector system, all neutron event data, including x_l, x_h, y_l, and y_h, are recorded in the data acquisition electronics. By using these event data, the total track length L(x,y), defined as the sum of the lengths of proton and triton, for each neutron event can be calculated as follows:

$$L(x,y) = sqrt\left((x_h - x_l)^2 + (y_h - y_l)^2\right) \tag{2}$$

The intrinsic error P_{err} of the position detection can be obtained from

$$P_{err}(x,y) = \frac{L_p}{L_p + L_t} L(x,y) - \frac{1}{2} L(x,y) \tag{3}$$

The errors are calculated from all recorded event data, and plotted in one-directions in Figure 5. In the figure, the horizontal and vertical axes represent the error and the number of events, respectively. The intrinsic spatial error of the detector is evaluated by fitting the data with a Gaussian function, and the full width at half maximum (FWHM) of the spatial error is 1.55 mm.

The incident neutrons are usually scattered by the MWPC itself because it has an aluminum (Al) window and the fill gas molecules act as scatterers. To estimate the effects of these scatterers, we calculated the amount of neutrons scattered by the MWPC. The calculation was performed by a Monte

Carlo simulation using PHITS (multi-purpose particle and heavy ion transport code system) [10]. This Monte Carlo code can simulate the transport and interaction process of neutrons, photons, electrons, and charged particles in an arbitrary three-dimension geometry system. In the simulation, a pencil beam of thermal neutron was irradiated perpendicularly to the detector, and the actual geometry of the MWPC such as the 7 mm window, the 20 mm conversion gap filled with gaseous mixture, and the 1 mm pitch detector element were taken as the parameters. The simulated result of neutrons scattered by the Al window and the fill gas is shown in Figure 6. The figure is the one-dimensional projection of the distribution of scattered neutrons. A simulation without the Al window (by setting the window thickness as 0 mm) is also plotted for comparison. It shows that the neutrons are scattered by the Al window and the fill gas. They were detected by the adjacent pixels relative to the incident position and the counting ratio of the pixel to the incident position was 10^{-4}. The ratio was also one order of magnitude smaller than that when the neutrons were scattered by the fill gas only. It is concluded that the effects of scattered neutrons to the intrinsic spatial error are negligible We continuously operated our MWPC for 500 min and the elapsed time dependence of measured counts is shown in Figure 7. It is observed that the signals remain constant during the operation.

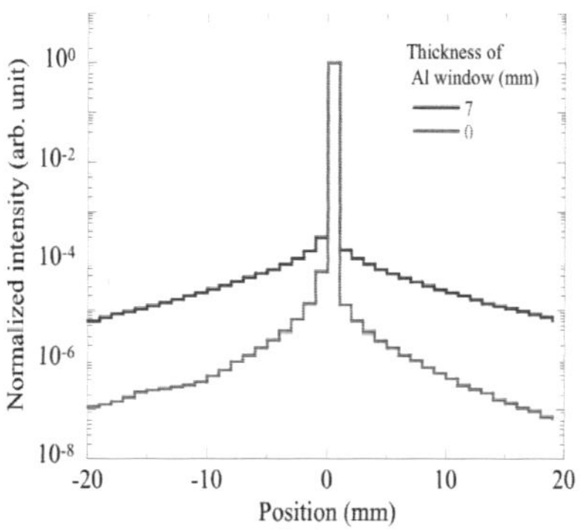

Figure 6. Distribution of scattered neutrons calculated by PHITS.

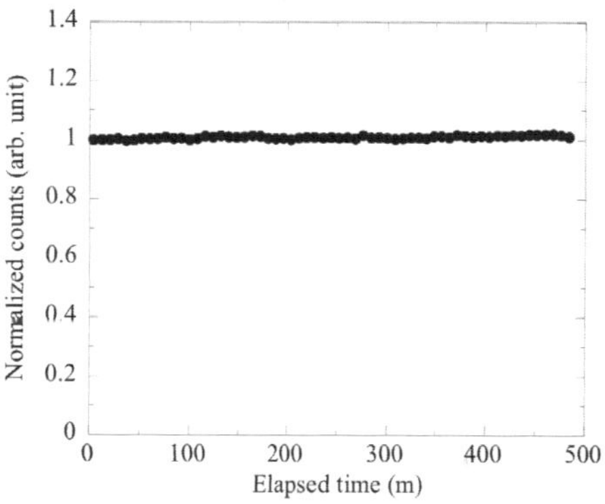

Figure 7. Normalized counts as a function of elapsed time.

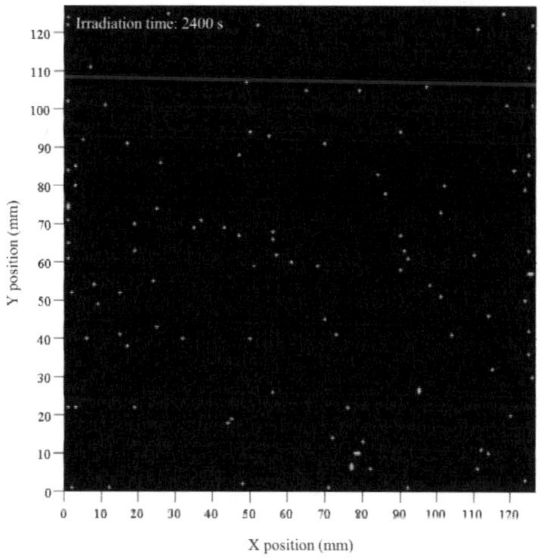

Figure 8. Measured image of gamma ray irradiation. The white dots indicate a detected signal.

The gamma-ray sensitivity of the detector was measured using a ^{60}Co gamma-ray source. The intensity of gamma-ray source was 8.9×10^5 Bq, and was placed on the surface of the detector window. A two-dimensional image generated by the ^{60}Co is shown in Figure 8 with an irradiation time of 2400 s. It is confirmed that the developed detector has low gamma-ray sensitivity of 8.2×10^{-9}.

4. Conclusions

A multiwire proportional chamber (MWPC) neutron detection system was developed. The detector had a neutron sensitive area of 128×128 mm^2 with a wire pitch of 1 mm. The system consisted of a 256 channel detector element (x: 128 lines, y: 128 lines) equipped with a pressure vessel, amplifier-shaper-discriminator boards, optical signal transmission devices, position encoders, and a data acquisition device. The estimated detection efficiency for thermal neutron was 74% determined from the gas condition and conversion gap. It exhibited an excellent uniformity of 7.9% with all pixels in the two-dimensional image measured using a Cf-252 neutron source. The average intrinsic spatial resolution was 1.55 mm FWHM in the sensitive region calculated by taking into account the track lengths of secondary particles with He/(20%)CF_4 at a pressure of 0.6 MPa. A gamma-ray sensitivity of 8.2×10^{-9} was obtained using a ^{60}Co gamma-ray source. After pulsed neutron irradiation experiments, this multiwire-type two-dimensional neutron detector system, which employs the individual line readout method, will be installed at J-PARC/MLF.

References

"J-PARC | Japan Proton Accelerator Research Complex," http://j-parc.jp/index-e.html.

H. Yamagishi, T. Nakamura, S. Masaoka, K. Soyama, and K. Aizawa, *Microstrip gas chamber with individual read-out using two-level discriminators for neutron detection, Nucl. Instrum. Meth. A* **529** (2004) 365.

T. Nakamura, H. Tanaka, H. Yamagishi, K. Soyama, K. Aizawa, A. Ochi, T. Tanimori, *A gas-based neutron imaging detector with individual read-outs, Nucl. Instrum. Meth. A* **573** (2007) 187.

H. Yamagishi, K. Toh, T. Nakamura, K. Sakasai, and K. Soyama, *Simulation program for multiwire-type two-dimensional neutron detector with individual readout, J. Instrum.* **6** (2011) C12025.

R. Allemand *et al., Position-sensitive detectors (p.s.d.) for neutron diffraction, Nucl. Instrum. Meth.* **126** (1975) 29.

N. Vellettaz, J.E. Assaf, and A. Oed, *Two-dimensional gaseous microstrip detector for thermal neutrons, Nucl. Instrum. Meth. A* **392** (1997) 73.

V. Radeka, N.A. Schaknowski, G.C. Smith, and B. Yu, *High-performance, imaging, thermal neutron detectors, Nucl. Instrum. Meth. A* **419** (1998) 642.

R. Kampmann, M. Marmotti, M. Haese-Seiller, V. Kudryashov, *2D-MWPC for the new reflectometer REFSANS/FRM-II: performance of the prototype, Nucl. Instrum. Meth. A* **529** (2004) 342.

J. F. Ziegler, J. P. Biersack, and U. Littmark, *The Stopping and Range of Ions in Solids. New York: Pergamon Press*, 1985.

K. Niita, N. Matsuda, Y. Iwamoto, H. Iwase, T. Sato, H. Nakashima, Y. Sakamoto and L. Sihver, *PHITS: Particle and Heavy Ion Transport code System, Version 2.23*, JAEA-Data/Code 2010-022 (2010).

AUTHOR INDEX

AUTHOR INDEX

Institute of Physics
Dirac House, Temple Back
Bristol BS1 6BE UK

CURRAN ASSOCIATES INC
proceedings
.com